Solid State Circuits Technologies

Edited by

Jacobus W. Swart

Solid State Circuits Technologies

Edited by Jacobus W. Swart

Republished by InTech

Janeza Trdine 9, 51000 Rijeka, Croatia

Technical Editor Teodora Smiljanic

Additional hard copies can be obtained from orders@intechopen.com

Solid State Circuits Technologies, Edited by Jacobus W. Swart

p. cm.

ISBN 978-953-307-045-2

For EU product safety concerns:
IN TECH d.o.o., Prolaz Marije Krucifikse Kozulić 3, 51000 Rijeka, Croatia,
info@intechopen.com or visit our website at intechopen.com.

Meet the editor

He graduated in Electrical Engineering and got a PhD at Escola Politécnica from the University of São Paulo in 1975 and 1981 respectively. He did pos-doc programs at Catholique University of Leuven, Belgium (1982-1983) and at Research Triangle Institute, USA (1991). He is a full professor at the State University of Campinas, since May 1997. After retiring in june 2013, he continued as collaborating professor at FEEC/UNICAMP. His experience is in the field of Electrical Engineering, with emphasis in materials and semiconductor devices, mainly on the following subjects: Si and GaAs integrated circuit technologies, CMOS, HBT, thin films, plasma etching and nanoelectronics. He was president of the Brazilian Microelectronics Society for two terms. He was member of CA-EE of CNPq for two terms and also member of the Deliberative Council of CNPq for 4 years. He is Distinquished Lecturer, chair of EDS Subcommittee for Regions/Chapters (SRC) for Region 9 and elected member of the Board of Governors of the Electron Device Society of IEEE. He is Fellow of IEEE. He was the coordinator of the IC-Brazil Program of MCTI from 2007 to 2013 and the coordinator of the NAMITEC research network since 2001, first under the Millennium Institute program and latter under the INCT program of CNPq.The proposal submitted in the Call INCT - MCTI / CNPq / CAPES / FAPs no. 16/2014 received funding recommendation (position 108) and was awarded with INCT Stamp certificate by CNPq. In 2013 he opened a company JW Swart Innovation in Semiconductors for which he acts as a consultant, mainly dedicated to work for imec from Belgium.

Preface

The evolution of solid-state circuit technology has a long history within a relatively short period of time. This technology has leaded to: the modern information society that connects us and tools; a large market; and, many types of products and applications. The solid-state circuit technology continuously evolves via breakthroughs and improvements every year. This book is devoted to review and present novel approaches for some of the main issues involved in this exciting and vigorous technology.

The book is composed of 22 chapters, written by authors coming from 30 different institutions located in12 different countries throughout the Americas, Asia and Europe. Thus, reflecting the wide international contribution to the book.

Low power consumption is becoming a paramount issue for modern integrated circuits, motivated by the huge integration level of modern electronics. In addition, the need for power-aware applications such as mobile electronics, RFIDs, implantable medical devices and smart sensor network motivates the development of low power consumption hardware. Circuit design techniques that aim for reduced power consumption are treated in the first two chapters. Accurate device modeling is essential for IC design and the models are constantly adapted to take into account smaller dimension effects. This subject is treated in chapter 3, focusing on the saturation mechanisms. Thermal noise and process variations affect the performance, yield and minimum bias voltage or power consumption of the circuits. These issues are the subjects of chapters 6 to 8.

The new and future CMOS technologies with constantly decreasing dimensions require new solutions to: reduce gate leakage; increase gate capacitance per area; reduce the sub-threshold slope; and increase transconductance, among other issues. These solutions have lead to new transistor structures, high-k dielectrics and metal gates. Critical technological innovations covering these solutions are presented in chapters 7 to 9.

Interconnects represents another critical issue in IC technology. A large part of the total die area is represented by interconnects having a large effect on the performance and reliability of the circuits. Carbon nanotubes are considered a promising material for interconnects. The modeling of interconnects as transmission lines and, in addition, the use of inductive-coupling links between chips are considered. Chapters 11 to 15 cover such important issues.

Microelectromechanical systems (MEMS) is a complementary field to integrated circuits. MEMS use similar materials and the same technology platforms. Furthermore, MEMS can be integrated in the same die of the electronic circuit for the case of smart sensors and actuators or MEMS can be integrated in the same package, as in a system in package approach. MEMS are essential for many existing applications. Moreover they are going through progressive evolution leading to new devices and new applications for all kind of automatization and sensor networks. Progress in materials, techniques, devices, interface circuits and packaging for MEMS are presented in the final 7 chapters of the book.

The broad range of subject presented in the book offers a general overview of the main issues in modern solid-state circuit technology. Furthermore, the book offers an in dept analysis on specific subjects for specialists. We believe the book is of great scientific and educational value for many readers.

I am profoundly indebted to the support provided by all of those involved in the work. First and foremost I would like to acknowledge and thank the authors that worked hard and generously agreed to share their results and knowledge. Second I would like to express my gratitude to the Intech team, that invited me to edit the book and give me their full support and a fruitful experience while working together to combine this book.

Editor

Jacobus W. Swart

Center of Technology for Information Renato Archer– CTI, Campinas, SP, Brazil

Contents

1

CMOS Voltage and Current Reference Circuits consisting of Subthreshold MOSFETs

— Micropower Circuit Components for Power-aware LSI Applications —

Ken Ueno
Hokkaido University
Japan

1. Introduction

The development of ultra-low power LSIs is a promising area of research in microelectronics. Such LSIs would be suitable for use in power-aware LSI applications such as portable mobile devices, implantable medical devices, and smart sensor networks [1]. These devices have to operate with ultra-low power, i.e., a few microwatts or less, because they will probably be placed under conditions where they have to get the necessary energy from poor energy sources such as microbatteries or energy scavenging devices [2]. As a step toward such LSIs, we first need to develop voltage and current reference circuits that can operate with an ultra-low current, several tens of nanoamperes or less, i.e., sub-microwatt operation. To achieve such low-power operation, the circuits have to be operated in the subthreshold region, i.e., a region at which the gate-source voltage of MOSFETs is lower than the threshold voltage [3; 4]. Voltage and current reference circuits are important building blocks for analog, digital, and mixed-signal circuit systems in microelectronics, because the performance of these circuits is determined mainly by their bias voltages and currents. The circuits generate a constant reference voltage and current for various other components such as operational amplifiers, comparators, AD/DA converters, oscillators, and PLLs. For this purpose, bandgap reference circuits with CMOS-based vertical bipolar transistors are conventionally used in CMOS LSIs [5; 6]. However, they need resistors with a high resistance of several hundred megaohms to achieve low-current, subthreshold operation. Such a high resistance needs a large area to be implemented, and this makes conventional bandgap references unsuitable for use in ultra-low power LSIs. Therefore, modified voltage and current reference circuits for lowpower LSIs have been reported (see [7]-[12], [14]-[17]). However, these circuits have various problems. For example, their power dissipations are still large, their output voltages and currents are sensitive to supply voltage and temperature variations, and they have complex circuits with many MOSFETs; these problems are inconvenient for practical use in ultra-low power LSIs. Moreover, the effect of process variations on the reference signal has not been discussed in detail. To solve these problems, I and my colleagues reported new voltage and current reference circuits [13; 18] that can operate with sub-microwatt power dissipation and with low sensitivity to temperature and supply voltage. Our circuits consist of subthreshold MOSFET circuits and use no resistors.

The following sections provide overviews of previous reported low-power reference circuits and a detailed explanation of our circuits. Section 2 describes the subthreshold current of MOSFETs and shows the temperature and process sensitivity of the current with a SPICE simulation. Section 3 describes the principle of conventional voltage and current reference circuits based on bandgap reference circuits. Sections 4 and 5 explain the operation principle of the reported voltage and current reference circuits and show the characteristics of prototype devices we made using 0.35-μm standard CMOS process technology. Finally, concluding remarks are presented in Sect. 6.

2. Subthreshold region (or weak inversion region) of MOSFETs

When the gate-source voltage of a MOSFET is lower than the threshold voltage, subthreshold current can be obtained. The subthreshold current through a MOSFET is an increasing exponential function of the gate-source voltage, and the current value is on the order of nanoamperes. Moreover, the subthreshold current is sensitive to temperature and process variations. The temperature and process characteristics of the subthreshold current are analyzed as follows.

Figure 1 shows the measured transfer curves of an nMOSFET in 0.35-μm CMOS process at different temperatures from -20 to 100°C. The drain-source voltage was set to 1 V. The threshold voltage is about 0.5 V in this device. The subthreshold drain current I_{DS} of a MOSFET is an exponential function of the gate-source voltage V_{GS} and the drain-source voltage V_{DS} and is given by

$$I_{DS} = KI_0 \exp\left(\frac{V_{GS} - V_{TH}}{\eta V_T}\right)\left(1 - \exp\left(-\frac{V_{DS}}{V_T}\right)\right), \quad (1)$$

$$I_0 = \mu C_{OX}(\eta - 1)V_T^2$$

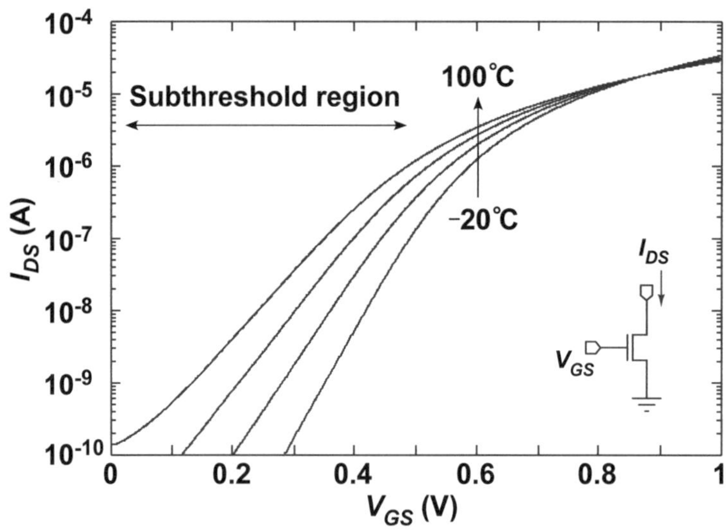

Fig. 1. Measured transfer curves of nMOSFET as a function of gate-source voltage V_{GS} at different temperatures.

where K is the aspect ratio (=W/L) of the transistor, μ is the carrier mobility, C_{OX} is the gate-oxide capacitance, V_T(=k_BT/q) is the thermal voltage, k_B is the Boltzmann constant, T is the absolute temperature, and q is the elementary charge, V_{TH} is the threshold voltage of a MOSFET, and η is the subthreshold slope factor [3], [19]. For V_{DS} > 0.1 V, current I_{DS} is independent of V_{DS} and is given by

$$I_{DS} = KI_0 \exp\left(\frac{V_{GS} - V_{TH}}{\eta V_T}\right). \tag{2}$$

The temperature dependence of the threshold voltage V_{TH} and the mobility μ of MOSFET can be given by

$$V_{TH} = V_{TH0} - \kappa T, \tag{3}$$

$$\mu(T) = \mu(T_0)(T/T_0)^{-m} \tag{4}$$

where $\mu(T_0)$ is the carrier mobility at room temperature T_0, m is the mobility temperature exponent, V_{TH0} is the threshold voltage at 0 K, and κ is the temperature coefficient of V_{TH} [20].

The temperature coefficient (T.C.) of the subthreshold current with fixed gate-source voltage is given by

$$\begin{aligned} T.C. &= \frac{1}{I_{DS}}\frac{dI_{DS}}{dT} \\ &= \frac{1}{\mu}\frac{d\mu}{dT} + \frac{1}{V_T^2}\frac{dV_T^2}{dT} + \frac{1}{\exp((V_{GS} - V_{TH})/\eta V_T)}\frac{d}{dT}\exp((V_{GS} - V_{TH})/\eta V_T) \\ &= \frac{2-m}{T} + \frac{\kappa - (V_{GS} - V_{TH})/T}{\eta V_T}. \end{aligned} \tag{5}$$

Process variations can be classified into two categories: i.e., within-die (WID) (intra-die) variation and die-to-die (D2D) (inter-die) variation [21]-[23]. The WID variation is caused by mismatches between transistor parameters within a chip and affects the relative accuracy of the parameters. In contrast, the D2D variation affects the absolute accuracy of transistor parameters between chips.

The process dependence of the subthreshold current can be expressed by

$$\frac{\Delta I_{DS}}{I_{DS}} = \frac{1}{I_{DS}}\left(\frac{\partial I_{DS}}{\partial \mu}\Delta\mu + \frac{\partial I_{DS}}{\partial V_{TH}}\Delta V_{TH}\right) = \frac{\Delta\mu}{\mu} - \frac{\Delta V_{TH}}{\eta V_T}. \tag{6}$$

The mobility variation $\Delta\mu$ is generally smaller than the threshold voltage variation ΔV_{TH}, so the current depends mainly on ΔV_{TH}.

Figure 2 shows the simulated subthreshold current with fixed gate-source voltages, obtained with a SPICE simulation with a set of 0.35-μm standard CMOS process. Current operating in the strong inversion region is also plotted for comparison. Fixed gate-source voltages were set to V_{TH}–0.2 V (weak inversion), and V_{TH}+0.2 V (strong inversion), respectively. Although

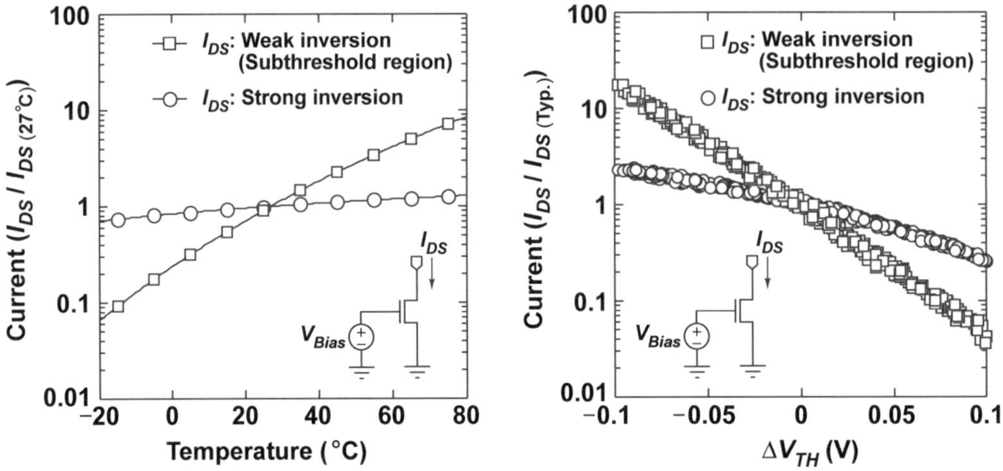

	T.C.	$\Delta I_{DS}/\Delta V_{TH}$
I_{DS} : Weak inversion ($V_{Bias}=V_{TH}-0.2$)	3%/°C	2.5%/mV
I_{DS} : Strong inversion ($V_{Bias}=V_{TH}+0.2$)	0.5%/°C	0.8%/mV

Fig. 2. (A). Simulated drain currents as a function of temperature. Fixed gate biases were set to V_{TH}–0.2 V (weak inversion), and V_{TH}+0.2 V (strong inversion). (B). Drain currents as a function of D2D threshold voltage variation ΔV_{TH}, as obtained from Monte Carlo simulation of 300 runs.

the current in the strong inversion region has a small temperature dependence (0.5%/°C), the subthreshold current has a large temperature dependence (3%/°C), as shown in Fig. 2-(A). Figure 2-(B) shows the simulated subthreshold current as a function of the threshold voltage variation ΔV_{TH}, as obtained from Monte Carlo simulation of 300 runs, assuming both die-to-die (D2D) variation (e.g., ΔV_{TH}, $\Delta \mu$, ΔT_{OX}, ΔL, ΔW) and within die (WID) variation (e.g., σ_{VTH}, σ_{μ}, σ_{TOX}, σ_L, σ_W) in transistor parameters [21; 22; 23]. Each open circle and square show I_{DS} for a run. The subthreshold current depends strongly on the threshold voltage variation (2.5%/mV) in comparison with the strong inversion current (0.8%/mV). Therefore, the subthreshold current is strongly dependent on temperature and process variations. In circuit designs, the process sensitivity of the subthreshold current has to be reduced by using large-sized transistors [23] and various analog layout techniques [24]. On the other hand, the exponential behavior and the high sensitivity to temperature of the subthreshold current can be used to compensate for temperature variation of a constant voltage, such as voltage reference circuits.

3. Voltage and current references based on bandgap reference circuits

Bandgap voltage reference circuits are widely used as voltage references. Figure 3 shows conventional bandgap voltage reference circuits [5],[6]. The circuits generate reference voltages independent of the process, supply voltage, and temperature, and consist of the MOSFET circuits, substrate pnp bipolar transistors, and resistors. The operation principles are as follows.

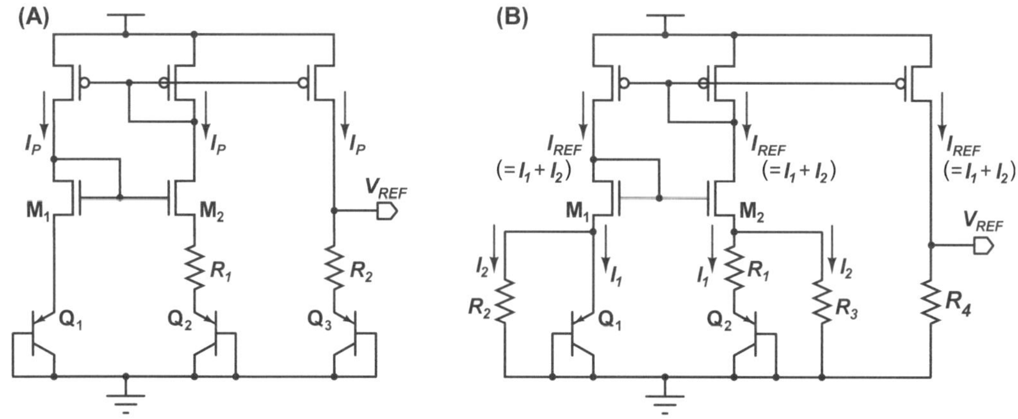

Fig. 3. (A). Conventional bandgap voltage reference circuit [5]. (B) Sub-1-V output bandgap voltage reference circuit [6] and current reference circuit [25].

3.1 Operation as voltage reference circuit

The collector current I_C of the bipolar transistor is given by

$$I_C = KI_S \exp\left(\frac{V_{BE}}{V_T}\right) \tag{7}$$

where K is the transistor size, I_S is the saturation current, and V_{BE} is the base-emitter voltage [5]. In the circuit in Fig. 3-(A), the operation current I_P is determined by the bipolar transistors Q_1 and Q_2 with different transistor sizes and the resistor R_1, and is given by

$$I_P = \frac{V_{BE1} - V_{BE2}}{R_1} = \frac{V_T \ln(K_2 / K_1)}{R_1}. \tag{8}$$

The current I_P is proportional to absolute temperature (PTAT). The resistor R_2 and the transistor Q_3 accept the current through the current mirror circuit and produce the output voltage, which is given by

$$V_{REF} = V_{BE3} + I_P R_2 = V_{BE3} + \frac{R_2}{R_1} V_T \ln(K_2 / K_1). \tag{9}$$

Equation (9) shows that V_{REF} can be expressed as a sum of the base-emitter voltage and thermal voltage scaled by the resistor ratio. Because V_{BE} has a negative T.C. and V_T has a positive T.C., output voltage V_{REF} with a zero T.C. can be obtained by adjusting the resistor ratio. The reference voltage is based on the bandgap energy of silicon, which is about 1.25 V. Banba *et al.* proposed a modified bandgap voltage reference circuit as shown in Fig. 3-(B). The circuit generates sub-1-V reference voltage. The operation currents I_1 and I_2 are given by

$$I_1 = \frac{V_{BE1} - V_{BE2}}{R_1} = \frac{V_T \ln(K_2 / K_1)}{R_1}, \qquad I_2 = \frac{V_{BE1}}{R_2}. \tag{10}$$

The resistor R_4 accepts the current $I_{REF}(=I_1+I_2)$ through a current mirror circuit and produces output voltage, so the output voltage can be expressed as

$$V_{REF} = I_{REF}R_4 = \frac{R_4}{R_2}V_{BE1} + \frac{R_4}{R_1}V_T \ln(K_2 / K_1). \quad (11)$$

Therefore, adjusting the resistor ratio, the circuit generates sub-1-V reference voltage that is independent of temperature.

3.2 Operation as current reference circuit

The circuit as shown in Fig. 3-(B) can be used as a current reference generator [25]. The temperature dependence of resistors is given by $R = R_0(1+ \alpha T)$, where R_0 is the resistance value at absolute zero temperature, and α is the temperature coefficient of the resistor. Because V_{BE} and $\Delta V_{BE}(=V_{BE1} - V_{BE2})$ have a negative and a positive temperature dependence, respectively, the temperature dependences can be expressed simply by $V_{BE}=V_{BE0}(1 - AT)$ and $\Delta V_{BE}=BT$, where A and B are the T.C. of V_{BE} and ΔV_{BE}, respectively, and V_{BE0} is the baseemitter voltage at absolute zero temperature. Therefore, the reference current $I_{REF}(=I_1+I_2)$ is given by

$$\begin{aligned} I_{REF} = I_1 + I_2 &= \frac{\Delta V_{BE}}{R_1} + \frac{V_{BE1}}{R_2} = \frac{BT}{R_{01}(1+\alpha T)} + \frac{V_{BE0}(1-AT)}{R_{02}(1+\alpha T)} \\ &= \frac{1}{R_{01}}(BT)(1-\alpha T) + \frac{V_{BE01}}{R_{02}}(1-AT)(1-\alpha T) \\ &\approx \frac{1}{R_{01}}(BT) + \frac{V_{BE01}}{R_{02}}(1-(A+\alpha)T). \end{aligned} \quad (12)$$

The left and right terms in Eq. (12) have negative and positive temperature dependence, respectively. Therefore, adjusting the appropriate resistor values, the circuit generates a reference current that is independent of temperature.

These circuits generate stable reference voltages and currents. However, the power dissipations of these circuits are too large (from 5 to 500 μW), so they need resistors with a high resistance of several hundred megaohms to achieve low-current, sub-microwatt operation. Such high resistance needs a large area to be implemented, and this makes conventional bandgap references unsuitable for use in ultra-low-power LSIs.

4. Overview of low-power voltage reference circuits

To achieve ultra-low-power operation and small area, modified voltage reference circuits without bipolar transistors have been reported (see [12]-[18]). These circuits consist of CMOS circuits that operate in the strong inversion and the subthreshold regions of MOSFET. The circuits generate a reference voltage that is independent of temperature and supply voltage. The next sections provide an overview of the reported low-power voltage reference circuits.

4.1 Voltage references based on ΔV_{GS}

Figure 4 shows voltage reference circuits based on the difference between the gate-source voltages of (A) two nMOS transistors, and (B) nMOS and pMOS transistors as reported by Song *et al.* [7] and Leung *et al.* [8], respectively. All MOSFETs operate in the strong inversion region.

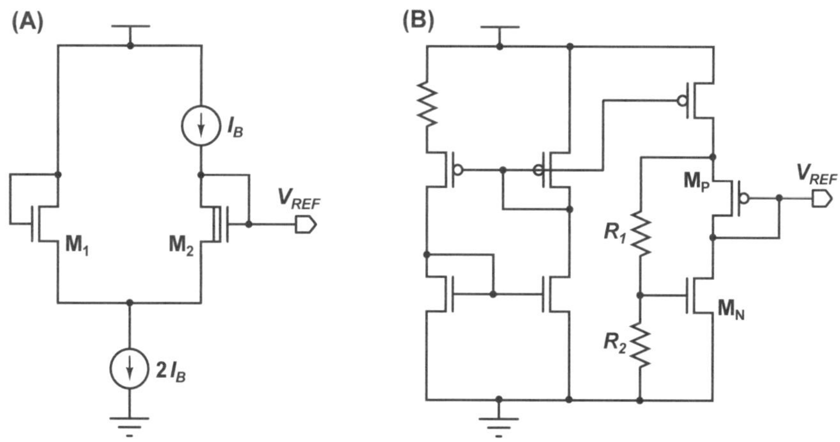

Fig. 4. Voltage reference circuits based on difference between gate-source voltages of (A) two nMOS transistors [7], and (B) nMOS and pMOS transistors [8].

The drain current I_{DS} that operates in the strong inversion, saturation region can be expressed as

$$I_{DS} = \frac{K\beta}{2}(V_{GS} - V_{TH})^2 \tag{13}$$

where K is the aspect ratio of the transistors, and $\beta(= \mu C_{OX})$ is the current gain factor.
The circuit in Fig. 4-(A) consists of M_1 and M_2 with different threshold voltage devices. The reference voltage is given by

$$\begin{aligned} V_{REF} &= V_{GS1} - V_{GS2} \\ &= (V_{TH01} - \kappa T) - (V_{TH02} - \kappa T) + \sqrt{\frac{2I_B}{\beta}}\left(\frac{1}{\sqrt{K_1}} - \frac{1}{\sqrt{K_2}}\right) \\ &\approx V_{TH01} - V_{TH02}. \end{aligned} \tag{14}$$

A low bias current I_B is used so that the temperature dependence of β can be ignored. Therefore, the reference voltage based on the difference between the threshold voltages can be obtained. However, the circuit requires a multiple-threshold voltage process, and, to cancel the temperature dependence of the reference voltage, the process must be controlled carefully so that the temperature coefficients κ of the two threshold voltages have the same value in each MOSFET.
Figure 4-(B) shows another voltage reference circuit based on the difference between the gate-source voltages of nMOS and pMOS transistors using a standard CMOS process. The reference voltage is given by

$$V_{REF} = \left(1 + \frac{R_1}{R_2}\right)V_{GSN} - V_{GSP}. \tag{15}$$

Therefore, adjusting the resistor ratio and the transistor sizes, the temperature dependence of the threshold voltages can be canceled, while the temperature dependence of the mobilities can be canceled only at room temperature. Consequently, the T.C. of the output voltage will be degraded for a wide temperature range. As reported in [8], a measured T.C. of 36.9 ppm/°C and a power dissipation of 30 μW were obtained. However, the power dissipation is still too large for use with sub-microwatt operation. To reduce the power dissipation, the circuit requires resistors with high resistance.

4.2 Voltage references operating in the strong inversion region of MOSFETs

Vita *et al.* proposed a voltage reference circuit consisting of transistors M_3–M_8 operating in the strong inversion region, and M_1 and M_2 operating in the subthreshold region as shown in Fig. 5-(A) [9]. In this circuit, the gate-source voltages for the four MOSFETs (M_1 through M_4) form a closed loop, so we find that $V_{GS3} + V_{GS1} = V_{GS2} + V_{GS4}$, i.e.,

$$\eta V_T \ln(K_2 / K_1) = \sqrt{2I_B / K_4\beta} - \sqrt{2I_B / K_3\beta}. \tag{16}$$

Therefore, the bias current I_B can be expressed by

$$I_B = \frac{K_4\beta}{2}\eta^2 V_T^2 \ln^2(K_2 / K_1)\left(\frac{\sqrt{K_3}}{\sqrt{K_3} - \sqrt{K_4}}\right)^2. \tag{17}$$

Transistors M_5–M_8 accept the current I_B and generate the output voltage. Most of the bias current I_B must flow through M_7 and M_8 rather than through M_5 and M_6 to compensate for the temperature dependence of the mobility μ. Therefore, the output voltage can be given by

$$V_{REF} = V_{GS8} + V_{GS5} - V_{GS7} = V_{TH} + \sqrt{\frac{2I_B}{\beta}}\left(\frac{1}{\sqrt{K_8}}\left(1 + \sqrt{\frac{K_6}{K_5}}\right) - \frac{1}{\sqrt{K_7}}\right)$$

$$= V_{TH} + \eta V_T \ln(K_2 / K_1)\frac{\sqrt{K_3K_4}}{\sqrt{K_3} - \sqrt{K_4}}\left(\frac{1}{\sqrt{K_8}}\left(1 + \sqrt{\frac{K_6}{K_5}}\right) - \frac{1}{\sqrt{K_7}}\right). \tag{18}$$

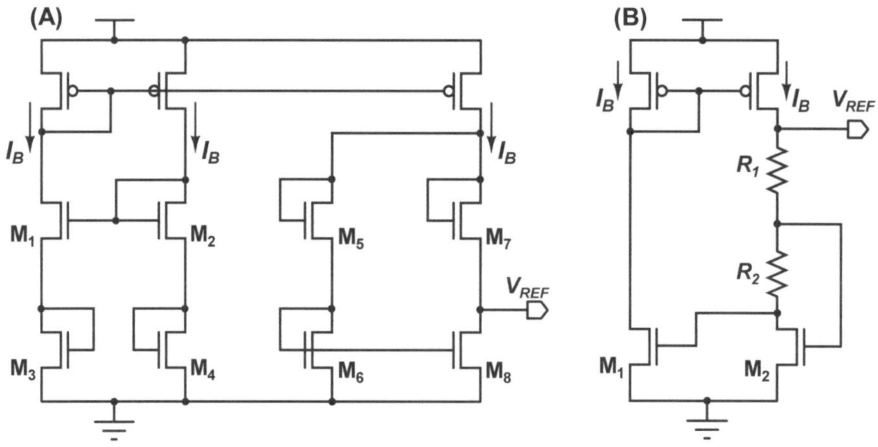

Fig. 5. Voltage reference circuit (A) operated in the strong inversion region [9], and (B) based on peaking current mirror circuit [10].

Because V_{TH} in Eq. (3) has a negative T.C. and V_T has a positive T.C., output voltage V_{REF} with a zero T.C. can be obtained by adjusting the size of the transistors.
As reported in [9], a measured T.C. of 12 ppm/°C and a power dissipation of 0.12 μW were obtained. Although the operation current of the circuit is on the order of nanoamperes, transistors M_3–M_8 operate in the strong inversion, saturation region. So, designs with careful transistor sizing are required for operation in each of the regions in MOSFETs.

4.3 Voltage references operating in the subthreshold region of MOSFETs

Cheng *et al.* developed a voltage reference using a peaking current mirror circuit as shown in Fig. 5-(B) [10]. All MOSFETs operate in the subthreshold region. The circuit forms a closed loop, i.e., $V_{GS1} = V_{GS2} - I_B R_2$, so the bias currents I_B can be expressed by

$$I_B = \frac{V_{GS2} - V_{GS1}}{R_2} = \frac{\eta V_T \ln(K_1 / K_2)}{R_2}. \tag{19}$$

The output voltage is given by

$$\begin{aligned} V_{REF} &= V_{GS2} + I_B R_1 \\ &= V_{GS2} + \frac{R_1}{R_2}\eta V_T \ln(K_1 / K_2). \end{aligned} \tag{20}$$

Because V_{GS} and V_T have a negative and a positive T.C., respectively, output voltage V_{REF} with a zero T.C. can be obtained by adjusting the resistor ratio. As reported in [10], a measured temperature coefficient of 62 ppm/°C and a power dissipation of 4.6 μW were obtained.
Huang *et al.* proposed a voltage reference circuit based on subthreshold MOSFETs [11] as shown in Fig. 6. The bias currents I_1 and I_2 are given by

$$I_1 = \frac{V_{GS8} - V_{GS9}}{R_2} = \frac{\eta V_T \ln(K_9 / K_8)}{R_2}, \qquad I_2 = \frac{V_{GS3}}{R_1} - \frac{K_5}{K_6} I_1. \tag{21}$$

Therefore, the output voltage can be expressed by

$$\begin{aligned} V_{REF} &= \left(\frac{K_{10}}{K_7} I_1 + \frac{K_{11}}{K_2} I_2 \right) R_3 \\ &= \frac{K_{11}}{K_2}\frac{R_3}{R_1} V_{GS3} + \left(\frac{K_{10}}{K_7} - \frac{K_{11}K_5}{K_2 K_6} \right) \frac{R_3}{R_2} \eta V_T \ln(K_9 / K_8). \end{aligned} \tag{22}$$

Because V_{GS} has a negative T.C. and V_T has a positive T.C., output voltage V_{REF} with a zero T.C. can be obtained by adjusting the resistor ratio and the transistor sizes. As reported in [11], a measured temperature coefficient of 271 ppm/°C and a power dissipation of 3.3 μW were obtained. In the circuits as shown in Figs. 5-(B) and 6, however, the power dissipations are still large. To achieve sub-microwatt operation, these circuits require resistors with a high resistance of several hundred megaohms.

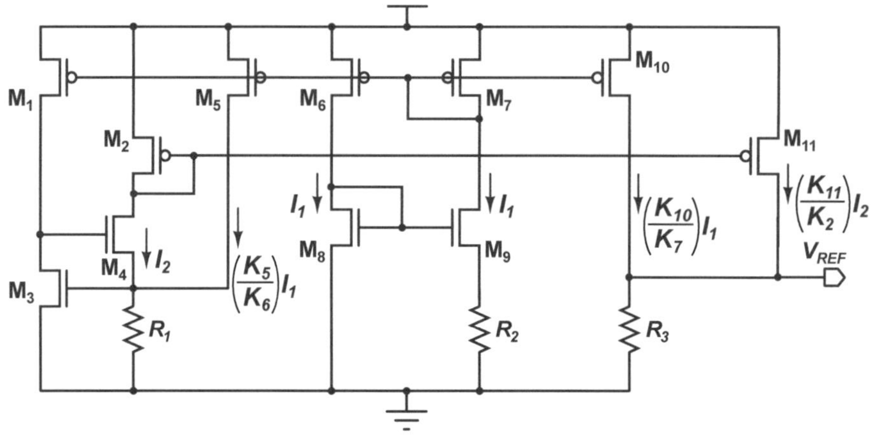

Fig. 6. Voltage reference circuit operated in the subthreshold region [11].

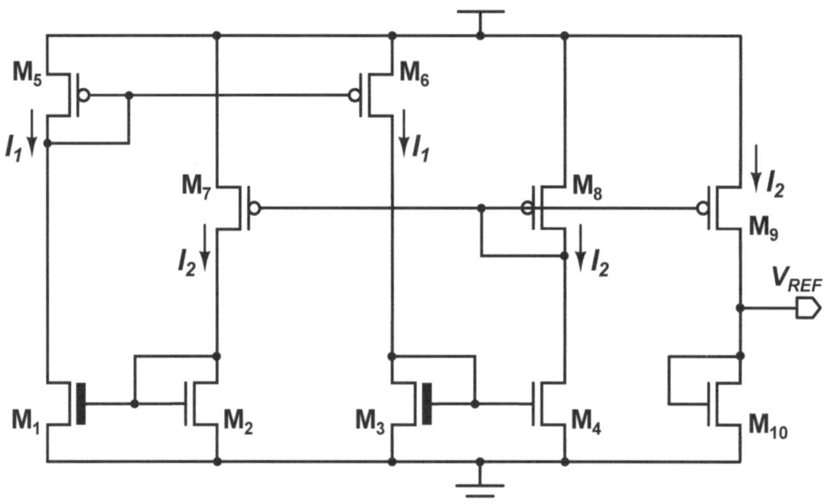

Fig. 7. Voltage reference circuit operated in the strong inversion and subthreshold regions using high-V_{TH} devices [12].

Vita *et al.* proposed a voltage reference circuit using two different threshold voltage devices as shown in Fig. 7 [12]. Transistors M_1 and M_3 with high-V_{TH} devices are operated in the subthreshold region, and M_2 and M_4 are operated in the strong inversion region. From $V_{GS1} = V_{GS2}$ and $V_{GS3} = V_{GS4}$, i.e.,

$$V_{TH_{HIGH}} + \eta V_T \ln\left(\frac{I_1}{K_1 I_0}\right) = V_{TH} + \sqrt{\frac{2I_2}{K_2\beta}}$$

$$V_{TH_{HIGH}} + \eta V_T \ln\left(\frac{I_1}{K_3 I_0}\right) = V_{TH} + \sqrt{\frac{2I_2}{K_4\beta}}. \tag{23}$$

Therefore, the output load current I_2 can be expressed as

$$I_2 = \frac{K_4\beta}{2(\sqrt{K_4/K_2}-1)^2}\eta^2 V_T^2 \ln^2(K_3/K_1). \tag{24}$$

Transistor M_{10} accepts the current I_2, and the output voltage can be given by

$$V_{REF} = V_{TH} + \sqrt{\frac{2I_2}{K_{10}\beta}}$$

$$= V_{TH} + \eta V_T \ln(K_3/K_1)\frac{\sqrt{K_4/K_{10}}}{\sqrt{K_4/K_2}-1}. \tag{25}$$

Because V_{TH} has a negative T.C. and V_T has a positive T.C., output voltage V_{REF} with a zero T.C. can be obtained by adjusting the size of the transistors.

As reported in [12], a measured T.C. of 10 ppm/°C and a power dissipation of 0.036 μW were obtained. However, the circuit requires a high-V_{TH} devices.

4.4 Voltage references consisting of subthreshold MOSFETs

Figure 8 shows our voltage reference circuit, which consists of a current source subcircuit and a bias-voltage subcircuit [13]. The current source subcircuit is a modified β multiplier self-biasing circuit that uses a MOS resistor M_{R1} instead of ordinary resistors. All the MOSFETs except for M_{R1} operate in the subthreshold region. MOS resistor M_{R1} is operated in a strong-inversion, deep-triode region. The circuit generates two voltages, one with a negative T.C. and one with a positive T.C., and combines them to produce a constant voltage with a zero T.C..

In the current source subcircuit, the current I_P is determined by two transistors M_1 and M_2, and the MOS resistor M_{R1}. The current I_P is given by

$$I_P = \frac{V_{DSR1}}{R_{M_{R1}}}$$

$$= K_{R1}\mu C_{OX}(V_{REF} - V_{TH})\eta V_T \ln(K_2/K_1). \tag{26}$$

In the bias-voltage subcircuit, the gate-source voltages (V_{GS3} through V_{GS7}) of the transistors form a closed loop [26], and the currents in M_4 and M_6 are $3I_P$ and $2I_P$. Therefore, we find that output voltage V_{REF} of the circuit is given by

$$V_{REF} = V_{GS4} - V_{GS3} + V_{GS6} - V_{GS5} + V_{GS7}$$

$$= V_{GS4} + \eta V_T \ln\left(\frac{2K_3K_5}{K_6K_7}\right)$$

$$= V_{TH} + \eta V_T \ln\left(\frac{3I_P}{K_4 I_0}\right) + \eta V_T \ln\left(\frac{2K_3K_5}{K_6K_7}\right) \tag{27}$$

where we assume that the mismatch between the threshold voltages of the transistors can be ignored. Equation (27) shows that V_{REF} can be expressed as a sum of the gate-source voltage V_{GS4} and thermal voltage V_T scaled by the transistor sizes. Because V_{TH} in Eq. (3) has a

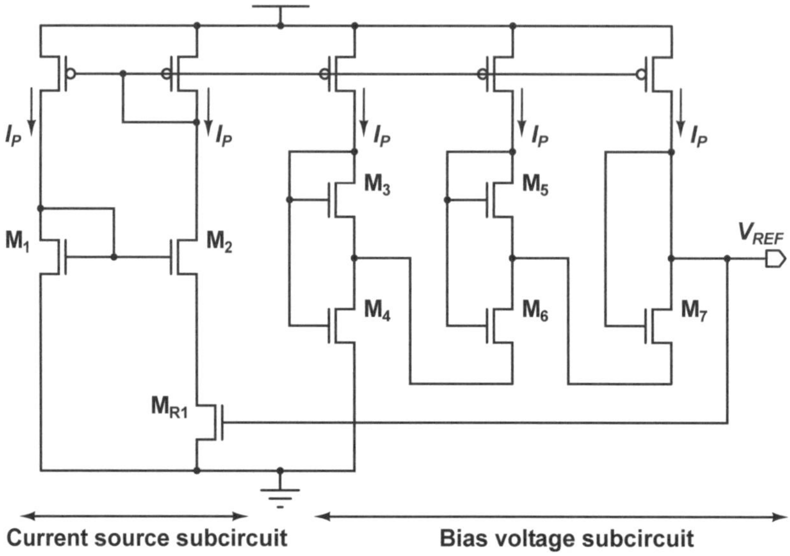

Fig. 8. Schematic of our voltage reference circuit [13]. All MOSFETs are operated in subthreshold region, except for MOS resistor M_{R1}, which is operated in strong-inversion, triode region.

negative T.C. and V_T has a positive T.C., output voltage V_{REF} with a zero T.C. can be obtained by adjusting the size of the transistors.
On the condition that $V_{REF} - V_{TH0} \ll \kappa T$ and $\eta V_T \ll \kappa T$, the T.C. of V_{REF} can be rewritten as

$$\frac{dV_{REF}}{dT} = -\kappa + \frac{\eta k_B}{q} \ln\left\{ \frac{6q\eta\kappa}{k_B(\eta - 1)} \frac{K_{R1}K_3K_5}{K_4K_6K_7} \ln\left(\frac{K_2}{K_1}\right)\right\}. \tag{28}$$

Therefore, a zero T.C. voltage can be obtained by setting the aspect ratios K_i in accordance with T.C.=0 (i.e., Eq. (28)=0). From Eqs. (27) and (28), we find that

$$V_{REF} = V_{TH0}. \tag{29}$$

This shows that the circuit generates a voltage equal to the threshold voltage of MOSFETs at 0 K. Using Eqs. (26) and (29), we can express current I_P as

$$I_P = K_{R1}\mu C_{OX}\kappa T \eta V_T \ln\left(\frac{K_2}{K_1}\right). \tag{30}$$

The current is determined only by the aspect ratios (K_1, K_2, and K_{R1}) and the temperature coefficient (κ) of the threshold voltage of MOSFETs, and it is independent of the threshold voltage V_{TH}, so the current I_P is less dependent on process variations as shown in the next section. The T.C. of the current can be given by

$$\frac{1}{I_P}\frac{dI_P}{dT} = \frac{1}{\mu}\frac{d\mu}{dT} + \frac{1}{T}\frac{dT}{dT} + \frac{1}{V_T}\frac{dV_T}{dT} = \frac{2-m}{T}. \tag{31}$$

The value of m is about 1.5 in standard CMOS process technologies, so current I_P has a positive T.C. and increases with temperature.

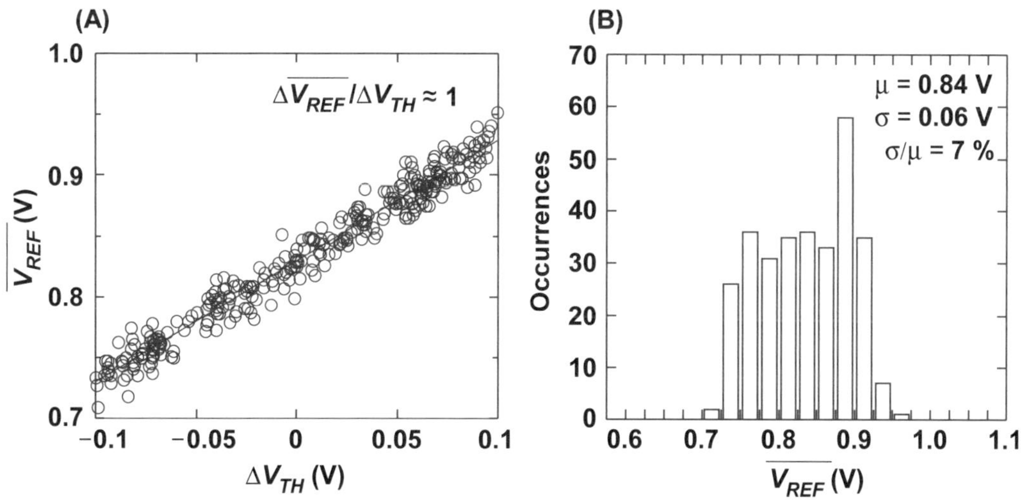

Fig. 9. (A). Average output voltage as a function of D2D variation ΔV_{TH} of threshold voltage, as obtained from Monte Carlo simulation of 300 runs. Output voltage shows a linear dependence on threshold voltage ($\Delta \overline{V_{REF}} / \Delta V_{TH} \approx 1$). (B). Distribution of output voltage, as obtained from Monte Carlo simulation.

4.4.1 Simulation and experimental results

We demonstrated the operation of our circuit with the aid of a SPICE simulation using a set of 0.35-μm standard CMOS parameters and assuming a 1.5-V power supply. To study the dependence of the output voltage on process variations, we performed Monte Carlo simulations assuming both D2D variation (e.g., ΔV_{TH}, $\Delta\mu$, ΔT_{OX}, ΔL, ΔW) and WID variation (e.g., σ_{VTH}, σ_{μ}, σ_{TOX}, σ_L, σ_W) in transistor parameters.

The results for 300 runs are depicted in Fig. 9. Figure 9-(A) shows the dispersion of V_{REF} from the average value ($\overline{V_{REF}}$) of V_{REF} from -20 to 80°C as a function of D2D threshold-voltage variation ΔV_{TH}. Each open circle shows $\overline{V_{REF}}$ for a run. As expected from Eq. (29), V_{REF} varies significantly with each run in a range from 0.75 to 0.95 V; this reflects the variation in transistor parameters for each run. The value of $\overline{V_{REF}}$ depends linearly on ΔV_{TH} because the circuit produces the voltage equal to the 0-K threshold voltage of MOSFETs. Figure 9-(B) shows the distribution of $\overline{V_{REF}}$. The average of $\overline{V_{REF}}$ was 840 mV, and the standard deviation was 60 mV. The coefficient of variation (σ/μ) was 7%, including D2D and WID variations.

We fabricated a prototype chip, using a 0.35-μm, 2-poly, 4-metal standard CMOS process. Figure 10-(A) shows measured output voltage V_{REF} as a function of temperature with supply voltage V_{DD} as a parameter. Almost constant voltage was achieved. The average of the output voltage was 745 mV. The temperature variation was 0.48 mV in a temperature range from -20 to 80°C, so the temperature coefficient was 7 ppm/°C. The line regulation was 20 ppm/V in the supply range of 1.4 to 3 V.

Figure 10-(B) shows measured current I_P as a function of temperature with power supply voltage as a parameter. The current I_P was about 36 nA at room temperature and reached the maximum of 39 nA at 80°C. The power dissipation of the circuit with a 1.5-V power supply was 0.32 μW at room temperature and varied from 0.28 to 0.35 μW at temperatures from -20 to 80°C. The temperature variation of the power dissipation was 0.2%/°C.

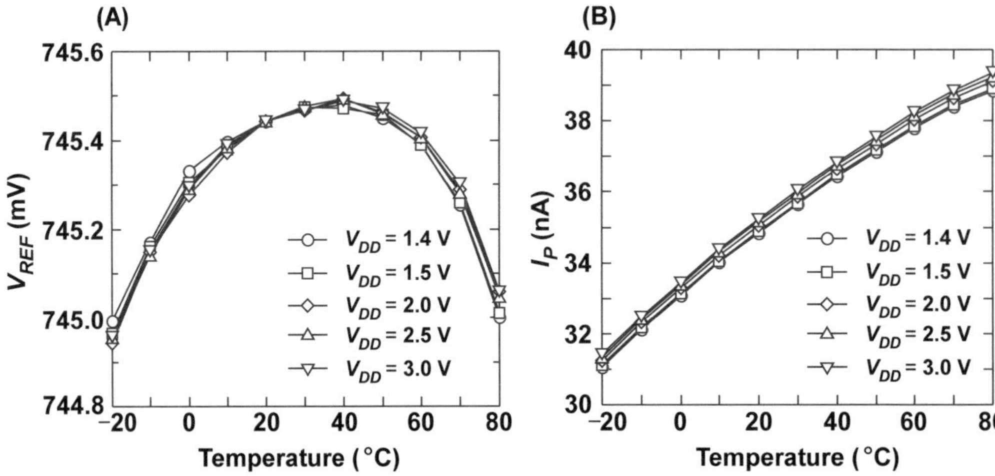

Fig. 10. (A). Measured output voltage V_{REF} as a function of temperature, with various supply voltages. Temperature coefficient was 7 ppm/°C and the supply regulation was 20 ppm/V. (B). Measured current I_P as a function of temperature for different supply voltages.

Table I summarizes the characteristics of our circuit [13] in comparison with other low-power CMOS voltage references reported in [8]-[12]. Our device is comparable to other circuits in power dissipation, PSRR, and chip area, and it is superior to others in T.C. and line sensitivity. Our circuit is therefore useful as a voltage reference for power-aware LSIs.

	This work [13]	JSSC '03 [8]	VLSI Symp. [9]
Process	0.35-μm, CMOS	0.6-μm, CMOS	0.35-μm, CMOS
Temperature range	-20 - 80°C	0 - 100°C	0 - 80°C
V_{DD}	1.4 - 3 V	1.4 - 3 V	1.5 - 4.3 V
$\overline{V_{REF}}$	745 mV	309 mV	891 mV
Power	0.3 μW(@1.4 V) Room temp.	29 μW(@3 V) Max. temp.	0.12 μW(@1.5 V) Room temp.
T.C.	7 ppm/°C	36.9 ppm/°C	12 ppm/°C
Line regulation	20 ppm/V	800 ppm/V	4600 ppm/V
PSRR	-45 dB(@100 Hz)	-47 dB(@100 Hz)	-59 dB(@100 Hz)
Chip area	0.055 mm^2	0.055 mm^2	0.015 mm^2

	Elec. Lett. '05 [10]	TCAS-II [11]	JSSC '07 [12]
Process	0.35-μm, CMOS	0.18-μm, CMOS	0.35-μm, CMOS
Temperature range	0 - 70°C	20 - 120°C	0 - 80°C
V_{DD}	1.4 - 3 V	0.85 - 2.5 V	0.9 - 4 V
$\overline{V_{REF}}$	579 mV	221 mV	670 mV
Power	4.6 μW(@2 V) N.A.	3.3 μW (@0.85 V) Average	0.036 μW (@0.9 V) Room temp.
T.C.	62 ppm/°C	271 ppm/°C	10 ppm/°C
Line regulation	6700 ppm/V	9000 ppm/V	2700 ppm/V
PSRR	-84 dB(@1 kHz)	N.A.	-47 dB(@100 Hz)
Chip area	0.126 mm^2	0.24 mm^2	0.045 mm^2

Table 1. Comparison of reported low-power CMOS voltage reference circuits

4.4.2 Discussion

Our circuit has several possible applications. The output voltage of our circuit can be used as a monitor signal for the D2D process variation in MOSFET threshold voltage because the output voltage is equal to the 0-K threshold voltage of MOSFETs in an LSI chip and is linearly dependent on the V_{TH} variation, as shown in Fig. 9-(A). This output voltage can be used to compensate for the threshold voltage variation in LSI chips. For example, consider the application to a reference current source. The process variation of the current I_P flowing in the circuit as shown in Fig. 8 (see Eq. (30)) can be expressed as

$$\frac{\Delta I_P}{I_P} = \frac{1}{I_P}\left(\frac{\partial I_P}{\partial \mu}\Delta\mu + \frac{\partial I_P}{\partial C_{OX}}\Delta C_{OX} + \frac{\partial I_P}{\partial \kappa}\Delta\kappa\right)$$

$$= \frac{\Delta\mu}{\mu} + \frac{\Delta C_{OX}}{C_{OX}} + \frac{\Delta\kappa}{\kappa}. \tag{32}$$

The current is independent of the threshold voltage variation. Although the current depends on the variation of the mobility $\Delta\mu/\mu$, gate-oxide capacitance $\Delta C_{OX}/C_{OX}$, and the temperature coefficient of the threshold voltage $\Delta\kappa/\kappa$, these variations are far smaller than the threshold voltage variation.

This way, the circuit can be used as an elementary circuit block for on-chip D2D process compensation systems, such as process- and temperature-compensated current references [27].

5. Overview of low-power current reference circuits

Current references with nanoampere-order currents are required to ensure circuit operation that is stable and highly precise, because power dissipation and performance of circuits are determined mainly by their bias currents. Nanoampere-current references for ultra-low-power LSIs have been reported in several papers [13]-[15]. The next sections provide an overview of the reported nanoampere current reference circuits.

5.1 Current references based on weak and strong inversion regions of MOSFETs

Sansen *et al.* developed a current reference circuit without resistors as shown in Fig. 11 [14]. Transistors M_2–M_{11} operate in the subthreshold region, and M_1 and M_{12} operate in the strong inversion region. The gate-source voltages of M_1–M_{12} form a closed loop, so we find that

$$V_{GS1} = V_{GS12}+V_{GS10}-V_{GS11}+V_{GS8}-V_{GS9}+V_{GS6}-V_{GS7}+V_{GS4}-V_{GS5}+V_{GS2}-V_{GS3}. \tag{33}$$

Assuming that the body effects of M_2–M_{10} are ignored, the output current I_{REF} is given by

$$I_{REF} = \frac{\beta}{2}\eta^2 V_T^2 \ln^2\left(120\cdot\frac{K_{11}K_9K_7K_5K_3}{K_{10}K_8K_6K_4K_2}\right)\left(\frac{K_1K_{12}}{K_{12}-K_1}\right). \tag{34}$$

The T.C. of the reference current is given by

$$T.C. = \frac{1}{I_{REF}}\frac{dI_{REF}}{dT} = \frac{1}{\mu}\frac{d\mu}{dT} + \frac{1}{V_T^2}\frac{dV_T^2}{dT} = \frac{2-m}{T}. \tag{35}$$

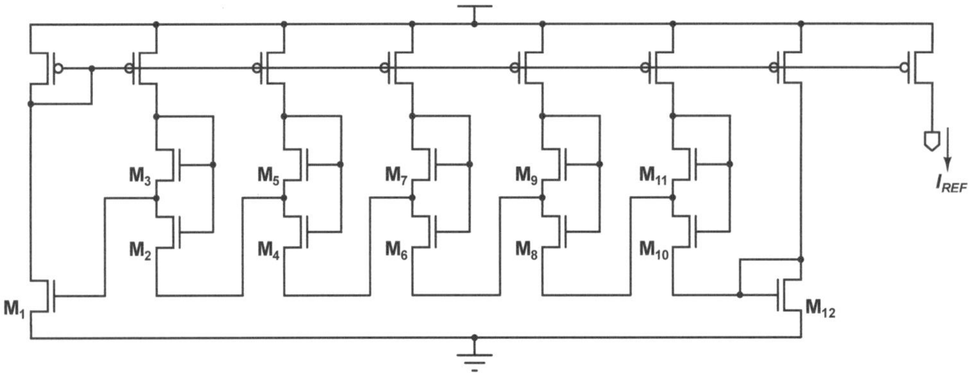

Fig. 11. Simplified schematic of current reference circuit without resistors [14]. Transistors M_2–M_{11} are operated in the subthreshold region, and M_1 and M_{12} are operated in the strong inversion region.

In a standard CMOS process, the mobility temperature exponent m is 1.5. Therefore, the output current has positive temperature dependence. As reported in [14], a measured temperature coefficient of 375 ppm/°C and a power dissipation of 10 μW were obtained, but the power dissipation is still large for use with sub-microwatt operation. Additionally, although the bias current of transistors M_2–M_{11} and M_1, M_{12} have the same value, nanoampere-order current, each transistor operates in a different region of the MOSFET. So, designs with careful transistor sizing and transistor matching are required.

5.2 Current references based on square root circuit

Lee *et al.* proposed a current reference circuit based on a square root circuit as shown in Fig. 12 [15]. Transistors M_1–M_4 operate in the subthreshold region, and other transistors operate in the strong inversion region. The gate-source voltages for the four MOSFETs (M_1 through M_4) form a closed loop, so we find that $V_{GS1} + V_{GS2} = V_{GS3} + V_{GS4}$. From the translinear principle [26], we can obtain

$$I_{REF} = \sqrt{\frac{K_3 K_4}{K_1 K_2} \cdot I_1 \cdot I_2}. \tag{36}$$

Current I_1 is determined by the gate-source voltages of M_5, M_6, and M_7. We find that $V_{GS7} + V_{GS6} = V_{DD} - V_{GS5}$, so current I_1 can be given by

$$I_1 = \frac{\beta (V_{DD} - 3V_{TH})^2}{2(1 + 2\sqrt{K_5 / K_6})^2} \tag{37}$$

where $K_6 = K_7$ is assumed.

The β-multiplier self-biasing circuit consisting of M_{16}–M_{19} and a resistor R generates current I_2. From $V_{GS18} = V_{GS16} + I_2 R$, current I_2 is given by

$$I_2 = \frac{2}{\beta R^2}\left(1 - \sqrt{K_{18} / K_{16}}\right)^2. \tag{38}$$

From Eqs. (36), (37), and (38), the output current can be rewritten as

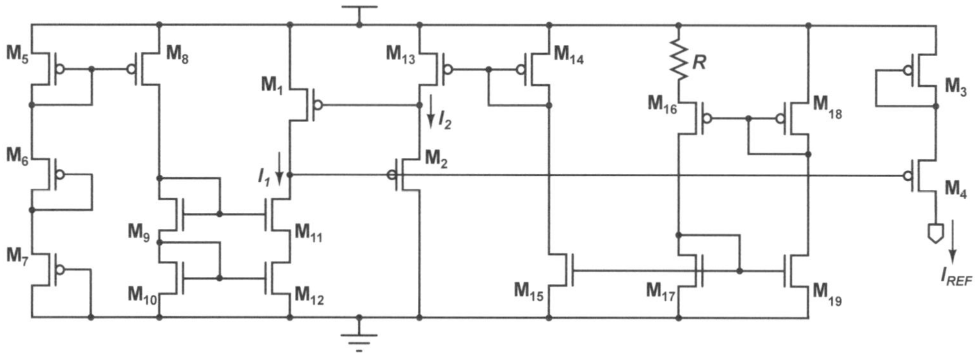

Fig. 12. Current reference circuit based on square root circuit [15]. Transistors M_5–M_{12} generate I_1, and transistors M_{13}–M_{19} generate I_2.

$$I_{REF} = \sqrt{\frac{K_3K_4}{K_1K_2}} \cdot \frac{1-\sqrt{K_{18}/K_{16}}}{1+2\sqrt{K_5/K_6}} \cdot \frac{V_{DD}-3V_{TH}}{R}. \tag{39}$$

Resistor R is an on-chip diffusion resistor, so the temperature dependence of the resistor is given by $R = R_0(1 + \alpha T)$, where R_0 is the resistance value at absolute zero temperature, and α is the temperature coefficient of the resistor. The T.C. of the output current can be expressed by

$$\begin{aligned} T.C. &= \frac{1}{I_{REF}}\frac{dI_{REF}}{dT} \\ &= (1+\alpha T)\frac{d}{dT}\left(\frac{1}{1+\alpha T}\right) + \frac{1}{(V_{DD}-3V_{TH})}\frac{d(V_{DD}-3V_{TH})}{dT} \\ &= -\frac{\alpha}{1+\alpha} + \frac{3\kappa}{V_{DD}-3V_{TH}}. \end{aligned} \tag{40}$$

As reported in [15], a measured T.C. of 230 ppm/°C was obtained. From Eqs. (39) and (40), however, the absolute value of the current and the T.C. depend strongly on the supply voltage V_{DD} and the threshold voltage V_{TH}.

5.3 Current references based on self-biasing technique without resistors

Figure13-(A) shows a β multiplier self-biasing circuit [31]. The circuit has a simple configuration and generates a PTAT current. However, the circuit requires large resistance of the resistor to reduce the operation current. To solve this problem, Oguey *et al.* developed a modified β multiplier self-biasing circuit that uses a MOS resistor, M_3, instead of ordinary resistors as shown in Fig. 13-(B) [16]. The gate-source voltage for MOS resistor M_3 is generated by a diode-connected transistor M_4. Transistors M_1 and M_2 operate in the subthreshold region. MOS resistor M_{R1} operates in a strong-inversion, deep-triode region, and the diode-connected transistor M_4 operates in the strong-inversion, saturation region. The drain currents I_3 and I_4 in M_3 and M_4 are given by

$$I_3 = K_3\beta(V_{GS} - V_{TH})V_{DS3}, \tag{41}$$

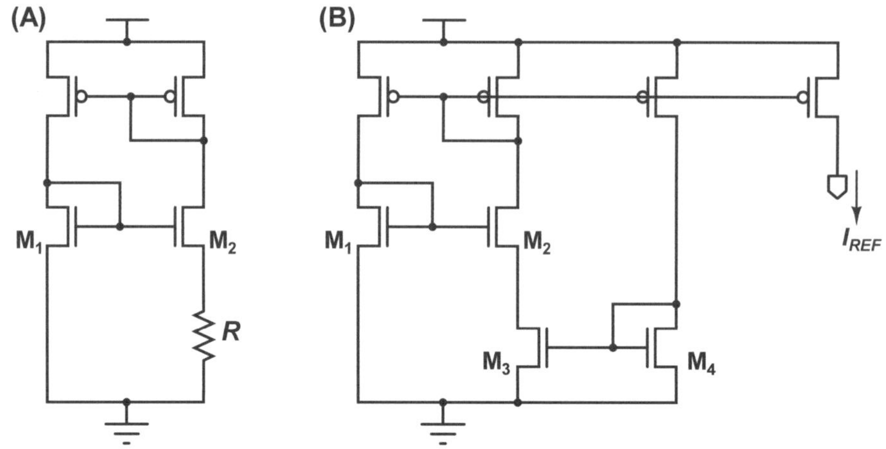

Fig. 13. (A). β-multiplier self-biasing circuit [31]. (B). Current reference circuit based on self-biasing circuit without resistors [16]. Transistors M_1 and M_2 operate in the subthreshold region, M_3 operates in the strong inversion, triode region, and M_4 is operated in the strong inversion, saturation region.

$$I_4 = \frac{K_4\beta}{2}(V_{GS} - V_{TH})^2. \tag{42}$$

The gate-source voltages of transistors M_3 and M_4 have the same value, so the output current can be expressed by

$$I_{REF} = K_3\beta\sqrt{\frac{2I_{REF}}{K_4\beta}}V_{DS} = \frac{2K_3^2\beta}{K_4}\eta^2 V_T^2 \ln^2(K_2 / K_1). \tag{43}$$

The temperature coefficient of the reference current is given by

$$T.C. = \frac{1}{I_{REF}}\frac{dI_{REF}}{dT} = \frac{1}{\mu}\frac{d\mu}{dT} + \frac{1}{V_T^2}\frac{dV_T^2}{dT} = \frac{2-m}{T}. \tag{44}$$

Therefore, the output current has positive temperature dependence. In other words, the T.C. of the current will never be zero. As reported in [16], a measured temperature coefficient of 1100 ppm/°C was obtained. Note that the transistors M_1–M_2, M_3 and M_4 operate in different regions of the MOSFET with the same current value, which is on the order of nanoamperes. So, designs with careful transistor sizing and transistor matching using large-sized transistors are required.

5.4 Current references consisting of subthreshold MOSFETs

Figure 14 shows the current reference circuit we proposed [18]. The circuit consists of a bias-voltage subcircuit and a current-source subcircuit. The bias-voltage subcircuit is a modified β multiplier self-biasing circuit as reported in [16]. Bias voltage V_B for MOS resistorM_3 is generated by a diode-connected transistor M_4. The current-source subcircuit accepts bias voltage V_B and generates reference current I_{OUT} that is independent of temperature and supply voltage. All MOSFETs operate in the subthreshold region except for M_3 and M_4.

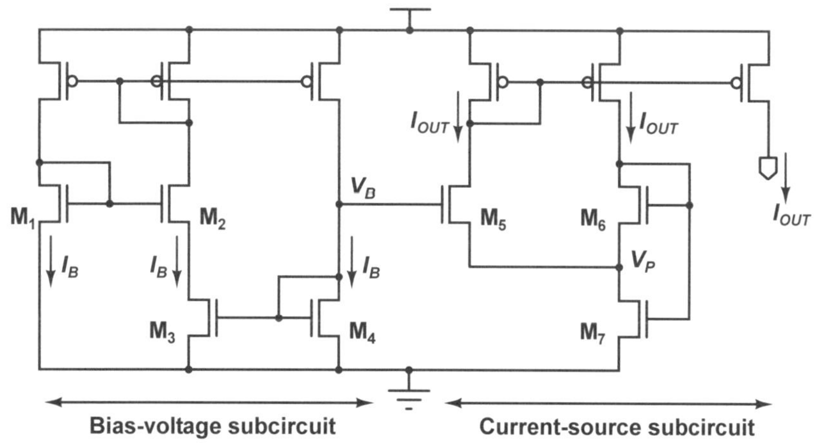

Fig. 14. Schematic of our current reference circuit [18]. All MOSFETs operate in the subthreshold region except for M_3 and M_4.

The current I_B is determined by the gate-source voltages of M_1 and M_2, and the drain-source voltage of M_3, so, we arrive at expression

$$I_B = \frac{V_{DS3}}{R_{M_3}}$$

$$= K_3 \mu C_{OX} (V_B - V_{TH}) \eta V_T \ln(K_2 / K_1) \tag{45}$$

for current I_B. Diode-connected transistor M_4 operates in the strong inversion and saturation regions. Its drain current I_B is given by

$$I_B = \frac{K_4 \mu C_{OX}}{2} (V_B - V_{TH})^2. \tag{46}$$

Because current I_B of M_3 is equal to I_B of M_4 (i.e., Eq. (45) = Eq. (46)), V_B is given by

$$V_B = V_{TH4} + \frac{2K_3}{K_4} \eta V_T \ln(K_2 / K_1). \tag{47}$$

Output current I_{OUT} through transistor M_5 can be given by

$$I_{OUT} = K_5 I_0 \exp\left(\frac{V_B - V_P - V_{TH5}}{\eta V_T} \right). \tag{48}$$

The source voltage V_P of transistor M_5 operated in the subthreshold region can be given by

$$V_P = V_{GS7} - V_{GS6}$$

$$= \eta V_T \ln(2K_6 / K_7) - \delta V_{TH76} \tag{49}$$

where δV_{TH76} is the difference between the threshold voltages of M_6 and M_7 with different transistor sizes (including the body effect in the transistors). From Eqs. (47), (48), and (49), we find that

$$I_{OUT} = I_0 \exp\left(\frac{\delta V_{TH}}{\eta V_T}\right)\frac{K_5K_7}{2K_6}\left(\frac{K_2}{K_1}\right)^{2K_3/K_4} \tag{50}$$

where $\delta V_{TH}(= V_{TH7} + V_{TH4} - V_{TH6} - V_{TH5})$ is the difference between the threshold voltages of transistors M_4–M_7. The value of δV_{TH} depends on the transistor sizes [28],[29]. This way, we can obtain a reference current with nanoampere-order.

The temperature coefficient (T.C.) of the output current I_{OUT} is given by

$$T.C. = \frac{1}{I_{OUT}}\frac{dI_{OUT}}{dT}$$

$$= \frac{1}{\mu}\frac{d\mu}{dT} + \frac{1}{V_T^2}\frac{dV_T^2}{dT} + \frac{1}{\exp\left(\frac{\delta V_{TH0}}{\eta V_T}\right)}\frac{d\exp\left(\frac{\delta V_{TH0}}{\eta V_T}\right)}{dT}$$

$$= \frac{2 - m - (\delta V_{TH0} / \eta V_T)}{T} \tag{51}$$

where $\delta V_{TH0}(= V_{TH07} + V_{TH04} - V_{TH06} - V_{TH05})$ is the difference between the threshold voltages at 0 K of transistors M_4–M_7. Therefore, the condition for a zero temperature coefficient can be given by

$$2 - m - (\delta V_{TH0} / \eta V_T) = 0. \tag{52}$$

Because the difference between the threshold voltages δV_{TH0} is insensitive to temperature, adjusting δV_{TH0} to an appropriate value will provide a zero T.C. at room temperature. Figure 15-(A) shows the calculated T.C. in Eq. (51) as a function of temperature with δV_{TH0} as a parameter. The mobility temperature exponent m was set to 1.5, and the subthreshold slope factor η was set to 1.3 [19; 30]. The T.C.s in the circuits reported in [13; 14; 16] are also plotted for comparison. The reported circuits [13; 14; 16] have a positive T.C. in a temperature range from -20 to 80°C, and these T.C.s will never be zero. On the other hand, our circuit can achieve a zero T.C. current at δV_{TH0}=17 mV and at room temperature.

In this way, we can obtain a zero T.C. current by setting an appropriate δV_{TH0}. The value of δV_{TH0} can be adjusted by the transistor sizes [28; 29].

Next, let us consider the effect of process variations on the output current. The process variations of the output current I_{OUT} can be expressed as

$$\frac{\Delta I_{OUT}}{I_{OUT}} = \frac{1}{I_{OUT}}\left(\frac{\partial I_{OUT}}{\partial \mu}\Delta\mu + \frac{\partial I_{OUT}}{\partial \delta V_{TH}}\Delta\delta V_{TH}\right)$$

$$= \frac{\Delta\mu}{\mu} + \frac{\Delta\delta V_{TH}}{\eta V_T}. \tag{53}$$

The mobility variation is generally smaller than the threshold voltage variation, so the output current depends mainly on $\Delta\delta V_{TH}/\eta V_T$, which is the variation of the threshold-voltage difference between transistors in a chip. Therefore, reducing WID variation is important in our device. The WID variation can be reduced by using large-sized transistors [23] and various analog layout techniques [24].

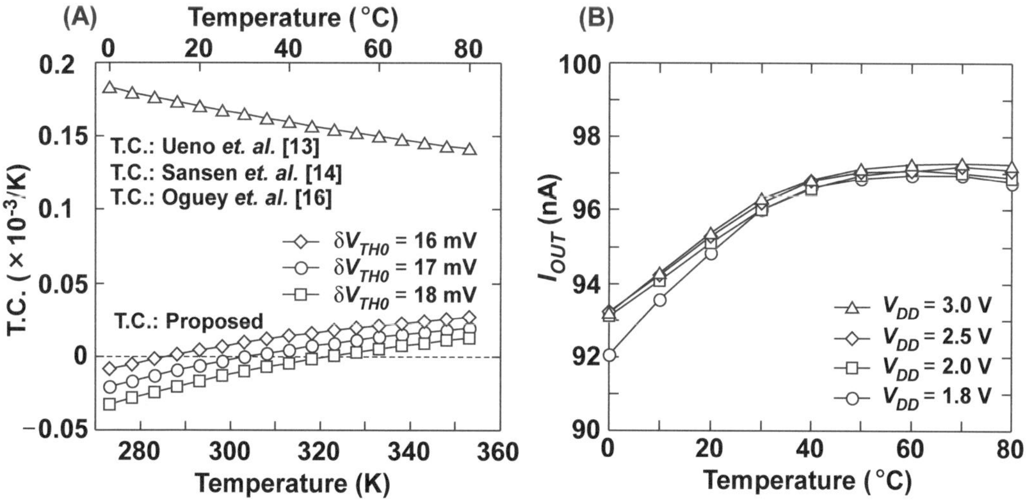

Fig. 15. (A). Calculated T.C.s of output currents as a function of temperature, with various δV_{TH0}; theoretical values obtained from Eqs. (31), (35), (44), and (51). (B). Measured output current I_{OUT} as a function of temperature with various supply voltages. T.C. was 520 ppm/°C.

5.4.1 Experimental results

We fabricated a prototype chip using a 0.35-μm, 2-poly, 4-metal standard CMOS process, and we designed the circuit so as to produce a 100-nA output current.

Figure 15-(B) shows measured output current I_{OUT} as a function of temperature with supply voltage V_{DD} as a parameter. The power supply voltage was set in a range from 1.8 to 3 V. The output current was about 96 nA and almost constant at temperatures in a range from 0 to 80°C. The temperature dependence and temperature coefficient were 50 pA/°C and 520 ppm/°C. An almost-constant reference current was obtained over a wide temperature range. The line regulation was 0.2%/V in a supply range of 1.8 to 3 V.

Table II summarizes the characteristics of our device in comparison with other low-power CMOS current references reported in [13]-[18]. Our device is superior to others in chip area. In the circuits reported in [13]-[18], there are trade-offs between the power dissipations and the T.C. of the reference currents. Our device achieved an acceptable trade-off. The power dissipation of our device was 1 μW at a 1.8-V power supply, and the load regulation was 0.02%/V.

6. Conclusion and discussion

In this chapter, overviews of previous reported low-power reference circuits and details of our circuits were provided. These circuits generate constant reference voltages and currents that are independent of supply voltage and temperature. However, to achieve sub-microwatt operation in circuits that consist of MOSFETs and resistors, they require resistors with a high resistance of several hundred mega ohms. Such a high resistance needs a large area to be implemented, and this is quite inconvenient for practical use in ultra-low power LSIs. Therefore, reference circuits for sub-microwatt operation have to be implemented without the use of resistors.

	This work [18]	JSSC '09 [13]	JSSC '88 [14]
Process	0.35-μm, CMOS	0.35-μm, CMOS	3-μm, CMOS
Temperature range	0 - 80℃	–20 - 80℃	0 - 80℃
V_{DD}	1.8 - 3 V	1.4 - 3 V	$\geq$3.5 V
I_{OUT}	96 nA	36 nA	774 nA
Power	1 μW(@1.8 V) Room temp.	0.3 μW(@1.5 V) Room temp.	10 μW(@5 V) N.A.
T.C.	520 ppm/℃	2200 ppm/℃	375 ppm/℃
Line regulation	0.2%/V	0.002%/V	0.015%/V
Load regulation	0.02%/V	N.A.	0.004%/V
Chip area	0.014 mm^2	0.06 mm^2	0.2 mm^2

	Elec. Lett. '96 [15]	JSSC '97 [16]	TCAS-II '05 [17]
Process	2-μm, CMOS	2-μm, CMOS	1.5-μm, CMOS
Temperature range	0 - 75℃	–40 - 80℃	–20 - 70℃
V_{DD}	5 V	$\geq$1.2 V	$\geq$1.1 V
I_{OUT}	285 nA	1 - 100 nA	0.41 nA
Power	N.A. N.A.	0.07 μW(@2.3 V) Room temp.	0.002 μW(@1.1 V) N.A.
T.C.	230 ppm/℃	1100 ppm/℃.	2500 ppm/℃
Line regulation	N.A.	10%/V	6%/V
Load regulation	N.A.	N.A.	N.A.
Chip area	N.A.	0.06 mm^2	0.046 mm^2

Table 2. Comparison of reported low-power CMOS current reference circuits

In the voltage reference circuits, reference voltages based on the difference between the threshold voltages (ΔV_{TH}), the difference between the gate-source voltages (ΔV_{GS}), and the threshold voltage at 0 K (V_{TH0}) have been proposed. However, the reference circuits based on ΔV_{TH} require a multiple-threshold voltage process, and the temperature dependence of the reference circuits based on ΔV_{GS} cannot be canceled for a wide temperature range. Therefore, these are unsuitable for practical use in ultra-low power LSIs. The voltage reference circuits based on V_{TH0} are promising circuit configurations because of their simple circuitries, sub-microwatt operation, and reference voltages that are insensitive to temperature over a wide temperature range. In our prototype, the T.C. and line regulation of the output voltage were 7 ppm/°C and 20 ppm/V and a power dissipation of 0.3 μW was obtained. However, because the absolute value of the reference voltages changes with the process variations of the threshold voltage, the circuit cannot be used as a reference voltage in conventional circuit systems. Therefore, the circuits require calibration techniques such as programmable MOS transistor arrays or adjustment of the bulk voltage of the MOSFET. Because the temperature dependence of the reference voltages can be canceled, one-point calibration techniques will enable us to compensate for process variations.

As other applications, because the output voltage shows a linear dependence on the threshold voltage variation, the reference voltage can be utilized as a D2D process variation signal for the techniques to compensate for the threshold voltage variation in an LSI chip.

Current reference circuits consisting of MOSFET circuits operating in the strong inversion region and the subthreshold region have been proposed. Because each MOSFET in the circuits operates in a different region with the same current value, which is on the order of nanoamperes, careful transistor sizing and reducing WID variation in the design are important. The WID variation can be reduced by conventional circuit design techniques. In our circuit, techniques such as using large-sized transistors and common centroid layout were used to reduce the effect of the WID variation.

From the theoretical results in the reported current references, the reference currents have a positive temperature dependence. Therefore, the circuits cannot be used as reference current circuits in environments with temperature changes. To solve this problem, we developed a temperature compensated current reference circuit with simple circuitry and a small area, and fabricated a prototype chip that generates a 100-nA output current. The T.C. and line regulation of the output current were 520 ppm/°C and 0.2%/V. A power dissipation of 1 μW was obtained.

These circuits will be useful as voltage and current reference circuits for subthreshold-operated, power-aware LSI applications such as RFIDs, mobile devices, implantable medical devices, and smart sensor networks.

7. References

[1] K. Ueno, T. Hirose, T. Asai, and Y. Amemiya, "CMOS smart sensor for monitoring the quality of perishables," IEEE Journal of Solid-State Circuits, vol. 42, no, 4, pp. 798-803, Apr. 2007.

[2] P. Fiorini, I. Doms, C. Van Hoof, R. Vullers, "Micropower energy scavenging," Proc. of the 34th European Solid State Circuits Conference (ESSCIRC), pp. 4 9, 2008.

[3] A. Wang, B.H. Clhoun, A.P. Chandracasan, Sub-threshold Design for Ultra Low-Power Systems, Springer, 2006.

[4] A. P. Chandrakasan, D. C. Daly, J. Kwong, Y. K. Ramadass, "Next Generation Micropower Systems," Proc. of IEEE Symposium on VLSI Circuits, pp. 2-5, 2008.

[5] P. R. Gray and R. G. Meyer, Analysis and Design of Analog Integrated Circuits, 3rd ed. New York: Wiley, 1993.

[6] H. Banba, H. Shiga, A. Umezawa, T. Miyaba, T. Tanzawa, S. Atsumi, and K. Sakui, "A CMOS bandgap reference circuit with sub-1-V operation," IEEE Journal of Solid-State Circuits, vol. 34, no. 5, pp. 670 - 674, May. 1999.

[7] B.-S. Song and P. R. Gray, "Threshold-voltage temperature drift in ion-implanted MOS transistors," IEEE J. Solid-State Circuits, vol. SC-17, no. 2, pp. 291-298, Apr. 1982.

[8] K. N. Leung, P. K. T. Mok, "A CMOS voltage reference based on weighted ΔV_{GS} for CMOS low-dropout linear regulators," IEEE Journal of Solid-State Circuits, vol. 38, no. 1, pp. 146 - 150, Jan. 2003.

[9] G. De Vita, G. Iannaccone, P. Andreani, "A 300 nW, 12 ppm/°C Voltage Reference in a Digital 0.35 μm CMOS Process," Dig. of Tech. Papers Symposium on VLSI Circuits, pp. 81-82, 2006.

[10] M.-H. Cheng, Z.-W. Wu, "Low-power low-voltage reference using peaking current mirror circuit," Electronics Letters, vol. 41, no. 10, pp. 572 - 573, 2005.

[11] P-H. Huang, H. Lin, Y-T. Lin, "A simple subthreshold CMOS voltage reference circuit with channel-length modulation compensation," IEEE Trans. Circuits Syst. II, Exp. Briefs, pp. 882 - 885, 2006.

[12] G. De Vita, G. Iannaccone, "A Sub-1-V, 10 ppm/°C, nanopower voltage reference generator" IEEE Journal of Solid-State Circuits, vol. 42, no. 7, pp. 1536 - 1542, Jul. 2007.

[13] K. Ueno, T. Hirose, T. Asai, Y. Amemiya, "A 300 nW, 15 ppm/°C, 20 ppm/V CMOS Voltage Reference Circuit Consisting of Subthreshold MOSFETs," IEEE J. Solid-State Circuits, vol. 44, no. 7, pp. 2047-2054, Jul. 2009.
[14] W.M. Sansen, F. O. Eynde, M. Steyaert, "A CMOS temperaturecompensated current reference," IEEE J. Solid-State Circuits, vol. 23, no. 3, pp. 821-824, Jun. 1988.
[15] C.-H. Lee, H.-J. Park, "All-CMOS temperature-independent current reference," Electronics Letters, vol. 32, pp. 1280-1281, Jul. 1996.
[16] H. J. Oguey and D. Aebischer, "CMOS current reference without resistance," IEEE J. Solid-State Circuits, vol. 32, no. 7, pp. 1132-1135, Jul. 1997.
[17] E. M. Camacho-Galeano and C. Galup-Montoro, "A 2-nW self-biased current reference in CMOS technology," IEEE Trans. Circuits Syst. II, Exp. Briefs, vol. 52, no. 2, pp. 61-65, Feb. 2005.
[18] K. Ueno, T. Asai, Y. Amemiya, "Current reference circuit for subthreshold CMOS LSIs," in Extended Abstract of Int. Conf. on Solid State Devices and Materials (SSDM), pp. 1000- 1001, 2008.
[19] Y. Taur, T.H. Ning, Fundamentals of Modern VLSI Devices, Cambridge University Press, 2002.
[20] I. M. Filanovsky, A. Allam, "Mutual compensation of mobility and threshold voltage temperature effects with applications in CMOS circuits," IEEE Trans. Circuits Syst. I, Fundam. Theory Appl, pp. 876-884, 2001.
[21] K. A. Bowman, S. G. Duvall, J. D. Meindl, "Impact of die-to-die and within-die parameter fluctuations on the maximum clock frequency distribution for gigascale integration," IEEE Journal of Solid-State Circuits, vol. 37, no. 2 pp. 183 - 190, Feb. 2002.
[22] H. Onodera, "Variability: Modeling and Its Impact on Design," IEICE Trans. Electron., Vol.E89-C, pp. 342 - 348, 2006.
[23] M. J. M. Pelgrom, A. C. J. Duinmaijer, A. P. G. Welbers, "Matching properties of MOS transistors," IEEE Journal of Solid-State Circuits, vol. 24, no. 5 pp. 1433 - 1439, Oct. 1989.
[24] A. Hastings, The Art of Analog Layout, Prentice Hall, 2001.
[25] J. Chen, B. Shi, "1 V CMOS current reference with 50 ppm/°C temperature coefficient," Electronics Letters, vol. 39, no. 2, pp. 209-210, Jan. 2003.
[26] B. Gilbert, "TRANSLINEAR CIRCUITS: A PROPOSED CLASSIFICATION," Electronics Letters, vol. 11, no. 1, pp. 15 - 16, 1975.
[27] K. Ueno, T. Hirose, T. Asai, Y. Amemiya, "A 46-ppm/°C temperature and process compensated current reference with on-chip threshold voltage monitoring circuit," Proc. of the IEEE Asian Solid-State Circuits Conference (A-SSCC), pp. 161-164, 2008.
[28] M. C. Hsu, B. J. Sheu, "Inverse-geometry dependence of MOS transistor electrical parameters", IEEE Trans. Computer-Aided Design, vol. CAD-6, pp. 582-585, July. 1987.
[29] Y. C. Cheng, M-C. Jeng, Z. Liu, J. H. Huang, M. Chen, K. Chen, P. K. Ko, C. Hu, "A physical and scalable IV model in BSIM3v3 for analog/digital circuit simulation.", IEEE Trans. Electron Devices, vol. 44, No. 2, pp. 277-287, Feb. 1997.
[30] S. M. Sze, Physics of Semiconductor Devices, 2nd ed, John Wiley & Son, 1981.
[31] Futaki H. A new type semiconductor (critical temperature resistor). Japan Journal of Applied Physics, vol. 4, no. 1, pp. 28-41, 1965.

2

Low-Power Analog Associative Processors Employing Resonance-Type Current-Voltage Characteristics

Trong Tu Bui[1] and Tadashi Shibata[2]
[1] *The University of Science-HCM City,*
[2]*The University of Tokyo,*
[1]*Vietnam*
[2]*Japan*

1. Introduction

Data-matching function plays an essential role in a number of information processing systems, such as those for voice/image recognition, codebook-based data compression, image coding, data search applications etc. In order to implement such functions effectively, both proper data representation algorithms and powerful search engines are essential. Concerning the former, robust image representation algorithms such as projected principle edge distribution (PPED) (Shibata et al., 1999; Yagi & Shibata, 2003; Yamasaki & Shibata, 2007) etc. have been developed on the basis of the edge information extracted from original images. Such an algorithm is robust against illumination, rotation, and scale variations, and has been successfully applied to various image recognition problems. Concerning the latter, because search operations are computationally very expensive and time-consuming, it would be better if these operations are carried out by dedicated VLSI associative processors rather than programs running on a general-purpose computer. In this regard, dedicated highly parallel associative processor chips have been developed for the purpose of real-time processing and low-power operation.

It has been demonstrated that associative processors can serve as the basis of humanlike flexible computation, and many examples of flexible pattern perception have been demonstrated that are based on analog and digital technologies as well as mixed signal technologies. Digital approaches are accurate in computation, but often require large chip real estate and often consume large power. Analog implementations are preferred in terms of low-power consumption and high-integration density. In this regard, various distance-calculating circuits, which are used to evaluate the similarity (or dissimilarity) between two vectors, have been proposed. Euclidean distance circuits (Tuttle et al., 1993) utilizing MOSFET square-law cells were employed in an 8-bit parallel analog vector quantization (VQ) chip. Konda et al. (1996) and Cauwenberghs & Pedroni (1997) proposed neuron MOSFET (νMOS)-based and charged-based Manhattan-distance evaluation cells, respectively. A νMOS-based Euclidean distance calculator used in a recognition system for handwritten digits was proposed (Vlassis et al., 2001). Kramer et al. (1997) also proposed an analog Manhattan-distance-based content-addressable memory (CAM) using the analog non-volatile memory technology. On the other hand, bell-shaped characteristics have been implemented in various analog associative processors (Ogawa & Shibata, 2001; Yamasaki & Shibata, 2003; Hasler et al., 2002; Peng et al., 2005). In such processors, bell-shaped current-voltage (*I-V*) characteristics, or resonance-type *I-V* characteristics, were utilized in building matching cells. This is because such resonance characteristics can represent the correlation between the input data and the template data in the sense that the output current becomes maximum when the input voltage coincides with the peak voltage. The resonance characteristics of single-electron transistors (SETs) were utilized to carry out associative processing for color classification (Saitoh et al., 2004). Since resonance characteristics are the typical nonlinear characteristics often observed in nano devices, such associative processors would be one of the most promising system applications in the coming era of nano devices. Although room-temperature SETs utilizing particular phenomena have been reported (Mastumoto et al., 1996; Uchida et al., 2002; Saitoh et al., 2004), all demonstrations have been reported at the device level or simple circuitry, rather than at realistic system levels. Numerous new developments are now being explored so as to make nano devices applicable to the next-generation integrated circuits. However, because these devices have a higher probability of being defective than conventional CMOS devices, designing reliable digital circuits with such devices is a major challenge. So far, CMOS-based associative processors are still dominant in practical applications. One of the drawbacks in analog implementation, however, is that the matching-cell behavior suffers from the problem of device mismatch. For this reason, architectures that are robust against such problems are desired.

In this chapter, a compact resonance-characteristics matching cell using only NMOS transistors in order to emulate the resonance-type *I-V* characteristics of nano devices and to build a small-area low-power associative processor will be described. In addition, a new calibration scheme (Bui & Shibata, 2008a) that can compensate for matching errors due to device mismatch is presented. System configuration of a single-core architecture and the major circuitries utilized in the prototype chip design as well as measurement results are presented in Section 2. In Section 3, a solution to how the system is hierarchically scaled up to a vast scale integration is presented. For a vast scale integrated system, a large number of template data can be implemented in multiple associative processors, making the recognition system more intelligent. In this regard, a fully-parallel multi-core/multi-chip scalable architecture of associative processors was developed (Bui & Shibata, 2008b; 2009). Moreover, the problem associated with inter-chip communication delay which is critical in the time-domain WTA operation was resolved by a newly-developed winner-code-decision scheme (Bui & Shibata, 2008b; 2009).

2. Single-core architecture of analog associative processor

2.1 System architecture

Figure 1 shows the block diagram of the single-core associative processor developed in our work (Bui & Shibata, 2008a). It consists of two main parts, the digital memory module and the proposed analog matching-cell module. The memory module employing SRAM is utilized to store template data that represent the past experience or knowledge. The similarity evaluation between the input data and the template data is carried out in parallel by vector-matching circuits in the matching-cell module. All data are represented as 64-dimension PPED vectors compatible with vectors generated from the vector-generation chip described in the study (Yamasaki & Shibata, 2007). Each vector-matching circuit itself consists of 64 vector-element matching cells (MCs) utilized to evaluate the similarity between vector elements. The matching score between vector elements is given as output current from the matching cell, which has bell-shaped *I-V* characteristics. Consequently, in the conventional manner, the matching scores between the input vector and template vectors are also currents obtained by taking the wired sum of element matching-cell output currents. Current memories are utilized to memorize the peak currents of the bell-shape characteristics and then to generate vector-matching scores by the calibration scheme proposed in Section 2.2.4. Utilizing these vector-matching scores, the winner-take-all (WTA) circuit (Ito et al., 2001) determines the maximum-likelihood template vector and identifies its location, namely, the code of the vector. Serial digital-to-analog converters (SDACs) are used to convert digital values to analog voltages prior to similarity evaluation processing. Once the template data are downloaded from the digital memory module to the matching-cell array via the digital-to-analog converters, the data are temporarily stored in all the matching cells as analog voltages and utilized for a number of parallel pattern matching operations that follow.

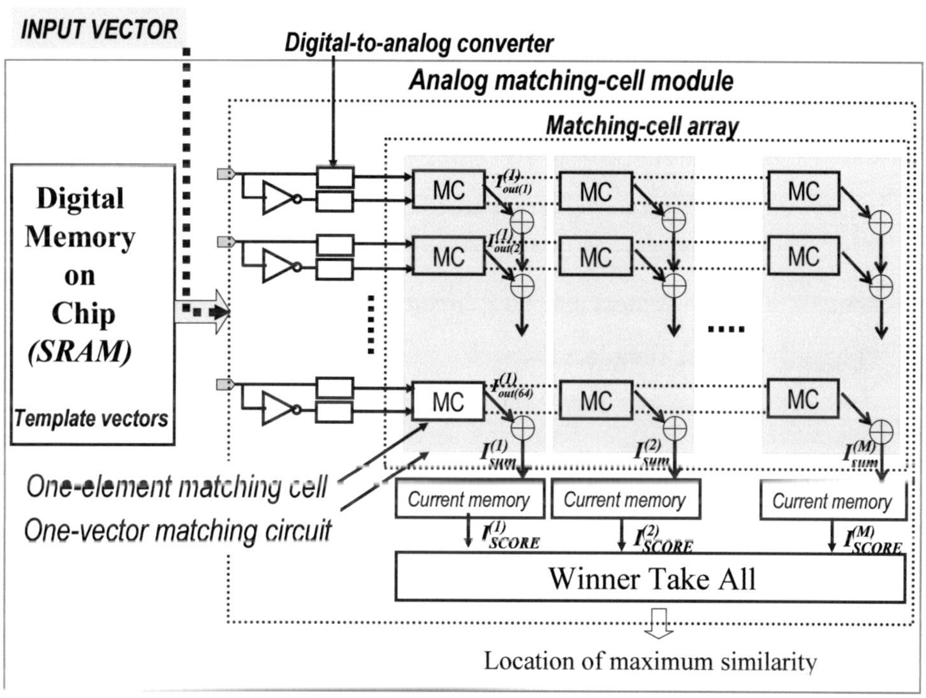

Fig. 1. Block diagram of single-core associative processor employing resonance-type current-voltage characteristics.

In analog associative processor implementations, the storing of analog template data is always a difficult issue. Analog nonvolatile memory technologies (Kramer et al., 1997; Yoon et al., 2000; Yamasaki et al., 2001; Kobayashi et al., 2005) have been developed for such purposes, but they are often very expensive to implement. In the proposed architecture, on the other hand, digital memories such as SRAM, DRAM, and flash can be employed to build a system that is inexpensive compared with analog nonvolatile memory technologies. By adding an analog matching-cell module to any existing memory system, an associative processor can be easily constructed in the architecture proposed in this work.

2.2 Circuit Implementation

2.2.1 Matching cell

Figure 2 shows the schematic of one element-matching cell, which is used to determine the similarity between each element of the input vector and the corresponding element of the template vector. The cell is composed of only NMOS transistors. This is advantageous in making the cell layout compact because extra areas for N-wells and PMOS transistors are not necessary. In this regard, the present cell is superior to the CMOS cell described in ref. (Yamasaki & Shibata, 2003) as well as the cell described in ref. (Konda et al., 1996).

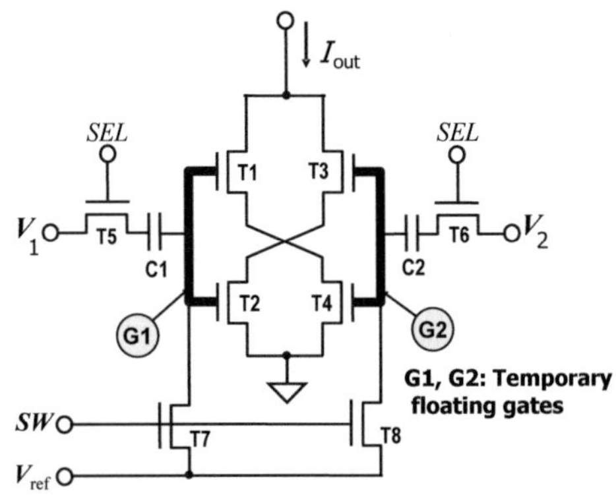

Fig. 2. Schematic of vector-element matching circuit (matching-cell circuit).

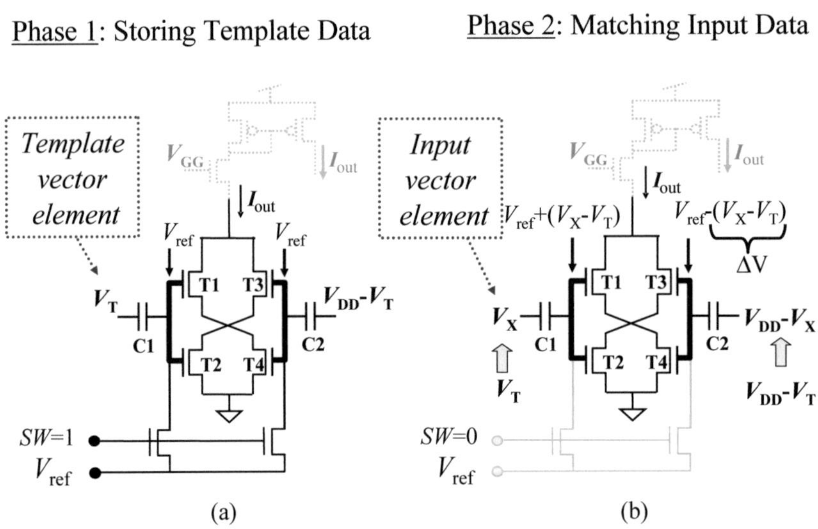

Fig. 3. Operation of matching cell, matching operation, is conducted in two phases. (a) Phase 1, the writing phase; template data are stored in matching cells. (b) Phase 2, the evaluation phase; similarities between template data and input data are evaluated.

Figure 3 illustrates two phases of the operation of the matching cell. In the figure, two NMOS switches (T_5 and T_6 in Fig. 2) connected to input terminals are omitted for simplicity of explanation. In the first phase, as shown in Fig. 3(a), template vector elements are stored temporarily inside matching cells. This phase is also called the writing phase, in which the template element voltage (V_T) and its complement (V_{DD}-V_T) are connected to two input terminals of the matching cell. The floating gates are first connected to the reference voltage, V_{ref}, and then disconnected from that voltage to make them electrically floating. After this phase, template vector elements are memorized as charges on the floating gates inside the corresponding matching cells. Phase 1 is repeated until all the necessary template vectors are downloaded from the memory module. In the second phase (also called the evaluation phase) shown in Fig. 3(b), the input element voltage (V_X) and its complement (V_{DD}-V_X) replace the positions of template elements. As a result, floating gate voltages of $V_{ref} + \Delta V$ and $V_{ref} - \Delta V$ are created. In the figure, ΔV is the difference voltage between the input vector element and the template vector element.

These two voltages create the bell-shaped *I-V* characteristics shown in Fig. 9. Indeed, since the gate voltages of the two serially connected transistors T_1 and T_4 are complementary analog signals, $V_{ref} + \Delta V$ and $V_{ref} - \Delta V$, respectively, they form bell-shaped *I-V* characteristics. Because of the back-gate effect occurring in T_1, these characteristics are slightly asymmetric. Similarly, the T_2-T_3 pair also creates asymmetric characteristics. By cross-coupling four transistors, as shown in Fig. 2, the asymmetry is removed.

The result of the evaluation from each matching cell is given as an output current (I_{out}). A higher current indicates greater similarity. The peak height of the output current I_{out} is also programmable by varying the reference voltage V_{ref} connected to the floating gates. The higher V_{ref} is, the higher the peak current becomes. These characteristics are described clearly in Section 2.3 and Fig. 9. In addition, it should be noted that once all the necessary template data are stored in the matching-cell array, only phase 2 is repeated for each new input vector.

The matching score between the input vector and the template vector is obtained by taking the wired sum of all I_{out}'s from 64 element-matching cells for one vector, as shown in Fig. 1 and eq. (1). In conventional approaches, a higher wired-sum current represents a greater similarity between two vectors.

$$I_{SCORE}^{(k)} = I_{SUM}^{(k)} = \sum_{i=1}^{64} I_{out(i)}^{(k)} \tag{1}$$

2.2.2 Winner-take-all circuitry

The block diagram of the winner-take-all circuit (WTA) is shown in Fig. 4. The matching scores from the vector-matching circuits are first converted to delay times by the current-to-delay-time converter (Yamasaki & Shibata, 2003). This is accomplished by using comparators that compare matching scores and a common ramp voltage signal. The shorter delay time corresponds to the larger matching score. The time-domain WTA circuit (Ito et al., 2001; Yamasaki & Shibata, 2003) utilizes an open-loop OR-tree architecture to sense the first up-setting signal and generates the binary address representing the location of the winner. In this manner, the maximum-likelihood template vector is identified.

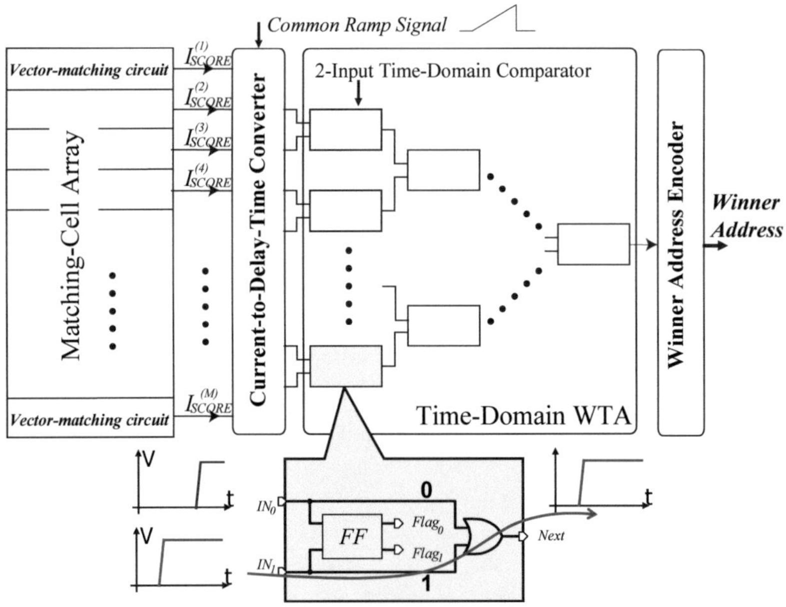

Fig. 4. Block diagram of the time-domain WTA, the flip-flop (FF) compares the timing difference between two input signals and senses the winner. The winner signal is also propagated to the next stage through the OR gate.

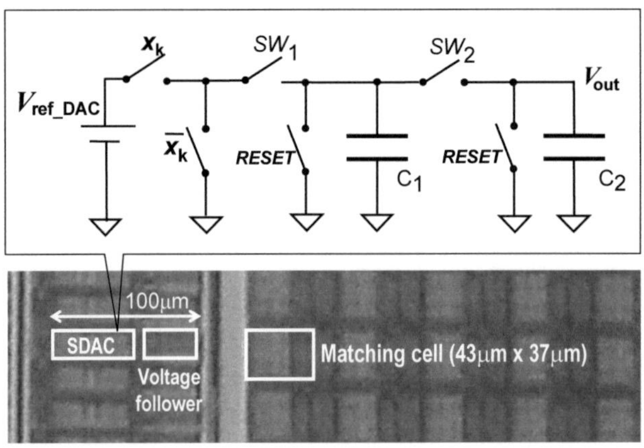

Fig. 5. Simplified schematic of SDAC and its layout area on the chip.

2.2.3 Serial digital-to-analog converter

As shown in Fig. 1, two digital-to-analog converters (DACs) are required for each of the vector elements since each matching cell requires two analog complementary signals; hence, 128 DACs are utilized in the system. Such an on-chip DAC needs to satisfy the requirement of small layout area, low-power dissipation, and small number of interconnects for data input. In this system, a serial digital-to-analog converter (SDAC) is utilized. The simplified schematic of the SDAC is shown in Fig. 5. The key feature of such a SDAC is its simplicity. It requires only two identical capacitors (C_1 and C_2) and a few switches. Basically, the operation of the SDAC is based on charging and sharing charges between two capacitors. The conversion is done sequentially; one clock cycle is required to convert one bit. Thus, N clock cycles would be required for an N-bit word. The output voltage, V_{out}, is proportional to the serial input data, as illustrated by eq. (2).

$$V_{out} = \frac{1}{2}\left\{ \cdots \frac{1}{2}\left[\frac{1}{2}(b_0 V_{ref\_DAC}) + b_1 V_{ref\_DAC} \right] + b_2 V_{ref\_DAC} + \cdots \right\}$$
$$V_{out} = V_{ref\_DAC}\left(\frac{b_0}{2^N} + \frac{b_1}{2^{N-1}} + \dots + \frac{b_{N-1}}{2} \right) \tag{2}$$

Because of its small size, the SDAC is a much better choice for the proposed architecture. Its layout area compared with the layout area of a matching cell is also shown in Fig. 5.

2.2.4 Calibration circuitry

Process variations influence device parameters, and hence matching-circuit behaviors. The matching result, therefore, may lead to errors. The new calibration scheme shown in Figs. 6 and 7 has been developed to mitigate the errors caused by device mismatch. According to the International Technology Roadmap for Semiconductors (ITRS-2008), transistor

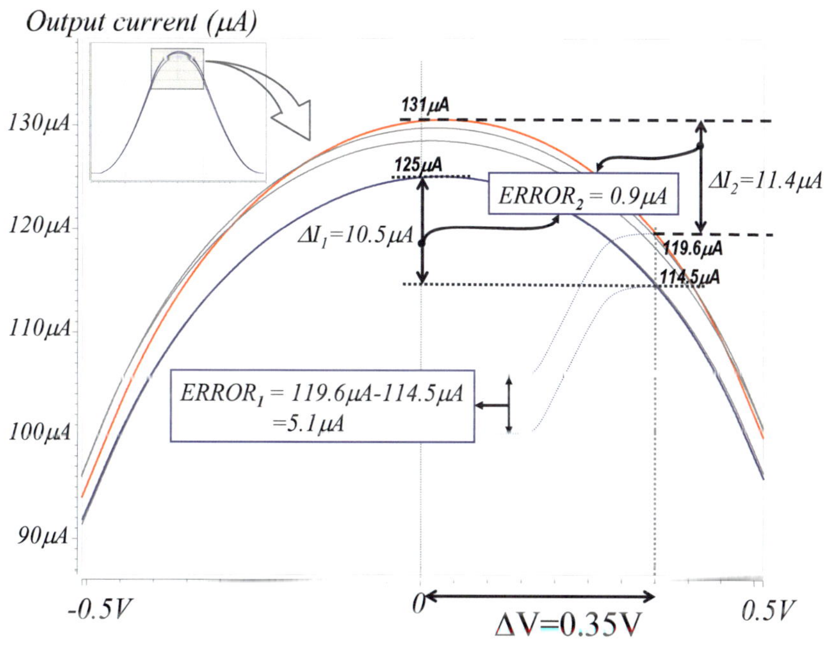

Fig. 6. Two distance-evaluating methods. Curves were generated by a 5-interation post-layout Monte Carlo simulation of a matching cell having random changes of 10% in transistors' length and width. The simulation was carried out at V_{DD} = 3.3 V and V_{ref} =1.65 V. Highest and lowest current curves were focused on. For the same distance between the input vector element and the template vector element, ΔV = 0.35 V, for example, the conventional distance-evaluating method and the proposed method are demonstrated.

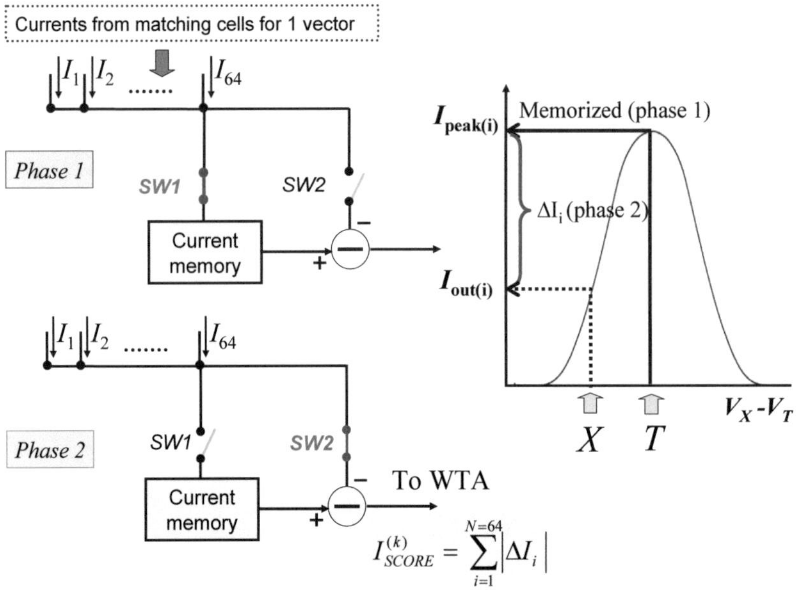

(a)

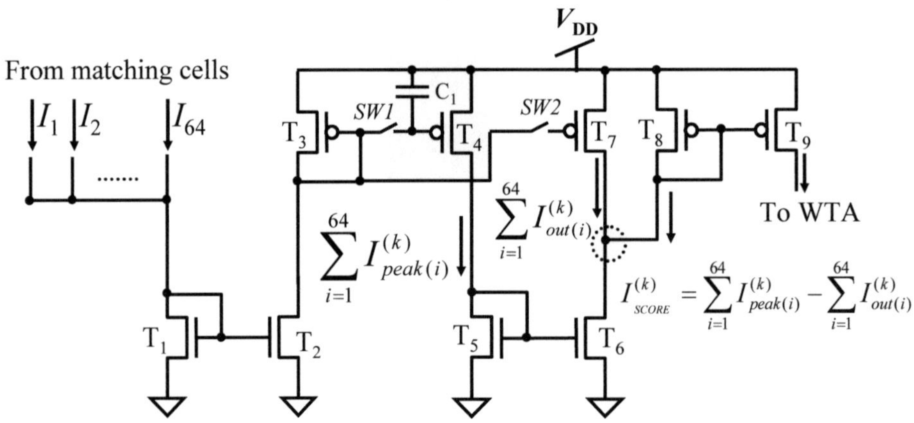

(b)

Fig. 7. Calibration scheme. (a) Calibration scheme operation. In phase 1, all peak output currents are memorized in current memories. In phase 2, the similarities between the input vector and the template vectors are evaluated. Only one current memory is required for one vector-matching circuit. (b) Circuit diagram of the current memory and subtractor.

dimensions may vary above 10%. The small figure at the top left of Fig. 6 illustrates matching-cell characteristics where the widths and the lengths of NMOS transistors of the matching cell vary randomly up to 10% as a result of process variations. These characteristics were obtained by a post-layout extracted circuit Monte Carlo simulation, and we focus on the highest and the lowest current curves. For the same distance between the input vector element and the template vector element, ΔV = 0.35 V, for example, two distance-evaluating methods are shown in the remaining part of Fig. 6, which is an enlarged image of the small rectangle at the top left. In the proposed method, the similarity is determined by the difference between the peak current and the output current at the moment of data matching. In the previous conventional approaches (Delbruck, 1991; Hasler et al., 2002; Yamasaki & Shibata, 2003; Ogawa & Shibata, 2001; Peng et al., 2005), the output current itself was utilized as the matching result. $ERROR_2$ (0.9 μA) and $ERROR_1$ (5.1 μA) in the figure refer to errors caused by the former method and the latter one, respectively. It is clearly shown that the proposed differential current method offers a better result. In order to implement this method, peak currents are stored in current memories in phase 1 (the writing phase), namely, at the time of template data download to matching cells. In phase 2 (the evaluation phase), differences between currents are obtained. Only phase 2 is repeated for each new input vector. This scheme is shown in Fig. 7(a), and the circuit diagram of the current memory and subtractor is presented in Fig. 7(b). The matching scores between input vector and template vectors are calculated by eq. (3).

$$I^{(k)}_{\text{SCORE}} = \sum_{i=1}^{64} \left| I^{(k)}_{\text{peak}(i)} - I^{(k)}_{\text{out}(i)} \right| = \sum_{i=1}^{64} I^{(k)}_{\text{peak}(i)} - \sum_{i=1}^{64} I^{(k)}_{\text{out}(i)} \tag{3}$$

According to this scheme, the greater similarity corresponds to the lower current rather than the higher one in the previous approaches.

2.3 Experimental results

2.3.1 Chip fabrication

The proof-of-concept chip was designed and fabricated using 0.35-μm 2P3M CMOS technology. The proposed matching-cell module includes 32 template vectors for the purpose of demonstration. The mechanism is preserved even in the case of a larger number of template vectors. The chip micrograph is shown in Fig. 8. The chip size is 4.9×4.9 mm², and the features of the chip are summarized in Table 1.

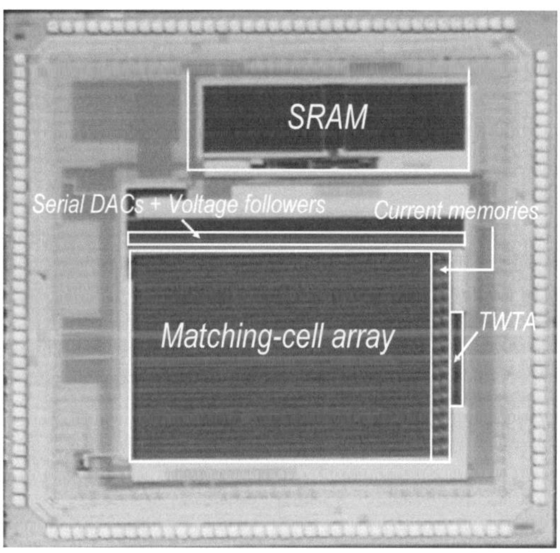

Fig. 8. Micrograph of the proof-of-concept chip fabricated using 0.35-μm CMOS process.

2.3.2 Measurement results and discussion

The measured characteristics of the vector element matching cell with various values of the reference voltage are illustrated in Fig. 9. Since the NMOS threshold voltage is around 0.6 V in the 0.35-μm CMOS technology in which the test chip was fabricated, it is shown that by varying V_{ref} from high to low values, the operation of the matching cell is altered from the above-threshold regime to the subthreshold regime, respectively. When operating in the subthreshold regime, the peak output current becomes as low as 80 nA at V_{ref} of 0.4 V. The results suggest an opportunity for building very low-power information processing systems.

Technology	2P3M 0.35-μm CMOS Process
Power supply (V)	3.3 (maximum)
Die size (mm²)	4.9 × 4.9
Number of vectors	32 vectors, 64 dimensions
Frequency (MHz)	33.3
Power consumption (mW)	21 at V_{ref} = 0.55 V, V_{DD} = 3 V, Clk = 33.3 MHz
Matching time (μs)	2.2 at 33.3 MHz

Table 1. Specifications of the proof-of-concept single-core chip.

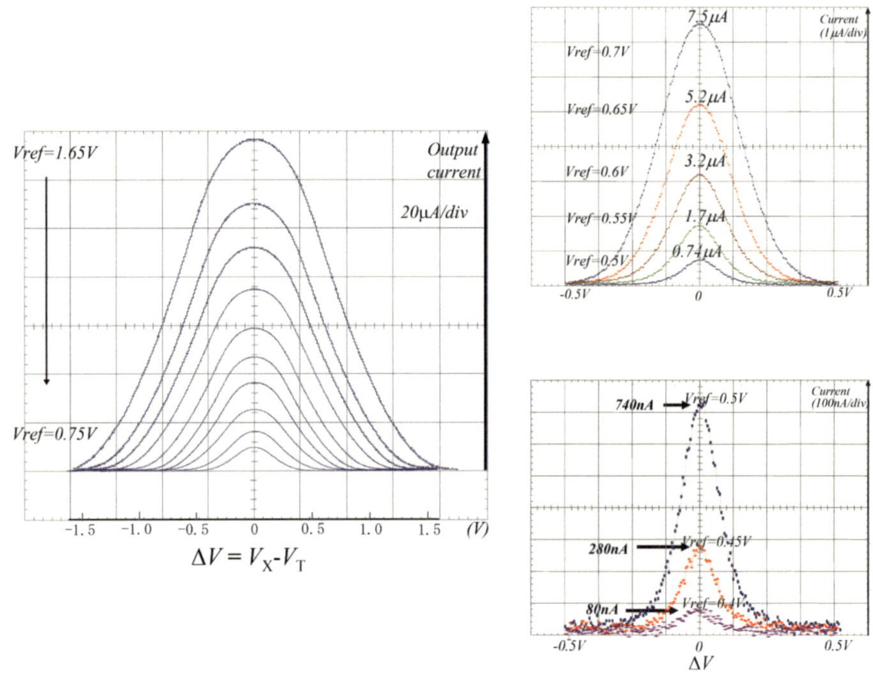

Fig. 9. Measured characteristics of the matching cell with various values of the reference voltage.

Figure 10 illustrates the experimental results for handwritten digit recognition utilizing the proposed architecture, as a simple demonstration. The digits "0"-"9" were converted to

PPED vectors so as to play the role of template vectors. The twenty-two other template vectors were dummy vectors. Then, the PPED vector of the handwritten digit "9" was employed as the input vector. The winner address shown in Fig. 10(a) corresponds to the location of the digit "9". This result verifies correct chip operation. Figures 10 (a) and 10 (b)

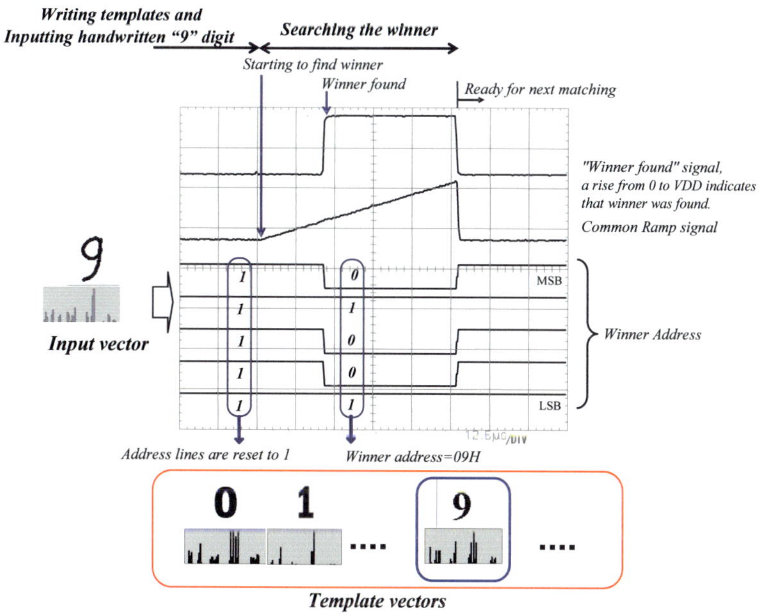

(a)

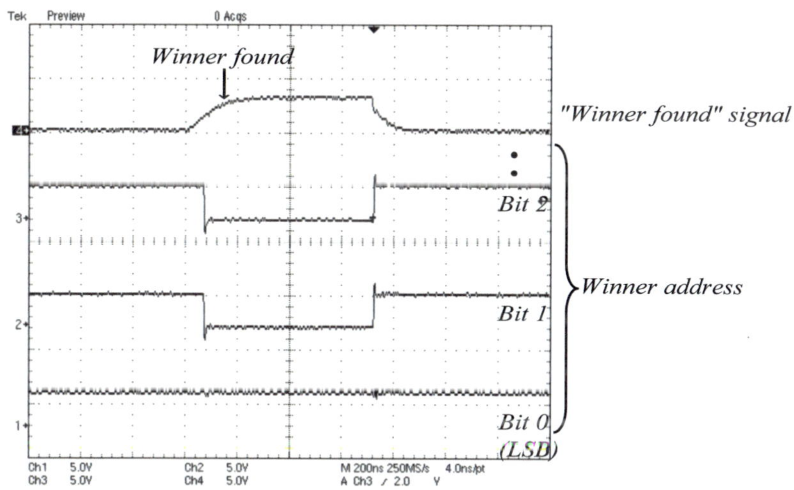

(b)

Fig. 10. Demonstration of the whole system operation. (a) Waveforms obtained with a logic scope describe the chip operation at 1 MHz for the purpose of illustration. The operating frequency is low because of the resolution limitation of the logic scope. (b) Waveforms obtained using an oscilloscope verify the chip operation at the frequency of 33.3 MHz show the waveforms captured from a logic scope and an oscilloscope, respectively. Since 72 clock cycles, comprising 8 cycles for SDAC and 64 cycles for an off-chip digital- to-analog converter utilized as the ramp-signal generator for the WTA circuit, are required to finish a template-matching cycle, the search time in this experiment is 2.2 μs at the frequency of 33.3 MHz and depends strongly on the speed of the ramp-up voltage signal employed in the current-to-delay-time converter. The system was set up to operate at the supply voltage of 3.0 V and the reference voltage of 0.55 V. As a result, the average power dissipation of the whole chip was about 21 mW.

Moreover, in Fig. 11, the average supply current of the whole chip, including the matching-cell array, the SRAM module, SDACs, voltage buffers, current memories, the WTA circuit, and I/O pads, measured with various V_{ref}'s is reported. The curves inherit the NMOS *I-V* characteristics owing to the NMOS-based matching-cell architecture. It can be observed that low supply currents are obtained with values of V_{ref} below the threshold voltage. These low reference voltages enable matching cells to operate in the subthreshold regime, in which the matching cell output currents drop exponentially with decreasing V_{ref}. As a result, the matching-cell array consumes very low power. Since the measured currents are for both the matching-cell array and the other parts, the supply currents in the subthreshold region remain at certain values rather than very low ones. These currents are mainly for the other parts whose power dissipations are reduced when lowering the supply voltage, and are independent of V_{ref}. Consequently, the supply currents are approximately constant values in the subthreshold region, as shown in Fig. 11.

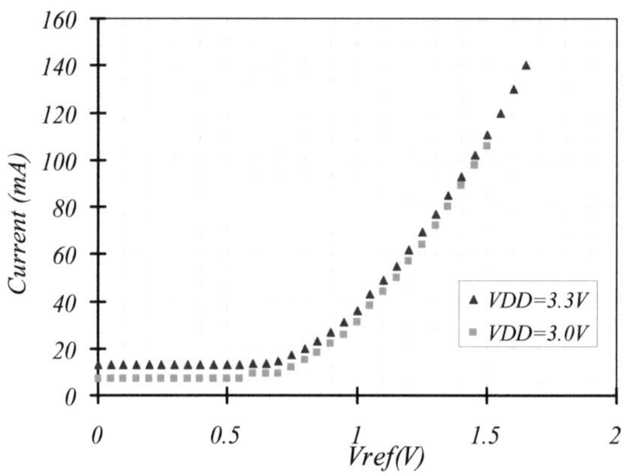

Fig. 11. Relationship between V_{ref} and supply current.

The performance of the associative processor is summarized with some others from the literature in Table 2. Because the time-domain WTA is utilized in this work because of its simple architecture, the search time is quite long compared with those of digital implementation (Nakata et al., 1999) and mixed signal implementation (Abedin et al., 2007). In addition to the matching-cell array, the WTA plays an important role in the power-saving scheme because the power consumption of the WTA increases significantly upon increasing the number of template vectors. In the present chip, the optimization of the speed and power dissipation of the WTA has not been considered. In order to make the proposed architecture practical and much better than digital approaches, a low power WTA would be considered in future studies. Furthermore, although analog flash implementation (Kramer et al., 1997) offers very low power consumption, such an implementation requires particular mechanisms in the template-writing phase, making the flash implementation difficult to control and hence, flexible programmability difficult to realize.

	Technol.	Power consumption (mW)	Matching time (μs)	Estimated power/MC (mW)
This work	Analog	21 (32 vectors, 64 elements)	2.2	0.01
Tuttle, *et al.* 1993	Analog	50*) (256 vectors, 16 elements)	2	0.012
Kramer, *et al.* 1997	Analog flash	195 (4K vectors, 64 elements)	4.6	0.00074
Oike, *et al.* 2004a	Digital	320.7 at V_{DD}=1.8V 15.1 at V_{DD}=0.9V (64 vectors, 32 elements)	2 ~8.12	0.157 0.0074
Nakada, *et al.* 1999	Digital	290 (256 vectors, 16 elements)	1.1	0.071
Abedin, *et al.* 2007	Mixed signal	195 (64 vectors, 16 elements)	0.16	0.19

Table 2. Performance comparison.
*)Not including power for memory and D/A converters.

3. Extension to a multi-core/multi-chip architecture of associative processors

3.1 Multi-core/Multi-chip configuration

In this session, a solution to how the system is hierarchically scaled up to a vast scale integration is presented. For a vast scale integrated system, a large number of template data can be implemented in multiple associative processors, thus making the recognition system more intelligent. In this regard, a multi-core/multi-chip architecture of associative processors has been developed (Bui & Shibata, 2008b; 2009).
In the literature, several multi-chip architectures based on all-digital technology have also been introduced (Nakata et al., 1999; Oike et al., 2004b). Although these systems offer accuracy, they occupy large chip real estate and usually have complicated structures. On the contrary, analog-technology-based system employing time-domain winner-take-all (WTA) is introduced in this study. The multi core/multi chip architecture inherits the architecture developed for the fully parallel single-core associative processor described in the previsous session. The problem associated with inter-chip communication delay which is critical in the time-domain WTA operation has been resolved by a newly-developed winner-code-decision scheme. In addition, switched-current technology has been utilized so as to further reduce the power consumption.
The block diagram of a multi-core/multi-chip associative system is shown Fig. 12. In general, the system includes many chips, and each chip itself has many cores. For the purpose of demonstration, the poof-of-concept system in this study is composed of four associative chips, namely, one master chip and three slave chips. Each chip consists of four

32-vector cores. (Each vector has 64 elements of 8-bit numbers.) As a result, a 512-vector associative system is constructed as a demonstration. The master chip and the slave chips are designed in the same configuration. They play master/slave roles when they are combined to form the whole system and operate in parallel. The master chip is distinguished from other slave chips by activating an additional majority-code-decision circuit described in the following section. Employing many cores on a single chip reduces the time required for downloading the information of template vectors stored in SRAMs to analog matching-cell arrays. In addition, four cores are activated separately, thus they can do matching operations independently or as a whole large system.
The 32-vector single-core architecture was already described in the previous section. In each core, template vectors are stored in on-chip digital memory, namely SRAM in the design. Employing digital memories is an inexpensive solution instead of using high-cost analog nonvolatile memory technologies. And compact serial digital-to-analog converters (SDACs) are used to convert digital values to analog voltages prior to similarity evaluation processing. The similarity evaluation between the input vector and template vectors is carried out in parallel by vector-matching circuits, each of which consists of 64 bell-shaped vector-element matching cells (MCs), a current memory, and a current subtractor as shown in Fig. 13. Signals *WR* and *RD* in Fig. 13 correspond to *WRITE* control signal and *READ* control signal, respectively. These signals permit to store matching results represented by currents into the current memories and to read out the matching scores from the subtractors. As mentioned in Section 2, current memory plays an important role in the device-mismatch calibration scheme in which the similarity is determined by the difference between the peak current and the output current at the moment of similarity evaluation. In the study, switched-current technology is employed to control *RD* and *WR* signals in order to cut-off currents flowing in the vector-matching circuits as well as the current memories except moments of downloading template vectors to the matching-cell arrays and evaluating similarities. As a result, the power dissipation is reduced further as compared with the design in Section 2.

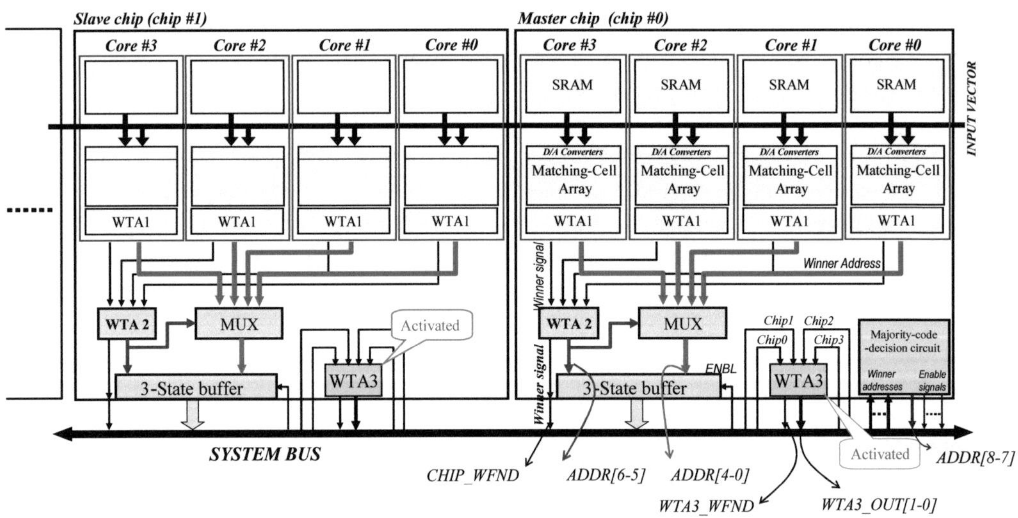

Fig. 12. Block diagram of the multi-core/multi-chip architecture.

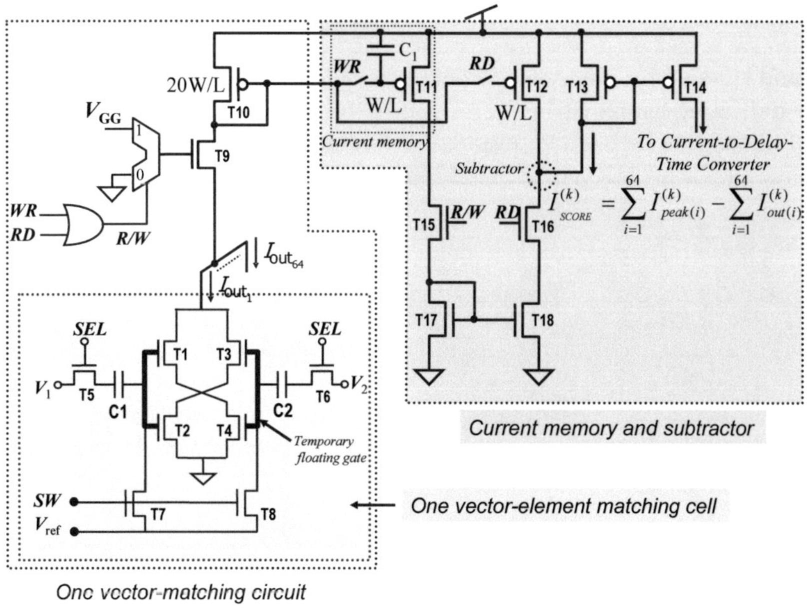

Fig. 13. Schematic of a vector-matching circuit.

Multi-core/Multi-chip configuration

The global winner, namely the template vector having the minimum distance to the input vector is searched for through a three-stage WTA circuit. Each WTA stage employing a time-domain WTA (Ito et al., 2001) senses the first up-setting signal among inputs and generates the binary address representing the location of the winner. The winner signal is also passed to the next WTA stage. In this manner, WTA1 searches for the local winner inside the 32-vector matching-cell array, WTA2 searches for the winner of one chip, and WTA3 searches for the global winner which is the winner when combining various chips together. All three WTA stages and the majority-code-decision circuit described below are layouted on each chip. The configuration is illustrated in Fig. 12.

However, when integrating several chips to form a larger system, signal propagation delays occurring in long inter-chip interconnects may lead to errors in time-domain signals. This will result in the decision error of the final WTA's (WTA3's). In order to deal with this problem, a balanced architecture should be satisfied to equalize delay times of inter-connection signals. However, even though with the balanced architecture, different propagation delays may still occur. Because of this problem, a redundant circuit following the final stage WTA, called the majority-code-decision circuit, has been developed. This circuit is only activated on the master chip. The circuit makes the decision based on the winner address codes generated by all WTA3's. The block diagram of the circuit is shown in Fig. 14. Basically, it consists of a binary counter, binary comparators, and a majority voting circuit (MVC). In the proof-of-concept chip, they are a 2-bit counter, 2-bit comparators, and a three-of-four MVC, respectively. As a result, the global result becomes more reliable than the architecture without a majority-code-decision circuit. In the case of a 2-bit 4-input majority-code-decision circuit like that in this study, the circuit can be constructed by combining two three-of-four MVCs whose outputs form the 2-bit majority winner code; but it is not the general case. It means that such architecture is not correct for other cases whose winner codes are larger than two bits. On the contrary, the method developed in this study is general and suitable for any case. The counter counts up from zero when it is activated; the winner-indicating-signal (*ADDR_FND*) indicates whether the majority winner code is found. This signal goes high when output of the counter coincides with the majority winner code.

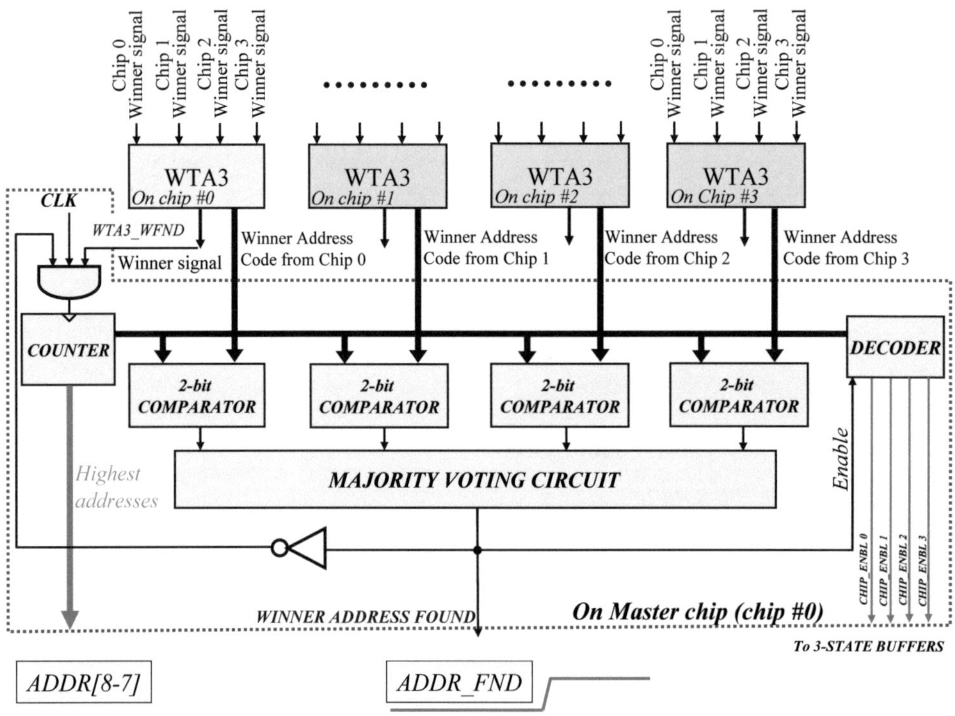

Fig. 14. WTA3 and the majority-code-decision circuitry.

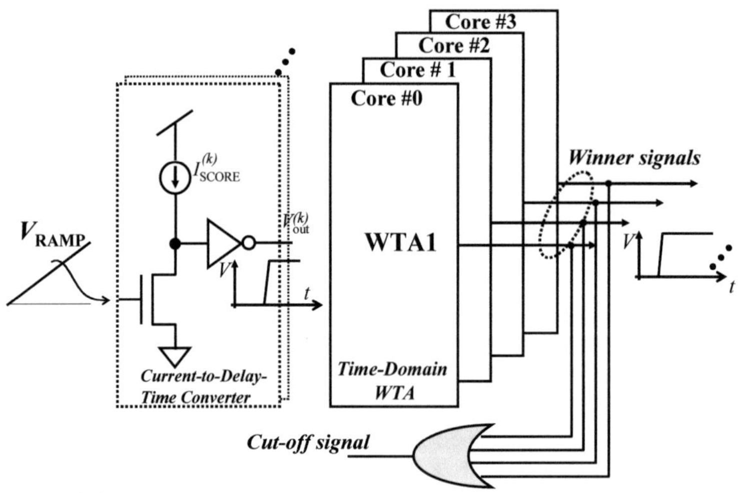

Fig. 15. Current-to-delay-time converter.

In addition, in order to further reduce the power dissipation, switched-current technology is also utilized in the current-to-delay-time converters by the method illustrated in Fig. 15. Winner signals obtained by WTA1's are combined by an OR-gate; the output signal is employed as a cut-off signal disconnecting both the common ramp voltage signal and score currents from current-to-delay-time converters. In this manner, once the winner signal is found by one of WTA1's, all current-to-delay-time converters are deactivated, thus further reducing the power consumption. This method can be applied to any large matching-cell array by dividing the array into several smaller blocks.

3.2 Experimental results

3.2.1 Chip fabrication

Measurement results obtained from the previous single-core chip fabricated in a 0.35-μm double-poly triple-metal CMOS technology have been discussed in Section 2. As an extended research, a proof-of-concept chip consisting of four cores was designed and fabricated in a 0.18-μm 5-metal CMOS technology. Figure 16 shows a micrograph of the test chip, and layout of a matching cell is shown in Fig. 17. Each core including a memory module and a matching-cell module occupies an area of 1760 μm × 570 μm. The size of matching cell is 19.7 μm × 7 μm. It should be noted again that the CMOS inverter-based matching cell presented in (Yamasaki & Shibata, 2003) is larger than the present cell due to the N-well region required for implementing PMOS transistors. This is an advantage of pure NMOS configuration. However, the present matching cell size is still large due to the large area required for capacitor layout. The specifications of the proof-of-concept chip are summarized in Table 3.

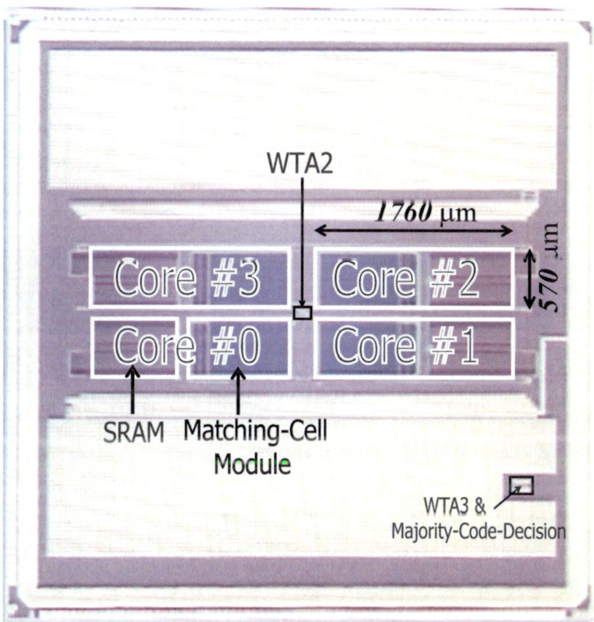

Fig. 16. Micrograph of the proof-of-concept chip fabricated using 0.18-μm CMOS process.

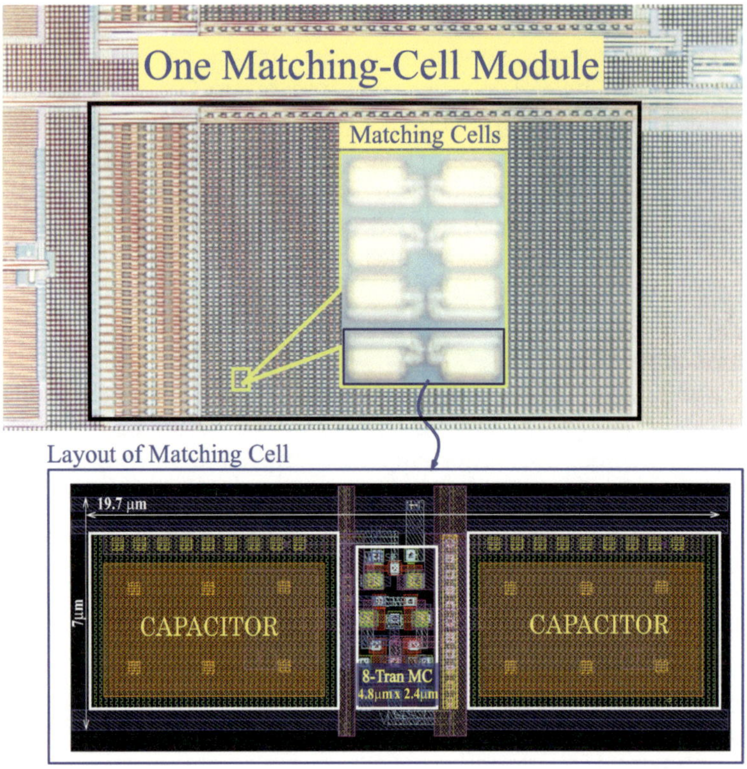

Fig. 17. Micrograph of a matching-cell module and layout of a matching cell (MC).

Technology	1P5M 0.18-μm CMOS
Power supply (V)	1.8
Core size (mm^2)	1.76 × 0.57
Matching cell size (μm^2)	19.7 × 7
Search time (μs)	8.16 at clock frequency = 16.7MHz (Incl. 8 clocks for SDAC and 128 clocks for the ramp voltage)
Power consumption (mW)	1.17 mW/32-vector matching-cell module; 6.48 mW/chip when operating in the subthreshold region with V_{DD}=1.8 V.
Function	128 vectors/chip, 512 vectors/4-chip system. Nearest match identification.

Table 3. Specifications of the proof-of-concept chip.

3.2.2 Measurement results

Figure 18(a) shows the characteristics of matching-cell measured with some small reference voltages. For the 0.18-μm CMOS technology in which the prototype chip has been fabricated, the threshold voltage of NMOS is around 0.45 V. As can be seen in the figure, in the subthreshold regime, the peak current of the matching cell characteristics is reduced to only several tens of nA. This is an important issue in power-saving schemes. The entire curve of peak output current with respect to the reference voltage shown in Fig. 18(b) has the shape of NMOS transistor characteristics.

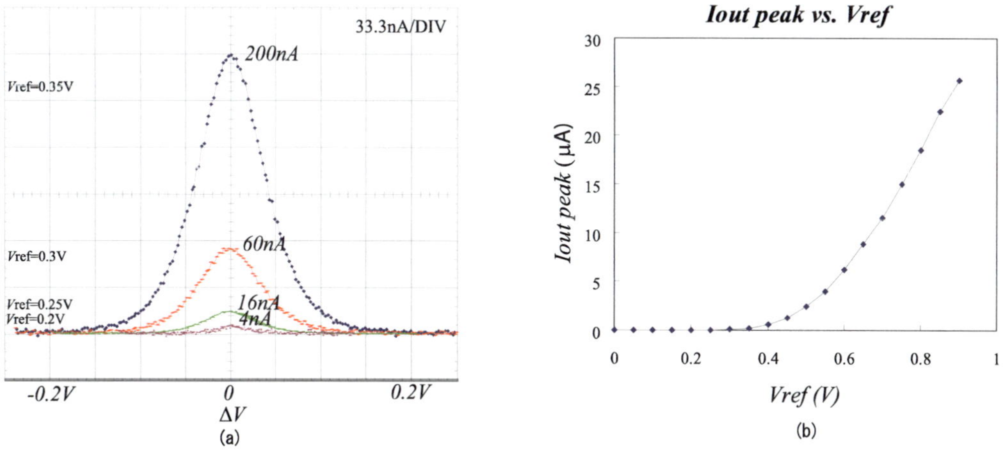

Fig. 18. Measured matching cell characteristics.

In Fig. 19, the average current of the whole chip including four cores and chip-I/O buffers and the current in a single 32-vector matching-cell module measured with various V_{ref}'s are reported. As can be seen from the figure, the curves have the shape of the NMOS *I-V* characteristics owing to the NMOS-based matching-cell architecture. In the subthreshold region, the current of the entire chip and that of one matching-cell module are 3.6 mA and 0.65 mA, respectively. As a result, the power consumption per matching cell is reduced to as small as 0.79 μW.

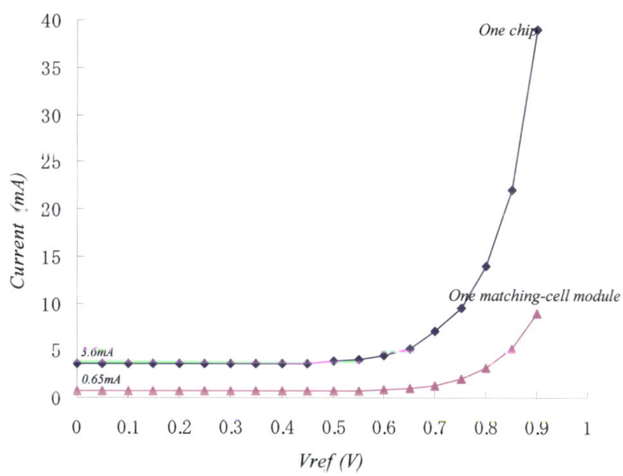

Fig. 19. Measured current as a function of the reference voltage V_{ref}.

Figure 20 (a) shows measured signals *CHIP_WFND* and *WTA3_WFND* generated by the WTA2 and WTA3 on the master chip, respectively. Waveforms at the output of the majority-code-decision circuit measured by an oscilloscope are shown in Fig. 20(b). The signal *WTA3_WFND* generated by WTA3 is employed as the control signal enabling the operation of the counter in Fig. 14. When the winner is found by the WTA3 on the master chip, the counter is activated, and begins to count up. When the counter output, *ADDR[8-7]*, coincides with the majority winner code, *ADDR_FND* signal goes high, indicating that the majority code was found and available on address lines *ADDR[8-7]*. This signal also stops the counter counting. In the demonstration, the majority winner code is 10_2 corresponding to chip #2. Majority-making-decision principle plays an important role not only in this design of a multi-chip architecture but also in miniscule-device-based designs where the device parameter variability is an important issue.

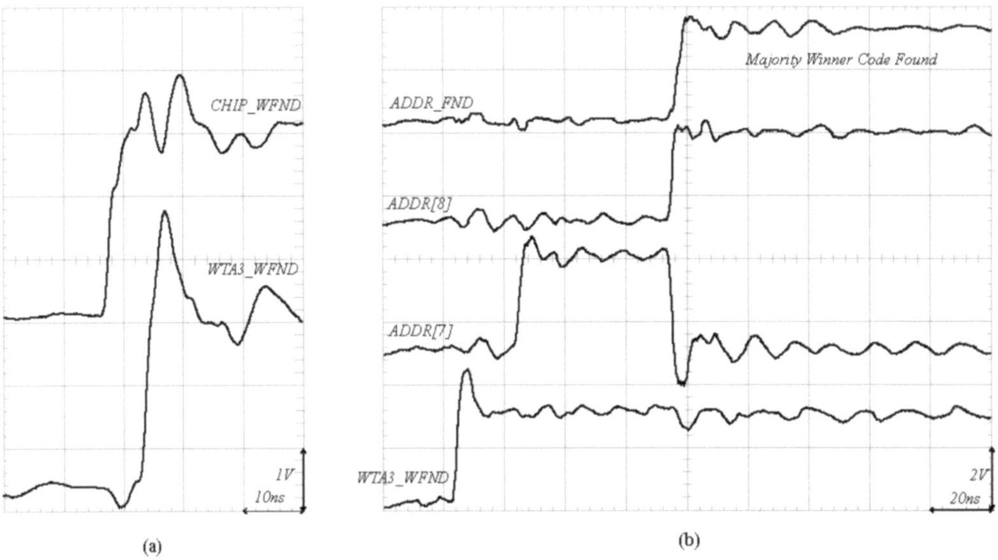

Fig. 20. Measured waveforms of the majority-code-decision circuit operating at a clock frequency of 20 MHz.

Demonstration of the whole system operation is illustrated in Fig. 21. All vectors of the test chip were assigned with given data. Required signals were connected to illustrate a system consisting four chips. After all template vectors were temporarily memorized inside matching-cell arrays, two input vectors were applied to the system input successively for matching. In Fig. 21, which is the measurement result captured from a logic scope, address lines *ADDR[4-0]*, *ADDR[6-5]*, and *ADDR[8-7]* represent winner address codes generated by WTA1, WTA2, and the majority-code-decision circuit, respectively. Namely, they are the winner template vector inside the winner core, the winner core inside the winner chip, and the winner chip of the multi-chip configuration, respectively. As a result, the global winner address is the combination of these three address codes. In this demonstration, the global winner addresses captured on the system bus are "100000101_2" representing the global winner is vector #5 (00101_2) of core #0 (00_2) in chip #2 (10_2) and "101010111_2" representing the global winner is vector #23 (10111_2) of core #2 (10_2) in chip #2 (10_2), respectively. *WTA_EVAL* signal enables the operation of the three-stage WTA circuitry. When this signal goes high, it also enables an off-chip ADC to generate the common ramp voltage used in current-to-delay-time converters. *GLOBAL_WFND* signal indicates that the winner template vector has been found and its address is available on the system bus. This signal also latches the global winner addresses on the system bus. The experimental results verify the correct operation of the system.

A searching cycle finishes in 136 clock cycles including eight clocks for on-chip D/A conversion of an input vector and 128 clocks for off-chip ramp voltage generation. In addition, employing many cores on a single chip reduces the time required for downloading the information of template vectors to analog matching-cell arrays.

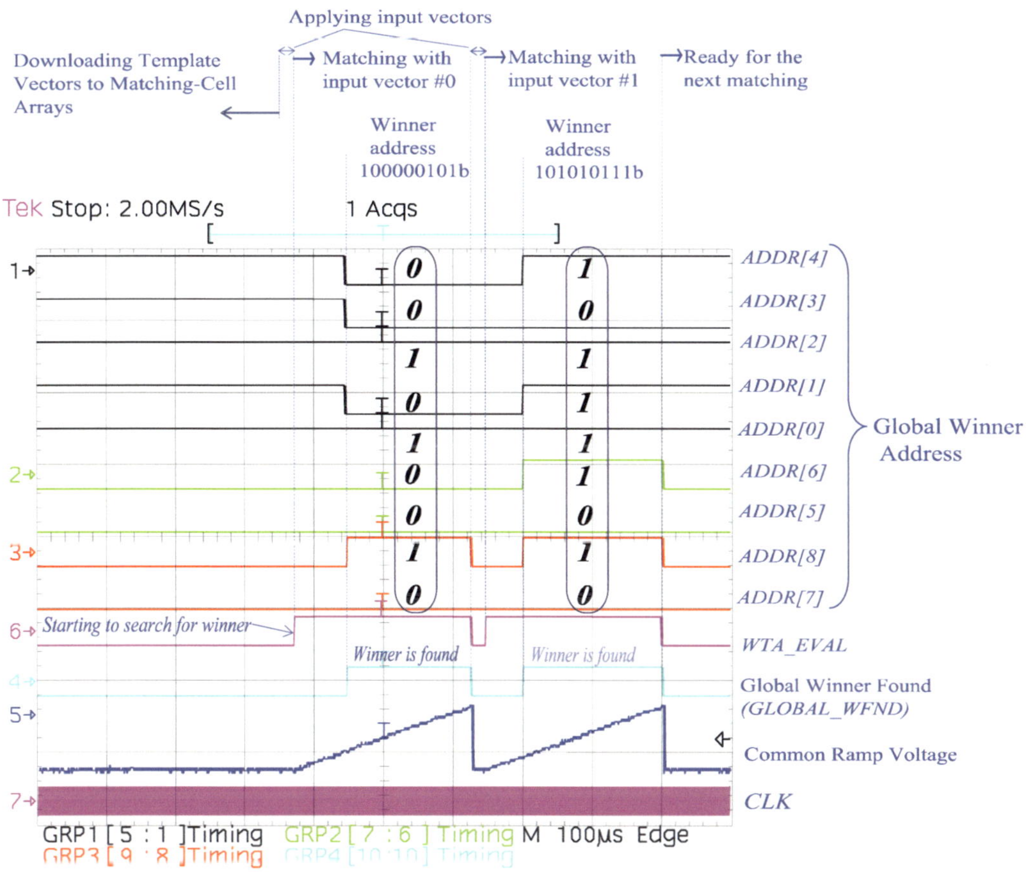

Fig. 21. Demonstration of the whole system operation by waveforms captured by a logic scope.

4. Conclusion

In this chapter, a methodology for building a low power high capacity associative system has been presented. Device mismatch problems as well as decision errors associated with inter-chip communication delays have been resolved by introducing the calibration scheme and the majority-code-decision circuit. Because of employing bell-shaped matching cell as similarity/dissimilarity-evaluation element, this study, therefore, provides an intermediary stage connecting CMOS designs and the coming era of nano devices. This is because such resonance-type current-voltage characteristics are typical characteristics often observed in nano-scale devices. The system also has the possibility of a large database capacity by employing the multi-core/multi-chip architecture. In principle, search time is independent of the number of cores as well as the number of chips. The operation of the systems as well as the concept of design have been verified by measurement results of the proof-of-concept chips designed in 0.35-μm and 0.18-μm CMOS processes.

The number of cores per chip as well as the number of chips in the system can be enlarged to a larger capacity without changing the methodology, making the system more intelligent by increasing the number of template vectors, i.e., giving a larger amount of knowledge in making decisions. Moreover, another method to enlarge the system is to employ WTA3's as extended WTA stages in a tree-like architecture. In principle, the system can be enlarged to any extent, making such an approach promising in giga-scale integration of nano devices.

In addition, as for associative processors developed in this chapter, template data are temporarily stored as voltages on capacitors. Due to design rules provided from the foundry, most of the matching-cell area is occupied by capacitors. If we assume a high-*k* MIM capacitance technology, such as the technology used in DRAM, to be available in the technologies used in this chapter's designs, the matching-cell array can be constructed in a very small area. It can be concluded that the combination of the high-*k* MIM capacitance technology with the architectures developed in this study would be a promising technology for analog implementations of associative processors in future.

5. Acknowledgements

The VLSI chips in this paper were fabricated as part of the chip fabrication program of the VLSI Design and Education Center (VDEC), the University of Tokyo, in collaboration with Rohm Corporation and Toppan Printing Corporation.

6. References

Abedin, Md. A.; Tanaka, Y.; Ahmadi, A.; Koide, T. & Mattausch, H. J. (2007). "Mixed Digital-Analog Associative Memory Enabling Fully Parallel Nearest Euclidean Distance Search", *Japanese Journal of Applied Physics (JJAP)*, Vol. 46, 2007, pp. 2231-2237.

Bui, T.T. & Shibata, T. (2008a). "Compact Bell-Shaped Analog Matching-Cell Module for Digital-Memory-Based Associative Processors," *Japanese Journal of Applied Physics (JJAP)*, vol. 47, No. 4, April 2008, pp. 2788–2796.

Bui, T.T. & Shibata, T. (2008b). "A Multi-core/Multi-chip Scalable Architecture of Associative Processors Employing Bell-Shaped Analog Matching Cells," *Proceedings of the 2008 9th International Conference on Solid-State and Integrated-Circuit Technology (ICSICT)*, pp. 1819-1822, Oct 20-23, 2008, Beijing.

Bui, T.T. & Shibata, T. (2009) "A Scalable Architecture of Associative Processors Employing Nano Functional Devices", *Proceedings of the 10th International Conference on Ultimate Integration of Silicon (ULIS)*, pp. 213-216, Mar.18-20, Aachen, Germany.

Cauwenberghs, G. & Pedroni, V. (1997). "A Low-Power CMOS Analog Vector Quantizer", *IEEE Journal of Solid-State Circuits,* Vol. 32, August 1997, pp. 1278-1283.

Delbruck, T. (1991). "Bump Circuits for Computing Similarity and Dissimilarity of Analog Voltages", *Proceedings of International Joint Conference on Neural Networks (IJCNN-91)*, pp. 475-479, June 1991, Seattle.

Hasler, P.; Smith, P.; Duffy, C.; Gordon, C.; Dugger, J. & Anderson, D. (2002). "A Floating-Gate Vector-Quantizer", *Proceedings of the 45th Midwest Symposium on Circuits and Systems (MWSCAS-2002)*, pp. 196-199, August 2002, Oklahoma.

Ito, K.; Ogawa, M.; & Shibata, T. (2001). "A High-Performance Ramp-Voltage-Scan Winner-Take-All Circuit in an Open Loop Architecture", *Japanese Journal Applied Physics (JJAP),* Vol. 41, 2001, pp. 2301-2305.

Kobayashi, D.; Shibata, T.; Fujimori, Y.; Nakamura, T. & Takasu, H. (2005). "A Ferroelectric Associative Memory Technology Employing Heterogate FGMOS Structure", *IEEE Transactions on Electron Devices,* Vol. 52, Oct. 2005, pp. 2188-2197.

Konda, M.; Shibata, T. & Ohmi, T. (1996). "Neuron-MOS Correlator Based on Manhattan Distance Computation for Event Recognition Hardware", *Proceedings of IEEE International Symposium on Circuits and Systems (ISCAS 1996),* pp. 217-220, May 1996, Atlanta, Georgia.

Kramer, A.; Canegallo, R.; Chinosi, M.; Doise, D.; Gozzini, G.; Navoni, L.; Rolandi, P. L. & Sabatini, M. (1997). "55GCPS CAM Using 5b Analog Flash", in *IEEE International Solid-State Circuits Conference(ISSCC) Digest of Technical Papers,* pp. 44-45, Feb. 1997, San Francisco.

Matsumoto, K.; Ishii, M.; Segawa, K. & Oka, Y. (1996). "Room Temperature Operation of a Single Electron Transistor Made by the Scanning Tunneling Microscope Nanooxidation Process for the TiOx/Ti System", *Applied Physics Letter,* Vol. 68, Jan. 1996, pp. 34-36.

Nakada, A.; Shibata, T.; Konda, M.; Morimoto, T. & and Ohmi, T. (1999). "A Fully Parallel Vector-Quantization Processor for Real-Time Motion-Picture Compression", *IEEE Journal of Solid-State Circuits,* Vol. 34, June 1999, pp. 822-830.

Ogawa, M. & Shibata, T. (2001). "NMOS-based Gaussian-Element-Matching Analog Associative Memory", *Proceedings of the 27th European Solid-State Circuits Conference ESSCIRC 2001,* pp. 257-260, Sept. 2001, Villach, Austria.

Oike, Y.; Ikeda, M. & Asada, K. (2004a). "A Word-Parallel Digital Associative Engine with Wide Search Range Based on Manhattan Distance", *Proceedings IEEE Custom Integrated Circuits Conference (CICC),* pp. 295-298, Oct. 2004, Orlando.

Oike, Y.; Ikeda, M.; & Asada, K. (2004b). "Hierarchical Multi-Chip Architecture for High Capacity Scalability of Fully Parallel Hamming-Distance Associative Memories", *IEICE Trans. Electron,* E87-C, Nov. 2004, pp. 1847-1855.

Peng, S.-Y.; Minch, B. A.; & Hasler, P. (2005). "A Programmable Floating Gate Bump Circuit with Variable Width", *Proceedings of IEEE International Symposium on Circuits and Systems (ISCAS 2005),* pp. 4341-4344, May 2005, Kobe.

Saitoh, M.; Harata, H. & Hiramoto, T. (2004). "Room-Temperature Demonstration of Integrated Silicon Single-Electron Transistor Circuits for Current Switching and Analog Pattern Matching", *Technical Digest IEEE International Electron Devices Meeting (IEDM),* pp. 187-190, Dec. 2004, San Francisco.

Shibata, T.; Yagi, M. & Adachi, M. (1999). "Soft-Computing Integrated Circuits for Intelligent Information Processing", *Proceedings of the Second International Conference on Information Fusion,* pp. 648-656, July 1999, Sunnyvale, California.

Tuttle, G. T.; Fallahi, S. & Adibi, A. A. (1993). "An 8b CMOS vector A/D converter", in *IEEE International Solid-State Circuits Conference (ISSCC) Digest of Technical Papers,* pp. 38-39, Feb. 1993, San Francisco.

Uchida, K.; Koga, J.; Ohba, R. & Toriumi, A. (2002). " Programmable Single-Electron Transistor Logic for Low-Power Intelligent Si LSI", *IEEE International Solid-State*

Circuits Conference (ISSCC) Digest of Technical Papers, pp. 206-207, Feb. 2002, San Francisco.

Vlassis, S.; Fikos, G. & Siskos, S. (2001). "A Floating Gate CMOS Euclidean Distance Calculator and Its Application to Hand-Written Digit Recognition", *Proceedings of International Conference on Image Processing*, pp. 350-353, Oct. 2001, Thessaloniki, Greece.

Yagi, M. & Shibata, T. (2003). "An Image Representation Algorithm Compatible with Neural-Associative-Processor-Based Hardware Recognition Systems", *IEEE Transactions on Neural Networks*, Vol. 14, Sept. 2003, pp. 1144-1161.

Yamasaki, H. & Shibata, T. (2007). "A Real-Time Image-Feature-Extraction and Vector-Generation VLSI Employing Arrayed-Shift-Register Architecture", *IEEE Journal of Solid-State Circuits*, Vol. 42, Sept. 2007, pp. 2046-2053.

Yamasaki, T. & Shibata, T. (2003). "Analog Soft-Pattern-Matching Classifier Using Floating-Gate MOS Technology", *IEEE Transactions on Neural Networks*, Vol. 14, Sept. 2003, pp. 1257-1265.

Yamasaki, T.; Suzuki, A.; Kobayashi, D.; & Shibata, T. (2001). "A Fast Self-Convergent Flash-Memory Programming Scheme for MV and Analog Data Storage", *Proceedings of IEEE International Symposium on Circuits and Systems (ISCAS 2001)*, pp. 930-933, May 2001, Sydney.

Yoon, S. M.; Tokumitsu, E.; & Ishiwara, H. (2000). "Ferroelectric Neuron Integrated Circuits Using $SrBi_2Ta_2O_9$-Gate FET's and CMOS Schmitt-Trigger Oscillators", *IEEE Transactions on Electron Devices*, Vol. 47, Aug. 2000, pp. 1630-1635.

3

The Evolution of Theory on Drain Current Saturation Mechanism of MOSFETs from the Early Days to the Present Day

Peizhen Yang[1], W.S. Lau[1], Seow Wei Lai[2], V.L. Lo[2], S.Y. Siah[2] and L. Chan[2]
[1]*Nanyang Technological University,*
[2]*Chartered Semiconductor Manufacturing*
Singapore

1. Introduction

Metal-oxide-semiconductor (MOS) digital logic is based on the enhancement-mode MOS transistors. During the past 40 years, the gate length of Si-based MOS transistors has been scaled down from about 10 μm to below 0.1 μm (100 nm). Currently, MOS transistors fabricated by 45 nm CMOS technology are readily available from various silicon foundries. Moreover, Taiwan Semiconductor Manufacturing Company (TSMC) has successfully developed 28 nm CMOS technology using the conventional silicon oxynitride as the gate insulator with polysilicon gate (Wu et al., 2009). IBM has demonstrated the use of high-K dielectric as the gate insulator with metal gate for their sub-22 nm CMOS technology (Choi et al., 2009). SEMATECH has developed their 16 nm CMOS technology using high-K/metal gate (Huang et al., 2009). Furthermore, several research groups have already reported on the development of 10 nm planar bulk MOS transistors (Wakabayashi et al., 2004; Wakabayashi et al., 2006; Kawaura et al., 2000). It has been reported using a hypothetical double-gate MOS transistor that a direct source-drain (S/D) tunneling sets an ultimate scaling limit for transistor with gate length below 10 nm (Jing & Lundstrom, 2002). Aggressive scaling brings about significant improvement in the integration level of Si-based MOS logic circuits. In addition, it also improves the switching speed because the drain current is increased when a smaller gate length and a smaller effective gate dielectric thickness are used. According to the conventional MOS transistor theory based on the constant electron mobility, the linear drain current (i.e. drain current at low drain voltage) will increase with the reduction of the gate length. Based on the classical concept of velocity saturation, the saturation drain current (i.e. drain current at high drain voltage) will not increase when the gate length is decreased. This theory is obviously contradictory to the experimental observation. Experimentally, we observe that the linear drain current and the saturation drain current are increased when the gate length is reduced. Hence, there is a need to investigate the drain current saturation mechanism in the nanoscale MOS transistors. First and foremost, we need to know the type of electrical conduction between the source and drain (S/D) regions for the state-of-the-art MOS transistors ($L \geq 32$ nm). Fig. 1 shows the various types of electrical conduction between the source and the drain of a n-channel MOS (NMOS) transistor (i) thermionic emission, (ii) thermally assisted S/D tunneling and (iii) direct S/D tunneling

(Kawaura & Baba, 2003). In the thermionic emission, carriers are thermally excited in the source, and then they go over the potential barrier beneath the gate. In the thermally-assisted S/D tunneling, carriers are thermally excited in the source, and then they tunnel slightly beneath the top of the potential barrier. Both thermionic emission and thermally-assisted S/D tunneling have strong temperature dependence. In contrast, the direct S/D tunneling does not need any thermal excitation and thus it has a weak dependence on temperature. Since the tunneling probability increases exponentially with decreasing potential barrier width, a decrease in the gate length will significantly increase the direct S/D tunneling and thus increase the subthreshold current (Kawaura & Baba, 2003). Fortunately, the tunneling current will only exceed the thermal current and degrade the subthreshold slope when the gate length is less than 5 nm (experimentally 4 nm and theoretically 6 nm) (Kawaura et al., 2000). Therefore, we only need to be concerned with thermionic emission between the source and the drain for the state-of-the-art MOS transistors ($L \geq 32$ nm). This chapter will discuss the evolution of theory on drain current saturation mechanism of MOS transistors from the early days to the present day. Section 2 will give an overview of the classical drain current equations that involve the concepts of velocity saturation and pinchoff. Section 3 will address the ambiguity involving the occurrence of velocity saturation and the presence of velocity overshoot in the nanoscale transistors. Section 4 will discuss the newer drain current transport concepts such as ballistic transport and quasi-ballistic transport. Section 5 will discuss the physics behind the apparent velocity saturation observed during transistor scaling and how it differs from the classical concept of velocity saturation within a transistor. Finally, Section 6 will discuss the actual mechanism behind the drain current saturation in nanoscale transistors.

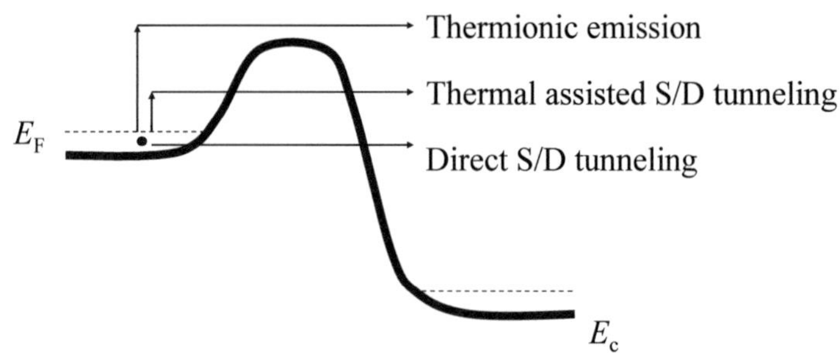

Fig. 1. Various types of electrical conduction between the source and the drain of a NMOS transistor (i) thermionic emission, (ii) thermally assisted S/D tunneling and (iii) direct S/D tunneling. Note that E_F refers to the Fermi level. E_c refers to the conduction band edge.

2. Classical drain current equations for MOS transistors

For long-channel MOS transistors (L = 10 μm), the drain current saturation is related to pinchoff (Hofstein & Heiman, 1963). A qualitative discussion of MOS transistor operation is useful, with the help of Fig.2. For NMOS transistor, a positive gate voltage (V_{GS}) will cause inversion at the Si/SiO_2 interface. When the drain voltage (V_{DS}) is small, the channel acts as a resistor and the drain current (I_{ds}) is proportional to V_{DS} (see Fig.3). This is known as the linear operation of the MOS transistor. The equation of the linear drain current is given by (Sah, 1991, b),

$$I_{ds} = \frac{\mu_{eff} W C_{ox}}{L_{eff}} \left[(V_{GS} - V_{th}) V_{DS} - 0.5 V_{DS}^2 \right] \quad (1)$$

where μ_{eff} is the low-field mobility. W is gate width. L_{eff} is the effective channel length. C_{ox} is the gate oxide capacitance per unit area. V_{th} is the threshold voltage.
By taking the partial derivative of equation (1) with respect to V_{DS}, the expression for the drain conductance (g_d) is as follows,

$$g_d = \left. \frac{\partial I_{DS}}{\partial V_{DS}} \right|_{V_{GS}} = \frac{\mu_{eff} W C_{ox}}{L_{eff}} (V_{GS} - V_{th} - V_{DS}) \quad (2)$$

Note that g_d decreases linearly with increasing V_{DS}. At $V_{DS} = V_{GS} - V_{th}$, g_d becomes zero and thus V_{DS} loses its influence on the number of electrons that can be injected by the source. This is because the depletion layer at the drain prevents the drain electric field from pulling out more electrons from source into the channel. Since V_{GS} can decrease the potential barrier of the source-to-channel pn junction, I_{ds} can be increased by using a bigger V_{GS}. Pinchoff point occurs when the electron density in the channel dropped to around zero. The current-saturation drain voltage (V_{Dsat}) is given by,

$$V_{Dsat} = V_{GS} - V_{th,sat} \quad (3)$$

where $V_{th,sat}$ is the saturation threshold voltage.
The saturation drain current (I_{ds}) is then given by (Sah,1991,b),

$$I_{ds} = \frac{\mu_{eff} W C_{ox}}{2 L_{eff}} (V_{GS} - V_{th})^2 \quad (4)$$

However, the constancy of I_{ds} at high V_{DS} is not maintained in the short-channel MOS transistors because the additional V_{DS} beyond (V_{GS} -V_{th}) will cause the pinchoff point to move slightly towards the source in order to deplete more electrons. This slight reduction in L_{eff} can be considered negligible for the long channel transistors but it becomes significant for the short channel transistors and thus results in a small g_d when $V_{DS} > V_{GS}$ - V_{th}

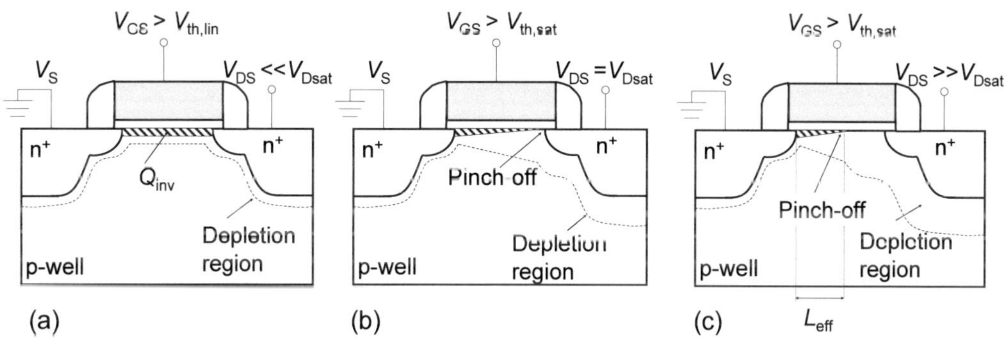

Fig. 2. NMOS transistor operating in (a) the linear mode, (b) the onset of saturation, and (c) beyond saturation where the effective channel length (L_{eff}) is reduced. $V_{th,lin}$ and $V_{th,sat}$ are the linear threshold voltage and the saturation threshold voltage , respectively. Q_{inv} is the inversion charge.

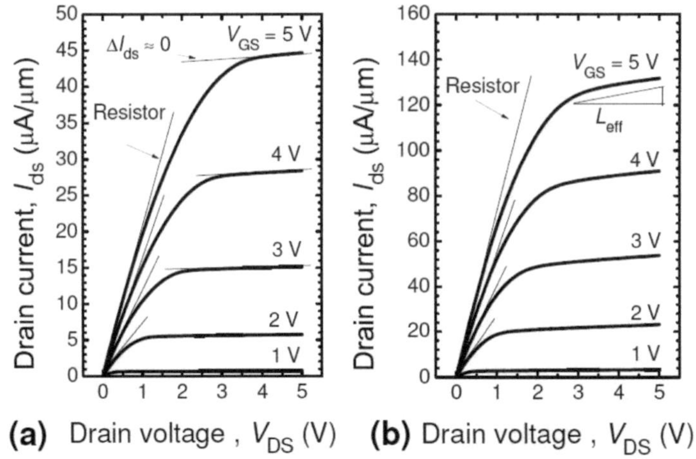

Fig. 3. Experimental I_{ds} versus V_{DS} characteristics of the NMOS transistor with physical gate oxide thickness of 300 Å (a) L =10 μm, W =10 μm, (b) L = 3 μm, W =10 μm.

For short-channel MOS transistors (L < 1 μm), (Taur et al., 1993) proposed that the drain current saturation, which occurs at V_{DS} smaller than the long-channel current-saturation drain voltage ($V_{Dsat} = V_{GS} - V_{th,sat}$), is caused by velocity saturation. From Fig.4, when the lateral electric field ($E_{lateral}$) is small (i.e. V_{DS} is low), the drift velocity (v_{drift}) is proportional to $E_{lateral}$ with μ_{eff} as the proportionality constant. When $E_{lateral}$ is further increased to the critical electric field ($E_{critical}$) that is around 10^4 V/cm, v_{drift} approaches a constant known as the saturation velocity (v_{sat}) (Thornber, 1980). Based on the time-of-flight measurement, at temperature of 300 K, v_{sat} for electrons in silicon is 10^7 cm/s while v_{sat} for holes in silicon is $6{\times}10^6$ cm/s (Norris & Gibbons, 1967).

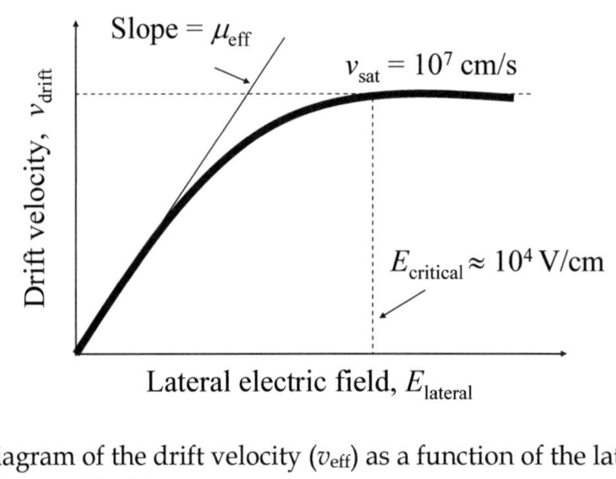

Fig. 4. Schematic diagram of the drift velocity (v_{eff}) as a function of the lateral electric field ($E_{lateral}$). Note that $E_{lateral} \approx V_{DS}/ L_{eff}$.

According to the velocity saturation model, the equation of the saturation I_{ds} for the nanoscale MOS transistor is given by (Taur & Ning, 1998, c),

$$I_{ds} = v_{sat} W C_{ox} \left(V_{GS} - V_{th,sat} \right) \quad (5)$$

In contrast with the theoretical predictions that v_{sat} is independent of μ_{eff} (Thornber, 1980), the experimental data show that the carrier velocity in the nanoscale transistor and the low-field mobility are actually related (Khakifirooz & Antoniadis, 2006). This can be better understood as follows. The effects of strain on μ_{eff} can be investigated qualitatively in a simple way through Drude model, $\mu_{eff} = q\tau / m^*$ where τ is the momentum relaxation time, m^* is the effective conductivity mass, and q is the electron charge (Sun et al., 2007). For <110> NMOS transistors that are fabricated on (100) Si substrate, there are four in-plane conduction band valleys (1, 2, 3, 4) and two out-of-plane conduction band valleys (5, 6), as shown in Fig. 5(a). The application of <110> uniaxial tensile stress will remove the degeneracy of the conduction band valleys such that the out-of-plane valleys (5, 6) will have a lower electron energy state that the in-plane valleys (1, 2, 3, 4). Since electrons will preferentially occupy the lower electron energy state, there will be more electrons in valleys (5, 6) compared to valleys (1, 2, 3, 4) and thus the effective in-plane mass becomes smaller. Besides the strain-induced splitting of the conduction band valleys, the strain-induced warping of the out-of-plane valleys (5, 6) in (100) silicon plane also plays a part in the electron mobility enhancement. In the absence of mechanical stress, the energy surface of the out-of-plane valleys (5, 6) is " circle" shaped and the effective mass of valleys (5,6) is m_T. When <110> tensile stress is applied, the effective mass of valleys (5, 6) along the stress direction ($m_{T,//}$) is decreased but the effective mass of valleys (5, 6) that is perpendicular to the stress direction ($m_{T,\perp}$) is increased (Uchida et al., 2005). By taking into account the change in the effective mass of the out-of-plane valleys (5, 6) and the strain-induced conduction subband splitting , the low-field mobility enhancement of the bulk <110> NMOS transistors under uniaxial <110> tensile stress can be modeled (Uchida et al., 2005).

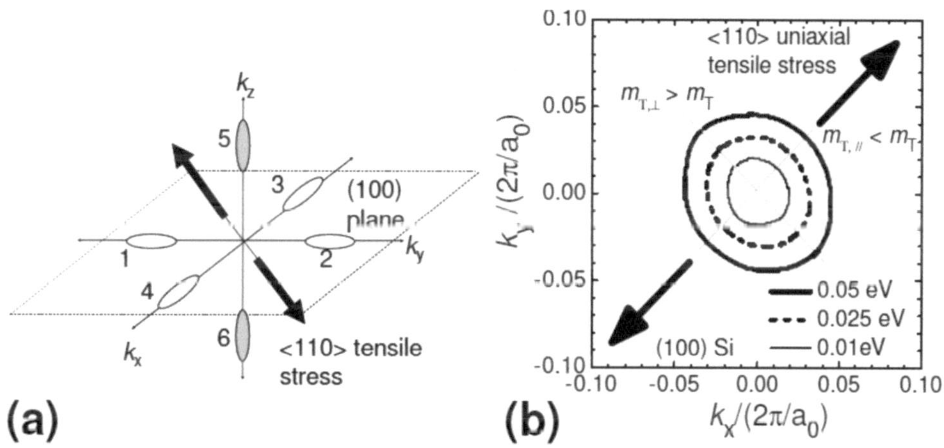

Fig. 5. Effects of <110> uniaxial tensile stress on the conduction band valleys of (100) silicon plane (a) Four in-plane valleys (1, 2, 3, 4) and two out-of-plane valleys (5,6), (b) Energy contours of the out-of-plane valleys (5, 6) , which is modified from (Uchida et al., 2005). Note that a_0 is the unstrained silicon lattice constant. k_x, k_y and k_z are the wave vectors along x direction, y direction and z direction , respectively. $m_{T,//}$ is the effective mass of valleys (5,6) along the stress direction ,and $m_{T,\perp}$ is the effective mass of valleys (5,6) in the direction that is perpendicular to the stress direction. m_T is the effective mass of valleys (5,6) in the absence of mechanical stress.

For <110> p-channel MOS (PMOS) transistors that are fabricated on (100) Si substrate, the lowest energy valence band edge has four in-plane wings (I1, I2, I3, I4) and eight out-of-plane wings (O1, O2, O3, O4). Fig.6, which is modified from (Wang et al., 2006), shows the effects of mechanical stress on the iso-energy contours of the valence band edge. In the absence of mechanical stress, the innermost contours are "star" shaped. When uniaxial compressive stress is applied along <110> channel direction, the innermost contours become oval shaped. In addition, the spacing between the contours increases for I1 and I3 wings while decreases for I2 and I4 wings. This indicates the hole energy lowering of I1 and I3 wings, and the hole energy rise of I2 and I4 wings. Since holes will preferentially occupy the lower hole energy state, there will be a carrier repopulation from I2 and I4 wings to I1 and I3 wings. As the channel length is along the direction of I2 and I4 wings, the hole mobility of <110> PMOS transistor will be improved. On the other hand, the application of uniaxial tensile stress along <110> channel direction leads to the opposite conclusion. The carriers are redistributed from I1 and I3 wings to I2 and I4 wings, leading to a hole mobility degradation in <110> PMOS transistor.

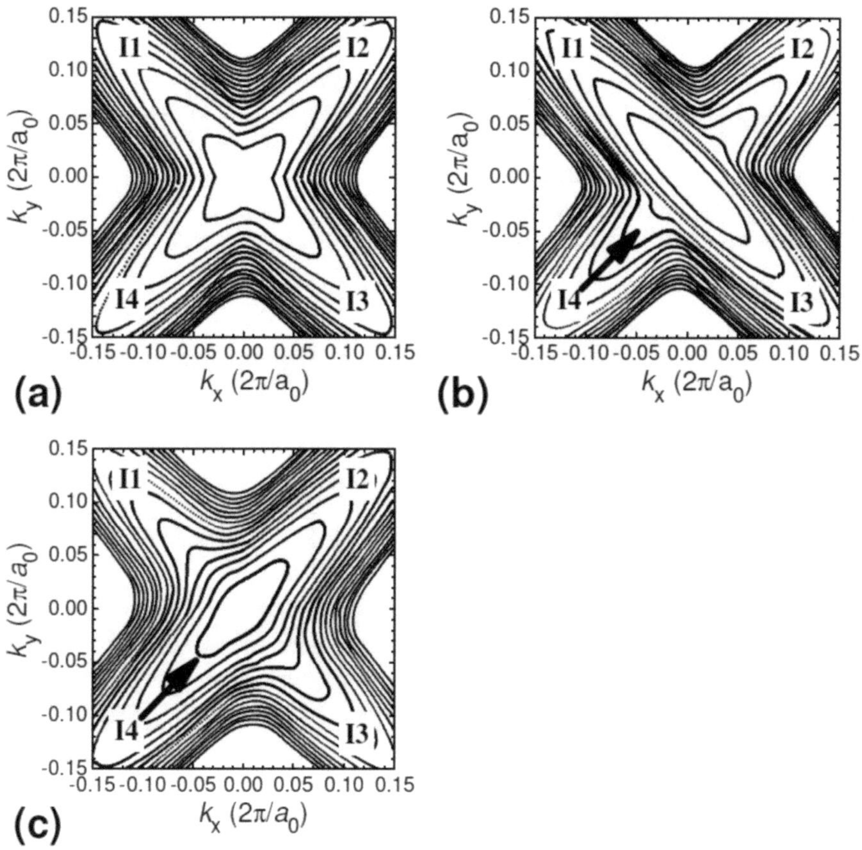

Fig. 6. Iso-energy contours separated by 25 meV in (100) silicon substrate for valence band edge, modified from (Wang et al., 2006). (a) No mechanical stress, (b) Uniaxial compressive stress along <110> direction, (c) Uniaxial tensile stress along <110> direction. Note that a_0 is the unstrained silicon lattice constant. k_x and k_y are the wavevectors along x direction and y direction, respectively. The arrow indicates the direction of the mechanical stress.

In addition to the simulation results of the strain-induced variation to the conduction band edge and the valence band edge, the change in the effective carrier mass by mechanical stress can also be studied by piezoresistance measurements. Device-level piezoresistance measurements in the channel plane can be readily done. From Table I, which is modified from (Chiang et al., 2007), the piezoresistance coefficient along the channel direction (π_L) is negative for NMOS transistor and is positive for PMOS transistor. This indicates that uniaxial tensile stress will decrease the effective carrier mass along the channel direction (m_x) for NMOS transistor but will increase m_x for PMOS transistor. In the other words, <110> tensile stress will increase the electron mobility of <110> NMOS transistor while <110> compressive stress will increase the hole mobility of <110> PMOS transistor. Since the on-state current (I_{on}) enhancement is observed in the nanoscale transistors with the implementation of various strain engineering techniques (Yang et al., 2004; C-H. Chen et al., 2004; Yang et al., 2008; Wang et al. , 2007), the carrier velocity in the nanoscale transistor must be related to the low-field mobility, and thus equation (5) needs to be modified so as to account for the strain-induced I_{on} enhancement.

Table I Device-level piezoresistance coefficients in the longitudinal direction (π_L), the tranverse direction (π_T), and the out-of-plane (π_{out}) direction for <110> channel MOS transistors that are fabricated on (100) Si substrate (Chiang et al., 2007). The units are in 10^{-11} m^2/N. Note that "longitudinal" means parallel to the direction of channel length in the channel plane, "transverse" means perpendicular to the direction of channel length in the channel plane, and "out-of-plane" means in the direction of the normal to the channel plane.

	NMOS transistor	PMOS transistor
π_L	-49	+90
π_T	-16	-46
π_{out}	+87	-44

However, for short channel transistors, the experimental V_{Dsat} is smaller than that predicted by equation (3) (Taur et al., 1993). Using the concept of velocity saturation, (Suzuki & Usuki, 2004) proposed an equation for V_{Dsat} that can account for the disparity between the experimental V_{DS} and the V_{Dsat} that is predicted by equation (3).

$$V_{Dsat} = \frac{V_{GS} - V_{th,sat}}{0.5 + \sqrt{0.25 + \dfrac{\mu_{eff}\left(V_{GS} - V_{th,sat}\right)}{v_{sat} L_{eff}}}} \quad (6)$$

Since velocity overshoot occurs in the nanoscale transistor (Kim et al., 2008; Ruch, 1972), equation (6) needs to be modified. In the physics-based model for MOS transistors developed by (Hauser, 2005), v_{sat} is treated as a fitting parameter that can be increased to 2.06×10^7 cm/s so as to fit the experimental I_{ds} versus V_{DS} characteristics of the nanoscale NMOS transistor (L = 90 nm). Although this approach is conceptually wrong, it serves as an easy way to avoid detailed discussion in velocity overshoot and quasi-ballistic transport. Hence, the resulting equation is as follows,

$$V_{Dsat} = \frac{V_{GS} - V_{th,sat}}{0.5 + \sqrt{0.25 + \frac{\mu_{eff}(L_{eff})}{v_{sat}(L_{eff})} \frac{V_{GS} - V_{th,sat}}{L_{eff}}}} \quad (7)$$

where μ_{eff} and v_{sat} are functions of L_{eff}. To avoid confusion, we introduce another parameter called the effective saturation velocity (v_{sat_eff}). According to (Lau et al., 2008, b), v_{sat_eff} is taken to be the average value of the carrier velocity (v_{eff}) when V_{GS} is close to the power supply voltage (V_{DD}). When uniaxial tensile stress is applied, both μ_{eff} and v_{sat_eff} of NMOS transistor will be increased. By replacing $v_{sat}(L_{eff})$ in equation (7) by $v_{sat\_eff}(\mu_{eff}, L_{eff})$,

$$V_{Dsat} = \frac{V_{GS} - V_{th,sat}}{0.5 + \sqrt{0.25 + \frac{\mu_{eff}(L_{eff})}{v_{sat\_eff}(\mu_{eff}, L_{eff})} \frac{V_{GS} - V_{th,sat}}{L_{eff}}}} \quad (8)$$

For long channel MOS transistors, the large L_{eff} will make the third term in the denominator of equation (8) negligible and thus $V_{Dsat} \approx (V_{GS} - V_{th,sat})$. For the short channel MOS transistors, the third term in the denominator of equation (8) must be considered and thus V_{Dsat} is expected to be smaller than $(V_{GS} - V_{th,sat})$. According to conventional MOS transistor theory (Taur & Ning, 1998, a), V_{Dsat} is given by $(V_{GS} - V_{th,sat})/m$ where the body effect coefficient (m) is typically between 1.1 and 1.4.

3. Does velocity saturation occur in the nanoscale MOS transistor?

For NMOS transistor, the electrons are accelerated by the lateral electric field ($E_{lateral}$) and thus the drift velocity (v_{drift}) increases. For (100) Si substrate, the optical phonon energy is bigger than 60 meV (Sah, 1991, a). When the kinetic energy of the electron exceeds 60 meV, the optical phonons are generated. However, the generation rate of optical phonon is very large and thus only a few electrons can have energy higher than 60 meV. An equilibrium is reached when the rate of energy gain from $E_{lateral}$ is equal to the rate of energy loss to phonon scattering. This corresponds to the maximum v_{drift} that occurs at $E_{lateral}$ around 10^4 V/cm. The maximum v_{drift} is known as the velocity saturation (v_{sat}). Based on the Monte Carlo simulation by (Ruch, 1972), the distance over which v_{drift} will overshoot the electron v_{sat} is less than 100 nm but this transient in velocity will only last for 0.8 ps before reaching its equilibrium value of 10^7 cm/s. According to (Mizuno, 2000), the amount of channel doping concentration (N_{ch}) will determine if velocity overshoot can be observed in bulk MOS transistors. For NMOS transistor with L = 80 nm, velocity overshoot can occur if $N_{ch} < 10^{17}$ cm^{-3}. For NMOS transistor with L = 30 nm, velocity overshoot can occur even if $N_{ch} \approx 10^{18}$ cm^{-3}. This can be attributed to the effective channel length (L_{eff}), which is a function of both the mask gate length (L) and N_{ch}. In fact, (Kim et al., 2008) has reported that the experimental findings of electron velocity overshoot in 36 nm bulk Si-based NMOS transistor at room temperature. Furthermore, the Monte Carlo simulation performed by (Miyata et al., 1993) show that electron velocity overshoot actually increases when the tensile stress is increased. This can account for the strain-induced I_{on} enhancement in the nanoscale NMOS transistors (Yang et al., 2004; C-H. Chen et al., 2004; Yang et al., 2008). Hence, it is more likely that velocity overshoot occur in the nanoscale transistor rather than velocity saturation.

Here, we will like to point out another misconception about the occurrence of velocity saturation in the nanoscale MOS transistors. Based on the classical concept of velocity saturation, the saturation I_{ds} of the short channel MOS transistor has a linear relationship with V_{GS} (see equation 5), and thus the saturation I_{ds} versus V_{DS} characteristics is expected to have constant spacing for equal V_{GS} step (Sze & Ng, 2007). On the other hand, the saturation I_{ds} of the long channel MOS transistor is controlled by pinchoff (Hofstein & Heiman, 1963). Based on the constant mobility assumption, equation 4 predicts that the saturation I_{ds} of long channel MOS transistor has a quadratic relationship with V_{GS} and thus the saturation I_{ds} versus V_{DS} characteristics is expected to have increasing spacing for equal V_{GS} step (Sze & Ng, 2007). However, constant spacing for equal V_{GS} step is often observed in the experimental I_{ds} versus V_{DS} characteristics of the long channel MOS transistor, as shown in Fig.3. This can be understood from the validity of the constant mobility assumption. Experimental data have shown that mobility is actually a function of V_{GS} (Takagi et al., 1994). From Fig.7, μ_{eff} first increases with increasing V_{GS} owing to Coulombic scattering and then decreases owing to phonon scattering and surface roughness scattering. To further investigate, we measured the I_{ds} versus V_{DS} characteristics and the I_{ds} versus V_{GS} characteristics of a long-channel NMOS transistor. Considering equal V_{GS} step, we observed an increasing spacing for 1 V$\leq V_{GS} \leq$ 3 V but constant spacing for 3 V $\leq V_{GS} \leq$ 5V in the saturation I_{ds} versus V_{DS} characteristics of the NMOS transistor (see Fig.8). Since the transconductance (g_m) is a measure of the low-field mobility (μ_{eff}) (Schroder, 1998), the g_m versus V_{GS} characteristics is expected to have the same features as the mobility versus V_{GS} characteristics. From Fig. 8(a), the drain current saturation of the NMOS transistor occurs at V_{DS} around 3 V. With reference to Fig. 8(b), when V_{DS} = 3 V and 0 V $\leq V_{GS} \leq$ 3 V, g_m increases monotonically with increasing V_{GS} owing to Coulombic scattering. When V_{GS} is further increased to beyond 3 V, surface roughness scattering will start to dominate and then g_m will decrease with increasing V_{GS}. Hence, for 1 V $\leq V_{GS} \leq$ 3 V, the saturation I_{ds} versus V_{DS} characteristics has increasing spacing for equal V_{GS} step. For 3 V $\leq V_{GS} \leq$ 5 V, the saturation I_{ds} versus V_{DS} characteristics has constant spacing for equal V_{GS} step. Since velocity saturation does not occur in long channel transistor, the constant spacing observed in the saturation I_{ds} versus V_{DS} characteristics at high V_{GS} cannot be used as an indicator of the onset of velocity saturation.

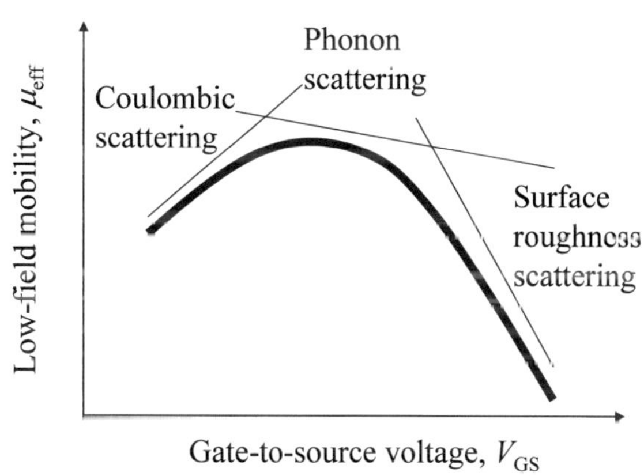

Fig. 7. Effects of the scattering mechanisms on the μ_{eff} versus V_{GS} characteristics of MOS transistor.

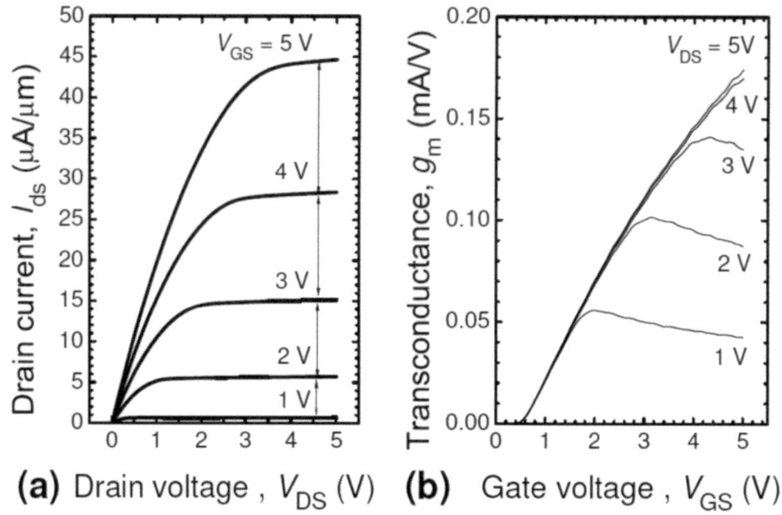

Fig. 8. Constant spacing is observed in the saturation I_{ds} versus V_{DS} characteristics of a NMOS transistor (L = 10 µm, W = 10 µm, physical gate oxide thickness of 300 Å) for equal V_{GS} step.

Here, it is interesting to note that it is common for the saturation I_{ds} versus V_{DS} characteristics of the zinc oxide thin-film transistors to have increasing spacing for equal V_{GS} step (Cheong et al., 2009; Yaglioglu et al., 2005). The mobility of these materials (~ 10 to 20 $cm^2/V.s$) is only one tenth of the mobility of silicon (~ 100 to 300 cm^2/Vs). In Fig.9, which is modified from (Cheong et al., 2009), the drain current saturation occurs at V_{DS} around 15 V. The increasing spacing observed in the saturation I_{ds} versus V_{DS} characteristics of the thin-

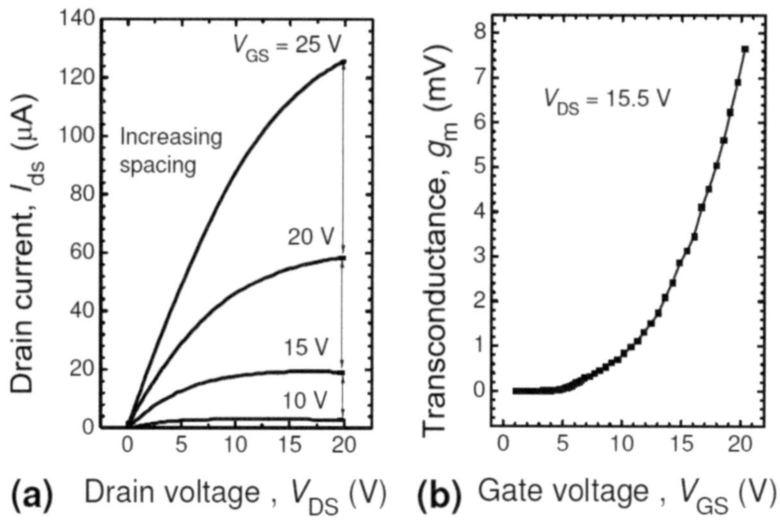

Fig. 9. Zinc oxide thin-film transistors with L = 20 µm and W = 40 µm (a) Increasing spacing observed in the experimental I_{ds} versus V_{DS} characteristics of, (b) Monotonically increasing g_m. Modified from (Cheong et al., 2009).

film transistor is related to the monotonically increasing g_m with increasing V_{GS}. Next, we will study the dependency of the saturation I_{ds} of the thin film transistor on V_{GS}. From Fig. 10, if I_{ds} and V_{GS} have linear dependency, $V_{th,sat}$ extracted by linear interpolation is around 17.5 V. If I_{ds} and V_{GS} have quadratic dependency, $V_{th,sat}$ extracted by extrapolating the linear portion of the $I_{ds}^{0.5}$ versus V_{GS} plot is around 10 V. As seen in the I_{ds} versus V_{DS} characteristics of the thin-film transistor (see Fig.9), the transistor is in cutoff mode when $V_{GS} \leq 10$ V. Hence, it is more appropriate to say that I_{ds} of thin-film transistor and V_{GS} have quadratic dependency rather than linear dependency.

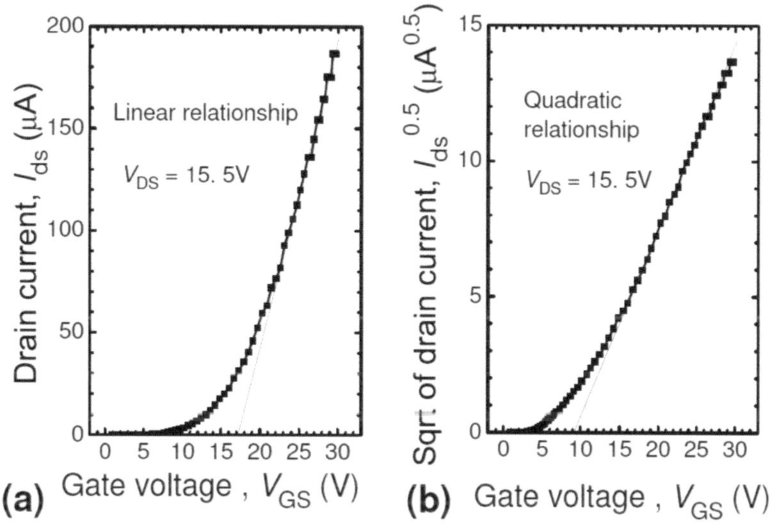

Fig. 10. Relationship between I_{ds} and V_{GS} of the zinc oxide thin-film transistors (L = 20 μm and W = 40 μm) (a) Linear dependency (b) Quadratic dependency. Modified from (Cheong et al., 2009).

4. Newer theories on the saturation drain current equations of the nanoscale MOS transistor

According to (Natori, 2008), the type of carrier transport in the MOS transistor depends on the relative dimension between the gate length (L) and the mean free path (λ), as illustrated in Fig. 11. Qualitatively, λ is the average distance covered by the channel carrier between the successive collisions. When L is much bigger than λ, the channel carriers will experience diffusive transport. When L is comparable to λ, the carriers undergo only a small number of scattering events from the source to the drain and thus the carriers will experience quasi-ballistic transport. Ballistic transport will only occur when $L < \lambda$. The experimentally extracted λ is in the range of 10 nm for the nanoscale transistor (M-J. Chen et al., 2004; Barral et al., 2009). Hence, the state of the art MOS transistor ($L \geq 32$ nm) is more likely to experience quasi-ballistic transport rather than ballistic transport. This section will discuss the main concepts of ballistic transport and then proceed to discuss about the existing quasi-ballistic theories. The emphasis of this section is to introduce a simplified equation for the saturation drain current of the nanoscale MOS transistor that is able to address quasi-ballistic transport while having electrical parameters that are obtainable from the standard device measurements. Here, we will introduce two equations that can satisfy the above criteria (i) Based on the concept of the effective saturation velocity (v_{sat_eff}) , which is a function of μ_{eff} and temperature (Lau et al. , 2008, b) and (ii) Based on the virtual source model (Khakifirooz et al., 2009).

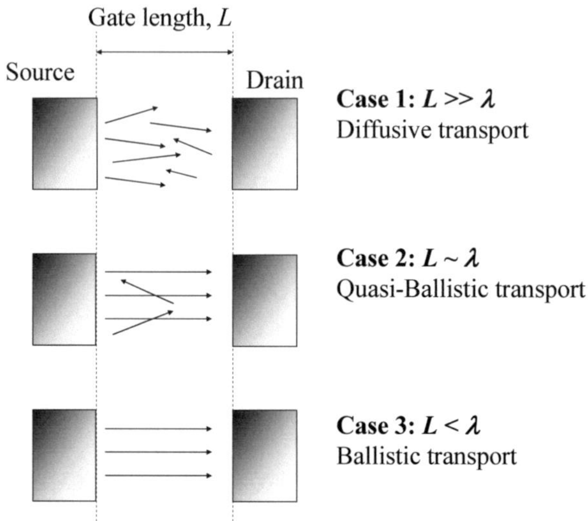

Fig. 11. Types of carrier transport in MOS transistors, which is modified from Fig. 1 in (Natori, 2008). Note that λ is the mean free path of the carrier.

4.1 Ballistic transport

In vacuum, electrons will move under the influence of electric field according to Newton's second law of motion,

$$F = m_e a = -qE \tag{9}$$

where F, m_e, a, q and E are the resultant force acting on the electron, the electron mass, the acceleration of the electron, the electronic charge , and the electric field ,respectively. Under such a situation, if the applied electric field is constant in both magnitude and direction, the electrons will accelerate in the direction opposite to that of the electric field. This type of transport is known as the ballistic transport. In the other words, if there is no obstacle to scatter the electrons, the electrons will experience ballistic transport (Heiblum & Eastman, 1987). Furthermore, (Bloch, 1928) postulated that the wave-particle duality of electron allows it to move without scattering in the densely packed atoms of a crystalline solid if (i) the crystal lattice is perfect and (ii) there is no lattice vibration. However, doping impurities such as boron, arsenic and phosphorus are added to the silicon crystal so as to tune the electrical parameters such as the threshold voltage and the off-state current (I_{off}). These dopants will disrupt the periodic arrangement of the crystal lattice and thus results in collisions with the impurity ions and the crystalline defects. Moreover, the atoms in crystals are always in constant motion according to the Particle Theory of Matter. These thermal vibrations cause waves of compression and expansion to move through the crystal and thus scatter the electrons (Heiblum & Eastman, 1987). Therefore, achieving ballistic transport in Si-based MOS transistors is only an ideal situation (Natori, 2008).

4.2 Quasi-ballistic transport

Having established that thermionic emission from the source to the channel is still relevant in the state-of-the-art MOS transistor ($L \geq 32$ nm) in Section 1, we will proceed to discuss the main concepts behind quasi-ballistic transport. (Lundstrom, 1997) derived an equation that relates the saturation I_{ds} of the nanoscale transistor to μ_{eff} as follows,

$$I_{ds} = \left[\frac{C_{ox} W}{\frac{1}{\upsilon_T} + \frac{1}{\mu_{eff}\ \varepsilon(0^+)}} \right] \left(V_{GS} - V_{th,sat} \right) \tag{10}$$

where the random thermal velocity of the carriers (υ_T) does not depend V_{GS}. The only variable in the υ_T equation is the temperature (T).

$$\upsilon_T = \upsilon_T(T) = \sqrt{(2k_B T)/(\pi\ m_t)} \tag{11}$$

where the transverse electron mass of silicon (m_t) is equal to 0.19 m_0 where the free electron mass (m_0) is equal to 9.11×10^{-31} kg (Singh, 1993). Using equation (11), υ_T is approximately equal to 1.2×10^7 cm/s at temperature of 25 °C. k_B is the Boltzmann constant. T is the absolute temperature. $\varepsilon(0^+)$ is defined as the average electric field within the length ℓ where a $k_B T/q$ potential drop occurs, as shown in Fig.12 in (Lundstrom & Ren, 2002). Despite the lack of equation for $\varepsilon(0^+)$ (Lundstrom, 1997; Lundstrom & Ren, 2002), Lundstrom has made an important contribution to relate the low-field mobility (μ_{eff}) to I_{on} of the deep submicron MOS transistors, and thus his theory is able to account for the strain-induced enhancement in I_{on} (Yang et al., 2004; C-H. Chen et al., 2004; Yang et al. 2008; Wang et al., 2007).

According to (Lundstrom, 1997), if a carrier backscatters beyond ℓ, it is likely to exit from the drain and is unlikely to return back to the source (see Fig. 12). For NMOS transistor, ℓ is the distance between the top of the conduction band edge and the point along the channel where channel potential drops by $k_B T/q$.

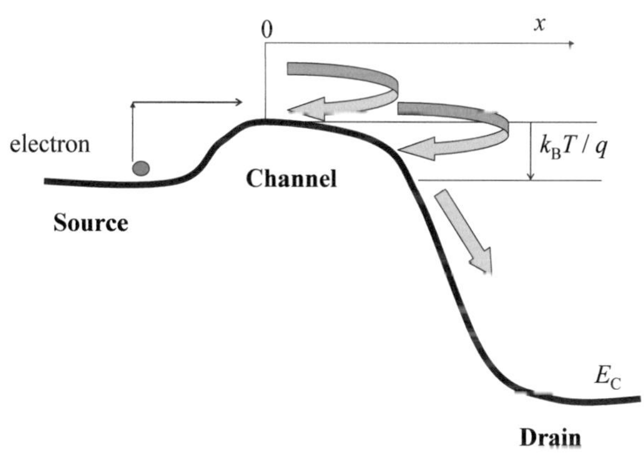

Fig. 12. Definition of the critical length (ℓ) for NMOS transistor. ℓ is defined to be the distance between the top of the conduction band edge and the point along the channel where channel potential drops by $k_B T/q$. Beyond ℓ, the carriers are unlikely to return to the source.

By inspection of equations (10) and (11), a loop-hole can be found in Lundstrom's 1997 theory. If equations (10) and (11) are correct, MOS transistors will function very poorly when the temperature is lowered from room temperature to very low temperature such as liquid helium temperature. However, there are numerous reports that MOS transistors and CMOS integrated circuits can function quite well at the liquid helium temperature (Chou et al., 1985; Ghibaudo & Balestra, 1997; Yoshikawa et al., 2005). Hence, there is a need to modify Lundstrom's 1997 theory. Indeed, (Lundstrom & Ren, 2002) made an attempt to incorporate Natori's 1994 theory into their theory. However, the resulting theory is very much not similar to equation (10) and has not been compared with real device performance.
Based on equation (24) in (Natori, 1994), the saturation drain current of the nanoscale MOS transistor is as follows,

$$I_{ds} = \frac{8\hbar W\left[C_{ox}\left(V_{GS}-V_{th,sat}\right)\right]^{3/2}}{3m_t\sqrt{q\pi M_v}} \tag{12a}$$

where $\hbar$ is the reduced Planck's constant. M_v is the product of the lowest valley degeneracy and the reciprocal of the fraction of the carrier population in the lowest energy level. For a NMOS transistor that is fabricated on (100) Si substrate, the fraction of the carrier population at the strong inversion is around 0.8 at 77 K but it decreases to around 0.4 at 300 K (Stern, 1972). In the other words, M_v is a function of temperature (T).
Rearranging equation (12a) results in,

$$I_{ds} = \left[\frac{C_{ox}\,W}{\dfrac{1}{v_{inj}(V_{GS},T)}}\right]\left(V_{GS}-V_{th,sat}\right) \tag{12b}$$

where the injection velocity (v_{inj}) is given by (Natori, 1994),

$$v_{inj}(V_{GS},T) = \frac{8\hbar\sqrt{C_{ox}\left(V_{GS}-V_{th,sat}\right)}}{3m_t\sqrt{q\pi M_v(T)}} \tag{12c}$$

With reference to Fig.8 in (Natori, 1994), v_{inj} increases with increasing temperature (T) and increasing V_{GS}. If Natori's theory is true, v_{inj} can be very high even though the temperature is very low. We propose that this feature of Natori's 1994 theory can be used to cover the shortcomings of Lundstrom's 1997 theory. However, there are some aspects of Natori's 1994 theory that contradict the experimental data. From Fig. 8 in (Natori, 1994), his theory, which disregards the channel scattering, predicted that the saturation I_{ds} of the nanoscale NMOS transistor will increase when temperature increases. However, this is contradictory to the experimental data. Fig. 13 shows that the experimental I_{ds} of a NMOS transistor (L= 60 nm) actually decreases when temperature increases. This can be explained by the increase in channel scattering when temperature increases (Takagi et al., 1994; Kondo & Tanimoto, 2001; Mazzoni et al., 1999). Moreover, equation (12b) cannot account for the strain-induced enhancement in I_{on} (Yang et al, 2004; C-H. Chen et al, 2004; Yang et al., 2008; Wang et al., 2007). Hence, without the help of Lundstrom's 1997 theory, Natori's 1994 theory is contradictory to the experimental data.

In addition, Natori's 1994 theory predicts that the saturation I_{ds} of the nanoscale MOS transistors will follow a $(V_{GS} - V_{th,sat})^{3/2}$ relationship. Fig. 14a shows the saturation $I_{ds}^{2/3}$ versus V_{GS} characteristics of a NMOS transistor (L = 60 nm). The threshold voltage extracted by the linear extrapolation is smaller than the threshold voltage of conduction. This shows that the saturation I_{ds} of the nanoscale MOS transistors does not follow a $(V_{GS} - V_{th,sat})^{3/2}$ relationship. Fig. 14b shows the saturation I_{ds} versus V_{GS} characteristics of the same NMOS transistor. In this case, the extracted threshold voltage is close to the threshold voltage of conduction. Hence, the saturation I_{ds} of nanoscale transistors is more likely to follow a $(V_{GS} - V_{th,sat})$ relationship.

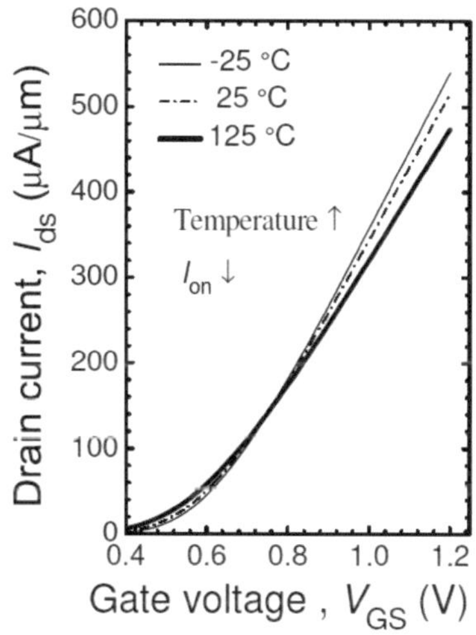

Fig. 13. Effects of temperature on the saturation I_{ds} versus V_{GS} characteristics of a NMOS transistor (L = 60 nm, W = 5 μm).

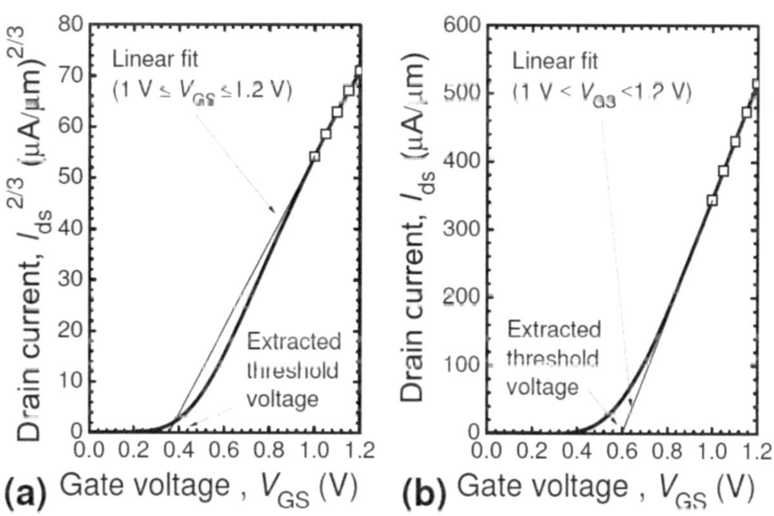

Fig. 14. As opposed to Natori's 1994 theory, the saturation I_{ds} of the short channel NMOS transistor does not follow a $(V_{GS} - V_{th,sat})^{3/2}$ relationship.

4.3 New equation that unifies Natori's 1994 theory and Lundstrom's 1997 theory

We propose a simplified equation that can unify both Natori's 1994 theory and Lundstrom's 1997 theory, as follows (Lau et al., 2008, b),

$$I_{ds} = \left[\frac{C_{ox}W}{\frac{1}{v_1(V_{GS},T)} + \frac{1}{v_1(V_{GS},T)}} \right] \left(V_{GS} - V_{th,sat} \right) \tag{13}$$

where

$$v_1(\, V_{GS},\, T) = v_{inj}(\, V_{GS},\, T) \tag{14}$$

$$v_2(\, V_{GS},\, T) = \mu_{eff}(\, V_{GS},\, T)\; \varepsilon\left(0^+\right) \tag{15}$$

(Lundstrom, 1997) proposed that v_1 is equal to v_T that is only dependent on T, as shown in equation (11). On the other hand, our theory proposed that v_1 is a function of both V_{GS} and T, and v_1 can be higher than v_T given by equation (11). (Natori, 1994) proposed that v_1 is equal to v_{inj}, which is a function of both V_{GS} and T. Recently, (Natori et al., 2003; Natori et al., 2005) simulated the v_{inj} characteristics using the multi-subband model (MSM). In weak inversion, v_{inj} is almost independent of V_{GS} and is approximately equal to 1.2 x 10^7 cm/s, which is equal to v_T. In strong inversion, v_{inj} will increase due to carrier degeneration but is confined within a narrow range from 1.2 x 10^7 cm/s to 1.6 x 10^7 cm/s.

Here, we would like to highlight that both Lundstrom's 1997 theory and Natori's 1994 theory did not consider the series resistance (R_{sd}). Although the conduction band edge (E_c) profile in the n-channel will be the same with or without R_{sd} (Martinie et al., 2008), the E_c within S/D regions will be different when the effects of R_{sd} is considered. If the effects of R_{sd} are disregarded, E_c within S/D regions will appear as a horizontal line, as illustrated in Fig. 12. However, the presence of R_{sd} will cause a potential drop in the S/D regions, resulting in a built-in electric field within the S/D regions (see Fig. 15). This electric field in the source region will accelerate the electrons. Since scattering decreases when temperature decreases (Takagi et al., 1994; Kondo & Tanimoto, 2001; Mazzoni et al., 1999), one would expect that there will be minimal scattering in the source when the temperature is very low. Hence, the presence of R_{sd} will allow the electrons to attain higher energy prior to thermionic emission into the channel. According to (M-J. Chen et al., 2004), the source series resistance (R_s) is about 75 Ω-μm. If the drain current (I_{ds}) is about 800 μA/μm, the voltage drop due to R_s is about 800 μA/μm x 75 Ω-μm = 60 mV. (Note that the thermal voltage, k_BT/q is approximately 26 meV at room temperature.) We proposed that the electrons are "heated" up by the 60 meV energy due to R_{sd} and thus their velocities can be significantly larger than 1.2 x 10^7 cm/s (as predicted by equation 12c). Moreover, this extra energy is expected to increase with increasing V_{GS} because higher V_{GS} implies a bigger I_{ds}. With this extra energy from electron heating in the R_{sd} region, the carriers can overcome the potential barrier at the liquid nitrogen temperature despite not being able to gain energy from the surrounding. The significance of v_2 term is that it establishes a link between I_{on} and μ_{eff}. This provides a better compatibility between theory and I_{on} enhancement in the nanoscale transistors by various stress engineering techniques (Yang et al., 2004; C-H. Chen et al., 2004; Yang et al.,

2008; Wang et al., 2007). However, there is no v_2 term in Natori's 1994 theory, as shown in equation (12b). Nevertheless, v_2 is covered by Lundstrom's 1997 theory, as shown in equation (10). Hence, we incorporate v_2 in Lundstrom's 1997 theory into equation (13).

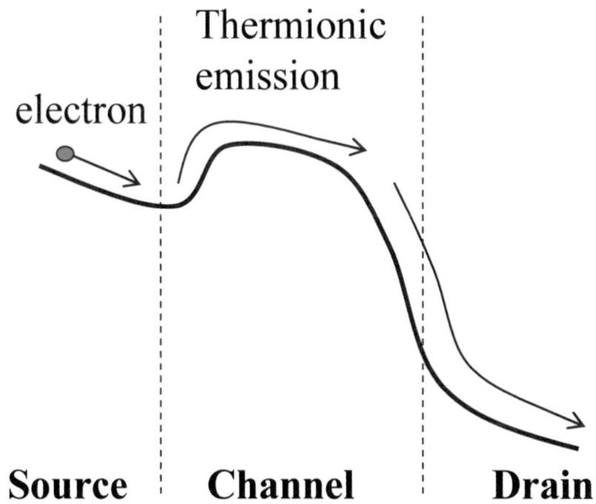

Fig. 15. The effects of S/D series resistance on the conduction band edge of a NMOS transistor in the saturation operation.

Another loop-hole in Lundstrom's 1997 theory is that there is no equation for $\varepsilon(0^+)$. From Fig.9 in (M-J. Chen et al., 2004), the slope of the near-source channel conduction band increases when V_{GS} increases. In the other words, the electric field near the top of potential barrier, $\varepsilon(0^+)$ increases with increasing V_{GS}. Hence, we deduce that $\varepsilon(0^+)$ is a function of both V_{GS} and V_{DS} such that $\varepsilon(0^+, V_{GS}, V_{DS} = V_{DD})$ is approximately equal to $\varepsilon(0^+, V_{GS}, V_{DS} = V_{Dsat})$. Note that V_{DD} is the power supply voltage. This is consistent with Fig. 5 in (Fuchs et al., 2005). Therefore, we propose that $\varepsilon(0^+)$ can be expressed as follows,

$$\varepsilon(0^+) = \frac{\alpha_1 V_{Dsat}}{L_{eff}} \tag{16a}$$

where the correction factor (α_1) is smaller than 1. Based on the conventional MOS transistor theory (Taur & Ning, 1998, a), V_{Dsat} is given by $(V_{GS} - V_{th,sat})/m$ where $1.1 \leq m \leq 1.4$. Furthermore, (Suzuki & Usuki, 2004) proposed a drain current model that shows that V_{Dsat} is smaller than $(V_{GS} - V_{th,sat})$ for the short-channel MOS transistors. This shows that the relationship of $V_{Dsat} = (V_{GS} - V_{th,sat})/m$ is still reasonably correct for very short MOS transistors. Therefore, $\varepsilon(0^+)$ can also be expressed by,.

$$\varepsilon(0^+) = \frac{\alpha_2 \left(V_{GS} - V_{th,sat}\right)}{L_{eff}} \tag{16b}$$

where the correction factor (α_2) is smaller than 1. The value of α_2 can be estimated from the effective carrier velocity (v_{eff}) versus V_{GS} characteristics and the μ_{eff} versus V_{GS} characteristics. Using the saturated transconductance method suggested by (Lochtefeld et al., 2002), v_{eff} was extracted as a function of V_{GS} as shown in Fig.16 (a). For the contact etch stop layer (CESL) with a tensile stress of 1.2 GPa, v_{sat_eff} of the NMOS transistor (L = 60 nm) was 7.3×10^6 cm/s. Using the constant current method with reference current, I_{ref}

$=0.1\mu A(W/L)$, the extracted $V_{th,sat}$ was about 0.3 V. Next, L_{eff} , which is extracted using the method proposed by (Guo et al., 1994) , was about 0.030 μm. Substituting L_{eff} = 3 x 10^{-6} cm, V_{GS} = 1.2 V , $V_{th,sat}$ = 0.3 V into equation (16b),

$$\varepsilon\left(0^{+}\right)=3\times10^{5}\alpha_2 \quad \text{(in units of V/cm)} \tag{16c}$$

Re-arranging $v_{sat\_eff} = \mu_{eff}\,\varepsilon(0^{+})$,

$$\varepsilon(0^{+})=\frac{v_{sat\_eff}}{\mu_{eff}} \tag{16d}$$

Next, μ_{eff} is extracted as a function of V_{GS} using a method described by (Schroder, 1998). From Fig. 16(b), when V_{GS} is 1.2 V, μ_{eff} was about 85 $cm^2V^{-1}s^{-1}$ at. Substituting v_{sat_eff} = 7.3 $\times10^6$ cm/s and μ_{eff} = 85 $cm^2V^{-1}s^{-1}$ into equation (16d),

$$\varepsilon(0^{+})=\frac{v_{sat\_eff}}{\mu_{eff}}=\frac{7.3\times10^{6}}{85}=8.588\times10^{4}\ \text{V/cm} \tag{16e}$$

According to (Lee et al., 2009), $\varepsilon(0^+)$of a PMOS transistor (L = 50 nm) is between 8 $\times10^4$ V/cm and 3 $\times10^5$ V/cm for various gate overdrives. By solving equations (16c) and (16e), α_2 is around 0.29. Note that α_2 is 0.5 for the conventional MOS transistor theory (Taur & Ning, 1998, a).

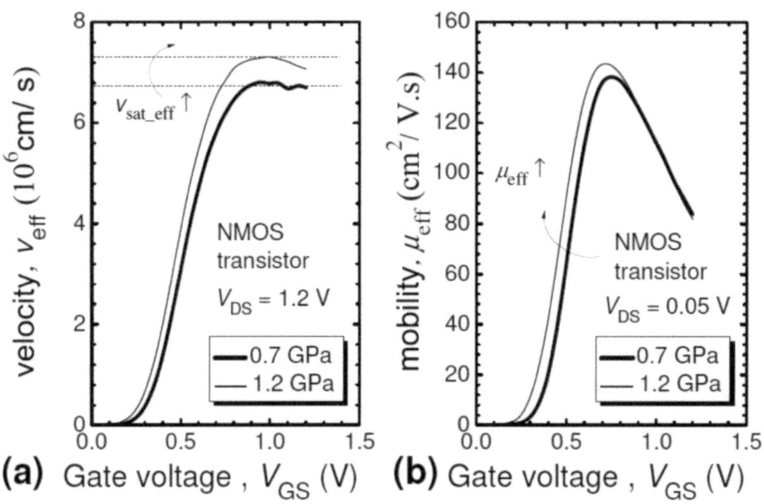

Fig. 16. Effects of uniaxial tensile stress on (a) the v_{eff} versus V_{GS} characteristics, (b) the μ_{eff} versus V_{GS} characteristics of a NMOS transistor (L = 60 nm, W = 0.12 μm). Note v_{sat_eff} is the average value of v_{eff} when V_{GS} is close to V_{DD}. The uniaxial tensile stress is induced by the contact etch stop layer (CESL). The film stress of the two CESL split are 0.7 GPa tensile stress and 1.2 GPa tensile stress.

Equation (13) is then modified by defining a new parameter called the effective carrier velocity (v_{eff}). The resulting equation is as follows (Yang et al., 2007; Lau et al., 2008, a; Lau et al., 2008, b),

$$I_{ds} = v_{eff}(\mu_{eff}, V_{GS}, T)WC_{ox}(V_{GS} - V_{th,sat}) \tag{17}$$

where v_{eff} is a function of μ_{eff}, V_{GS} and T at a constant V_{DS} (see Fig.16a and Fig.17). Furthermore, v_{eff} is also related to v_1 and v_2, as follows,

$$v_{eff}(\mu_{eff}, V_{GS}, \mathrm{T}) = \left(\frac{1}{v_1(V_{GS}, T)} + \frac{1}{v_2(\mu_{eff}, V_{GS}, T)} \right)^{-1} \tag{18}$$

When temperature decreases, v_{inj} decreases (Natori, 1994). Since v_1 is related to v_{inj} (see equation 14), v_1 is expected to decrease with decreasing temperature. On the other hand, mobilities due to Coulombic scattering, phonon scattering and surface roughness scattering will increase with decreasing temperature (Takagi et al., 1994; Kondo & Tanimoto, 2001; Mazzoni et al., 1999). As v_2 is related to μ_{eff} (see equation 15), we expect v_2 to increase when temperature decreases. Fig. 17 shows that the experimental v_{eff} increases when temperature decreases, and hence v_2 dominates over v_1.

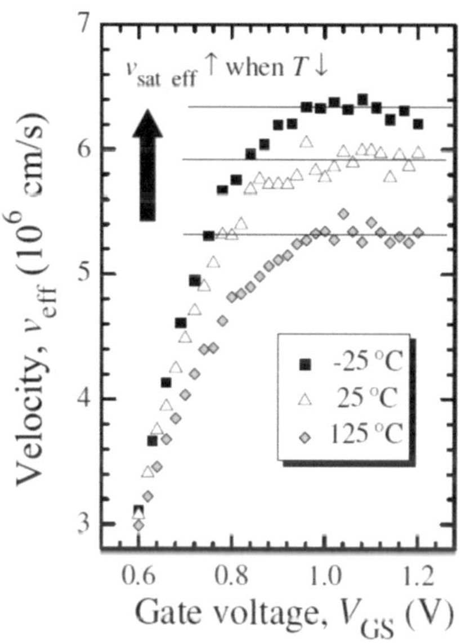

Fig. 17. The effect of temperature on v_{sat_eff}. Note that v_{sat_eff} corresponds to the average value of v_{eff} when V_{GS} is close to V_{DD} (L = 60 nm, W = 5 μm, V_{DS} = V_{DD} = 1.2 V).

Another evidence to illustrate the importance of v_2 over v_1 is through their behavior with V_{GS}. Fig.18 shows the behavior of v_1, v_2, v_{eff} with V_{GS}. Since v_1 is related to v_{inj}, v_1 is expected to increase when V_{GS} increases (Natori, 1994). On the other hand, v_2 is related to μ_{eff}, as shown in equation (15). Hence, the v_2 versus V_{GS} characteristics will tend to follow that of the μ_{eff} versus V_{GS} characteristics (see Fig. 7). When V_{GS} is low, v_2 is expected to increase with increasing V_{GS} owing to the screening of the Coulombic scattering centres. When V_{GS} is high, an increase in V_{GS} will decrease μ_{eff} owing to the surface roughness scattering. From equation (15), v_2 is the product of μ_{eff} and $\varepsilon(0^+)$. From equation (16b), $\varepsilon(0^+)$ is expected to increase with increasing V_{GS}. Hence, v_2 is expected to approach a constant at high V_{GS} owing to the opposing effects of μ_{eff} and $\varepsilon(0^+)$.

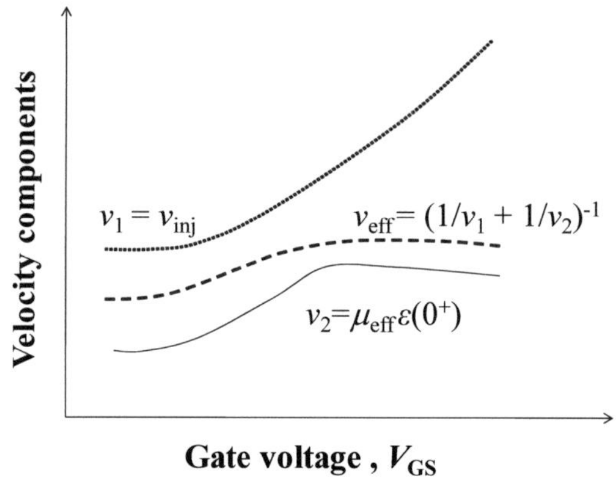

Fig. 18. A schematic diagram showing the relationship of v_1, v_2 and v_{eff} with V_{GS}.

Since v_{eff} approaches a constant when V_{GS} close to V_{DD} (see Fig. 16a and Fig. 17), it is more appropriate to replace v_{eff} in equation (17) can be replaced by v_{sat_eff}, resulting in (Yang et al., 2007; Lau et al., 2008,a; Lau et al., 2008,b),

$$I_{ds} = v_{sat\_eff}(\mu_{eff}, T)WC_{ox,inv}\left(V_{GS} - V_{th,sat\_IV}\right) \tag{19}$$

where v_{sat_eff} is the average value of v_{eff} when V_{GS} is close to V_{DD}. In Fig. 16(a), v_{sat_eff} increases when tensile stress increases, and thus leads to I_{on} enhancement in the short channel NMOS transistor. This shows that equation (19) is able to account for the strain-induced I_{on} enhancement by various strain engineering techniques (Yang et al., 2004; C-H. Chen et al., 2004; Yang et al. 2008; Wang et al., 2007). As shown in Fig.17, v_{sat_eff} increases when temperature decreases, resulting in a better I_{on} performance at very low temperature. This shows that equation (19) is able to explain the I_{on} enhancement at liquid helium temperature (Chou et al., 1985; Ghibaudo & Balestra, 1997; Yoshikawa et al., 2005).

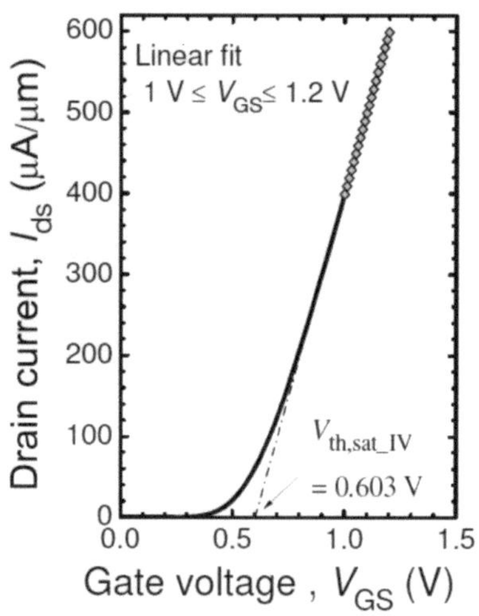

Fig. 19. Extraction of V_{th,sat_IV} from the saturation I_{ds} versus V_{GS} characteristics of a NMOS transistor (L = 60 nm, W = 2 µm, V_{DS} = 1.2 V).

Moreover, $V_{th,sat}$ in equation (17) needs to be replaced by V_{th,sat_IV}. As illustrated in Fig. 19, V_{th,sat_IV} can be extracted from the saturation I_{ds} versus V_{GS} characteristics. First, a best-fit line is performed on the saturation I_{ds} versus V_{GS} characteristics when V_{GS} is close to V_{DD}. For our transistors, v_{eff} approaches a constant when 1 V ≤ V_{GS} ≤ 1.2 V. V_{th,sat_IV} can be found by the interception between the best-fit line and the V_{GS} axis. In this example, V_{th,sat_IV} was 0.603 V. For comparison, we extracted $V_{th,sat}$ using the Constant Current (CC) method with the reference drain current (I_{ref}) defined as 0.1μA(W/L). The extracted $V_{th,sat}$ was 0.351 V, which is much smaller than V_{th,sat_IV}. Moreover, we also observed that V_{th,sat_IV} is also bigger than the linear threshold voltage ($V_{th,lin}$). In Fig. 20(a), $V_{th,lin}$ extracted using CC method was 0.484 V. In Fig. 20(b), $V_{th,lin}$ extracted using maximum g_m method was 0.557 V. We believe that V_{th,sat_IV} is bigger than $V_{th,lin}$ and $V_{th,sat}$ because it accounts for the additional V_{GS} that is required to produce electrons to screen the Coulombic scattering centres, as shown in Fig. 21. On the other hand, $V_{th,lin}$ and $V_{th,sat}$ indicate the onset of inversion. Furthermore, polysilicon depletion and quantum mechanical effects will make the gate oxide appears thicker, and thus C_{ox} in equation (17) has to be replaced by $C_{ox,inv}$, which is the gate oxide capacitance per unit area at inversion.

4.4 Virtual source model for nanoscale transistors in saturation mode

(Khakifirooz et al., 2009) proposed a semi-empirical model for the saturation drain current of the nanoscale transistor. This model is based on the location of the "virtual source", which is the top of the conduction band profile for NMOS transistor, as shown in Fig. 22. Based on the "charge-sheet approximation", the saturation I_{ds} of the nanoscale transistor can be described by the product of the local charge density and the carrier velocity, as follows (Khakifirooz & Antoniadis, 2008).

$$I_{ds} = WQ_{ixo}v_{xo} \tag{20}$$

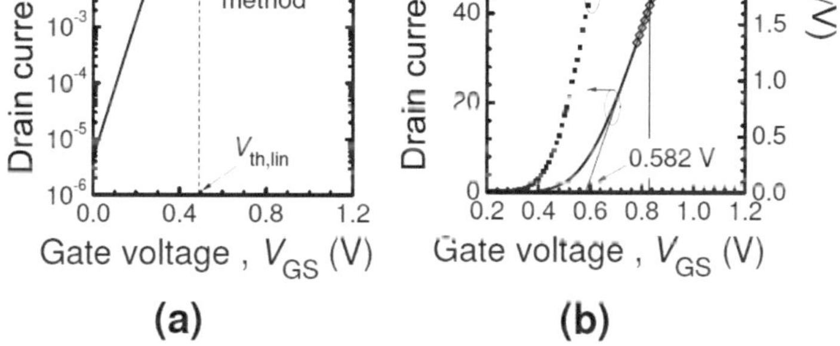

Fig. 20. Extraction of $V_{th,lin}$ of a NMOS transistor in the linear operation (L = 60 nm, W = 2 μm, V_{DS} = 0.05 V) (a) Using constant current method with I_{ref} = 0.1 μA W/L, $V_{th,lin}$ = 0.484 V, (b) Using maximum g_m method ($V_{th,lin}$ = 0.582 - $V_{DS}/2$ = 0.557 V).

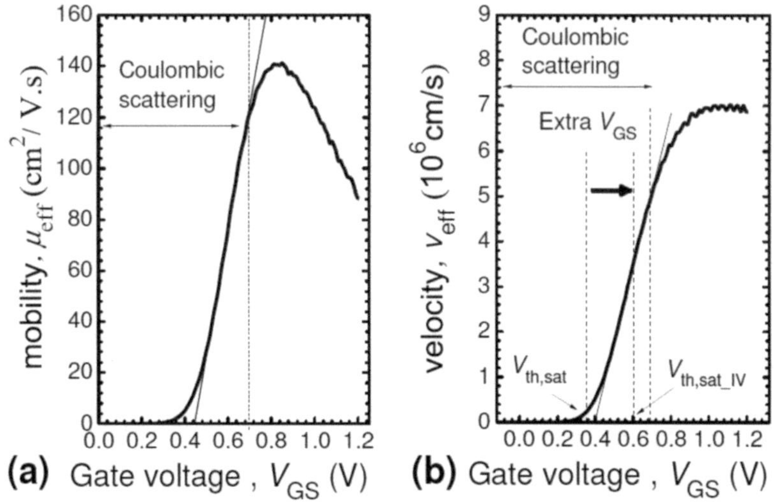

Fig. 21. V_{th,sat_IV} includes a component to overcome the Coulombic scattering by "screening".

The virtual source charge density (Q_{xio}) is given by (Khakifirooz et al., 2009),

$$Q_{xio} = C_{ox}\frac{k_B T}{q} ln\left[1 + \exp\left(\frac{V_{GS} - I_{ds}R_s - V_{th,sat}}{mk_B T / q}\right)\right] \tag{21}$$

where R_s is the source series resistance. The body-effect coefficient (m) can be expressed as (Taur & Ning, 1998,b),

$$m = 1 + \frac{\sqrt{\varepsilon_0 \varepsilon_{Si} q N_{ch} / (4\psi_B)}}{C_{ox}} \tag{22}$$

where ε_0 is the permittivity of free space. ε_{Si} is the dielectric constant of silicon. N_{ch} is the channel doping concentration. ψ_B is the difference between the Fermi level in the channel region and the intrinsic Fermi level.

The virtual source velocity (v_{xo}) is the average velocity of the channel carriers at the potential barrier near the source.

$$v_{x0} = \frac{v}{1 - C_{ox} R_s W (1 + 2\delta) v} \tag{23}$$

where δ is the drain-induced-barrier lowering (DIBL) with units of V/V. The carrier velocity can be extracted as follows,

$$v = \frac{I_{ds} / W}{C_{ox}\left(V_{GS} - V_{th,sat}\right)} \tag{24}$$

According to (Khakifirooz et al., 2009), the above model has a reasonably good fit to the experimental I_{ds} versus V_{GS} characteristics and the experimental I_{ds} versus V_{DS} characteristics of nanoscale Si-based MOS transistors fabricated using poly-SiON gate stack as well as high-

K metal gate stack. The extracted v_{xo} for NMOS transistor (L = 35 nm) is around 1.4×10^7 cm/s. Since v_{sat} for electrons in silicon is 10^7 cm/s (Norris & Gibbons, 1967), this shows that velocity saturation does not occur in the nanoscale Si-based MOS transistor.

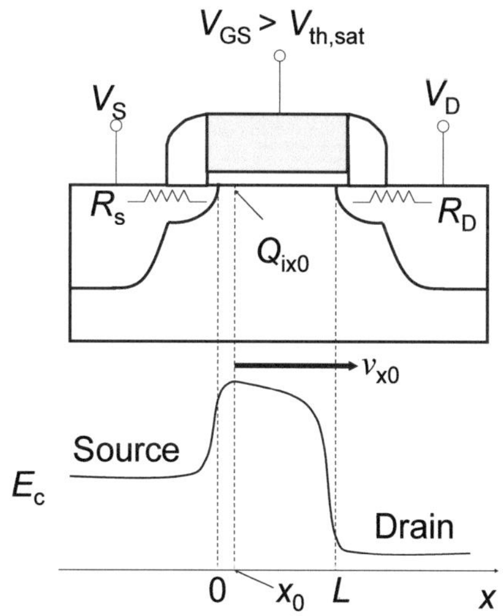

Fig. 22. Illustration of the virtual source point (x_0) in a NMOS transistor. The carrier charge density (Q_{ixo}) and the virtual velocity (v_{xo}) are defined at the top of the conduction band profile along the channel direction. R_s is the source series resistance. R_D is the drain series resistance.

5. Apparent velocity saturation in the nanoscale MOS transistor

Fig. 23 shows the maximum μ_{eff} versus L characteristics and the v_{sat_eff} versus L characteristics of a bulk NMOS transistor. μ_{eff} is extracted from the linear I_{ds} versus V_{GS} characteristics (Schroder, 1998). v_{eff} is extracted using the saturation transconductance method (Lochtefeld et al., 2002). R_{sd} correction to v_{eff} has to be done as described by (Chou & Antoniadis, 1987) . R_{sd} is extracted using a modified version of the method according to (Chern et al., 1980). Note that v_{sat_eff} is the average value of v_{eff} when V_{GS} is close to V_{DD}. By taking the maximum μ_{eff} to be independent of the gate length, v_{sat_eff} = constant x L_{eff}^{-1}, based on equation (16b) and equation (16d). However, the experimental v_{sat_eff} = constant x $L_{eff}^{-\beta}$ where β is less than 1 despite the uncertainty in R_{sd} measurements (see Fig. 24). This indicates that the carrier velocity tends to saturate when L decreases (see Fig. 23b).

Since the relationship between the carrier velocity and the low-field mobility is well-established (Khakifirooz & Antoniadis, 2006), we can have a better understanding of the apparent velocity saturation in the nanoscale MOS transistors by looking at the mobility. A strong reduction of mobility is typically observed in the silicon-based MOS transistors when the gate length is scaled (Romanjek et al., 2004; Cros et al., 2006; Cassé et al., 2009; Huet et al., 2008, Fischetti & Laux, 2001). The reason of this degradation is still not clearly understood. It is first attributed to the halo implants as its contribution to the channel doping concentration increases with decreasing gate length (Romanjek et al., 2004). However, this mobility degradation is also observed in the undoped double gate MOS

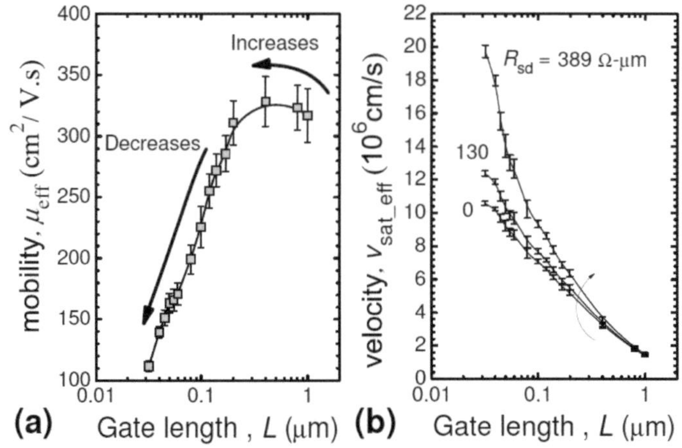

Fig. 23. Effects of scaling on bulk NMOS transistors ($W = 1$ μm) (a) the μ_{eff} versus L characteristics, (b) the v_{sat_eff} versus L characteristics. Note that v_{sat_eff} increases with increasing μ_{eff} . $R_{sd} = 0$ Ω-μm refers to the case where R_{sd} correction is not performed.

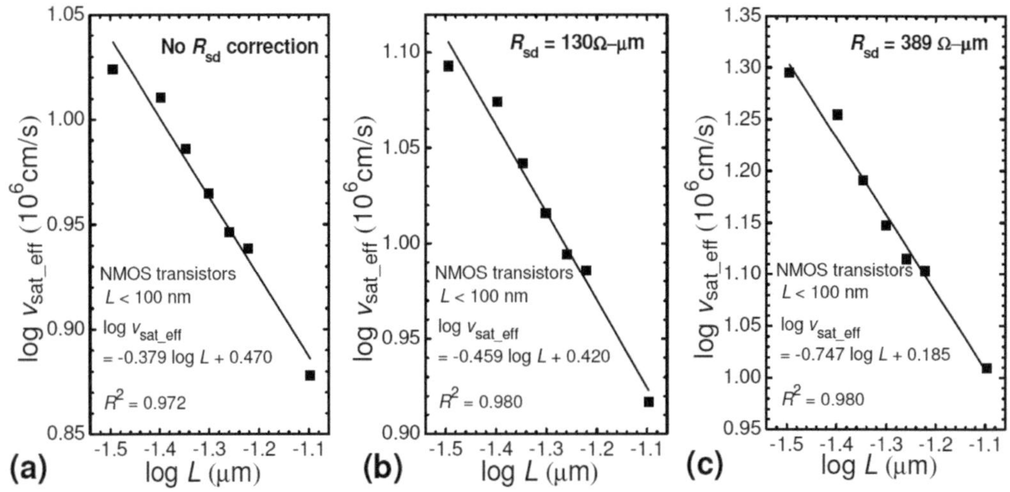

Fig. 24. Validity of v_{sat_eff} = constant x $L_{eff}^{-\beta}$ where β is less than 1 despite the uncertainty in R_{sd} measurements. Note that log $v_{sat\_eff} = -\beta \log L_{eff}$ + log constant.

transistors (Cros et al., 2006) and the undoped fully-depleted silicon-on-insulator (FD-SOI) MOS transistors (Cassé et al., 2009). This indicates that the halo implant is not the dominant factor involved in the degradation. Another limiting transport mechanism expected to be non-negligible in the short-channel MOS transistor is the presence of crystalline defects induced by S/D extension implants (Cros et al., 2006). Furthermore, Monte Carlo studies shows that ballistic transport has significant impact on the mobility degradation (Huet et al., 2008). Another explanation is that the increase in the long-range Coulombic scattering interactions between the high-density electron gases in the S/D regions and the channel electrons for very short channel MOS transistors (Fischetti & Laux, 2001). In an attempt to clarify the mobility degradation mechanism, (Cassé et al., 2009) used the differential magnetoresistance technique for mobility extraction to eliminate the effects of series resistance (R_{sd}) and the ballisticity introduced by L-independent resistance. However, strong mobility degradation is still observed in the undoped FD-SOI MOS transistors (L < 100 nm) at 20 K and thus the mobility degradation is likely to be caused by (i) the long-range Coulombic scattering interactions between the electron gases in the S/D regions and the channel electrons, (ii) the charged defects at the S/D regions, and (iii) the neutral defects at the S/D regions (Cassé et al., 2009). The apparent saturation of carrier velocity when L decreases can be understood as follows. As discussed in section 4.3, the effects of v_2 dominates over the effects of v_1 such as $v_{eff} \approx v_2$. From equation (15), v_2 is the product of μ_{eff} and $\varepsilon(0^+)$. From equation (16b), $\varepsilon(0^+)$ increases when L_{eff} decreases. In short, when L decreases, μ_{eff} decreases but $\varepsilon(0^+)$ increases. Hence, v_{eff} is expected to approach a constant when L decreases. Since v_{sat_eff} is the average value of v_{eff} when V_{GS} is close to V_{DD}, v_{sat_eff} is expected to approach a constant when L decreases. This is probably why (Hauser, 2005) is able to use the velocity saturation model (see equation 5) to fit the experimental I_{ds} versus V_{DS} characteristics of the nanoscale NMOS transistor (L = 90 nm). Note that Hauser used v_{sat} as a fitting parameter. In his physics-based model, v_{sat} is taken to be 2.06×10^7 cm/s rather than 1×10^7 cm/s (saturation velocity of electrons in silicon at room temperature). Therefore, the physics behind the apparent saturation of the carrier velocity is different from that of velocity saturation (the rate of energy gain from the lateral electric field is equal to the rate of energy loss to the surroundings by phonon scattering).

6. Drain current saturation mechanism of the nanoscale MOS transistors

As mentioned in section 2, the two well-known mechanisms for drain current saturation in MOS transistors are pinch off and velocity saturation. However, we have shown that velocity saturation is unlikely to occur in the nanoscale MOS transistors. In addition, (Kim et al., 2008) reported that the experimental observation of velocity overshoot in the nanoscale bulk NMOS transistor (L = 36 nm) at room temperature. In section 5, we have unveiled that the apparent velocity saturation that occurs during scaling is caused by (i) the long-range Coulombic scattering interactions between the electron gases in the S/D regions and the channel electrons, (ii) the charged defects at the S/D regions and (iii) the neutral defects at the S/D regions (Cassé et al. 2009). Since velocity saturation involves the tradeoff between the rate of energy gain from lateral electric field and the rate of energy loss to the surroundings by phonon scattering, we believe that velocity saturation does not occur in the nanoscale transistors. Hence, it is possible that the drain current saturation mechanism in nanoscale MOS transistor is caused by pinch off rather than velocity saturation. In fact, several groups of researchers have developed compact models for the pinch-off region of the nanoscale MOS transistors (Navarro et al., 2005; Weidemann et al., 2007). For V_{DD} = 1 V, the pinch-off point is less than 10 nm from the drain side (Navarro et al., 2005). This shows that the pinch-off point will always remain within the channel even though this point tends to shift towards the source side with increasing V_{DS}.

Our previous work gives the experimental evidence that the drain current saturation in the nanoscale NMOS transistor is caused by pinchoff (Lau et al., 2009). By simply changing the polarity of the drain bias (V_D), it is possible to create a situation whereby pinchoff is unlikely to occur. As shown in Fig. 25, the normal biasing involves the application of a positive V_D to the drain terminal of a NMOS transistor. On the other hand, the unusual biasing involves the application of a negative V_D to the drain terminal of a NMOS transistor. The most obvious implication of such biasing is the direction of the electron flow. For the normal biasing condition, the electrons are injected from source terminal to drain terminal. For the unusual biasing condition, the electrons are injected from drain terminal to source terminal. In the other words, the effective source terminal for the unusual biasing is actually the drain terminal. To avoid confusion, we define V_{GS}^* as the potential difference between the gate terminal and the terminal that injects electrons into the channel. V_{DS}^* is the potential difference between the source terminal and drain terminal. From equation (8), the condition for pinchoff to occur is as follows,

$$V_{DS}^* \geq \frac{V_{GS}^* - V_{th,sat}}{m} \tag{25}$$

where m is between 1.1 and 1.4 (Taur & Ning, 1998,a). For our NMOS transistors, V_{DD} is 1.2 V. Under the normal biasing, V_{GS}^* is 1.2 V and V_{DS}^* is 1.2 V (see Fig. 25a). Under the unusual biasing, V_{GS}^* is 2.4 V and V_{DS}^* is 1.2 V (see Fig. 25b). Hence, normal biasing will be able to satisfy the condition for pinchoff and thus pinchoff can occur. However, the condition for pinchoff cannot be satisfied under the unusual biasing because V_{GS}^* is much bigger than V_{DS}^*. From Fig.26, the nanoscale NMOS transistor (L = 45 nm) used in our study does not suffer from punchthrough. Note that negative V_D will forward bias the p-well-to-n$^+$drain junction. To minimize the amount of forward biased p-n junction current in NMOS transistor under the unusual biasing, we limited the V_D to be -0.4 V (see Fig. 27). As shown

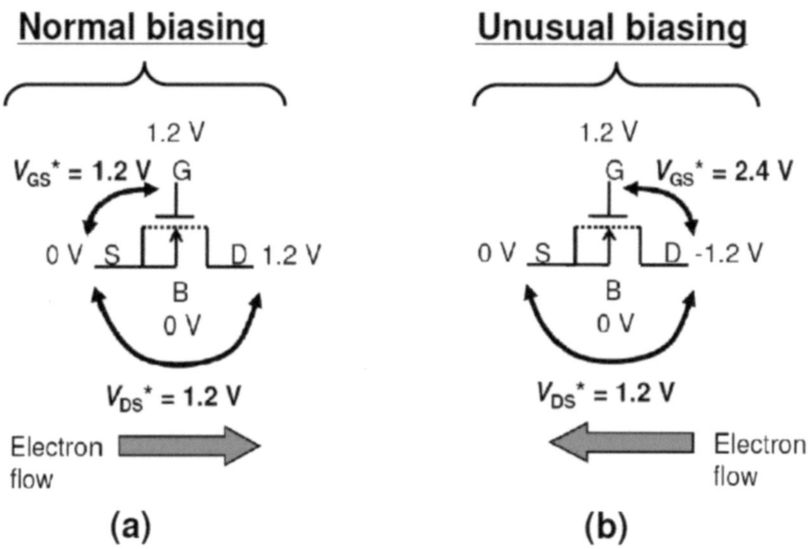

Fig. 25. Biasing conditions of the NMOS transistor (a) Under the normal biasing, a positive V_D of 1.2 V is applied to the drain terminal. The p-well-to-n$^+$ drain junction is reversed biased. (b) Under the unusual biasing condition, a negative V_D of -1.2 V is applied to the drain terminal. The p-well-to-n$^+$ drain junction is forward biased.

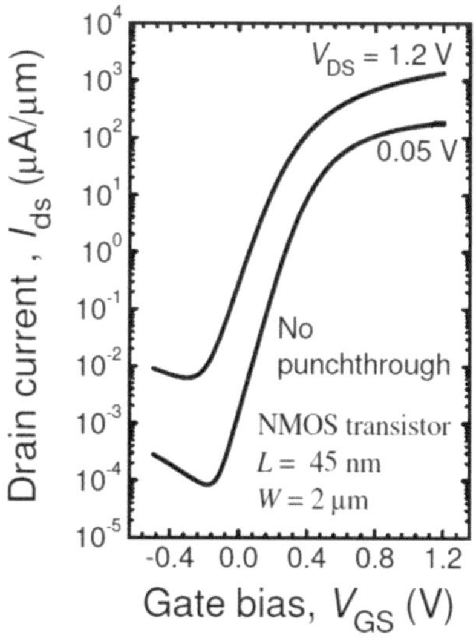

Fig. 26. The I_{ds} versus V_{GS} characteristics of the nanoscale NMOS transistor (L = 45 nm, W = 2 μm).

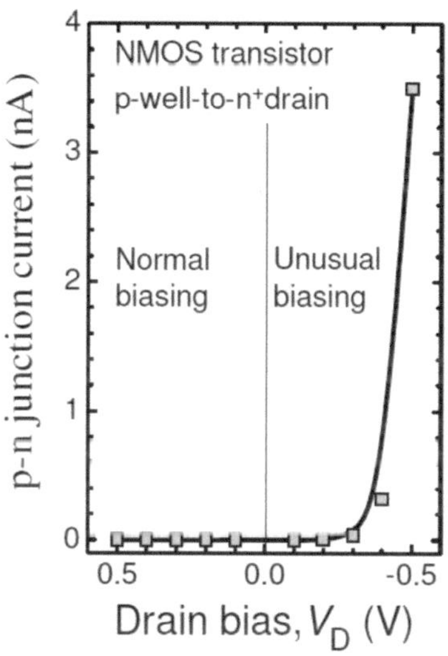

Fig. 27. Selection of the unusual V_D biasing condition for NMOS transistor.

in Fig. 28, the application of V_D= -0.4 V to the NMOS transistor will shift the I_{ds} versus V_{GS} characteristics towards the left. If drain current saturation mechanism is caused by velocity saturation, we will expect drain current saturation to occur in both normal V_D biasing and unusual V_D biasing. If drain current saturation mechanism is caused by pinchoff, we will expect drain current saturation to occur in the normal V_D biasing but not in the unusual V_D biasing. Fig. 29 shows that there is no obvious current saturation in the experimental I_{ds} versus V_{DS} characteristics of the NMOS transistor under the unusual biasing (negative V_D).

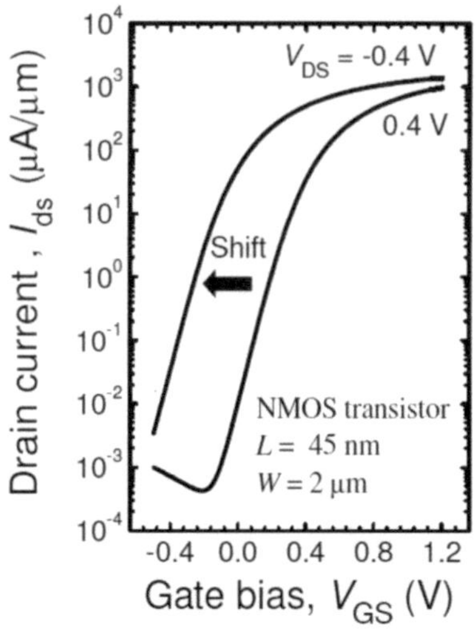

Fig. 28. Effects of the negative V_D on I_{ds} versus V_{GS} characteristics of NMOS transistor.

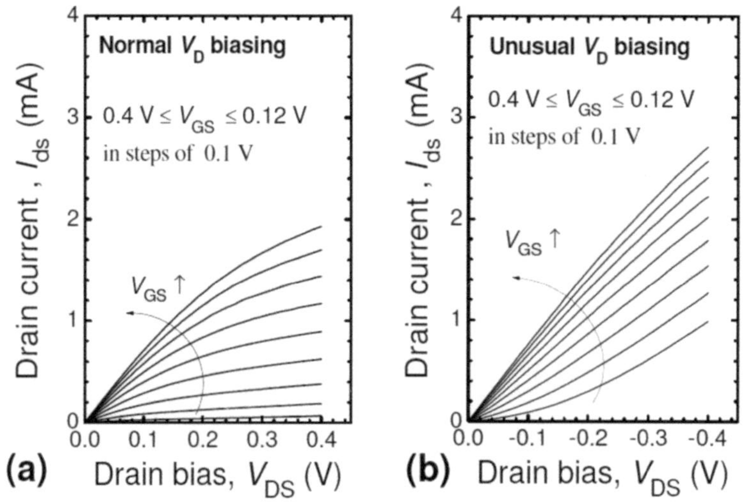

Fig. 29. The I_{ds} versus V_{DS} characteristics of a bulk NMOS transistor (L = 45 nm, W = 2 μm) (a) Normal V_D biasing (positive V_D), (b) Unusual V_D biasing (negative V_D).

7. References

Barral, V.; Poiroux, T.; Munteanu, D.; Autran, J-L. & Deleonibus, S. (2009). Experimental Investigation on the Quasi-Ballistic Tranport: Part II - Backscattering Coefficient Extraction and Link With the Mobility. *IEEE Trans. Electron Dev.*, Vol. 56, No. 3., (Mar 2009) pp.420-430, ISSN: 0018-9383.

Bloch, F. (1928). Uber die Quantenmechanik der Elektronen in Kristalgittern. *Zeitschrift fur Physik*, Vol. 52, No. 7-8, pp. 555-600. (Note: This paper was in German. The title after translation into English is "Quantum mechanics of electrons in crystal lattices".)

Cassé, M. ; Rochette, F.; Thevenod, L.; Bhouri, N.; Andrieu, F.; Reimbold, G.; Boulanger, F.; Mouis, M.; Ghibaudo, G. & Maude, D.K. (2009). A comprehensive study of magnetoresistance mobility in short channel transistors: Application to strained and unstrained silicon-on-insulator field-effect transistors. *J. Appl. Phys.*, Vol. 105, No. 8, (April 2009) pp. 084503-1 to 084503-9, ISSN: 0021-8979.

Chen, C-H.; Lee, T.L.; Hou, T.H.; Chen, C.L.; Chen, C.C.; Hsu, J.W.; Cheng, K.L.; Chiu, Y.H.; Tao, H.J.; Jin, Y.; Diaz, C.H.; Chen, S.C. & Liang, M.-S. (2004). Stress Memorization Technique (SMT) by Selectively Strained-Nitride Capping for sub-65 nm High-Performance Strained-Si Device Application. Symposium on VLSI Technology Digest of Technical Papers, pp. 56-57, ISBN-10: 0 7803 8289 7, Honolulu, USA, Jun 2004, Widerkehr and Associates, Gaithersburg.

Chen, M-.J.; Huang, H.-T. ; Chou, Y.-C.; Chen, R.-T.; Tseng, Y.-T.; Chen, P.-N. & Diaz, C.H. (2004). Separation of channel backscattering coefficients in nanoscale MOSFETs. *IEEE Trans. Electron Dev.*, Vol. 51, No. 9, (Sept 2004) pp. 1409-1415, ISSN: 00189383.

Cheong, W-S.; Yoon, S-M.; Yang, S. & Hwang, C-S. (2009). Optimization of an Amorphous In-Ga-Zn-Oxide Semiconductor for Top-Gate Transparent Thin-Film Transistor. *Journal of the Korean Physical Society*, Vol. 54, No. 5, (May 2009) pp. 1879-1884, ISSN: 0374-4884.

Chern, J. G. J.; Chang, P.; Motta, R. F. & Godinho, N. (1980). A new method to determine MOSFET channel length. *IEEE Electron Device Lett.*, Vol. 1, No. 9, (Sept 1980) pp. 170-173, ISSN: 0741-3106.

Chiang, W.T.; Pan, J.W.; Liu, P.W.; Tsai, C.H. & Ma, G.H. (2007). Strain Effects of Si and SiGe Channel on (100) and (110) Si Surfaces for Advanced CMOS Applications. Symposium on VLSI Technology, Systems and Application (VLSI-TSA), pp. 84- 85, ISBN-10: 1424405858 , Hsinchu, Taiwan, April 2007, Institute of Electrical and Electronics Engineers Inc., United States.

Choi, K.; Jagannathan, H.; Choi, C.; Edge, L.; Ando, T.; Frank, M.; Jamison, P.; Wang, M.; Cartier, E.; Zafar, S.; Bruley, J.; Kerber, A.; Linder, B.; Callegari, A.; Yang, Q.; Brown, S.; Stathis, J. ; Iacoponi, J.; Paruchuri, V. & Narayanan, V. (2009). Extremely Scaled Gate-First High-k/Metal Gate Stack with EOT of 0.55 nm Using Novel Interfacial Layer Scavenging Techniques for 22nm Technology Node and Beyond. Symposium on VLSI Technology Digest of Technical Papers, pp. 138-139, ISBN: 978-1-4244-3308-7, in Kyoto, Japan, Jun 2009.

Chou, S.Y.; Antoniadis, D.A. & Smith, H.I. (1985). Observation of electron velocity overshoot in sub-100-nm-channel MOSFET's in Silicon. *IEEE Electron Dev. Lett.*, Vol. 6, No. 12, (Dec 1985) pp. 665-667, ISSN: 0741-3106.

Chou S.Y. & Antoniadis, D.A. (1987). Relationship between measured and intrinsic transconductances of FET's. *IEEE Trans. Electron Dev.*, Vol. 34, No. 2, (Feb 1987) pp. 448 - 450, ISSN: 0018-9383.

Cros, A.; Romanjek, K.; Fleury, D.; Harrison, S.; Cerutti, R.; Coronel, P.; Dumont, B.; Pouydebasque, A.; Wacquez, R.; Duriez, B.; Gwoziecki, R.; Boeuf, F.; Brut, H.; Ghibaudo, G. & Skotnicki, T. (2006). Unexpected mobility degradation for very short devices: A new challenge for CMOS scaling. *IEEE Electron Devices Meeting* (IEDM), pp. 1-4 , ISBN-10: 1424404398, San Francisco, CA, United States, Dec 2006, Institute of Electrical and Electronics Engineers Inc., Piscataway, NJ, United States.

Fischetti, M.V. & Laux, S.E. (2001). Long-range Coulomb interactions in small Si devices. Part I: Performance and reliability. *J. Appl. Phys*, Vol. 89, No. 2, (Jan 2001) pp. 1205-1231, ISSN: 00218979.

Fuchs, E.; Dolfus, P.; Carval, G.L.; Barraud, S.; Villanueva, D.; Salvetti, F.; Jaouen, H. & Skotnicki, T.(2005). A new backscattering model giving a description of the quasi-ballistic transport in nano-MOSFET. *IEEE Trans. Electron Dev.*, Vol. 52, No. 10, (Oct 2005) pp. 2280-2289, ISSN: 0018-9383.

Ghibaudo, G. & Balestra, F. (1997). Low temperature characterization of silicon CMOS devices. *Microelectron. Reliab.*, Vol. 37, No. 9, (Sept 1997) pp. 1353- 1366, ISSN: 0026-2714.

Guo, J.-C.; Chung, S.S.-S. & Hsu, C.C.-H. (1994). A new approach to determine the effective channel length and the drain-and-source series resistance of miniaturized MOSFET's . *IEEE Trans. Electron Dev.*, Vol. 41, No. 10, (Oct 1994) pp. 1811-1818, ISSN: 0018-9383.

Hauser, J.R. (2005). A New and Improved Physics-Based Model for MOS Transistors. *IEEE Trans. Electron Dev.*, Vol. 52, No. 12, (Dec 2005) pp. 2640- 2647, ISSN: 00189383.

Heiblum, M. & Eastman, L. F. (1987). Ballistic Electrons in Semiconductors. *Scientific American*, Vol. 256, No. 2, (Feb 1987) pp. 64- 73, ISSN: 00368733.

Hofstein, S.R. & Heiman, F.P. (1963). Silicon insulated-gate field-effect transistor. *IEEE Proceedings*, Vol. 51, No. 9, (Sept 1963) pp. 1190- 1202.

Huang, J.; Heh, D.; Sivasubramani, P.; Kirsch, P. D.; Bersuker, G.; Gilmer, D. C.; Quevedo-Lopez, M.A.; Hussain, M. M.; Majhi, P.; Lysaght, P.; Park, H.; Goel, N.; Young, C.; Park, C.S.; Park, C.; Cruz, M.; Diaz, V.; Hung, P. Y.; Price, J.; Tseng, H.-H. & Jammy, R. (2009) Gate First High-k/ Metal Gate Stacks with Zero SiO_x Interface Achieving EOT = 0.59 nm for 16 nm Application. Symposium on VLSI Technology Digest of Technical Papers, pp. 34-35, ISBN: 978-1-4244-3308-7, in Kyoto, Japan.

Huet, K.; Querlioz, D.; Chaisantikulwat, W.; Saint-Martin, J.; Bournel, A.; Moulis, M. & Dollfus, P. (2008). Monte Carlo study of apparent magnetoresistance mobility in nanometer scale metal oxide semiconductor field effect transistors. *J. Appl. Phys.*, Vol. 104, No. 4, (Aug 2008) pp. 044504-1 to 044504-7, ISSN: 00218979.

Jing, W. & Lundstrom, M. (2002). Does source-to-drain tunneling limits the ultimate scaling of MOSFETs ? *IEEE Electron Devices Meeting* (IEDM), pp. 707-710, ISBN-10: 0 7803 7462 2, San Francisco, CA, USA, Dec 2002, Institute of Electrical and Electronics Engineers, Piscataway, NJ, USA.

Kawaura, H.; Sakamoto, T. & Baba, T. (2000). Observation of source-to-drain direct tunneling in 8 nm gate electrically variable shallow junction metal-oxide-semiconductor field-effect transistors. *Appl. Phys. Lett.*, Vol. 76, No. 25, (Jun 2000) pp. 3810-3812, ISSN: 00036951.

Kawaura, H. & Baba, T. (2003). Direct Tunneling from Source to Drain in Nanometer-Scale Silicon Transistors. *Jpn. J. Appl. Phys.*, Vol. 42, No. 2A, (Feb 2003) pp. 351-357, ISSN: 00214922.

Khakifirooz, A. & Antoniadis, D.A. (2006). Transistor Performance Scaling: The Role of Virtual Source Velocity and Its Mobility Dependence. *IEEE Electron Devices Meeting* (IEDM), pp. 1-4 , ISBN-10: 1424404398, San Francisco, CA, United states, Dec 2006, Institute of Electrical and Electronics Engineers, Piscataway, NJ, United States.

Khakifirooz, A. & Antoniadis, D.A. (2008). MOSFET Performance Scaling – Part I: Historical Trends. IEEE Trans. Electron Dev., Vol. 55, No. 6 (Jun 2008) pp. 1391-1400, ISSN 0018-9383.

Khakifirooz, A.; Nayfeh, O.M. & Antoniadis, D.A. (2009). A simple semiempirical short-channel MOSFET current: voltage model continuous across all regions of operation and employing only physical parameters. IEEE Trans. Electron Dev., Vol. 56, No. 8 (Aug 2009) pp. 1674-1680, ISSN 0018-9383.

Kim, J.; Lee, J.; Yun, Y.; Park, B-G.; Lee, J.D. & Shin, H. (2008). Extraction of Effective Carrier Velocity and Observation of Velocity Overshoot in Sub-40 nm MOSFETs. *Journal of Semiconductor Technology and Science*, Vol. 8, No. 2, (Jun 2008) pp. 115-120, ISSN: 1598-1657. Note that this is a Korean journal.

Kondo, M. & Tanimoto, H. (2001). Accurate Coulomb mobility model for MOS inversion layer and its application to NO-oxynitride devices. *IEEE Trans. Electron Dev.*, Vol. 48, No. 2, (Feb 2001) pp. 265-270, ISSN: 00189383.

(a) Lau, W.S.; Yang, Peizhen; Eng, C.W.; Ho, V.; Loh, C.H.; Siah, S.Y.; Vigar, D. & Chan, L. (2008). A study of the linearity between I_{on} and $\log I_{off}$ of modern MOS transistors and its application to stress engineering. *Microelectronics Reliab.*, Vol. 48, No. 4, (April 2008) pp. 497-503, ISSN: 0026-2714.

(b) Lau, W.S.; Yang, Peizhen; Ho, V.; Toh, L.F.; Liu, Y.; Siah, S.Y. & Chan, L. (2008). An explanation of the dependence of the effective saturation velocity on gate voltage in sub-0.1 μm metal-oxide-semiconductor transistors by quasi-ballistic transport theory. *Microelectronics Reliab.* , Vol. 48, No. 10, (Oct 2008) pp. 1641-1648, ISSN: 0026-2714.

Lau, W.S.; Yang, Peizhen; Chian, Jason Zhiwei; Ho, V.; Loh, C.H.; Siah, S.Y. & Chan, L. (2009). Drain current saturation at high drain voltage due to pinch off instead of velocity saturation in sub-100 nm metal-oxide-semiconductor transistors. *Microelectronics Reliab.*, Vol. 49, No. 1, (Jan 2009) pp. 1-7, ISSN: 00262714.

Lee, W.; Kuo, J. J.-Y; Chen, W. P.-N; Su, P. & Jeng, M-C. (2009). Impact of Uniaxial Strain on Channel Backscattering Characteristics and Drain Current Variation for Nanoscale PMOSFETs. *Symposium on VLSI Technology Digest of Technical Papers*, pp. 112-113, ISBN: 978-1-4244-3308-7, Kyoto, Japan, Jun 2009.

Lochtefeld, A.; Djomehri, I.J.; Samudra, G. & Antoniadis, D.A. (2002). New insights into carrier transport in n-MOSFETs. *IBM J. Res. & Dev.*, Vol. 46, No. 2-3 , (March-May 2002) pp. 347-357 , ISSN: 00188646.

Lundstrom, M. (1997). Elementary Scattering Theory of the Si MOSFET. *IEEE Electron Device Lett.*, Vol. 18, No. 7, (July 1997) pp. 361-363, ISSN: 0741-3106.

Lundstrom, M. & Ren, Z. (2002). Essential physics of carrier transport in nanoscale MOSFETs. *IEEE Trans. Electron Dev.*, Vol. 49, No. 1, (Jan 2002) pp. 133 -141, ISSN: 00189383.

Martinie, S.; Le Carval, G.; Munteanu, D.; Soliveres, S & Autran, J.-L. (2008). Impact of ballistic and quasi-ballistic transport on performances of double-gate MOSFET-Based Circuits. IEEE Trans. Electron Dev., Vol. 55, No. 9, (Sept 2008) pp. 2443-2453, ISSN 0018-9383.

Mazzoni, G.; Lacaita, A.L.; Perron, L.M. & Pirovano, A. (1999). On surface roughness-limited mobility in highly doped n-MOSFET's. *IEEE Trans. Electron Dev.*, Vol. 46, No. 7, (July 1999) pp.1423-1428, ISSN: 0018-9383.

Miyata, H.; Yamada, T. & Ferry, D.K. (1993). Electron transport properties of a strained Si layer on a relaxed $Si_{1-x}Ge_x$ substrate by Monte Carlo simulation. *Appl. Phys. Lett.*, Vol. 62, No. 21, (May 1993) pp. 2661- 2663, ISSN. 00036951.

Mizuno, T. (2000). New Channel Engineering for Sub-100 nm MOS Devices Considering Both Carrier Velocity Overshoot and Statistical Performance Fluctuations. *IEEE Trans. Electron Dev.*, Vol. 47, No. 4, (April 2000) pp.756-761, ISSN: 00189383.

Natori, K. (1994). Ballistic metal-oxide-semiconductor field effect transistor. *J. Appl. Phys.*, Vol. 76, No. 8, (Oct 1994) pp. 4879-4890, ISSN: 0021-8979.

Natori, K.; Shimizu, T. & Ikenobe, T. (2003). Multi-subband effects on performance limit of nanoscale MOSFETs. *Jpn. J. Appl. Phys.*, Vol. 42, No. 4B, (April 2003) pp. 2063-2066, ISSN: 00214922.

Natori, K.; Wada, N. & Kurusu, T. (2005). New Monte Carlo simulation technique for quasi-ballistic transport in ultrasmall metal oxide semiconductor field-effect transistors. *Jpn. J. Appl. Phys.*, Vol. 44, No. 9A, (Sept 2005) pp.6463-6470, ISSN: 00214922.

Natori, K. (2008). Ballistic / quasi-ballistic transport in nanoscale transistors. *Applied Surface Science*, Vol. 254, No. 19, (July 2008) pp. 6194-6198, ISSN: 01694332.

Navarro, D.; Mizuoguchi, T.; Suetake, M.; Hisamitsu, K.; Ueno, H.; Miura-Mattausch, M.; Mattausch, H. J.; Kumashiro, S.; Yamaguchi, T.; Yamashita, K. & Nakayama, N. (2005). A Compact Model of the Pinch-off Region of 100 nm MOSFETs Based on the Surface-Potential. *IEICE Trans. Electron.*, Vol. E88-C, No. 5, (May 2005) pp. 1079-1086, ISSN: 09168524.

Norris, C.B. Jr.; & Gibbons, J.F. (1967). Measurement of high-field carrier drift velocities in silicon by a time-of-flight technique. *IEEE Trans. Electron Dev.*, Vol. 14, No. 1, (Jan 1967) pp. 38-42.

Romanjek, K.; Andrieu, F.; Ernst, T. & Ghibaudo, G. (2004). Improved Split C-V Method for Effective Mobility Extraction in sub-0.1 μm Si MOSFETs. *IEEE Electron Dev. Lett.*, Vol. 25, No. 8, (Aug 2004) pp. 583-585, ISSN: 0741-3106.

Ruch, J.G. (1972). Electron Dynamics in Short Channel Field-Effect Transistors. *IEEE Trans. Electron Dev.*, Vol. 19, No. 5, (May 1972) pp. 652-654, ISSN: 00189383.

(a) Sah, C.T. (1991). *Fundamentals of Solid-State Electronics*, 1st ed., World Scientific, ISBN: 9810206372, Singapore, p.245.

(b) Sah, C.T. (1991). *Fundamentals of Solid-State Electronics*, 1st ed., World Scientific, ISBN: 9810206372, Singapore, p.554.

Schroder, Dieter K. (1998). *Semiconductor Material and Device Characterization*, 2nd ed., John Wiley & Sons, ISBN: 0471241393, New York, pp. 548- 549.

Singh, J. (1993). *Physics of semiconductor and their heterostructures*, McGraw-Hill, ISBN: 0070576076, New York, p. 161.

Stern, F. (1972). Self-Consistent Results for n-Type Si Inversion Layers. *Phys. Rev.* B, Vol. 5, No. 12, (Jun 1972) pp. 4891-4899, ISSN: 0556-2805.

Sun, Y; Thompson, S.E. & Nishida, T. (2007). Physics of strain effects in semiconductors and metal-oxide-semiconductors field-effect transistors. *J. Appl. Phys.*, Vol. 101, No. 10, (May 2007) pp. 104503-1 to 104503-22, ISSN: 0021-8979.

Suzuki, K. & Usuki, T. (2004). Metal oxide semiconductor field effect transistor (MOSFET) based on a physical high-field carrier-velocity model. *Jpn. J. Appl. Phys.*, Vol. 43, No. 1, (Jan 2004) pp. 77-81, ISSN: 00214922.

Sze, S.M. & Ng, Kwok K. (2007). *Physics of Semiconductor Devices*, 3rd ed., Wiley-Interscience, Hoboken, N.J., 0471143235, p. 309.

Takagi, S.; Toriumi, A.; Iwase, M. & Tango, H. (1994). On the Universality of Inversion Layer Mobility in Si MOSFET's Part I – Effects of Substrate Impurity Concentration. *IEEE Trans. Electron Dev.*, Vol. 41, No. 12, (Dec 1994) pp. 2357- 2362, ISSN: 00189383.

Taur, Y.; Hsu, C.H.; Wu, B.; Kiehl, R.; Davari, B. & Shahidi, S. (1993). Saturation transconductance of deep-submicron-channel MOSFETs. *Solid-State Electronics*, Vol. 36, No. 8, (Aug 1993) pp. 1085-1087, ISSN: 0038-1101.

(a) Taur, Y & Ning, T. (1998). *Fundamentals of modern VLSI devices*. 1st ed. New York: Cambridge University Press; ISBN: 0521550564, New York, p.121.
(b) Taur, Y & Ning, T. (1998). *Fundamentals of modern VLSI devices*. 1st ed. New York: Cambridge University Press; ISBN: 0521550564, New York, p.128.
(c) Taur, Y & Ning, T. (1998). *Fundamentals of modern VLSI devices*. 1st ed. New York: Cambridge University Press; ISBN: 0521550564, New York, p. 152.

Thornber, K.K. (1980). Relation of drift velocity to low-field mobility and high-field saturation velocity. *J. Appl. Phys.*, Vol. 51, No. 4, (April 1980) pp. 2127-2136, ISSN: 0021-8979.

Uchida, K.; Krishnamohan, T.; Saraswat, K.C. & Nishi, Y. (2005). Physical Mechanisms of Electron Mobility Enhancement in Uniaxial Stressed MOSFETs and Impact of Uniaxial Stress Engineering in Ballistic Regime. *IEEE Electron Devices Meeting* (IEDM), pp. 129-132, ISBN-10: 078039268X, Washington, DC, MD, United states, Dec 2005, Institute of Electrical and Electronics Engineers, Piscataway, NJ, United States.

Wang, E.X.; Mantagne, P.; Shifren, L.; Obradovic, B.; Kotlyar, R.; Cea, S.; Stettler, M. & Giles, M.D. (2006). Physics of Hole Transport in Strained Silicon MOSFET Inversion Layers. *IEEE Trans. Electron Dev.*, Vol. 53, No. 8, (Aug 2006) pp. 1840-1851, ISSN: 00189383.

Wang, J. ; Tateshita, Y.; Yamakawa, S.; Nagano, K.; Hirano, T.; Kikuchi, Y.; Miyanami, Y.; Yamaguchi, S.; Tai, K. ; Yamamoto, R.; Kanda, S.; Kimura, T.; Kugimiya, K.; Tsukamoto, M.; Wakabayashi, H.; Tagawa, Y.; Iwamoto, H.; Ohno, T.; Saito, M.; Kadomura, S. & Nagashima, N. (2007). Novel Channel-Stress Enhancement Technology with eSiGe S/D and Recessed Channel on Damascene Gate Process. *Symposium on VLSI Technology Digest of Technical Papers*, pp. 46-47, ISBN-13: 978-4-900784-03-1, Kyoto, Japan, Jun 2007, Institute of Electrical and Elecronics Engineers, Piscataway, NJ, United States.

Wakabayashi, H.; Ezaki, T.; Hane, M.; Ikezawa, T.; Sakamoto, T.; Kawaura, H.; Yamagami, S.; Ikarashi, N.; Takeuchi, K.; Yamamoto, T. & Mogami, T. (2004). Transport Properties of Sub-10-nm Planar-Bulk-CMOS Devices. *IEEE Electron Devices Meeting* (IEDM), pp. 429-432 , ISBN-10: 07803 8684 1, San Francisco, CA, United states, Dec 2004, Institute of Electrical and Electronics Engineers, Piscataway, NJ, United States.

Wakabayashi, H.; Ezaki, T.; Sakamoto, T.; Kawaura, H.; Ikarashi, N.; Ikezawa, N.; Narihiro, M.; Ochiai, Y. ; Ikezawa, T.; Takeuchi, K.; Yamamoto, T.; Hane, M. & Mogami, T. (2006). Characteristics and Modeling of Sub-10 nm Planar Bulk CMOS Devices Fabricated by Lateral Source/Drain Junction Control. *IEEE Trans. Electron Dev.*, Vol. 53, No. 9, (Sept 2006) pp. 1961-1970, ISSN: 00189383.

Weidemann, M. ; Kloes, A. & Iniguez B. (2007). Compact Model for Electric Field at Pinch-Off and Channel Length Shortening in Bulk MOSFET. *IEEE Electron Devices and Solid-State Circuits* (EDSSC), pp. 1147-1150, ISBN-10: 1424406374, Tainan, Taiwan, Dec 2007, Institute of Electrical and Electronics Engineers Inc., Piscataway, NJ, United States.

Wu, S. Y.; Liaw, J.J.; Lin, C.Y.; Chiang, M.C.; Yang, C.K.; Cheng, J.Y., Tsai, M.H.; Liu, M.Y.; Wu, P.H.; Chang, C.H.; Hu, L.C.; Lin, C.I.; Chen, H.F.; Chang, S.Y.; Wang, S.H.; Tong, P.Y.; Hsieh, Y.L.; Pan, K.H.; Hsieh, C.H.; Chen, C.H.; Yao, C.H.; Chen, C.C; Lee, T.L.; Chang, C.W.; Lin, H.J.; Chen, S.C.; Shieh, J.H.; Tsai, M.H.; Jang, S.M.;

Chen, K.S.; Ku, Y.; See, Y.C. & Lo, W.J. (2009). A Highly Manufacturable 28nm CMOS Low Power Platform Technology with Fully Functional 64Mb SRAM Using Dual/Tripe Gate Oxide Process. *Symposium on VLSI Technology Digest of Technical Papers*, pp. 210-211, ISBN: 978-1-4244-3308-7, in Kyoto, Japan.

Yaglioglu, B.; Yeom, H-Y.; Beresford, R. & Paine, D.C. (2005). The Fabrication and Characterization of Amorphous Indium Zinc Oxide (In_2O_3:10wt%ZnO) based Thin Film Transistors. *Materials Research Society Symposium Proceedings*, pp. 19-23, ISBN-10: 1558998608 , Boston, MA, United states , Nov 2005, Materials Research Society, Warrendale, PA, United States.

Yang, B.Frank; Takalkar, R.; Ren, Z.; Black, L.; Dube, A.; Weijtmans, J.W.; Li, J.; Johnson, J.B.; Faltermeier, J.; Madan, A.; Zhu, Z.; Turansky, A.; Xia, G.; Chakravarti, A.; Pal, R.; Chan, K.; Reznicek, A.; Adam, T.N. ; Yang, B.; de souza, J.P.; Harey, E.C.T. ; Greene, B.; Gehring, A.; Cai, M.; Aime, D.; Sun, S.; Meer, H.; Holt, J.; Theodore, D.; Zollner, S.; Grudoswki, P.; Sadana, D.; Park, D.-G.; Mocuta, D.; Schepis, D.; Maciejewski, E.; Luning, S. ; Pellerin, J. & Leobandung, E. (2008). High-performance nMOSFET with in-situ Phosphorus-doped embedded Si:C (ISPD eSi:C) source-drain stressor. *IEEE Electron Devices Meeting* (IEDM), pp. 51-54, ISBN-13: 978-1-4244-2377-4, San Francisco, CA, USA, Dec 2008, Institute of Electrical and Electronics Engineers, Piscataway, NJ, United States.

Yang, H. S; Malik, R.; Narasimha, S.; Li, Y.; Divakaruni, R.; Agnello, P.; Allen, S.; Antreasyan, A.; Arnold, J. C.; Bandy, K.; Belyansky, M.; Bonnoit, A.; Bronner, G.; Chan, V.; Chen, X.; Chen, Z.; Chidambarrao, D.; Chou, A.; Clark, W.; Crowder, S. W.; Engel, B.; Harifuchi, H.; Huang, S. F.; Jagannathan, R.; Jamin, F. F.; Kohyama,Y.; Kuroda, H.; Lai, C.W.; Lee, H.K.; Lee, W.-H.; Lim, E.H.; Lai, W.; Mallikarjunan, A.; Matsumoto, K.; McKnight, A.; Nayak, J. ; Ng, H.Y.; Panda, S.; Rengarajan, R.; Steigerwalt, M.; Subbanna, S.; Subramanian, K.; Sudijono, J.; Sudo, G.; Sun, S.-P.; Tessier, B.; Toyoshima, Y.; Tran, P.; Wise, R.; Wong, R.; Yang, I.Y.; Wann, C.H.; Su, L.T.; Horstmann, M.; Feudel, Th.; Wei, A.; Frohberg, K.; Burbach, G.; Gerhardt, M.; Lenski, M.; Stephan, R.; Wieczorek, K.; Schaller, M.; Salz, H.; Hohage, J.; Ruelke, H.; Klais, J.; Huebler, P.; Luning, S.; van Bentum, R.; Grasshoff, G. ; Schwan, C. ; Ehrichs, E.; Goad, S.; Buller, J.; Krishnan, S.; Greenlaw, D.; Raab, M. & Kepler, N. (2004). Dual Stress Liner for High Performance sub-45nm Gate Length SOI CMOS Manufacturing. *IEEE Electron Devices Meeting* (IEDM), pp. 1075-1077, ISBN-10: 0 7803 8684 1, San Francisco, CA, USA, Dec 2004, Institute of Electrical and Electronics Engineers, Piscataway, NJ, United States.

Yang, Peizhen; Lau, W.S.; Ho, V.; Loh, C.H.; Siah, S.Y. & Chan, L. (2007). A comparison between the quasi-ballistic transport model and the conventional velocity saturation model for sub-0.1 μm MOS transistors. *IEEE Electron Devices and Solid-State Circuits* (EDSSC), pp. 99-102, ISBN-10: 1424406374, Tainan, Taiwan, Dec 2007, Institute of Electrical and Electronics Engineers Inc., Piscataway, NJ, United States.

Yoshikawa, N.; Tomida, T.; Tokuda, M.; Liu, Q.; Meng, X.; Whiteley, S.R. & Van Duzer, T. (2005). Characterization of 4 K CMOS devices and circuits for hybrid Josephson-CMOS systems. *IEEE Trans. Appl. Supercond.*, Vol. 15, No. 2, (Jun 2005) pp. 267-271, ISSN: 1051-8223.

4

Thermal Noise in Modern CMOS Technology

Chih-Hung Chen
McMaster University
Canada

1. Introduction

Because of the high cut-off frequency (f_T) in hundreds of gigahertz resulting from the aggressive reduction of physical size and the enhancement of carrier mobility, metal-oxide-semiconductor field effect transistors (MOSFETs) become widely used in radio-frequency (RF) and high-speed integrated circuits (ICs). However, when working at high frequencies and high speed, thermal noise becomes a critical issue preventing these circuits from their anticipated performance. This chapter presents how thermal noise is characterized, how it is modeled, and what is its trend in future CMOS technology.

2. Noise characterization

Because the thermal noise is overwhelmed by the $1/f$ noise in devices at low frequencies, it is usually evaluated at high frequencies, at least above the $1/f$ corner frequency. Different from the low-frequency noise characterization, which can be directly conducted using a spectrum analyzer, thermal noise characteristics has to be evaluated by its noise factor (or noise figure in dB) and/or its four noise parameters, namely minimum noise factor (F_{min}) or minimum noise figure in dB, (NF_{min}), equivalent noise resistance (R_n), and optimized source admittance ($Y_{opt} = G_{opt} + j \cdot B_{opt}$). We describe how noise factors and noise parameters are measured in 2.1, how to remove the parasitic effects of probe pads and metal interconnections in a device-under-test (DUT) in 2.2, and how to extract the noise sources of interest in 2.3.

2.1 Noise measurement

Noise parameters are commonly used parameters in the microwave noise characterization of a linear noisy two-port network. One of its applications is to calibrate a noise measurement system (Chen et al., 2007), and another example is to remove the parasitic effects of metal interconnections in a DUT (Chen & Deen, 2001). They are also used to extract the noise sources of interest in devices (Chen & Deen, 2001; Chen & Deen, 2000; Asgaran et al., 2007), which assist in device noise modeling (Chen & Deen, 2002; Asgaran, Deen & Chen, 2004; Deen et al., 2006). In this section, we present the setup of a noise measurement system and different algorithms to improve the measurement accuracy.

The conventional set of noise parameters are based on Rothe and Dahlkes' work (Rothe & Dahlke, 1956). In this work, a noisy two-port network is represented by voltage or current sources connected to the noiseless network. Haus et al. expanded this concept and developed the four well-known noise parameters - minimum noise factor F_{min}, equivalent noise resistance R_n, optimal source conductance G_{opt} (i.e., the real pat of Y_{opt}), and optimal source susceptance B_{opt} (i.e., the imaginary part of Y_{opt}) (Haus et al., 1960). This representation allows easy calculation of noise figures for a noisy two-port network. The intuitive, impedance-based representation of the noisy two-port network also demonstrates the dependence of noise factors on the source admittances attached to the input of the network. Since the introduction of this two-port noise representation, many measurement and extraction methods have been introduced (Lane, 1969; Mitama & Katoh, 1979; Vasilescu, Alquie & Krim, 1988; O'Callaghan & Mondal, 1991).

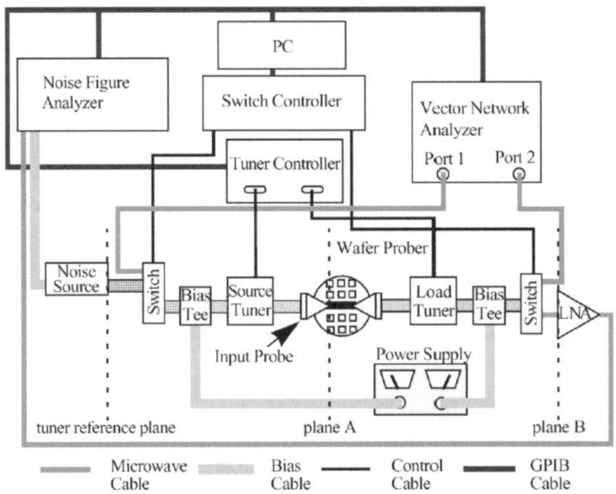

Fig. 1. System configuration for microwave noise measurements (Chen et al., 2007).

For a commonly-used a noise measurement system (see Fig. 1), it consists of a noise source, a vector network analyzer, a noise figure analyzer (NFA), microwave impedance tuners, a low-noise amplifier (LNA), and other peripheral components (e.g., PC). This system can be furthered simplified using a PNA-X (with noise option) from Agilent to replace both VNA and NFA (Simpson, 2009). The noise source is to generate two noise outputs with different equivalent noise temperatures, namely hot (T_h) and cold (T_c) temperatures during the noise measurements. The source tuner is to provide different source admittances for the receiver, and the load tuner is to match the output of the DUT for a maximum power transfer. The LNA is to boost the weak noise signal to increase the accuracy of the measured noise power. It also helps to reduce the noise factor of the receiver to increase the noise factor accuracy of the DUT, especially when Friis's equation (Friis, 1944) is applied to remove receiver's noise contribution. However, the gain of the LNA has to be carefully selected in order not to saturate the receiver in the NFA.

For noise parameter measurements, in general, they can be divided into two different categories. The first category involves the forward and reverse noise measurements based on the concept of noise wave. It was first introduced by Penfield in 1962 (Penfield, 1962). Instead of representing the internal noise of the two-port network by voltage or current sources, Penfield's method uses noise waves. Such wave-based representation allows the use of the scattering parameters, which are widely used in the microwave frequency range. Unlike the conventional noise parameters, wave-based noise parameters represent the intrinsic noise behavior of a noisy two-port network. They do not necessarily depend on the reflection coefficient seen by the input of the two-port. Since a noisy two-port can also be represented by other combinations of voltage or current sources, Hillbrand and Russer provided a more general treatment using waves to replace these sources (Hillbrand & Russer, 1976). Using the wave-based noise analysis technique, Meys developed a measurement method to characterize a linear two-port's noise properties (Meys, 1978). With Meys' formulation of the wave-based noise parameters, Valk et al. developed a method to de-embed the two-port noise parameters from a cascaded two-port network (Valk et al., 1988). Using different noise-wave definitions, Hecken developed a different set of noise parameters for noisy multi-ports (Hecken, 1981). Wedge developed a set of two-port noise parameters by modeling the intrinsic noise as noise waves leaving each port (Wedge & Rutledge, 1992; Wedge & Rutledge, 1991). From a practical point of view, the purpose of a set of noise parameters should help designers decide how to terminate a noisy two-port for optimal noise or power performance. Engen and Wait presented a set of noise parameters with physical meanings for the ease of this application (Engen, 1970; Wait & Engen, 1991). Based Wedge's noise parameters and using a similar approach to Engen's work, Randa presented a method in which the available noise temperature for the input port of the device can be also obtained (Randa & Walker, 2007; Randa, 2002). A reverse measurement is still necessary, but this more generalized approach removes the assumption that the reverse available power gain of the two-port is negligible (Chen, Wang & Bakr, 2008). The major issue stopping the wave-based approach from the on-wafer noise measurements in practice is the requirement of the reverse measurements.

In the second category, however, only the forward measurements are conducted to obtain the noise powers (or noise factors) at different source admittances/impedances. Under this catagory, there are two approaches to obtain these crucial noise parameters. In the first approach, four (or more) noise factors are obtained first using the Y-factor method (Agilent Application Note 57-1). The four noise parameters are then calculated by solving the linearized noise factor equations with algorithms or methods to take care of the experimental errors in the noise factors and the source admittances (IRE Subcommittee 7.9 on Noise, 1963; Lange, 1967; Lane, 1969; Gupta, 1970; Caruso & Sannino, 1978; Mitama & Katoh, 1979; Sannino, 1979; Pospieszalski, 1986; Vasilescu, Alquie & Krim, 1988; Davidson et al., 1989; O'Callaghan & Mondal, 1991; Archer & Batchelor, 1992; Boudiat & Laporte, 1993; Tiemeijer et al., 2005; Wiatr & Walker, 2005). The second approach, on the other hand, directly solves the noise parameters using the power equation (Adamian & Uhlir, 1973; Tutt, 1994). This method leads to the so-called "cold-only" method in which the noise power in the hot state is only measured during the system calibration, but not in the measurement of the DUT (Adamian & Uhlir, 1973; Tutt, 1994; Meierer & Tsironis, 1995; Kantanen et al., 2003). Recently, two methods to improve the measurement accuracy by taking care of the impedance difference between the hot and cold states are presented by Kantanen (Kantanen et al., 2003) for Y-factor based approach and by Chen (Chen et al., 2007) for power equation based approach, respectively.

2.2 Noise parameter de-embedding

Because the physical size of devices is small, probe pads and interconnections are usually designed to access devices when performing noise measurements. With the continuous downscaling of the device dimensions, the impact of the surrounding parasitics on the device characteristics has steadily gained importance in the a.c. and noise measurements of a DUT, which includes a transistor, probe pads, and metal interconnections between the probe pads and the transistor. Since the probe pads and interconnections introduce additional parasitics including resistances, inductances, and capacitances, de-embedding procedures for both measured scattering and noise parameters must be performed prior to analyzing the performance of an intrinsic transistor to isolate the intrinsic performance from that due to extrinsic parasitic effects for on-wafer measurements.

In 1987, van Wijnen et al. presented a method to remove the capacitive parasitics of probe pads from the on-wafer s-parameter measurements by measuring an additional "OPEN" dummy structure (van Wijnen, Claessen & Wolsheimer, 1987). In 1991, Koolen et al. improved the de-embedding procedure with the consideration of the influence of the interconnections by measuring another "SHORT" dummy structure (Koolen, Geelen & Versleijen, 1991). Lee et al., in 1994, modified the "SHORT' structure and the de-embedding method presented by Koolen et al. so as to extract the parasitic inductances of the interconnections (Lee, Ryum & Kang, 1994). In 2000, Kolding presented a procedure to predict the series losses and coupling parasitics (Kolding, 2000). In these de-embedding methods, in general, a parallel-series configuration which assumes that the impedance of interconnections is in series with the transistor, and the admittance of probe pads is in parallel with the interconnections and the transistor is used to model the DUT.

In the parallel-series configurations, it is assumed that the capacitive effect of interconnections is negligible, and the inductive and resistive effects are dominant at the frequencies of interest. However, this might not be true for the designs with long interconnections or at operating frequencies at several tens of GHz, where the distributive effects of the interconnections become important. Therefore, the interconnections cannot be modeled as an inductor in series with a resistor, and the DUT has to be modeled as probe pads, interconnections, and the transistor connected in a cascade configuration. The de-embedding procedure presented by Biber in 1998 (Biber et al., 1998) is based on cascade configurations, but it still neglects the capacitive effect of the interconnections. In addition, it requires specific equivalent circuit models for both probe pads and the interconnections. In 2002, Chen and Deen presented a general de-embedding procedure based on the cascade configurations without the requirement of any equivalent circuit models for probe pads and interconnections (Chen & Deen, 2001). Cho et al. improved Chen and Deens' method by presenting a scalable noise de-embedding technique for the characterization of devices in various sizes without designing their corresponding dummy structures (Cho et al., 2005). This can save a lot of wafer space in designing microwave test structures.

2.3 Noise source extraction

Behavior of physical noise sources in MOSFETs, namely channel thermal noise, induced gate noise, and their correlation, is needed when we develop any physics-based compact noise model. Obtaining the spectral densities of these noise sources of interest as a function of frequency, bias condition, and device geometry directly from the intrinsic noise and s-parameters is the key step in the noise modeling. The extracted noise spectral densities of these desired noise sources can provide important insights on the noise characteristics of devices and serve as a useful guide for noise modeling. There are several methods for the noise source extraction (Chen & Deen, 2000; Knoblinger, Klein & Baumann, 2000; Chen et al., 2001; Knoblinger, 2001). Both methods by Chen (Chen & Deen, 2000) and Knoblinger

(Knoblinger, Klein & Baumann, 2000) published in year 2000 only extract the channel noise. In 2001, both of them presented new methods to extract channel noise, induced gate noise, and their correlation (Chen et al., 2001; Knoblinger, 2001). Fig. 2 shows the extracted channel noise, induced gate noise, correlation noise, and cross-correlation coefficient as a function of frequencies for devices with different channel lengths (Chen et al., 2001). It is observed that the channel noise is about frequency independent, the induced gate noise and the correlation term are proportional to f^2 and f, respectively, where f is the operating frequency (solid lines in the figures). In addition, when the channel length decreases, both induced gate noise and its correlation with the channel thermal noise also decrease because of the reduction of the gate-to-source capacitance.

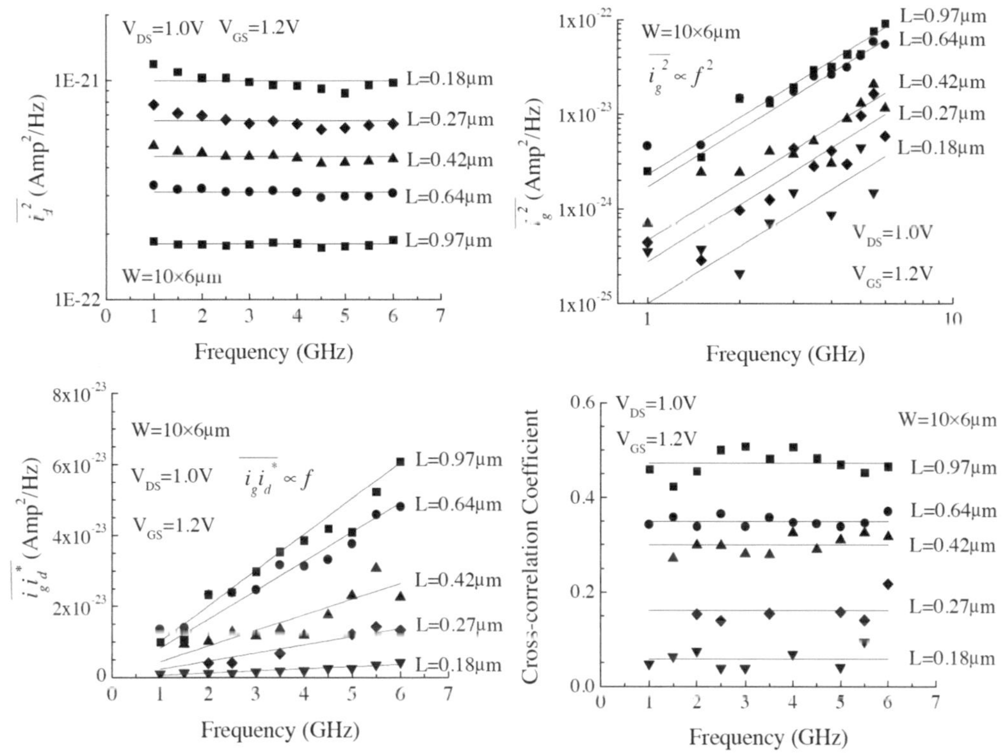

Fig. 2. Extracted channel noise, induced gate noise, correlation noise, and cross-correlation coefficient as a function of frequencies for devices with different channel lengths (Chen et al., 2001).

According to Chen and Deens' analytical equation (Chen & Deen, 2000), it is shown that among these noise parameters - NF_{min}, R_n, and Γ_{opt}, only the equivalent noise resistance R_n extrapolated at low frequencies provides a direct insight for the channel noise. Therefore, any proposed channel noise model should compare the calculated and measured R_n versus frequency characteristics. It is not sufficient to verify the channel noise model by just comparing the NF_{min} only, which is affected by the induced gate noise as well. Fig. 3 shows the measured (symbols) and calculated (lines) NF_{min} and r_n (R_n normalized to 50Ω) versus frequency characteristics for an n-type MOSFET with L = 0.97 μm and W = 10 × 6 μm (ten 6 μm fingers connected in parallel) biased at V_{DS} = 1.0 V and V_{GS} = 1.2 V using different combination of these noise sources. It is shown that the induced gate noise has strong impact on the NF_{min}, especially for long channel devices, but little influence on r_n. In addition, it seems that the correlation noise has little impact on NF_{min} and r_n.

¶

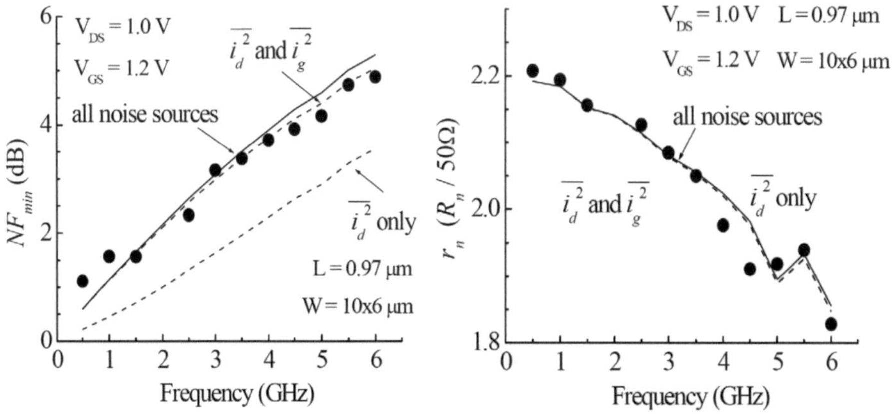

Fig. 3. Impacts of noise sources on NF_{min} and r_n (R_n normalized to 50Ω).

Another example to illustrate this idea is shown in Fig. 4. The solid lines are obtained using the extracted channel noise, induced gate noise and their correlation, and the dashed lines are obtained by replacing the extracted channel thermal noise with the thermal noise models commonly used by analog IC designers (i.e., $8kTg_{do}/3$ and $8kTg_m/3$). It is shown that although the conventional channel noise models agree well with the measured NF_{min}, they predict lower r_n.

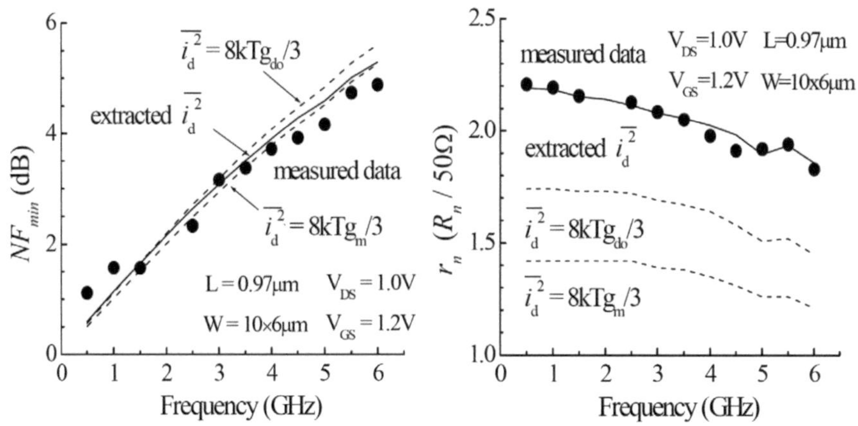

Fig. 4. Verification of channel thermal noise based on NF_{min} and r_n (R_n normalized to 50Ω).

3. Noise modeling

Physics-based noise models for channel thermal noise, induced gate noise, and their correlation are important when examining experimental results. It also provides circuit designers some guidelines in designing low-power, low-noise ICs. This section presents thermal noise modeling in 3.1 and how they can be implemented into commercial circuit simulators in 3.2.

3.1 Thermal noise modeling

Due to the enhancement in CMOS technology, new noise phenomena emerge. This section discusses the impacts of channel length modulation effects, hot carrier effects, and velocity saturation effects down to 65 nm technology node.

Starting from van der Ziel, Jordan, and Jordans' pioneering work back to 1962 and 1965 (van der Ziel, 1962; Jordan & Jordan, 1965), the modeling of the channel thermal noise in field-effect transistors has been continuous since then. In 1967, Klaassen and Prins derived an equation to calculate the power spectral density of the channel thermal noise as (Klaassen & Prins, 1967)

$$S_{i_d^2} = \frac{4kT}{L_{eff}^2} \int_0^{L_{eff}} g(V(x))dx = \frac{4kT}{I_D L_{eff}^2} \int_{V_{seff}}^{V_{deff}} g(V)dV \tag{1}$$

where k is Boltzmann's constant, T is the lattice temperature, L_{eff} is the effective channel length, $g(V(x))$ is the channel conductance at the position x, and I_D is the d.c. drain current. Here V_{seff} and V_{deff} are the effective source and drain voltages, respectively. In order to treat a MOSFET as a resister-like element, van der Ziel presented a simpler equation as (van der Ziel, 1970)

$$S_{i_d^2} = 4kTg_{do}\gamma \tag{2}$$

where g_{do} is the output conductance at $V_{DS} = 0$, and the value of γ is 1 and 2/3 in the triode and saturation regions, respectively. The parameter γ in (2) is widely used later in the literature to demonstrate the enhanced channel noise in short-channel transistors. Another frequently used equation for the channel thermal noise proposed by Tsividis was given by (Tsividis, 1987)

$$S_{i_d^2} = \frac{4kT}{L_{eff}^2} \mu_{eff} \, |Q_{inv}| \tag{3}$$

where μ_{eff} is the effective mobility and Q_{inv} is the total inversion layer charge. These models, (1) to (3) are considered as the conventional models which worked well for long-channel transistors. In 1986, Adibi reported that the γ value of a 0.7 μm transistor is higher than 2/3 when working in the saturation region (Abidi, 1986). Different theories were then proposed to discover the origin of the enhanced channel thermal noise.

- Chen and Deens' model

Before Chen and Deen proposed their model in 2002 (Chen & Deen, 2002), all of the theories (Triantis, Birbas & Kondis, 1996; Klein, 1999; Scholten et al., 1999; Jin, Chan & Lau, 2000; Park & Park, 2000; Knoblinger, Klein & Tiebout, 2001) attributed the enhanced channel thermal noise to the hot carrier effect, following the similar arguments for the excess noise in field-effect transistors (Klassen, 1970; Baechtold, 1971; Takagi & Matsumoto, 1977; Jindal, 1986). Chen and Deen, however, considered the channel length modulation (CLM) effect and proposed the spectral density of the channel noise as (Chen & Deen, 2002)

$$S_{i_d^2} = \frac{4kT}{L_{elec}^2}\mu_{eff}\,|Q_{inv}| + \delta\frac{4kTI_D}{L_{elec}^2 E_{crit}^2}V_{DSsat} \tag{4}$$

where E_{crit} is the critical electrical field, Q_{inv} is the total inversion charge in the gradual channel region, and L_{elec} is the electrical channel length of the device ($L_{elec} = L_{eff} - \Delta L$, where ΔL is the channel length of the velocity saturated region). The second term in (4) is used to account for the carrier heating effect. However, in the experimental verification (see Fig. 5), very good agreements with measured data are achieved without including the hot carrier effect (i.e., $\delta = 0$) (Chen & Deen, 2002). Based on this observation, it was argued that no carrier heating is needed to model the channel thermal noise, and that the lattice temperature should be used for the temperature T in (4). In addition, the noise generated from the velocity saturated region in the channel, measured at the drain terminal, is assumed to be negligible.

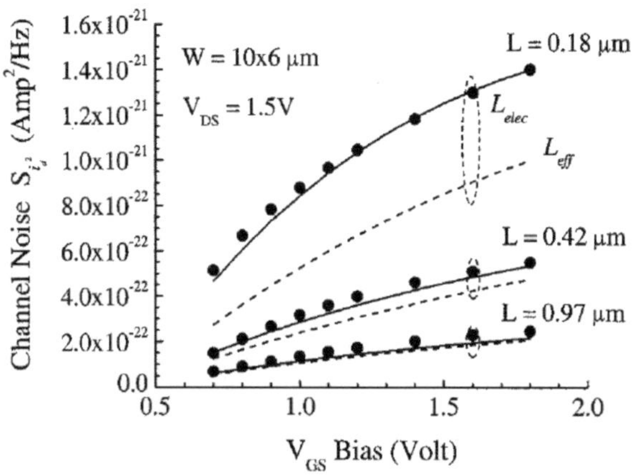

Fig. 5. Extracted (symbols) and calculated (lines) spectral densities of the channel thermal noise of n-type MOSFETs in a 0.18 μm COMS technology (Chen & Deen, 2002).

- Paasschens, Scholten & van Langeveldes' model

As indicated by Paasschens et al. (Paasschens, Scholten & van Langevelde, 2005), the limitation in (1) is that it cannot be applied to those devices whose channel conductance is a function of both position and voltage, i.e., $g = g(x, V)$ like LDMOS. In this case, Paasschens et al. separated the position and voltage dependence for the channel conductance as

$$g(x,V) = \frac{h(V)}{r(x)}. \tag{5}$$

Then, the channel thermal noise can be obtained by

$$S_{i_d^2} = 4kT\frac{\int_0^{L_{elec}} h(V(x))r(x)dx}{(\int_0^{L_{elec}} r(x)dx)^2} = \frac{4kT}{I_D}\frac{\int_{V_{seff}}^{V_{deff}} h(V)^2 dV}{(\int_0^{L_{elec}} r(x)dx)^2}. \tag{6}$$

Fig. 6 shows the normalized channel thermal noise for a ring-MOSFET biased at $V_{DS} = V_{GS} - V_{TH} = 1$ V. Both conventional Klaassen-Prins equation in (1) (i.e., $g = g(V)$) and thermal noise equation for resisters (i.e., $g = g(x)$) predict wrong results in the case of ring-MOSFETs.

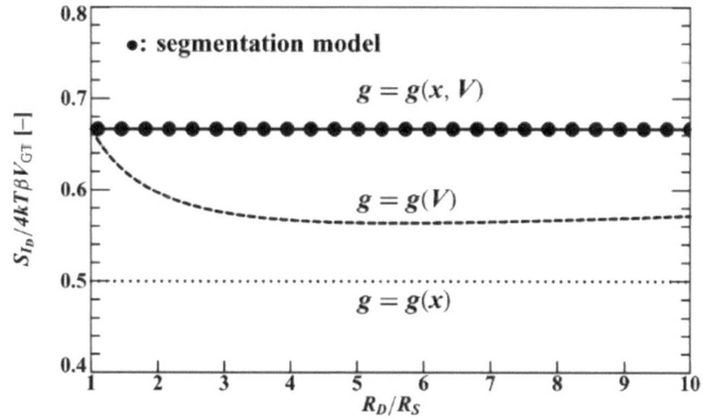

Fig. 6. Normalized channel thermal noise for a ring-MOSFET based on the classical Klaassen-Prins equation (dashed), the thermal noise equation for resistors (dotted), and the modified Klaassen-Prins equation (solid) (Paasschens, Scholten & van Langevelde, 2005).

For the velocity saturation effect due to the lateral electrical field, Paasschens et al. proposed a modified Klaassen-Prins equation as (Paasschens, Scholten & van Langevelde, 2005)

$$S_{i_d^2} = 4kT\frac{\int_0^{L_{elec}} (g_o / g)^{2p} g dx}{(\int_0^{L_{elec}} (g_o / g)^p dx)^2} = \frac{4kT}{I_D L_{vsat}^2} \int_{V_{seff}}^{V_{deff}} g_c^2 dV \quad (7)$$

with g_o and g_c defined in Paasschens et al.'s paper. Here p is the parameter to include the velocity saturation effect. Fig. 7 shows the calculated channel thermal noise with and without the velocity saturation effect. We can see that the velocity saturation effect reduces the channel thermal noise appeared at the drain terminal of the transistor.

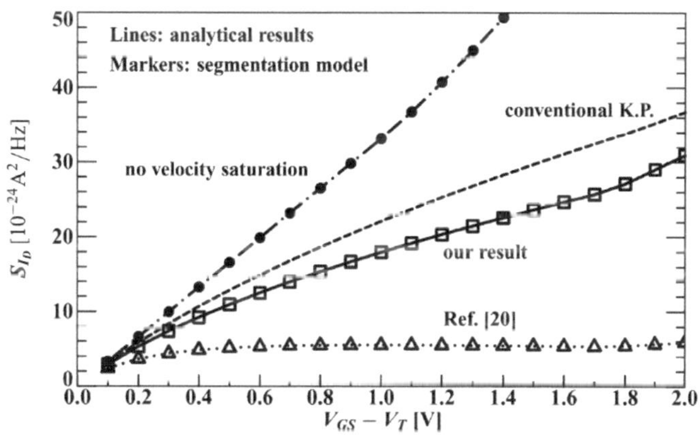

Fig. 7. Calculated channel thermal noise with and without the velocity saturation effect (Paasschens, Scholten & van Langevelde, 2005).

- Roy and Enzs' model

The carrier heating and mobility degradation are the two major concerns after Chen and Deens' model. Roy and Enz proposed a model for the channel noise as (Roy & Enz, 2005)

$$S_{i_d^2} = \frac{4kT_l W_{eff} \int_0^{L_{eff}} \frac{\mu_o |Q_{inv}|}{\mu_{eff} + \mu_{eff}^{'} \cdot E} dx}{L_{elec}^2 (1 + \frac{1}{L_{elec}} \int_{V_s}^{V_{deff}} \frac{u_{eff}^{'}}{\mu_{eff} + \mu_{eff}^{'} \cdot E})^2 dV} \tag{8}$$

where

$$\mu_{eff} = \frac{\mu_o}{\sqrt{1 + (\frac{dV / dx}{E_c})^2}} \sqrt{\frac{T_c}{T_l}}, \quad \mu_{eff}^{'} = \partial \mu_{eff} / \partial E, \tag{9}$$

T_c and T_l are the carrier and the lattice temperatures, respectively, E is the lateral electrical field in the gradual channel region, W_{eff} is the effective channel width, μ_o is the mobility without the velocity degradation result from E. Roy and Enz reported in their paper that the carrier heating and mobility reduction have an opposite effect on the power spectral density of the channel thermal noise. The mobility reduction decreases it, whereas the carrier heating enhances it. They believed that as Chen and Deens' model does not consider the carrier heating, the effect of mobility reduction gets largely compensated. They also believed that this is why Scholten et al. (Scholten et al., 2003) were able to match the experimental result without considering carrier heating. Jeon et al. also reported that the hot-carrier effect should be taken into account when modeling their 0.13 μm transistors (Jeon et al., 2007). On the contrary, Schenk et al. used device simulators to calculate the hot-electron effect for 0.25 μm transistors and concluded that the hot-electron effect on the channel thermal noise is not important under the normal operation conditions (Schenk et al., 2003). Fig. 8 shows the calculated γ value as a function of gate bias for models proposed by Chen (Chen & Deen, 2002), Han (Han, Shin & Lee, 2004), and Roy (Roy & Enz, 2005), respectively.

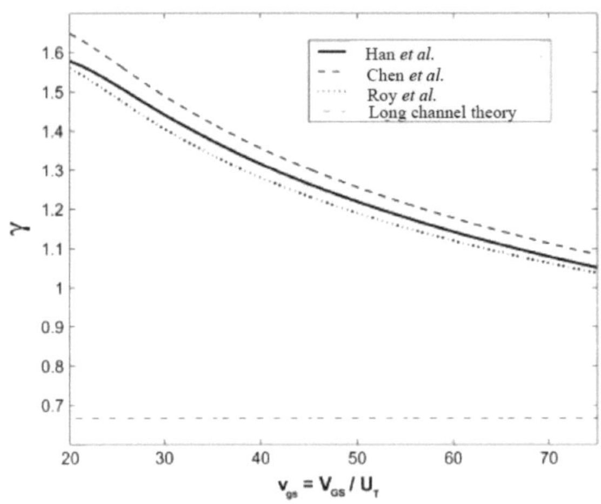

Fig. 8. Simulated γ values versus normalized v_{gs} bias for different models (Roy & Enz, 2005).

For the range of γ values, Fig. 9 shows the measured γ values from different technologies published in the literature (Dronavalli & Jindal, 2006). In general, the majority of the published γ values vary between 2/3 and 3 for channel length down to 120 nm.

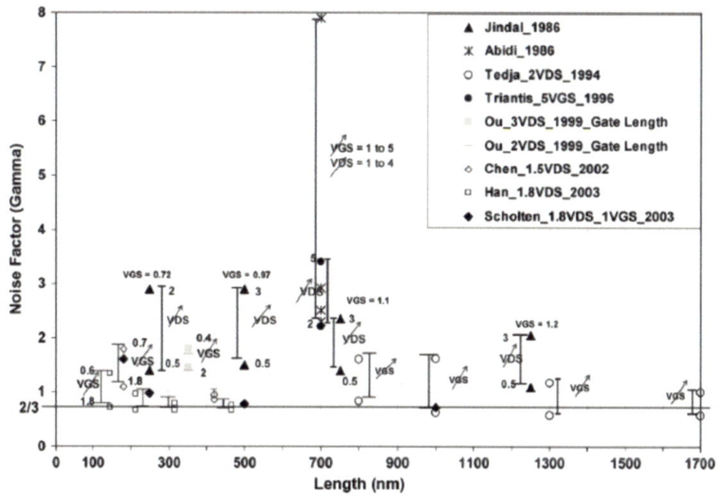

Fig. 9. Measured γ values for different technologies reported in the literature (Dronavalli & Jindal, 2006).

For the state-of-the-art 65 nm CMOS technology, Fig. 10 shows the measured (symbols) and simulated (lines) γ values for transistors fabricated by United Microelectronics Corporation (UMC) with W = 32×4 μm, and L = 60 nm, 80 nm, 120 nm, and 180 nm, respectively biased at V_{DS} = 1.2 V (Chen et al., 2008). We can see that the γ value could be as high as 4 for the 65 nm CMOS technology now.

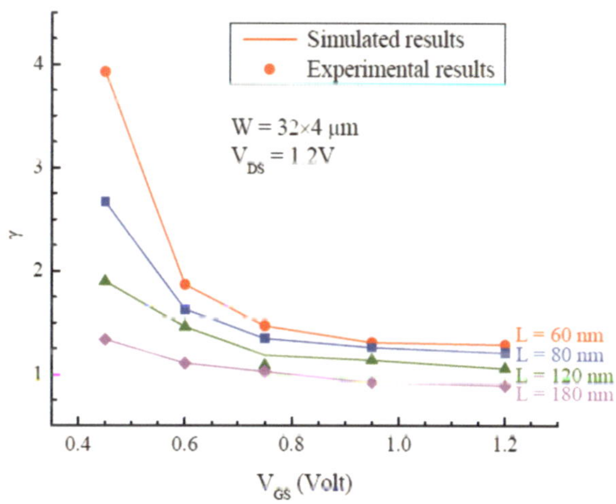

Fig. 10. Measured (symbols) and simulated (lines) γ values versus V_{GS} characteristics for transistors with W = 32×4 μm, and L = 60 nm, 80 nm, 120 nm, and 180 nm, respectively biased at V_{DS} = 1.2 V (Chen et al., 2008).

3.2 Thermal noise implementation

Any physics-based noise model has to be implemented into the circuit simulators before they can be used by IC designers. This is usually done through the software vendor or model developers, which might take long time to accomplish. Assuming that the noise spectral densities of channel thermal noise and induced gate noise are obtained from either theoretical calculation based on any noise model mentioned in previous sections or experimental results, Chen et al. provided a simple method to implement the enhanced channel noise and the induced gate noise for RF IC applications using a subcircuit approach (Chen, Li & Cheng, 2004). This approach is general and can work with any compact model (e.g., BSIM, MOS 11 or EKV model) and circuit simulator (e.g., SpectreRF or ELDO). Fig. 11 shows an equivalent circuit to demonstrate the implementation method. Because most of the circuit simulators cannot handle correlated noise sources, the correlation noise is not implemented at this point.

- Enhanced channel thermal noise

The enhanced channel thermal noise i_{de} shown in Fig. 11 is implemented using a Current Controlled Current Source (CCCS), and its value is determined by the noise current generated by the reference resistance R_{de} as shown in Fig. 12(a). Its resistance value is determined by (Chen, Li & Cheng, 2004)

$$R_{de} = \frac{4kT}{\overline{i_d^2} - \overline{i_{dcom}^2}}, \tag{10}$$

where k is Boltzmann's constant, T is the absolute temperature, and $\overline{i_{dcom}^2}$ is the channel thermal noise generated by the compact model.

- Induced gate noise

The induced gate noise can be naturally generated by using the segmentation method as presented in Scholten's paper (Scholten et al., 2003). However, the disadvantage of this approach is that it increases the number of transistors and therefore the simulation complexity, especially for the distortion analysis. In Chen's paper, the induced gate noise i_g shown in Fig. 11 is implemented by using another Current Controlled Current Source (CCCS), whose power spectral density is generated by the noise reference circuit shown in Fig. 12(b) with C_{ind} and R_{ind} selected by (Chen, Li & Cheng, 2004)

$$C_{ind} = 100 \cdot \frac{P_{ind} \cdot f_{max}}{8kT\pi} \tag{11}$$

and

$$R_{ind} = \frac{P_{ind}}{16kT\pi^2 C_{ind}^2} \tag{12}$$

where f_{max} is the maximum frequency up to which the simulation will be valid, and P_{ind} represents the coefficient in the induced gate noise vs. frequency characteristics, i.e.,

$$P_{ind} = \frac{\overline{i_g^2}}{f^2}. \tag{13}$$

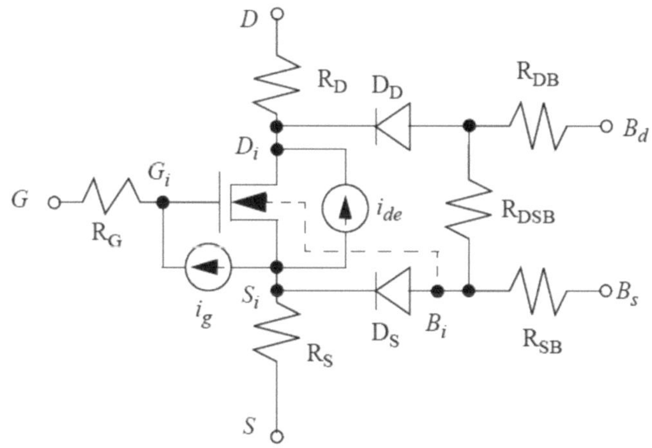

Fig. 11. Noise equivalent circuit of a MOSFET including parasitic resistance (R_D, R_G, and R_S), substrate network (D_D, D_S, R_{DB}, R_{SB}, and R_{DSB}), enhanced channel noise (i_{de}), and induced gate noise (i_g) for RF IC applications (Chen, Li & Cheng, 2004).

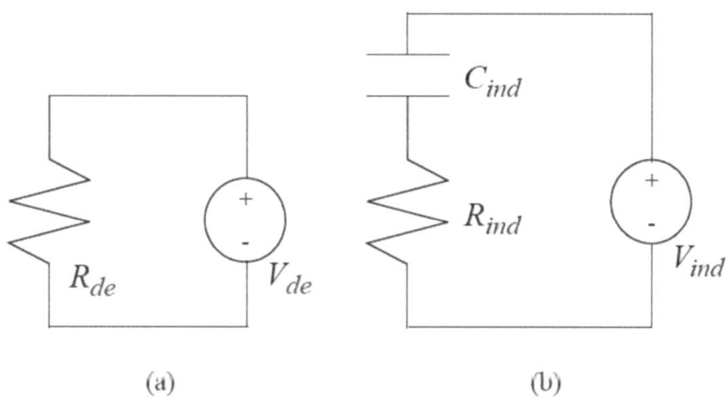

Fig. 12. Noise reference circuits to generate the noise currents for (a) the enhanced channel noise i_{de} and (b) the induced gate noise i_g shown in Fig. 11 (Chen, Li & Cheng, 2004).

4. Future work

As presented in the paper, all of the recent channel noise models focus on the noise from the gradual channel region, and how to characterize the noise contribution from the velocity saturation region in the nanometer MOSFETs is an area for future research. On the other hand, the design of integrated circuits with low power consumption is the trend for future circuit designs. In some cases, transistors might work in the moderate or weak inversion region. Therefore, the channel noise models for transistors working in these regions will become important for low-power applications. Finally, the scaling issues and the temperature characteristics of the active noise sources in the transistor are other research areas for future studies.

5. References

Abidi, A. A. (1986). High-frequency noise measurements on FET's with small dimensions. *IEEE Trans. on Electron Devices*, Vol. ED-33, 1986, pp. 1801-1805.

Adamian, V. & Uhlir, A. (1973). A novel procedure for receiver noise characterization. IEEE *IEEE Trans. Instrum. Meas.*, Vol. IM-22, No. 2, June 1973, pp. 181-182.

Agilent Application Note 57-1. Fundamentals of RF and Microwave Noise Figure Measurements.

Archer, J. W. & Batchelor, R. A. (1992). Fully automated on-wafer noise characterization of GaAs MESFET's and HEMT's. *IEEE Trans. Microwave Theory Tech.*, Vol. 40, No. 2, Feb. 1992, pp. 209-216.

Asgaran, S. et al. (2007). Analytical extraction of MOSFET's high frequency noise parameters from NF50 measurements and its application in RFIC design. *IEEE J. Solid-State Circuits*, Vol. 42, No. 5, May 2007, pp. 1034-1043.

Asgaran, S., Deen, M. J., & Chen, C. H. (2004). Analytical Modeling of MOSFET's Channel Noise and Noise Parameters. *IEEE Trans. Electron Devices*, Vol. 51, No. 12, Dec. 2004, pp. 2019-2114.

Baechtold, W. (1971). Noise temperature in silicon in the hot electron region. *IEEE Trans. Electron Devices*, vol. ED-18, No. 12, Dec. 1971, pp. 1186-1187.

Biber, C. E., et al. (1998). Technology Independent degradation of minimum noise figure due to pad parasitics. *Proceedings of 1998 IEEE MTT-S International Microwave Symposium*, Vol. 1, 1998, pp. 145-148.

Boudiaf, A. & Laporte, M. (1993). An accurate and repeatable technique for noise parameter measurements. *IEEE Trans. Instrum. Meas.*, Vol. 42, No. 2, Apr. 1993, pp. 532-537.

Caruso, G. & Sannino, M. (1978). Computer-aided determination of microwave two-port noise parameters. *IEEE Trans. Microwave Theory Tech.*, Vol. MTT-26, No. 9, 1978, pp. 639-642.

Chen, C. H. et al. (2001). Extraction of the induced gate noise, channel thermal noise and their correlation in sub-micron MOSFETs from RF noise measurements. *IEEE Trans. Electron Devices*, Vol. 48, No. 12, Dec. 2001, pp. 2884-2892.

Chen, C. H. et al. (2007). Novel noise parameter determination for on-wafer microwave noise measurements. *IEEE Trans. Instrum. Meas.*, Vol. 57, Issue 11, May 2007, pp. 2462-2471.

Chen, C. H. et al. (2008). Thermal noise performance in recent CMOS technologies. *Proceedings of the 9th International Conference on Solid-State and Integrated-Circuit Technology*, pp. 476-479, Beijing, China, Oct. 20-23, 2008.

Chen, C. H. & Deen, M. J. (2000). Direct extraction of the channel thermal noise in metal-oxide-semiconductor field effect transistor from measurements of their RF noise parameters. *J. Vacuum Science and Technology A*, Vol. 18(2), March/April 2000, pp. 757-760.

Chen, C. H. & Deen M. J. (2001). A general noise and s-parameter de-embedding procedure for on-wafer high-frequency noise measurements of MOSFETs. *IEEE Trans. Microwave Theory Tech.*, Vol. 49, No. 5, May 2001, pp. 1004-1005.

Chen, C. H. & Deen, M. J. (2002). Channel noise modeling of deep sub-micron MOSFETs. *IEEE Trans. Electron Device*, Vol. 49, No. 8, Aug. 2002, pp. 1484-1487.

Chen, C. H., Li, F. & Cheng, Y. (2004). MOSFET drain and induced-gate noise modeling and experimental verification for RF IC design. *Proceedings of IEEE International*

Conference on Microelectronics Test Structures (*ICMTS 2004*), pp. 51-56, Awaji, Japan, March 22-25, 2004.

Chen, C. H., Wang, Y. L., & Bakr, M. (2008). Wave-based approach for microwave noise characterization. *Fluctuation and Noise Letters*, Vol. 8, Issue 1, March 2008, pp. R1-R14.

Cho, M.-H. et al. (2005). A scalable noise de-embedding technique for on-wafer microwave device characterization. *IEEE Microwave Wireless Comp. Lett.*, Vol. 15, No. 10, Oct. 2005, pp. 649-651.

Davidson, A. C. et al. (1989). Accuracy improvements in microwave noise parameter measurements. *IEEE Trans. Microwave Theory Tech.*, Vol. 37, No. 12, Dec. 1989, pp. 1973-1978.

Deen, M. J. et al. (2006). High frequency noise of modern MOSFETs: compact modeling and measurement issues. *IEEE Trans. Electron Devices*, Vol. 53, No. 9, Sep. 2006, pp. 2062-2081.

Darfeuille, S. et al. (2007). Novel design and analysis procedures for differential-based filters on silicon. *Int. J. RF Microwave CAE*, Vol. 17, Jan. 2007, pp. 49–55.

Dobrowolski, J. A. (1989). A CAD-oriented method for noise figure computation of two-ports with any internal topology. *IEEE Trans. Microwave Theory Tech.*, Vol. 37, No. 1, Jan. 1989, pp. 15–20.

Dobrowolski, J. A. (1991). Noise power sensitivities and noise figure minimization of two-ports with any internal topology. *IEEE Trans. Microwave Theory Tech.*, Vol. 39, Jan. 1991, pp. 136–140.

Dronavalli, S. & Jindal, R. P. (2006). CMOS device noise considerations for terabit lightwave systems. *IEEE Trans. Electron Devices*, Vol. 53, No. 4, Apr. 2006, pp. 623-630.

Engbert, J. & Larsen, T. (1991). Extended definitions for noise temperatures of linear noisy one- and two-ports. *IEE Proc. H Microwaves, Antennas and Propagation*, Vol. 138, Feb. 1991, pp. 86–90.

Engen, G. F. (1970). A new method of characterizing amplifier noise performance. *IEEE Trans. Instrum. Meas.*, Vol. IM–19, Nov. 1970, pp. 344–349.

Friis, H. T. (1944). Noise figures of radio receivers. *Proc. of the IRE*, Vol. 32, No. 7, July 1944, pp. 419-422.

García-García, Q. (2004). Noise in lossless microwave multiports. *Int. J. RF Microwave CAE*, Vol. 14, Mar. 2004, pp. 99–110.

Grosch, T. O. & Carpenter, L. A. (1993). Two-port to three-port noise-wave transformation for CAD applications. *IEEE Trans. Microwave Theory Tech.*, Vol. 41, No. 9, Sep. 1993, pp. 1543–1548.

Gupta, M. S. (1970). Determination of the noise parameters of a linear 2-port. *Electron. Lett.*, Vol. 6, No. 17, 20th Aug. 1970, pp. 543-544.

Han, K., Shin, H. & Lee, K. (2004). Analytical drain thermal noise current model valid for deep submicron MOSFETs. *IEEE Trans. Electron Devices*, Vol. 51, No. 2, Feb. 2004, pp. 261-269.

Haus, H. A. et al. (1960). IRE standards on methods of measuring noise in linear two-ports, 1959. *Proc. IRE*, Vol. 48, Jan. 1960, pp. 60-68.

Haus, H. A. et al. (1960). Representation of noise in linear two-port. *Proc. IRE*, Jan. 1960, pp. 69–74.

Hecken, R. P. (1981). Analysis of linear noisy two-ports using scattering waves. *IEEE Trans. Microwave Theory Tech.*, Vol. MTT–29, Oct. 1981, pp. 997–1004.

Hillbrand, H. & Russer, P. H. (1976). An efficient method for computer aided noise analysis of linear amplifier networks. *IEEE Trans. Circuits Syst.*, Vol. 23, No. 4, Apr. 1976, pp. 235–238.

IRE Subcommittee 7.9 on Noise (1963). Description of the noise performance of amplifiers and receiving systems. *Proc. of the IEEE*, Vol. 51, Issue 3, Mar. 1963, pp. 436-442.

Jeon, J. et al. (2007). An analytical channel thermal noise model for deep-submicron MOSFETs with short channel effects. *Solid State Electron.*, Vol. 51, Issue 7, July 2007, pp. 1034-1038.

Jin, W., Chan, P. C. H. & Lau, J. (2000). A physical thermal noise model for SOI MOSFET. *IEEE Trans. Electron Devices*, Vol. 47, Issue 4, Apr. 2000, pp. 768-773.

Jindal, R. P. (1986). Hot-electron effects on channel thermal noise in fine-line NMOS field-effect transistors. *IEEE Trans. Electron Devices*, Vol. ED-33, Issue 9, Sep. 1986, pp. 1395-1397.

Jordan, A. G. & Jordan, N. A. (1965). Theory of noise in metal oxide semiconductor devices. *IEEE Trans. Electron Devices*, Vol. ED-12, March 1965, pp. 148-156.

Kanaglekar, N. G., McIntosh, R. E. & Bryant, W. E. (1987). Wave analysis of noise in interconnected multiport networks. *IEEE Trans. Microwave Theory Tech.*, Vol. 35, No. 2, Feb. 1987, pp. 112-116.

Klassen, F. M. (1970). On the influence of hot carrier effects on the thermal noise of field-effect transistors, *IEEE Trans. Electron Devices*, Vol. ED-17, No. 10, Oct. 1970, pp. 858-862.

Kantanen, M. et al. (2003). A wide-band on-wafer noise parameter measurement system at 50-75 GHz. *IEEE Trans. Microwave Theory Tech.*, Vol. 51, No. 5, May 2003, pp. 1489-1495.

Klaassen, F. M. & Prins, J. (1967). Thermal noise of MOS transistors. *Philips Res. Rep.*, Vol. 22, 1967, pp. 504-514.

Klein, P. (1999). An analytical thermal noise model of deep submicron MOSFET's. *IEEE Electron Device Lett.*, Vol. 20, Aug. 1999, pp. 399-401.

Knoblinger, G. (2001). RF-Noise of Deep-Submicron MOSFETs: Extraction and Modeling. *Proceedings of European Solid-State Device Research Conference* (*ESSDERC 2001*), pp. 331-334, 2001.

Knoblinger, G., Klein, P. & Baumann, U. (2000). Thermal channel noise of quarter and sub-quarter micron NMOSFETs. *Proceedings of IEEE International Conference on Microelectronic Test Structures*, pp. 95 - 98, 2000.

Knoblinger, G., Klein, P. & Tiebout, M. (2001). A new model for thermal channel noise of deep-submicron MOSFETs and its application in RF-CMOS design. *IEEE J. Solid-State Circuits*, Vol. 36, Issue 5, May 2001, pp. 831-837.

Kolding, T. E. (2000). A four-step method for de-embedding gigahertz on-wafer CMOS measurements. *IEEE Trans. Electron Devices*, Vol. 47, No. 4, Apr. 2000, pp. 734-740.

Koolen, M. C. A. M., Geelen, J. A. M. & Versleijen, M. P. J. G. (1991). An improved de-embedding technique for on-wafer high-frequency characterization. *Proceedings of IEEE Bipolar Circuits and Technology Meeting*, pp. 188-191, 1991.

Lane, R. Q. (1969). The determination of device noise parameters. *Proc. of the IEEE*, Vol. 57, No. 8, Aug. 1969, pp. 1461-1462.

Lange, J. (1967). Noise characterization of linear twoports in terms of invariant parameters. *IEEE J. Solid-State Circuits*, Vol. SC-2, No. 2, June 1967, pp. 37-40.

Lee, S., Ryum, B. R. & Kang, S. W. (1994). A new parameter extraction technique for small-signal equivalent circuit of polysilicon emitter bipolar transistors. *IEEE Trans. Electron Devices*, Vol. 41, No. 2, Feb. 1994, pp. 233-238.

Meierer, R. & Tsironis, C. (1995). An on-wafer noise parameter measurement technique with automatic receiver calibration. *Microwave J.*, Vol. 38, No. 3, Mar. 1995, pp. 22-37.

Meys, R. P. (1978). A wave approach to the noise properties of linear microwave devices. *IEEE Trans. Microwave Theory Tech.*, Vol. MTT–26, Jan. 1978, pp. 34–37.

Mitama, M. & Katoh, H. (1979). An improved computational method for noise parameter measurement. *IEEE Trans. Microwave Theory Tech.*, Vol. MTT-27, No. 6, June 1979, pp. 612-615.

O'Callaghan, J. M. & Mondal, J. P. (1991). A vector approach for noise parameter fitting and selection of source admittances. *IEEE Trans. Microwave Theory Tech.*, Vol. 28, No. 8, Aug. 1991, pp. 1376-1382.

Park, C. H. & Park, Y. J. (2000). Modeling of thermal noise in short-channel MOSFETs at saturation. *Solid State Electron.*, Vol. 44, Issue 11, Nov. 2000, pp. 2053-2057.

Paasschens, J. C. J., Scholten, A. J. & van Langevelde, R. (2005). Generalizations of the Klaassen-Prins equation for calculating the noise of semiconductor devices. *IEEE Trans. Electron Devices*, Vol. 52, No. 11, Nov. 2005, pp. 2463-2472.

Penfield, P. (1962). Wave representation of amplifier noise. *IRE Trans. Circuit Theory*, Vol. CT–9, Mar. 1962, pp.84–86.

Pospieszalski, M. W. (1986). On the measurement of noise parameters of microwave two-ports. *IEEE Trans. Microwave Theory Tech.*, Vol. 34, No. 4, Apr. 1986, pp. 456-458.

Randa, J. (2002). Noise-parameter uncertainties: a monte carlo simulation. *J. Res. Nat. Inst. Stand. Tech.*, Vol. 107, No. 5, 2002, pp. 431–444.

Randa, J. (2001). Noise characterization of multiport amplifiers. *IEEE Trans. Microwave Theory Tech.*, Vol. 49, No. 10, Oct. 2001, pp. 1757–1763.

Randa, J. & Walker, D. K. (2007). On-wafer measurement of transistor noise parameters at NIST. *IEEE Trans. Instrum. Meas.*, Vol. 56, No. 2, Apr. 2007, pp. 551–554.

Rothe, H. & Dahlke, W. (1956). Theory of noisy fourpoles. *Proc. IRE*, Vol. 44, June 1956, pp. 811-818.

Roy, A. S. & Enz, C. C. (2005). Compact modeling of thermal noise in the MOS transistor. *IEEE Trans. Electron Devices*, Vol. 52, No. 4, Apr. 2005, pp. 611-614.

Sannino, M. (1979). On the determination of device noise and gain parameters. *Proc. of the IEEE*, Vol. 67, No. 9, Sep. 1979, pp. 1364-1366.

Schenk, A. et al. (2003). Simulation of RF nnoise in MOSFETs using different transport models. *IEICE Trans. Electron.*, Vol. E86-C, No. 3, March 2003, pp. 481-489.

Scholten, A. J. et al. (1999). Accurate thermal noise model for deep-submicron CMOS. IEDM Tech. Dig., 1999, pp. 155-158

Scholten, A. J. et al. (2003). Noise modeling for RF CMOS circuit simulation. *IEEE Trans. Electron Devices*, Vol. 50, No. 3, Mar. 2003, pp. 618-632.

Simpson, G. (2009). Setting up ultra-fast noise parameters using the Agilent PNA-X. Maury Microwave application notes 5C-084, June 2009.

Takagi, K. & Matsumoto, K. (1977). Noise in silicon and FET's at high electric fields. *Solid-state Electron.*, Vol. 20, Jan. 1977, pp. 1-3.

Tiemeijer, L. F. et al. (2005). Improved Y-factor method for wide-band on-wafer noise parameter measurements. *IEEE Trans. Microwave Theory Tech.*, Vol. 53, No. 9, 2005, pp. 2917-2925.

Triantis, D. P. , Birbas, A. N. & Kondis, D. (1996). Thermal noise modeling for short-channel MOSFETs. *IEEE Trans. Electron Devices*, Vol. 43, Issue 11, Nov. 1996, pp. 1950-1955.

Tsividis, Y. P. (1987). *Operation and Modeling of the MOS Transistor*, New York: Wiley, 1987.

Tutt, M. N. (1994). Low and high frequency noise properties of heterojunction transistors. Ph.D. thesis, EE&CS, The University of Michigan, Ann Arbor, MI, 1994.

Valk, E. C. et al. (1988). De-embedding two-port noise parameters using a noise wave model. *IEEE Trans. Instrum. Meas.*, Vol. 37, No. 2, June 1988, pp. 195–200.

van der Ziel, A. (1962). Thermal noise in field-effect transistors. *Proc. IRE*, Vol. 56, 1962, pp. 1808-1812.

van der Ziel, A. (1970). *Noise Sources, Characterization, Measurement*, NJ: Prentice-Hall, 1970.

van Wijnen, P. J., Claessen, H. R. & Wolsheimer, E. A. (1987). A new straight forward calibration and correction procedure for "on wafer" high frequency s-parameter measurements (45 MHz - 18 GHz). *Proceedings of IEEE Bipolar Circuits and Technology Meeting*, pp. 70-73, 1987.

Vasilescu, G. I., Alquie, G. & Krim, M. (1988). Exact computation of two-port noise parameters. *Electron. Lett.*, Vol. 25, No. 4, Feb. 1988, pp. 292-293.

Wait, D. & Engen, G. F. (1991). Application of radiometry to the accurate measurement of amplifier noise. *IEEE Trans. Instrum. Meas.*, Vol. 40, No. 2, Apr. 1991, pp. 433–437.

Wedge, S. W. & Rutledge, D. B. (1992). Wave techniques for noise modeling and measurement. *IEEE Trans. Microwave Theory Tech.*, Vol. 40, No. 11, Nov. 1992, pp. 2004–2012.

Wedge, S. W. & Rutledge, D. B. (1991). Noise waves and passive linear multiports. *IEEE Microwave and Guided Wave Lett.*, Vol. 1, No. 5, May 1991, pp. 117-119.

Wiatr, W. & Walker, D. K. (2005). Systematic errors of noise parameter determination caused by imperfect source impedance measurement. *IEEE Trans. Instrum. Meas.*, Vol. 54, No. 2, Apr. 2005, pp. 696-700.

5

Statistical Prediction of Circuit Aging under Process Variations

Wenping Wang[1], Vijay Reddy[2], Varsha Balakrishnan[1], Srikanth Krishnan[2] and Yu Cao[1]

[1]*Department of Electrical Engineering, Arizona State University, Tempe, AZ 85287,*
[2]*External Development and Manufacturing, Texas Instruments, PO Box 650311, MS 3740, Dallas TX 75243, USA*

1. Introduction

With relentless scaling of CMOS technology, circuit timing uncertainty due to temporal degradation and static process variations poses a dramatic challenge to IC design (*International Technology Roadmap for Semiconductors*, 2008; Reddy et al., 2002; Nassif, 2001; Lin et al., 1998). The deterioration of circuit performance over time, i.e., aging, is usually caused by several physical mechanisms such as channel-hot-carrier (CHC), negative-bias-temperature-instability (NBTI), and time-dependent-dielectric-breakdown (TDDB) (Schroder & Babcock, 2003; Alam & Mahapatra, 2005; Wang et al., 2007; Vattikonda et al., 2006; Ogawa et al., 2003). Among these effects, NBTI is the leading mechanism that is responsible for the majority part of circuit aging (Kimizuka et al., 1999; Wang et al., 2007). In (Wang et al., 2007), the authors show that for 65nm technology, CHC degradation is much smaller than NBTI degradation, almost one order lower in the degradation magnitude. NBTI primarily increases the threshold voltage (V_{th}) of PMOS devices. Such parameters shift significantly affects circuit lifetime and performance (e.g., power, speed and failure rate), and in the worst case, may even result in a complete parametric failure of a system (Borkar, 2006; Alam & Mahapatra, 2005; Wang et al., 2007; Vattikonda et al., 2006; Bhardwaj et al., 2006; Kumar et al., 2006; Paul et al., 2006). To cope with this threat and guarantee circuit lifetime, it is critical to include NBTI into circuit analysis and adaptively develop design techniques to effectively mitigate its negative impact on performance.

For a VLSI design, an accurate prediction of circuit performance degradation under NBTI remains as a tremendous challenge. As shown in (Wang et al., 2007), NBTI has a strong dependence on dynamic operation conditions, such as supply voltage (V_{dd}), temperature (T) and input signal probability (α_s). Usually these parameters are not spatially or temporally uniform, but vary significantly from gate to gate and from time to time. Similar to the burn-in process, we may use high voltage and high temperature to guardband the worst case condition. However, the search for the worst case α_s is computationally inhibitive due to the extremely large space of signal probabilities for each input node. A practical method is proposed to predict the upper bound of each gate under all possible input α_ss (Agarwal et al., 2008).

Besides these uncertainties, static process variation poses another challenge that leads to the variance of circuit aging. Fig. 1 illustrates an example that shows the statistical measurement of switching frequency and the leakage (*IDDQ*) of ring oscillators (ROs) before the reliability test. In this 65nm technology, more than 3X and 25% variability are observed in circuit leakage and the speed, respectively. These variabilities are attributed to statistical distributions of device parameters that are caused by the manufacturing process (Nassif, 2001). Examples include dopant concentration, channel length (*L*), oxide thickness (t_{ox}), etc. Their impact on device and circuit performance is usually investigated through a reduced set of device parameters - V_{th} is the most important one among them, as the interface between process and electrical studies. In (Liu et al., 2007), we have identified that the leading variation sources are *L*, V_{th} and carrier mobility (μ). By including the extracted variations of these three sources in the nominal model file, we are able to accurately predict the change of IV characteristics in all regions. Other variations, such as that in t_{ox}, are included into these variations (e.g., V_{th} is a function of t_{ox}) and indirectly affect device performance.

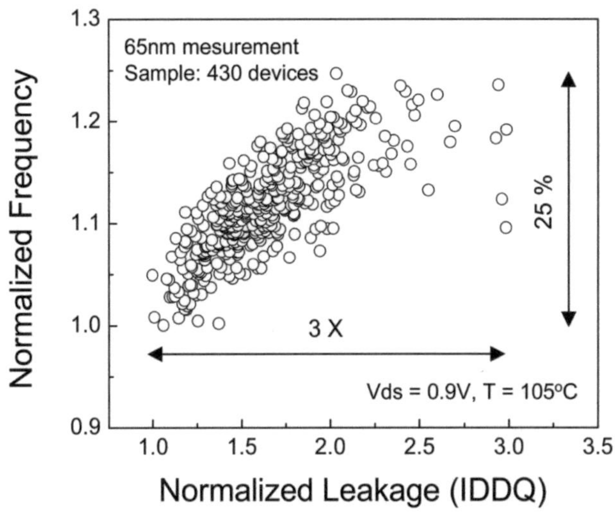

Fig. 1. Measured RO leakage and frequency variations before the stress.

Since NBTI effect has an exponential dependence of V_{th} (Wang et al., 2007; Alam & Mahapatra, 2005; Wang et al., 2007), circuit aging strongly interacts with process variations, significantly shifting both the mean and the variance of the circuit performance. By focusing on this primary variation source - V_{th}, we are therefore able to gain key insights and project the first-order trend. Other secondary process sources are neglected in this work. With the availability of more detailed data in the future, we will incorporate more sources. Fig. 2 shows the measured speed degradation from the same set of ROs as those in Fig. 1. Circuit performance and its variability not only depend on static process variations, but also change over the period of dynamic operation because of the effect of circuit aging (Schroder & Babcock, 2003; Wang et al., 2007). Therefore, accurate prediction of circuit performance distribution during its life time should consider the impact of static variations, primary reliability mechanisms, and more importantly, their interactions. This prediction is essential for designers to safely guardband the circuit for a sufficient life time. Otherwise, we have to either use an overly pessimistic bound, or resort to expensive stress tests in order to collect enough statistical information.

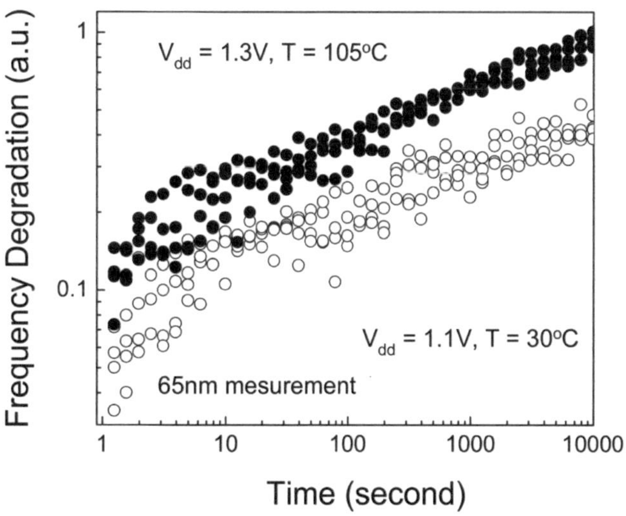

Fig. 2. Measured frequency degradation of a 11-stage ring oscillator under different stress conditions (four selected samples each condition).

A few works have been published in the literature to estimate the statistical variations in temporal NBTI degradation (Rauch, 2002; 2007; Rosa et al., 2006; Kang et al., 2007). Their assumption is the number of broken bonds in the channel is a Poisson random variable, and correspondingly V_{th} follows the Poisson distribution. With technology scaling, additional V_{th} variations, such as random dopant fluctuation and short channel effects, need to be considered. The measurement data show that the distribution of V_{th} variations follows the Gaussian distribution (Liu et al., 2007). In addition, the correlations between process variation and NBTI are ignored in previous work. In this work, we begin with the assumption that process variation induced V_{th} change is Gaussian random variable. We leverage compact models of transistor degradation and circuit performance to achieve accurate and efficient reliability prediction. Dynamic NBTI effect is incorporated into the analytical framework to account for the aging of circuit speed and the leakage (Schroder & Babcock, 2003; Wang et al., 2007). Based on our initial observation with the available data, the specific contributions and conclusions of this work include:

- A statistical predictive methodology of circuit aging is proposed. In this analytical approach, only five model parameters need to be extracted from fresh data (i.e., before the stress). With the initial information of the transistor and circuit topology, these models provide accurate prediction of circuit performance degradation and the variability.
- The degradation rate of circuit speed and its standard deviation follows a power law of 1/6. While the mean of circuit timing goes up with the stress time, the variance actually declines due to the interaction between NBTI effect and process variations. The degradation rate of both values is independent on the amount and the type of variations in the circuit.
- The mean and the standard deviation of logarithmic *IDDQ* reduce with the stress time as $t^{1/6}$, with the variance more sensitive to global variations.
- A hierarchical statistical aging analysis methodology is proposed to efficiently predict circuit aging under both process variations and operation uncertainties.

The outline of the paper is as follows: In Section 2, the statistical modeling for both transistor and circuit performance degradation is described, and the proposed models are comprehensively verified with silicon data from industrial 65nm technology, as well as SPICE simulation results. Section 3 presents a hierarchical statistical aging analysis methodology for circuit performance prediction. Finally Section 4 concludes this work.

2. Statistical modeling of circuit aging

NBTI is the dominant effect of circuit aging in advanced CMOS technology (Schroder & Babcock, 2003; Kimizuka et al., 1999). We propose a hierarchical solution to bridge the underlying device physics with efficient circuit analysis. Based on the reaction-diffusion mechanism, we developed the model of Vth shift under NBTI effect. In the presence of process variations and aging, using Gradual Change Approximation (GCA), this model is further expanded as a linear function of Vth shift to efficiently predict the performance degradation.

To characterize the change of circuit performance under the stress, the Alpha-power law based delay model and the leakage model are calibrated for a given gate. Under the condition that the amount of NBTI-induced V_{th} shift is much smaller than the nominal value of V_{th}, both models are simplified to extract the dependence of circuit performance to V_{th} change.

Finally, the gate-level models are integrated into various circuit paths to analytically predict the aging of path timing and the leakage. Both the mean value and the variance of these important metrics are derived as a function of static performance variability, the nominal sensitivity of circuit performance, and other operation conditions, such as supply voltage and temperature.

2.1 Gradual change approximation

NBTI manifests itself in a gradual increase in the magnitude of PMOS threshold voltage, resulting in the degradation of circuit performance over time. A set of publications have shown that NBTI is only pronounced in a long term, i.e., the performance degradation of transistors and circuits is a gradual aging process (Kumar et al., 2006; Wang et al., 2007). ΔV_{th} in different devices due to variation and NBTI is still a relative small portion compared to the nominal V_{th} value. Thus, when we derive the statistical degradation model, the first order Taylor expansion (i.e., linear expression) is applicable to transistor and circuit metrics, such as Equations (1) and (2), as well as gate delay and leakage analysis. This is the Gradual Change Approximation (GCA), which can be applied to simplify the following model derivations.

$$e^x = \sum_{n=0}^{\infty} \frac{x^n}{n!} = 1 + x + \frac{x^2}{2!} + \frac{x^3}{3!} + \cdots \qquad \text{for all } x \tag{1}$$

$$(1+x)^p = \sum_{n=0}^{\infty} \binom{p}{n} x^n \qquad \text{for all } |x| < 1, \text{and all complex } p \tag{2}$$

2.2 Transistor degradation model

NBTI occurs when a negative gate bias is applied to the PMOS devices and it has two phases: stress and recovery (Alam & Mahapatra, 2005; Wang et al., 2007). In stress phase, the

holes in the channel weaken the Si-H bonds, which results in the generation of the positive interface charges and hydrogen species. During recovery phase, the interface traps are annealed by the hydrogen species and thus, V_{th} degradation ($\Delta V_{th-nbti}$) is partially recovered.
Two main theories are proposed to interpret NBTI degradation: Reaction-Diffusion (RD) (Alam & Mahapatra, 2005; Alam et al., 2007; Krishnan et al., 2005) and hole Trapping/Detrapping (T/DT) (Shen et al., 2006; Parthasarathy et al., 2006; Huard et al., 2006). R-D model naturally explains the long-term power law time exponent of NBTI (n ~1/6) (Alam et al., 2007; Wang et al., 2007; Bhardwaj et al., 2006). However, the experimental data obtained by using on-the-fly (OTF) measurement (Parthasarathy et al., 2006; Rangan et al., 2003), or ultra-fast (UF) measurement (Reisinger et al., 2006) show that the power law dependence with a time exponent of n > 1/6. R-D model cannot explain well the fast transient in the beginning of recovery of NBTI. Thus, we introduce an experimental term to capture the behavior of the fast response in recovery (Bhardwaj et al., 2006; Wang et al., 2007), and the long-term $\Delta V_{th-nbti}$ is given by,

$$\Delta V_{th-nbti} = \left(\sqrt{K_v^2 \cdot T_{clk} \cdot \alpha_s} / \left(1 - \beta_t^{1/2n}\right)\right)^{2n} \tag{3}$$

where
$$\beta_t = 1 - \frac{2\xi_1 \cdot t_e + \sqrt{\xi_2 \cdot C \cdot (1-\alpha_s) \cdot T_{clk}}}{2t_{ox} + \sqrt{C \cdot t}} \tag{4}$$

$$K_v = \left(\frac{q t_{ox}}{\epsilon_{ox}}\right)^3 K^2 C_{ox}(V_{gs} - V_{th})\sqrt{C} exp\left(\frac{2E_{ox}}{E_0}\right) \tag{5}$$

where T_{clk}, α_s, and n are clock period, input signal probability, and time exponential constant (1/6), respectively. K, ξ_1, ξ_2 and E_0 are fitting parameters. K_v describes the dependence of the bias voltage, T, t_{ox} and other technology parameters associated with NBTI degradation (Bhardwaj et al., 2006; Wang et al., 2007). For more details about the meaning and value of the parameters, please refer to (Bhardwaj et al., 2006; Wang et al., 2007). The dependence of NBTI effect on V_{th} (which is a parameter lumping many process details) is still under the debate. For instance, reference (Alam & Mahapatra, 2005) has discussed several possibilities. A general observation is that NBTI is strongly affected by the electric field. Under the stress condition, this field is proportional to $(V_{gs} - V_t)/t_{ox}$, which leads to our model. By so far, this model matches data from 180nm down to 45nm, making it a generic model to describe NBTI effect in technology scaling. We are further collecting data from multi-V_{th} technology at the same technology node, with the target to verify it with more process details. Fig. 3 verifies this model with one set of experimental data under different stress conditions. Besides this long term prediction model, both static and real time dynamic models are also available in (Bhardwaj et al., 2006; Wang et al., 2007). The models well capture NBTI recovery effect (Agarwal et al., 2008).
This model assumes nominal degradation without considering the statistical process variations. If there are global and local process variations, V_{th} in Equation (5) should be expressed as.

$$V_{th} = V_{th0} + \Delta V_{th-g} + \Delta V_{th-l} \tag{6}$$

where V_{th0} is the nominal threshold voltage, ΔV_{th-g} and ΔV_{th-l} represent the change of threshold voltage due to global and local variations, respectively. Equation (6) shows that

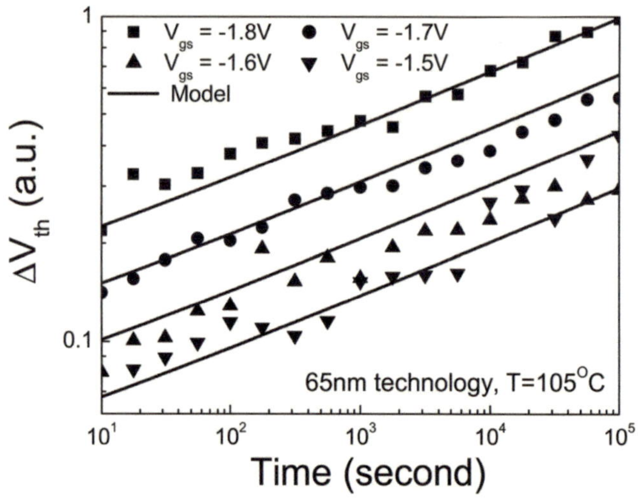

Fig. 3. Threshold voltage degradation under different stress conditions.

positive variation results in V_{th} increase, which correspondingly leads to smaller V_{th} degradation (according to Equations (3) and (5)), while negative variation results in larger V_{th} degradation. Fig. 4 shows V_{th} degradation over time for three different transistors. Due to process variations, Device 1 starts with a larger V_{th} and Device 3 starts with a smaller V_{th}. Substitute their fresh V_{th} to Equations (3) - (5), ΔV_{th} for these three devices is shown as Fig. 4. At the beginning, the difference in V_{th} degradation between Device 1 and Device 3 is 20.97%. With the increase of stress time, the difference becomes smaller and smaller. After 10^5s stress, it decreases to 15.57%. Such a compensation between process variations and aging is well captured by our model.

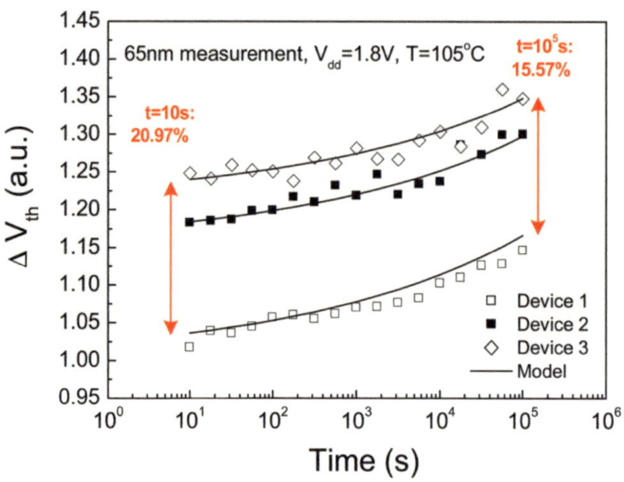

Fig. 4. Threshold voltage degradation over time for different devices

Since the degradation rate of different circuit paths is pronouncedly different due to different switching activities and circuit topologies (Wang et al., 2007), in (Agarwal et al., 2008), we introduce a Maximum Dynamic Stress (MDS) simulation technique with α_s

approaching to 1, which gives a simple and realistic estimation of the upper limit of gate delay degradation under dynamic NBTI. By using GCA and MDS described in (Agarwal et al., 2008), the long term prediction model is further simplified as a variation dependent model, i.e.,

$$\Delta V_{th-nbti} = A\big(1 - S_v(\Delta V_{th-g} + \Delta V_{th-l})\big)t^n \tag{7}$$

where the value of A depends on both technology parameters and operating conditions; S_v is the nominal sensitivity of NBTI degradation to V_{th} shift. In this work, $A = 2.5 \times 10^{-3} V/s^{1/6}$ and $S_v = 7V^{-1}$. Fig. 5 validates this simplified model (Equation (7)) with the long term predictive model (Equations (3) - (5)) under different process variations. Within $\pm 30mV$, it provides accurate prediction of V_{th} degradation.

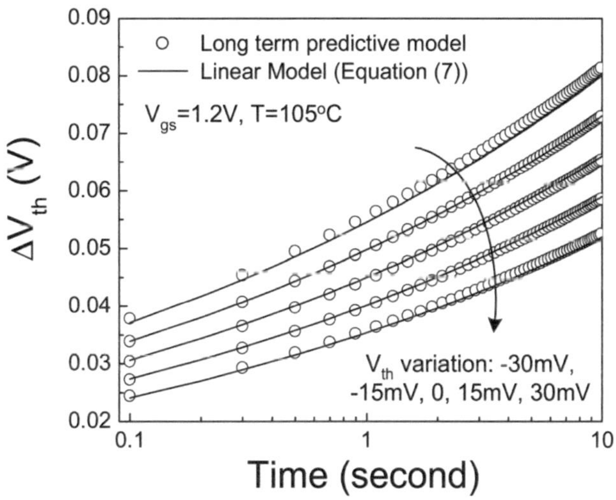

Fig. 5. Verification of process variation dependent NBTI model.

Besides V_{th} change, carrier mobility also degrades with increasing stress time (Wang et al., 2007). According to the universal effective mobility model, the dominant components of carrier scattering are phonon scattering, surface roughness scattering, and coulomb scattering. Aging effects induced interface charges result in stronger column scattering, which affects the carrier mobility mostly in low V_{gs} region. Because of this reason, mobility degradation is more important for analog circuit aging, but not for digital circuits. This behavior has been confirmed by the 65nm data (Wang et al., 2007). Thus, in this work, since the analysis is focused on digital circuit in which the devices operate at saturation region, the mobility degradation is ignored.

2.3 Gate delay degradation model

A widely used gate delay (T_{di}) model is based on the Alpha-power law that was proposed in (Sakurai & Newton, 1990),

$$T_{di} = (C_{li}V_{dd})/(\beta_i(V_{dd} - V_{thi})^{\alpha}) \tag{8}$$

where C_{li} is the effective load capacitance of the gate; β_i is a parameter depending on gate size. Under both process variations and NBTI effect, V_{thi} of PMOS is given by

$$V_{thi} = V_{th0} + \Delta V_{thi} \tag{9}$$

$$\Delta V_{thi} = \Delta V_{thi-g} + \Delta V_{thi-l} + \Delta V_{thi-nbti} \tag{10}$$

Substituting V_{thi} into Equation (8) and using the GCA of $1/(1-x)_p \approx 1 + p \cdot x$ (for $x \ll 1$), we obtain:

$$T_{di} = \frac{C_{li}V_{dd}}{\beta_i(V_{dd} - V_{th0} - \Delta V_{thi})^{\alpha}} \approx \frac{C_{li}V_{dd}}{\beta_i(V_{dd} - V_{th0})^{\alpha}}\left(1 + \frac{\alpha \Delta V_{thi}}{(V_{dd} - V_{th0})}\right) \tag{11}$$

We define $T_{d0i} = (C_{li}V_{dd})/(\beta_i(V_{dd} - V_{th0})^{\alpha})$, which is the gate delay without process variations and NBTI degradation ($\Delta V_{thi} = 0$), and $S_{ti} = \alpha/(V_{dd} - V_{th0})$, which is the nominal sensitivity of gate delay to PMOS V_{th} shift. These two parameters rely on the process technology and the circuit structure. They can be conveniently extracted from SPICE simulation at the nominal condition. Thus, Equation (11) becomes:

$$T_{di} = T_{d0i}(1 + S_{ti}\Delta V_{thi}) \tag{12}$$

Substitute Equations (7) and (10) into (12),

$$T_{di} = T_{d0}(1 + S_{ti}(At^n + (1 - AS_v t^n)\Delta V_{thi-g} + (1 - AS_v t^n)\Delta V_{thi-l})) \tag{13}$$

Since 65nm measurement data show that the distribution of V_{th} variation is Gaussian distribution (Nassif, 2001; Liu et al., 2007), in this work, we assume ΔV_{thi-g} and ΔV_{thi-l} are Gaussian random variables. Their mean (μ_g and μ_l) are 0 and their standard deviations (σ_g and σ_l) depend on the manufacturing process (Borkar et al., 2003). Since gate delay is linearly proportional to the threshold voltage change, the probability distribution function (PDF) of gate delay also follows the normal distribution $N \sim (\mu_{T_{di}}, \sigma^2_{T_{di}})$.

At $t = 0$, $\Delta V_{th-nbti} = 0$. Assuming global and local variations are uncorrelated random variables (Boning & Nassif, 2001), $\mu_{T_{di}}(0)$ and $\sigma^2_{T_{di}}(0)$ are given by:

$$\mu_{T_{di}}(0) = T_{d0i}, \quad \sigma_{T_{di}}(0) = T_{d0i}S_{ti}\sqrt{\sigma_g^2 + \sigma_l^2} \tag{14}$$

At $t > 0$, from Equation (13), we get

$$\mu_{T_{di}}(t) = \mu_{T_{di}}(0)(1 + AS_{ti}t^n), \quad \sigma_{T_{di}}(t) = \sigma_{T_{di}}(0)(1 - AS_v t^n) \tag{15}$$

Given the initial conditions of the process and timing information for the transistor and the gate, Equation (15) predicts the mean and standard deviation with increasing time. From these equations, we have four observations:

1. The mean of gate delay increases with the stress time, while the variance decreases. Since a lower-V_{th} transistor has a faster degradation rate and thus, larger V_{th} increase, this phenomenon compensates static process variations and reduces the variance during the stress period.
2. As the stress time increases, the aging of both mean and standard deviation follows the same power law of $t^{1/6}$.
3. The degradation rate of gate delay and its variance are independent of the amount and the type of variations.

4. The degradation rate is determined by the sensitivities to V_{th} shift. Process variations only affect the fresh variability, but not the degradation rate.

2.4 Circuit performance degradation model

2.4.1 Path timing

The PDF of a path comprising n gates corresponds to the linear combination of the n PDFs of gate delays. The mean and the variance of the path delay (T_d) are given by

$$\mu_{T_d} = \sum_{i}^{n} \mu_{T_{di}}, \quad \sigma_{T_d}^2 = \sum_{i}^{n}\sum_{j}^{n} \sigma_{T_{di}} \cdot \rho_{ij} \cdot \sigma_{T_{dj}} \tag{16}$$

where ρ_{ij} is the correlation coefficient between two gates. For the simplicity of the demonstration, we assume the inter-gate correlation is the same for all the gates, i.e., $\rho_{ij} = \rho$, while this methodology is general enough for all statistical conditions. Thus, the variance of path delay is derived as

$$\sigma_{T_d}^2 = \sum_{i}^{n}\sum_{j}^{n} \sigma_{T_{di}} \cdot \rho \cdot \sigma_{T_{dj}} + \sum_{i}^{n}(1-\rho) \cdot \sigma_{T_{di}}^2 \tag{17}$$

In the case of local variations, $\rho = 0$, i.e., the variations between two gates are uncorrelated. The case of $\rho = 1$ describes global variations, i.e., the variations between two gates are correlated. With both local and global variations, $\sigma_{T_d}^2$ is given by the linear combination of the variance of the local and global variations. Fig. 6 shows the delay distribution of ROs due to circuit aging. The distribution of gate delay becomes increasingly narrower under the stress as indicated by Equation (15).

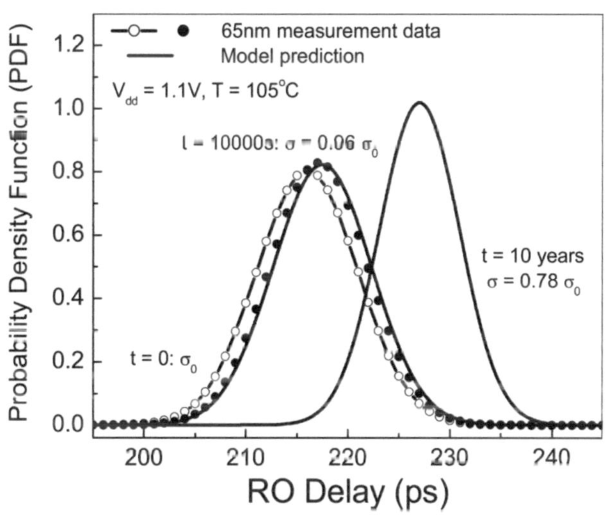

Fig. 6. The PDF's of 65nm RO delay during aging.

The proposed predictive methodology is generated for a path consisting of various types of gates. Fig. 7 shows such a circuit example. By stressing the path for different years, Fig. 8 compares the model prediction with SPICE simulation results of gate delay. Under different types and amount of variations, the model provides accurate prediction of the mean and standard deviation.

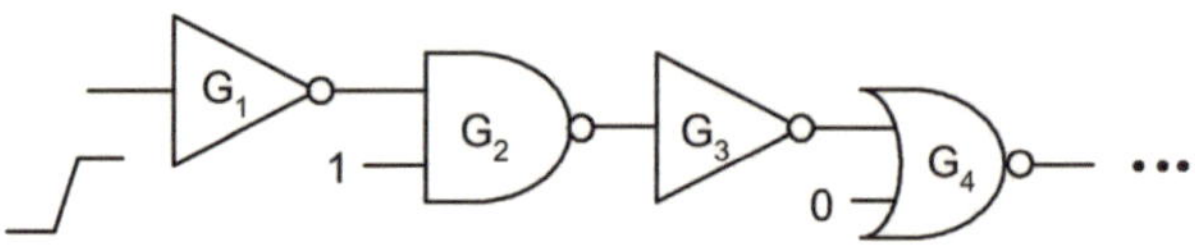

Fig. 7. Circuit example for path timing analysis.

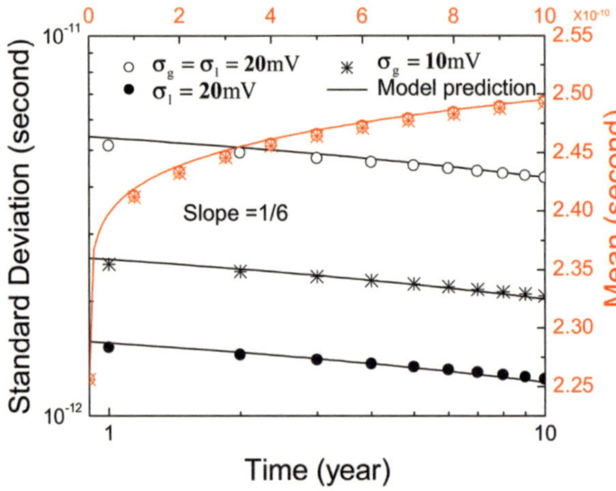

Fig. 8. The temporal increase of the mean and standard deviation of circuit speed (path is shown in Fig. 7).

2.4.2 Leakage

IDDQ of a circuit is defined as the total amount of leakage current at the standby. It has an exponential dependence on V_{th}:

$$IDDQ = \sum_{i}^{n} I_{0i} \cdot e^{-\frac{V_{thi}}{mv_T}} = \sum_{i}^{n} I_{0i} \cdot e^{-\frac{(V_{th0i}+\Delta V_{thi})}{mv_T}} \tag{18}$$

where $I_{0i} = \beta_i(m-1)(1-e^{-V_{ds}/v_T})$, m is the body effect coefficient and v_T is the thermal voltage (kT/q). Substitute Equation (10) into (18), we get

$$\begin{aligned} IDDQ &= \sum_{i}^{n} I_{0i} e^{-\frac{V_{th0i}}{mv_T}} e^{-\frac{\Delta V_{thi-g}+\Delta V_{thi-l}+\Delta V_{thi-nbti}}{mv_T}} \\ &= \sum_{i}^{n} IDDQi(0) e^{-\frac{\Delta V_{thi-g}+\Delta V_{thi-l}+\Delta V_{thi-nbti}}{mv_T}} \end{aligned} \tag{19}$$

IDDQi(0) is the gate leakage at t = 0, i.e., $\Delta V_{thi\text{-}g} = \Delta V_{thi\text{-}l} = \Delta V_{thi\text{-}nbti} = 0$. Taking the natural logarithms on both sides of Equation (19), we have

$$Ln(IDDQ) = Ln\left(\sum_{i}^{n} IDDQi(0) e^{-\frac{\Delta V_{thi-g}+\Delta V_{thi-l}+\Delta V_{thi-nbti}}{mv_T}}\right) \tag{20}$$

Under global variations, using Equation (7), Equation (20) becomes the following

$$Ln(IDDQ) = -\frac{At^n + (1 - AS_v t^n)\Delta V_{thi-g}}{mv_T} + Ln\Big(\sum_i^n IDDQi(0)\Big) \tag{21}$$

The mean and standard deviation of circuit leakage are

$$\mu_{Ln(IDDQ)}(t) = \mu_{Ln(IDDQ)}(0) - A \cdot t^n/(mv_T) \tag{22}$$

$$\sigma_{Ln(IDDQ)}(t) = (1 - AS_v t^n)/(mv_T) \cdot \sigma_g \tag{23}$$

where $\mu_{Ln(IDDQ)}(0) = Ln(\sum_i^n IDDQi(0))$

Under local variations, Equation (20) is approximated as

$$\begin{aligned} Ln(IDDQ) &\approx Ln\Big(e^{-\frac{At^n}{mv_T}} e^{-\frac{(1-AS_v t^n)\Delta V_{thi-l}}{(mv_T)\cdot\eta}} \sum_i^n IDDQi(0)\Big) \\ &= -\frac{At^n}{mv_T} + \frac{(1 - AS_v t^n)\Delta V_{thi-l}}{(mv_T)\cdot \eta} + Ln\Big(\sum_i^n IDDQi(0)\Big) \end{aligned} \tag{24}$$

where η has the value between 0 and 1, depending on the circuit structure. The mean and standard deviation of circuit leakage are

$$\mu_{Ln(IDDQ)}(t) = \mu_{Ln(IDDQ)}(0) - A \cdot t^n/(mv_T) \tag{25}$$

$$\sigma_{Ln(IDDQ)}(t) = (1 - AS_v t^n)/(mv_T) \cdot (\sigma_l/\eta) \tag{26}$$

Akin to path timing, logarithmic *IDDQ* has the same time dependence under either global or local variation. Their impact is only different by a factor of η, which is derived from the circuit structure. Fig. 9 shows that the logarithmic mean and the standard deviation degradation of leakage current follow the power law of $t^{1/6}$. The mean is relatively independent of the type of variations, while standard deviation is more sensitive to the global variation.

3. Statistical aging analysis

The analysis above generates the statistics of each gate under the aging effect. In reality, the distributions of logic gates in a circuit are correlated depending on their statistical properties. To obtain the information of path timing degradation, we can incorporate statistical timing analysis techniques to handle the correlation. In this section, we propose a hierarchical method to extract related model parameters and prepare the framework for the integration.

3.1 Statistical static timing analysis flow

Fig. 10 shows the statistical circuit performance analysis flow which is used to predict the circuit performance. This flow incorporates conventional static timing analysis with the statistical properties of the circuit and the process technology. The primary components include:

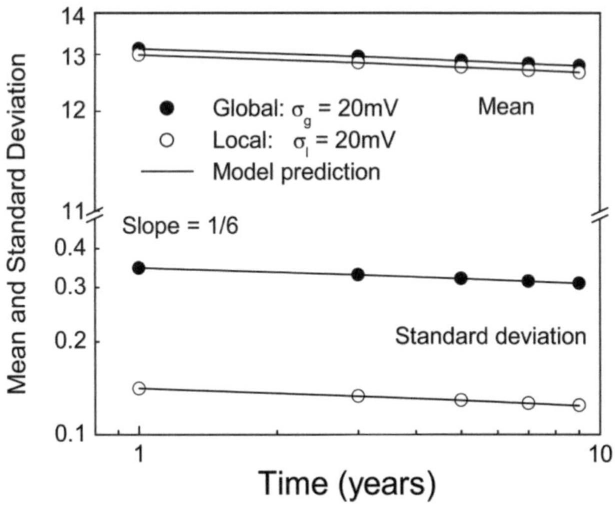

Fig. 9. The mean and standard deviation degradation of leakage.

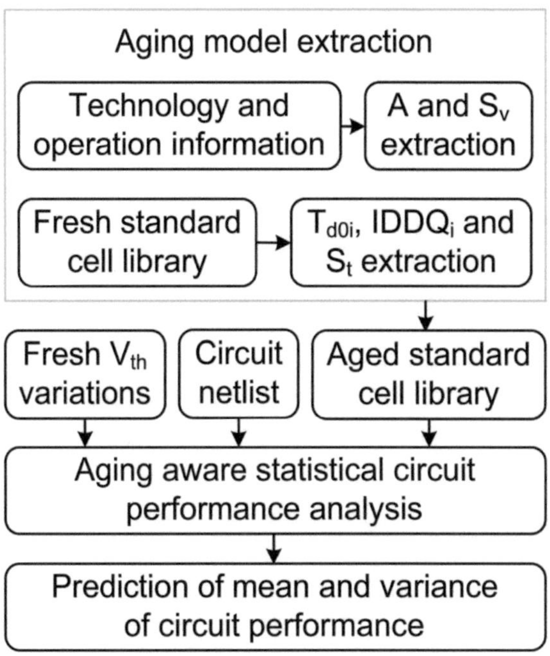

Fig. 10. Statistical circuit performance analysis flow.

1. Model parameter extraction. Given technology and operating condition information, A and S_v are extracted. Given standard cell performance library, T_{d0i}, $IDDQ_i$ and S_t are extracted. For more details about parameter extraction, please refer to (Wang et al., 2007) and Section 3.2.1.
2. Substituting all the parameters into Equations (13), (21) and (24), we obtain the aged standard cell performance library. The cell timing and leakage are functions of transistor global and local V_{th} variations and stress time.

3. Generating two statistical distribution of transistor fresh V_{th} variations, one is for global variation and the other is for local variation. Their means are 0, and standard deviations are determined by the manufacturing process.
4. Given circuit netlist, random assign V_{th} variations from the generated distribution to each transistor. With specified V_{th} variations, the cell timing and leakage of the aged standard cell library are only function of stress time.
5. Given stress time, using circuit performance analyzer with the aged standard cell library, we get the circuit performance degradation. For example, we want to do timing analysis for given circuit. Since the aged library provides each cell timing information, the path timing is obtained by adding up all the individual cell timings in the path. We can do leakage analysis in the same way.
6. For circuit performance distribution prediction, given initial V_{th} variation distribution, Equations (16), (17), (22), (23), (25), (26) directly give the mean and standard derivation of circuit performance distribution.

3.2 Model prediction and silicon validation
3.2.1 Model parameter extraction

In order to accurately predict the circuit performance degradation, there are five parameters need to be characterized at $t = 0$, including: A, S_v, S_t, T_{d0i} and $IDDQi$.

A: Parameter of long term V_{th} degradation under nominal conditions. Its value is extracted from Equations (3) - (5), with the dependence of temperatures, V_{dd}, and switching activity.

S_v: the sensitivity of NBTI-induced transistor degradation to the nominal value of V_{th}.

S_t: the sensitivity of gate delay to PMOS V_{th} shift.

T_{d0i}: nominal gate delay, without V_{th} variation and aging.

$IDDQi$: nominal gate leakage, without V_{th} variation and aging.

Note these parameters are all extracted from the nominal condition, but not affected by process variations. Variations only change the distribution of T_d and $IDDQ$, which can be obtained from the statistics at $t = 0$ (i.e., before the stress). During the stress, the interaction between variability and reliability follows the prediction of our new models. These statistical reliability models improve the predictability in the design stage, avoiding expensive reliability test at the circuit level.

3.2.2 Test chip and measurement

To validate the statistical models for the circuit performance, an inverter ring oscillator was designed as the prototype circuit (Reddy et al., 2002). Fig. 11 shows the simplified schematic of the ring oscillator. The channel length of the transistors in the ring oscillator is drawn at the minimum design rule. There is a NAND gate that allows the RO oscillation enable/disable (OE pin). When the oscillation is disabled, i.e., OE = 0V, IDDQ is measured through V_{RING}.

When the oscillation is enabled, i.e., OE = V_{dd}, the RO oscillates at full speed (several GHz) and Ring Oscillator Frequency (Fosc) is measured periodically without any interruptions. Similar as the on-the-fly measurement at the device level, this dynamic stress measurement is nonstop. In this work, the ROs are stressed under various supply voltage and temperature conditions. The temperature was the same for both stress and measurement and no electrical stress was applied until the temperature was at the proper value.

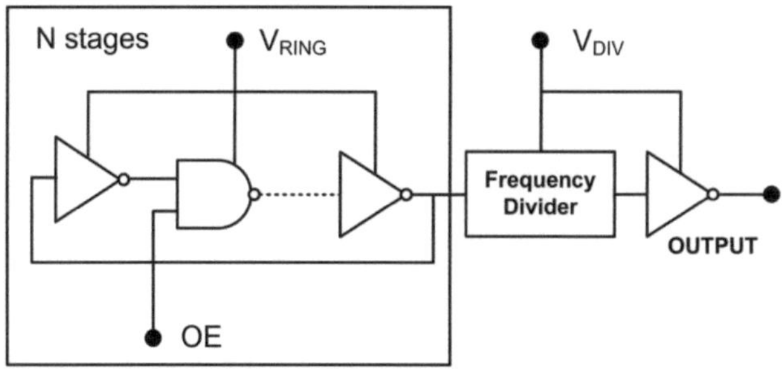

Fig. 11. Simplified schematic of inverter ring oscillator circuit. Oscillation is disabled by grounding the Oscillation Enable (OE) pin.

3.2.3 Model validation with silicon data

The proposed statistical model is verified by 65nm technology available silicon data under a few operating conditions. Fig. 12 shows the delay degradation of ROs. The dots present the mean changes; the error bars are scaled delay distribution of the sample circuits; and the lines are the model predictions. While the mean value increases as $t^{1/6}$, as a signature of the dominance of NBTI effect in circuit aging, the distribution declines with stress time. Fig. 13 evaluates the change of both the active current (*IDDA*), and the leakage current (*IDDQ*). Since *IDDA* ≈*CV*/ *f*, for given switching frequency, *IDDA* is easily extracted. Our predictive models only require the sensitivities of transistor and circuit aging, as well as the statistics before the stress. Then the degradation of circuit performance is fully predicted toward the end of the life time.

3.3 Simulation setup of circuit benchmarks

The statistical aging analysis has been implemented in C to predict the circuit performance degradation of ISCAS85, 89 and ITC99 benchmark circuits (, n.d.; *ITC99 Benchmark*, n.d.). We

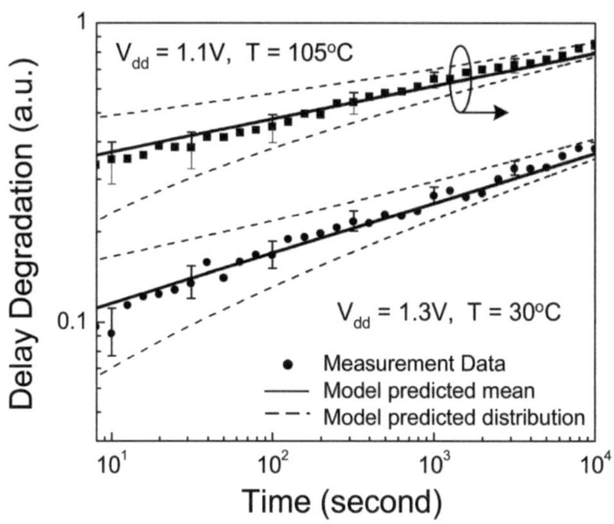

Fig. 12. Delay degradation of ROs under various stress conditions (four selected samples each condition).

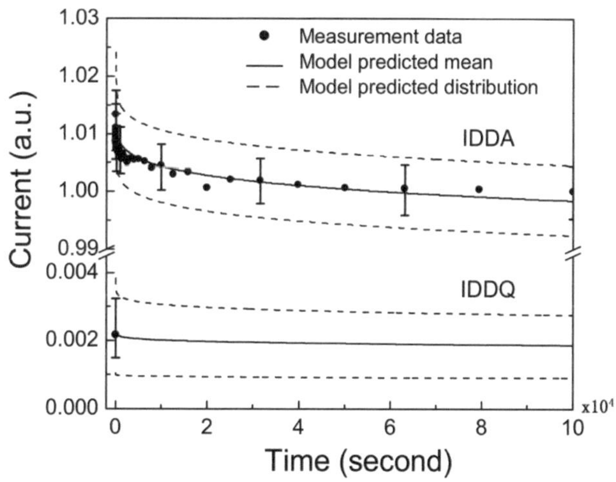

Fig. 13. The prediction of *IDDA* and *IDDQ* degradation.

obtain an aging aware library by running SPICE simulation using PTM 65nm technology (Zhao & Cao, 2006). This library consists of 5 different cells: INVERTER, 2 input NAND, 3 input NAND, 2 input NOR, 3 input NOR. The benchmark circuits in the BLIF format are synthesized by SIS (E. M. Sentovich, K. J. Singh, L. Lavagno, C. Moon, R. Murgai, A. Saldanha, H. Savoj, P. R. Stephan, and R. K. Brayton & Sangiovanni-Vincentelli, 1992) using this standard library. Random local and global variations are generated from a Gaussian distribution with mean 0 and standard deviation which is determined by the manufacturing process. The delay of each gate is calculated at time = 0 (with the variations) and after 10 years (with variations and NBTI degradation) using the aging aware library. Timing analysis is then performed, and the path delay for each path is calculated as the sum of the individual gate delays. After the run, the path with the maximum delay is identified as the critical path in the design.

3.4 Results and discussions

Figures 14 and 15 show the path delay distribution under both NBTI effect and process variation with time increasing. From these two figures, it can be seen that the mean of the path delay increases, while the standard deviation of the path delay decreases. Both of them follows the power law dependence of $t^{1/6}$. As shown in the figure, at t = 0, the delay difference between maximum and minimum is 10.18%, while after 10 years stress, it reduces to 5.29%. The Path delay degradation caused by NBTI effect for 10 years stress is 14.06%, which is more than the delay difference caused by process variations at t = 0. Thus, for the circuit designers, it is critical to consider both NBTI effect and process variation at the early design stage.
Table 1 further shows the simulation results for different benchmark circuits at *Time* = 0 and *Time* = 10 years. For a given circuit, *Path* represents the number of the total path in the circuit. T_{d0} and T_{d10} are the critical path delay of the circuit without process variations at *Time* =0 and *Time* = 10 years, respectively. The *Max*, *Mean* and *Min* columns correspond to the maximum, mean and minimum of the sets of the critical path delay under different simulation iteration. Δ is the delay difference between maximum and minimum. Δ_{nbti} is the NBTI-induce path delay degradation. *m* indicates the margin need to be add during the circuit design stage.

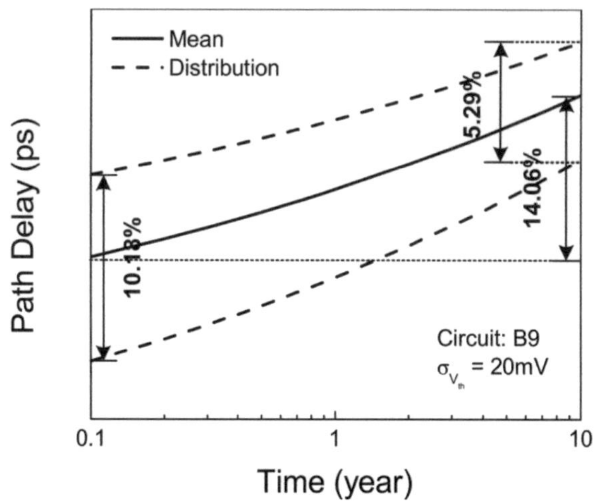

Fig. 14. Path delay mean change under both NBTI effect and process variations.

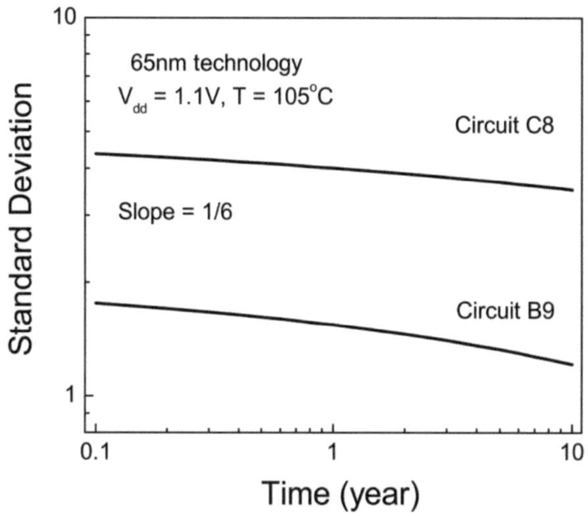

Fig. 15. Path delay standard deviation change under both NBTI effect and process variations.

Based on the simulation results of Table 1, we have three observations.

1. For most circuits, NBTI-induced delay degradation is comparable or larger than the process variations-induced delay difference. For example, circuit K2, after 10 years stress, NBTI-induced delay degradation is 15.65%, while the delay difference caused by process variation at *Time* = 0 is 13.22%. Thus, for circuit designers, it is necessary to add a guardband for NBTI in addition to guardband for process variations.
2. The impact of process variations on circuit delay strongly depends on the circuit structure. For instance, at *Time* = 0, process variation-induced delay difference for circuit Frg2 is 33.09%, while it is 4.97% for circuit Apex6. This effect is seen at *Time* = 10 years also.
3. The mean of the circuit delay increases with time, while the standard deviation decreases. Both of them follow the NBTI power law of $t^{1/6}$ and are independent of process variations (Fig. 14 and Fig. 15).

Circuit	Path	*Time* = 0(*ps*)					*Time* = 10*years*(*ps*)					$\Delta_{nbti}(\%)$	$m(\%)$
		$\sigma_l = 0$	$\sigma_l = 20mV$				$\sigma_l = 0$	$\sigma_l = 20mV$					
		T_{d0}	*Max*	*Mean*	*Min*	$\Delta(\%)$	T_{d10}	*Max*	*Mean*	*Min*	$\Delta(\%)$		
k2	10404	44458.02	47136.9	44435.39	41261.71	13.22	51416.18	52806.87	51404.44	49756.88	6.86	15.65	18.78
des	99845	34743.06	36887.09	34850.87	33244.59	10.48	39558.86	40568.06	39552.43	38624.65	5.59	13.86	16.77
b9	206	144.51	152.49	145.45	137.76	10.18	164.94	169.03	165.29	161.39	5.29	14.06	16.89
c8	201	227.79	240.24	227.78	216.57	10.39	263.04	269.48	263.02	256.75	5.59	15.47	18.30
comp	2016	4349.09	4452.62	4347.67	4326.27	2.91	4960.90	5014.65	4960.15	4902.26	2.58	14.07	15.30
frg2	4792	1438.80	1859.83	[illegible]	1383.74	33.09	1652.40	2006.68	1897.89	1661.26	24.01	14.85	39.47
term1	649	1710.81	1786.63	[illegible]	1630.49	9.13	1968.15	2007.51	1968.98	1926.45	4.74	15.04	17.34
apex6	1499	991.08	1016.69	991.31	967.41	4.97	1136.71	1150.01	1136.82	1124.43	2.58	14.69	16.04
count	368	827.46	855.96	827.62	799.14	6.87	948.16	962.96	948.25	933.46	3.56	14.59	16.38
example2	1307	443.47	467.57	445.37	428.27	8.86	505.71	518.22	506.62	497.83	4.60	14.04	16.86
f51m	174	342.17	359.84	342.11	324.35	10.37	391.52	400.69	391.49	382.26	5.38	14.42	17.10
frg1	127	785.78	805.38	786.00	763.64	5.31	900.42	910.59	900.50	888.93	2.76	14.59	15.89
lal	188	160.47	176.76	160.92	149.43	17.03	185.74	194.19	185.80	178.34	9.88	15.74	21.01
ldd	447	619.40	638.88	620.29	598.31	6.55	709.44	719.53	709.90	698.49	3.40	14.54	16.17
mux	189	1074.41	1112.05	1075.01	1033.14	7.34	1236.53	1256.07	1236.84	1215.11	3.81	15.09	16.91

Table 1. Simulation results for ISCAS and ITC99 circuit benchmark

By integrating the gate-based dynamic stress bound and process variation prediction model into circuit timing analysis, we verify that our hierarchical method provide a safe and tight bound of circuit timing degradation. Fig. 16 is the netlist of benchmark circuit C17. Fig. 17 shows the delay distribution of circuit C17 under various input vectors and process variations. As shown in the figure, the proposed method provides a safe and tight bound in the estimation of the circuit timing degradation, which helps to improve design predictability and avoid pessimistic guardbanding under NBTI effect.

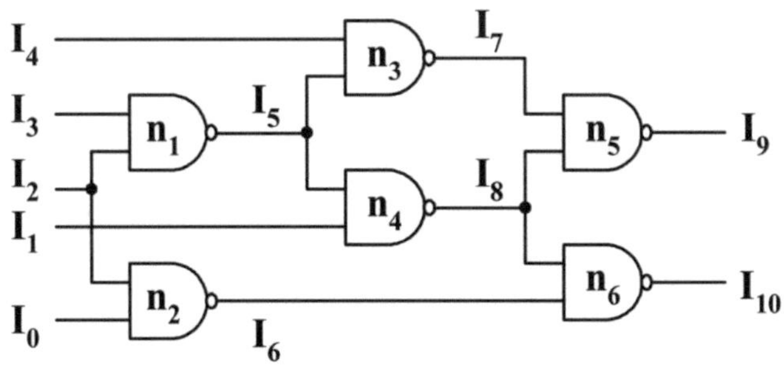

Fig. 16. Circuit C17 netlist.

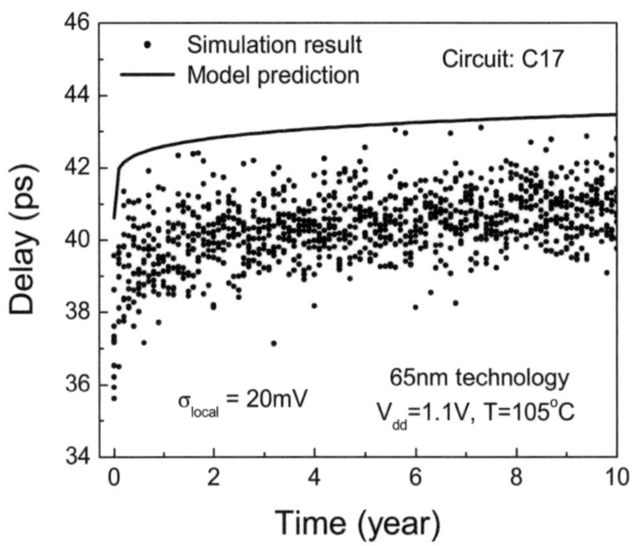

Fig. 17. Delay distribution of circuit C17 with various input vectors and process variations.

3.5 Impact of statistical variations on SRAM

The six transistor SRAM (6T SRAM) design is highly sensitive to process variations, especially local random variations. The mismatch between neighboring transistors reduces the cell Static Noise Margin (SNM) (Krishnan et al., 2006; Lin et al., 2006; Rosa et al., 2006). The impact of statistical variations on SRAM SNM is mainly divided into Read, Write, and Hold stability. Under the NBTI effect, a weaker PMOS transistor increases Read failures, but improves the Write operation (Krishnan et al., 2006; Lin et al., 2006). Fig. 18 shows the PDF of 6T SRAM Read noise margin during the aging. In the SRAM cell, only one side of PMOS is stressed (i.e., with the gate biased at the ground), and the PMOS in the other side remains in the recovery. The mismatch in the stress mode reduces Read noise margin on one side, with no impact on the other side. As shown in Fig. 18, the mean value of Read noise margin decreases with longer aging time. Since Read SNM is determined only by the stressed PMOS side, NBTI induced SNM degradation is not cancelled out between stressed PMOS side and unstressed PMOS side. Therefore, the tail of Read SNM is determined by the stressed PMOS side. This behavior is different from the aging effect in a logic path, where random variability in each stage is averaged in the path timing. Further studies on statistical aging modeling are needed to predict the nonlinear behavior in SRAM.

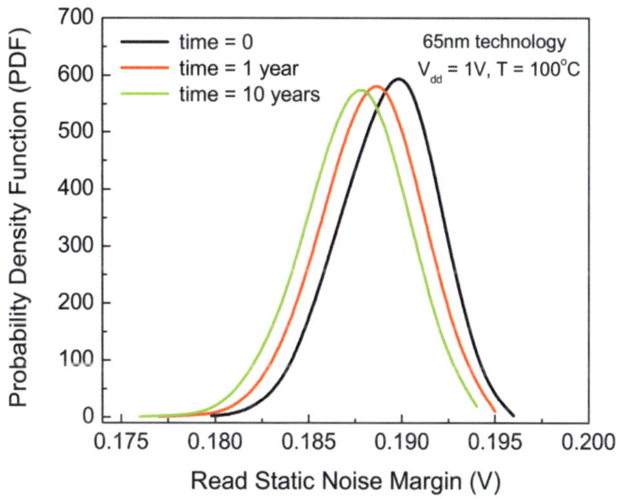

Fig. 18. The probability density function of SRAM read noise margin during aging.

4. Conclusions

In this work, a statistical methodology is developed to predict circuit performance degradation under both NBTI effect and process variations. These analytical solutions reveal that the degradation rate and its standard deviation are independent on the type and the amount of process variations. In order to predict the mean and the variance of circuit aging, only the characteristics of transistor degradation and circuit performance sensitivity to aged parameters are required. The aging of circuit speed and *IDDQ* shows a power law dependence on the stress time, as an evidence of the dominance of NBTI effect. The proposed method is implemented into SPICE simulation and timing analysis tools. With verification with 65nm silicon data, it supports statistical aging analysis in standard design flow, improving design predictability and helping avoid pessimistic guardbanding under the increasingly severe aging effect. We are collecting a larger volume of statistical data to further verify these conclusions.

5. Acknowledgements

This project is supported by SRC, Task 1609. It is also partially supported by Gigascale System Research Center (GSRC), one of five research centers funded under the Focus Center Research Program, a Semiconductor Research Corporation Program.

6. References

(n.d.). *Collaborative Benchmarking and Experimental Algorithmics Lab.* http://www.cbl.ncsu.edu/.

Agarwal, M., Balakrishnan, V., Bhuyan, A., Paul, B. C., Wang, W., Yang, B., Yu, C. & Mitra, S. (2008). Optimized circuit failure prediction for aging: Practicality and promise, *IEEE International Test Conference* pp. 1–10.

Alam, M. A., Kufluoglu, H., Varghese, D. & Mahapatra, S. (2007). A comprehensive model for pmos nbti degradation: Recent progress, *Microelectronics Reliability* Vol. 47(No. 6): 853–862.

Alam, M. A. & Mahapatra, S. (2005). A comprehensive model of pmos nbti degradation, *Microelectronics Reliability* Vol. 45: 71–81.

Bhardwaj, S., Wang, W., Vattikonda, R., Cao, Y. & Vrudhula, S. (2006). Predictive modeling of the nbti effect for reliable design, *IEEE Custom Integrated Circuits Conference* pp. 189- 192.

Boning, D. & Nassif, S. (2001). *Chapter 6: Models of process variations in device an interconnect in Book Design of High-Performance Microprocessor Circuits*, IEEE Press.

Borkar, S. (2006). Electronics beyond nano-scale CMOS, *ACM/IEEE Design Automation Conference* pp. 807–808.

Borkar, S., Karnik, T., Narendra, S., Tschanz, J., Keshavarzi, A. & De, V. (2003). Parameter variations and impact on circuits and microarchitecture, *ACM/IEEE Design Automation Conference* pp. 338–342.

E. M. Sentovich, K. J. Singh, L. Lavagno, C. Moon, R. Murgai, A. Saldanha, H. Savoj, P. R. Stephan, and R. K. Brayton & Sangiovanni-Vincentelli, A. (1992). SIS: A system for sequential circuit synthesis, *Technical report.* URL: *citeseer.ist.psu.edu/sentovich92sis.html*

Huard, V., Parthasarathy, C. R., Guerin, C. & Mammase, M. (2006). Physical modeling of negative bias temperature instabilities for predictive extrapolation, *IEEE International Reliability Physics Symposium Proceedings* pp. 733–734.

International Technology Roadmap for Semiconductors (2008). http://www.itrs.net/Links/2008ITRS/Home2008.htm.

ITC99 Benchmark (n.d.). http://www.cerc.utexas.edu/itc99-benchmarks/bench. html.

Kang, K., Park, S. P., Roy, K. & Alam, M. A. (2007). Estimation of statistical variation in temporal nbti degradation and its impact on lifetime circuit performance, *IEEE/ACM International Conference on Computer-Aided Design* pp. 730–734.

Kimizuka, N., Yamamoto, T., Mogami, T., Yamaguchi, K., Imai, K. & Horiuchi, T. (1999). The impact of bias temperature instability for direct-tunneling ultra-thin gate oxide on mosfet scaling, *VLSI Symposium on Technology* pp. 73–74.

Krishnan, A. T., Chancellor, C., Chakravarthi, S., Nicollian, P. E., Reddy, V., Varghese, A., Khamankar, R. B., Krishnan, S. & Levitov, L. (2005). Material dependence of hydrogen diffusion: Implications for nbti degradation, *IEEE International Electron Devices Meeting* pp. 688–691.

Krishnan, A. T., Reddy, V., Aldrich, D., Raval, J., Christensen, K., Rosal, J., O'Brien, C., Khamankar, R., Marshall, A., Loh, W.-K., McKee, R. & Krishnan, S. (2006). Sram cell static noise margin and v_{MIN} sensitivity to transistor degradation, *IEEE International Electron Devices Meeting* pp. 1–4.

Kumar, S. V., Kim, C. H. & Sapatnekar, S. S. (2006). An analytical model for negative bias temperature instability, *International Conference on Comuter-Aided Design* pp. 493–496.

Lin, J. C., Oates, A. S., Tseng, H. C., Liao, Y. P., Chung, T. H., Huang, K. C., Tong, P. Y., Yau, S. H. & Wang, Y. F. (2006). Prediction and control of nbti - induced sram v_{ccmin} drift, *IEEE International Electron Devices Meeting* pp. 1–4.

Lin, Z., Spanos, C., Milor, L. & Lin, Y. (1998). Circuit sensitivity to interconnect variation, *IEEE Trans. on Semiconductor Manufacturing* Vol. 11: 557–568.

Liu, W. Z. F., Agarwal, K., Acharyya, D., Nassif, S., Nowka, K. & Cao, Y. (2007). Rigorous extraction of process variations for 65nm cmos design, *European Solid-State Circuits Conference* pp. 89–92.

Nassif, S. R. (2001). Modeling and analysis of manufacturing variations, *IEEE Custom Integrated Circuits Conference* pp. 223–228.

Ogawa, E. T., Kim, J., Haase, G. S., Mogul, H. C. & McPherson, J. W. (2003). Leakage, breakdown and tddb characteristics of porous low-k silica, *IEEE International Reliability Physics Symposium* pp. 166–172.

Parthasarathy, C. R., Denais, M., Huard, V., Ribes, G., Vincent, E. & Bravaix, A. (2006). New insights into recovery characteristics post nbti stress, *IEEE International Reliability Physics Symposium Proceedings* pp. 471–477.

Paul, B. C., Kang, K., Kufluoglu, H., Alam, M. A. & Roy, K. (2006). Temporal performance degradation under NBTI: Estimation and design for improved reliability of nanoscale circuits, *ACM/IEEE Design, Automation, and Test Europe* pp. 780–785.

Rangan, S., Mielke, N.&Yeh, E. C. C. (2003). Universal recovery behavior of negative bias temperature instability [pmosfets], *IEEE International Electron Devices Meeting* pp. 14.3.1-14.3.4.

Rauch, S. E. (2002). The statistics of nbti induced vt and β mismatch shifts in pmosfets, *IEEE Trans. on Device Material Reliability* pp. 89–93.

Rauch, S. E. (2007). Review and reexamination o freliability effects related to nbti-indued statistical variations, *IEEE Transactions on Device and Materials Reliability* Vol. 7(No. 4): 524–530.

Reddy, V., Krishnan, A. T., Marshall, A., Rodriguez, J., Natarajan, S., Rost, T. & Krishnan, S. (2002). Impact of negative bias temperature instability on digital circuit reliability, *IEEE International Reliability Physics Symposium* pp. 248–254.

Reisinger, H., Blank, O., Heinrigs,W., Muhlhoff, A., Gustin,W. & Schlunder, C. (2006). Analysis of nbti degradation- and recovery- behavior based on ultra fast vt-measurements, *IEEE International Reliability Physics Symposium Proceedings* pp. 448–453.

Rosa, G. L., Ng, W. L., Rauch, S., Wong, R. & Sudijono, J. (2006). Impact of nbti induced statistical variation to sram cell stability, *IEEE International Reliability Physics Symposium Proceedings* pp. 274–282.

Sakurai, T. & Newton, A. R. (1990). Alpha-power law mosfet model and its application to cmos logics, *IEEE Journal of Solid-State Circuits* Vol. 25(No. 2): 584–594.

Schroder, D. K. & Babcock, J. A. (2003). Negative bias temperature instability: Road to cross in deep submicron silicon semiconductor manufacturing, *Journal of Applied Physics* Vol. 94(No. 1): 1–18.

Shen, C., Li, M. F., Foo, C. E., Yang, T., Huang, D. M., Yap, A., Samudra, G. S. & Yeo, Y. C. (2006). Characterization and physical origin of fast v_{th} transient in nbti of pmosfets with sion dielectric, *IEEE International Electron Devices Meeting* pp. 12.5.1–.

Vattikonda, R.,Wang,W. & Cao, Y. (2006). Modeling and minimization of pmos nbti effect for robust nanometer design, *ACM/IEEE Design Automation Conference* pp. 1047–1052.

Wang, W., Reddy, V., Krishnan, A. T., Vattikonda, R., Krishnan, S. & Cao, Y. (2007). Compact modeling and simulation of circuit reliability for 65nm cmos technology, *IEEE Transactions on Device and Materials Reliability* Vol. 7(No. 4): 509–517.

Zhao, W. & Cao, Y. (2006). New generation of predictive technology model for sub-45nm early design explorations, *IEEE Tran. on Electron Devices* Vol. 53(No. 11): 2816–2823. http://www.eas.asu.edu/$\sim$ptm.

6

Standby Supply Voltage Minimization for Reliable Nanoscale SRAMs

Jiajing Wang and Benton H. Calhoun
University of Virginia
United States

1. Introduction

Increased leakage current and device variability are posing major challenges to CMOS circuit designs in deeply scaled technologies. Static Random Accessed Memory (SRAM) has been and continues to be the largest component in embedded digital systems or Systems-on-Chip (SoCs). It is expected to occupy over 90% of the area of SoC by 2013 (Nakagome et al., 2003). As a result, SRAM is more vulnerable to those challenges. To effectively reduce SRAM leakage and/or active power, supply voltage (V_{DD}) is often scaled down during standby operation (e.g. (Qin et al., 2004; Flautner et al., 2002; Bhavnagarwala et al., 2004; Wang et al., 2007)) and/or active operation (e.g. (Morita et al., 2006; Joshi et al., 2007)). For ultra-low-energy applications, SRAMs operating with V_{DD} near/below the threshold voltage (V_T) are also proposed (e.g. (Calhoun & Chandrakasan, 2007; Verma & Chandrakasan, 2008)). However, all SRAM functions, including read stability, write ability, access performance, and hold stability, are less reliable at lower voltage, which leads to the reduction of yield. The minimum supply voltage (Vmin) is limited by the lowest acceptable yield and determines the maximum achievable power reduction. Applying an underestimated Vmin will cause intolerable failures and decrease SRAM yield. On the other hand, applying an overestimated Vmin will waste power and energy. However, finding the optimum Vmin becomes difficult in the presence of global and local variations.

In this chapter, we particularly explore SRAM Vmin during standby mode, i.e. data retention voltage (DRV). We first analyze the impacts of local/random and global/systematic variations on DRV, and then present new statistical and adaptive design methods to address those impacts. The goal of this chapter is to develop effective methods for achieving the best leakage power savings while maintaining the desired yield under variations.

2. Variation impact on data retention voltage

2.1 Data Retention Voltage (DRV)

Fig. 1 shows the structure of the conventional 6T SRAM cell. The cell consists of two cross-coupled inverters ((PL,NL) & (PR,NR)) and the pass-gate transistor XL/XR on each side. Q and QB are the internal nodes storing the data. During standby mode, the WL signal remains low. BL/BLB signals are often precharged to either high or low. Although floating bitline is also proposed to further reduce BL leakage current (Wang et al., 2007), we assume that the BLs remain high in this chapter. Fig. 1 also illustrates the paths of the major leakage

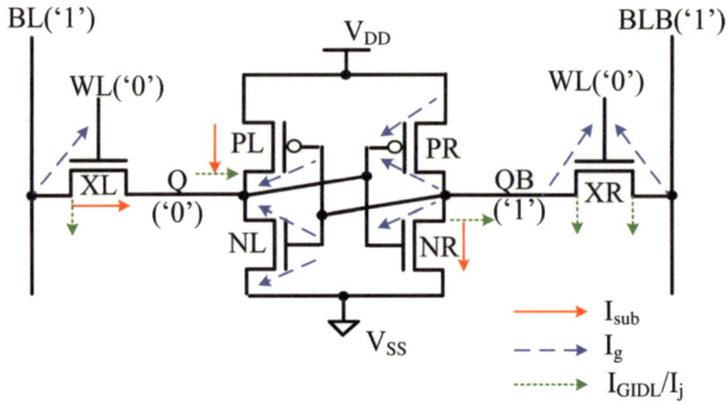

Fig. 1. 6T SRAM cell and the path of the major leakage currents.

current components during standby mode for nanometer technologies. They are sub-threshold leakage current (I_{sub}), gate leakage current (I_g), gate induced drain leakage (I_{GIDL}), and junction leakage current (I_j). I_{sub} is the drain-to-source current when the transistor operates in weak inversion. It decreases exponentially with the reduction of the drain-to-source voltage (V_{DS}) due to the drain induced barrier lowering (DIBL) effect (Ferre & Figueras, 2005). I_g is the direct tunneling current through the gate oxide to the channel as well as to the overlap region between gate and source/drain extension. Since it grows exponentially with the scaling of the gate oxide thickness, I_g becomes the dominant leakage source for CMOS technologies beyond 45nm. Recent new high-k metal gate device option provides large reduction in gate leakage (Mistry et al., 2007). In addition, a lower V_{DD} exponentially reduces I_g. I_{GIDL} is caused by the high electric field under the gate-to-drain overlap region, and I_j is caused by the reverse-biased pn junction (Roy et al., 2003). Both I_{GIDL} and I_j also decrease dramatically with V_{DD}. Therefore, V_{DD} scaling can effectively reduce the total cell leakage current, $I_{lk,total}$. Fig. 2 shows that $I_{lk,total}$ can be reduced by more than 10× for a cell in 45nm. Due to the direct effect of V_{DD}, the cell leakage power, which is equal to $I_{lk,total} \cdot V_{DD}$, can be further reduced with a lower V_{DD}.

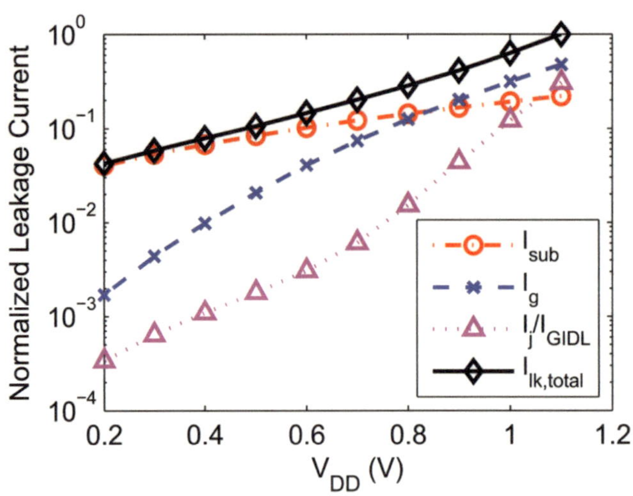

Fig. 2. The normalized cell leakage current versus V_{DD}.

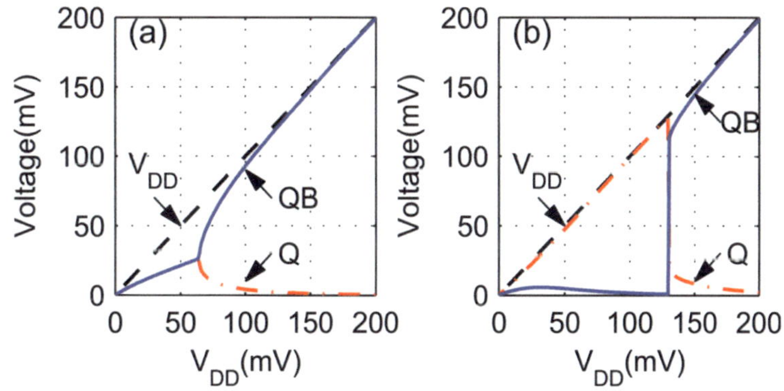

Fig. 3. The voltage of the storage nodes against V_{DD} for (a) a balanced cell and (b) a imbalanced cell (© 2007 IEEE).

However, the drawback of a scaled V_{DD} is the degradation of the cell stability. Fig. 3 shows that excessive V_{DD} scaling results in the loss of the stored data ('0' in this example). Fig. 3(a) particularly shows the balanced case when there is no mismatch between the transistors on the left side (PL/NL/XL in Fig. 1) and those on the right side (PR/NR/XR in Fig. 1). Q and QB converge to a metastable point as a result of the degraded gain. Fig. 3(b) shows the other case when the cell is imbalanced by some mismatch in V_T. In this case, Q and QB flip to the more stable state ('1' here). The data retention voltage (DRV) defines the minimum V_{DD} below which the SRAM cell can not preserve its data (Qin et al, 2004). So DRV is the fundamental limiter of the lower V_{DD} operation and prohibits additional power savings. We define DRV0 and DRV1 as the minimum V_{DD} for preserving '0' and '1' respectively. For the balanced case as in Fig. 3(a), DRV0=DRV1; for the imbalanced case, one increases while the other decreases (e.g. DRV0>>DRV1 for the example in Fig. 3(b)). To ensure the cell can safely hold both '0' and '1', the actual DRV is the maximum value of DRV0 and DRV1. Fig. 3 thereby implies that DRV increases when any mismatch occurs.

Unfortunately, device variability increases with technology scaling. In order to predict the maximum achievable power savings from lowering V_{DD}, we must evaluate the impact of device variability on DRV. All the variations can be categorized into two groups: *global/systematic* variation and *local/random* variation. Global variations influence all the transistors on the chip. On the other hand, local variations have a different effect on individual transistors, and thus cause mismatch between adjacent devices. Next, we will examine the impact of these variations on DRV.

2.2 Impact of local/random variation

Variations occur in a variety of physical parameters, mainly including the threshold voltage (V_T), the gate oxide thickness (Tox), the channel effect length (L_{eff}), and the channel effect width (W_{eff}). Among these parameters, DRV is most sensitive to V_T (Qin et al., 2004). In addition, the variation of L_{eff} can cause V_T variation due to the short channel effect. Therefore, we mainly consider the impact of V_T variation on DRV. Random doping fluctuation (RDF) is the dominant source of local V_T variation, and it deteriorates with continuous device scaling. The RDF induced random V_T variation can be modeled as a normal distribution with its standard deviation (σ_{V_T}) inversely proportional to the square root of the channel area as below (Asenov et al., 2003).

$$\sigma_{V_T} \propto \frac{1}{\sqrt{W_{eff}L_{eff}}} \quad (1)$$

SRAM cells commonly use transistors with smaller geometry for higher density. Thus they are naturally more susceptible to random variations due to a larger value of σ_{V_T}.
Given the statistics of parametric variations, we can use Monte Carlo (MC) simulation to investigate the impact of variations on the figure of merit. Fig. 4 is the histogram of the cell DRV values with a 5000-point MC simulation in a commercial 90nm CMOS process. The DRV exhibits a non-Gaussian distribution with a longer tail on the right side. The tail value of the distribution is the lowest supply voltage that can be applied to the whole SRAM array without losing any data. We call it the standby Vmin for an SRAM. Vmin determines the maximum achievable power reduction for the entire SRAM array. Therefore, the estimation of the tail value becomes crucial. Modern SRAMs often contain millions of cells, thus the tail event only occurs once out of millions of cell simulations. For such a rare event, the Monte Carlo method requires at least millions of runs, thereby becoming prohibitively expensive. To speed up the estimation of these rare events, various methods arise and fall into the following two major categories.

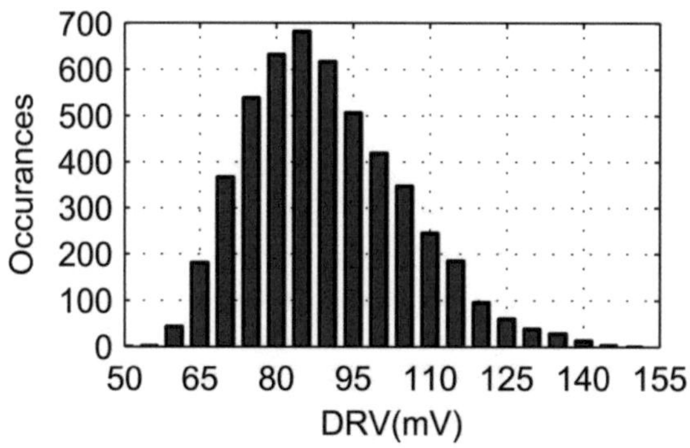

Fig. 4. The histogram of DRV from Monte Carlo simulation with 5000 samples (© 2007 IEEE).

- *Non-Monte-Carlo (non-MC) methods*
  The first non-MC method is to develop a comprehensive analytical model. Although Qin et al. (2004) proposed a theoretical model to approximate the DRV of a single cell, they did not address the statistical characteristics of DRV. The question of how variations impact the long tail of the DRV distribution is not answered. The second and more generic non-MC method is the boundary searching approach, which intends to find the boundaries in the parameter space that correspond to success/failure of the circuit without using MC sampling (Gu & Roychowdhury, 2008). The authors demonstrated its efficiency for estimating SRAM read access yield when considering only two major design parameters. However, the real access yield is also determined by other design parameters that have a minor impact on read access. When all the parameters are searched, this method becomes quite expensive.
- *Improved Monte-Carlo (MC) methods*
  The huge expense of MC for rare event estimation is mainly due to the inefficiency of the rare event sampling. Importance sampling (Kanj et al., 2006) and the Statistical Blockade (SB) tool (Singhee & Rutenbar, 2007) are two interesting techniques to hasten the generation of the rare events. However, their efficiency highly relies on the

goodness of the sampling distribution and the tail filter respectively. Extrapolation is an alternative way to avoid a full MC simulation. We can run a relatively small number of samples and fit them into a known distribution. After that, we can quickly acquire the estimates in the extreme tail region by simply calculating with the fitting distribution. Although it is much simpler, its accuracy is dependent on how good the fitting distribution is. For non-Gaussian variables like DRV, it is hard to find a proper known distribution that can well fit the skewed tail region. Fitting a normal and log-normal distribution either underestimates or overestimates the tail values, respectively. The SB tool proposes to use the generalized Pareto distribution (GPD) to particularly fit the tail samples. Its accuracy is dependent on the number of tail samples, which also requires fast Monte Carlo methods like the tail filter in the SB tool to accelerate its generation.

In this chapter, we propose a new fast method to predict the tail of the DRV distribution. We use the extrapolation method so that only a small number of Monte Carlo samples is required. High accuracy is achieved by using a dedicated statistical model for DRV (Wang, Singhee et al., 2007). We will describe the details of this method in section 3.

2.3 Impact of global/systematic variation

Global variations include manufacturing related process variations, voltage supply fluctuations, and temperature changes (i.e. PVT variations). We assume the temperature range is [0°C, 105°C] and the voltage fluctuation range is [-25mV, 25mV]. Fig. 5 shows the DRV histogram of a 5-Kb SRAM array at three PVT cases: typical, best-case, and worst-case. The typical case is at the TT (typical-N and typical-P) process corner, 25°C, and zero voltage fluctuation; the best case for the technology we use is at the SS (Slow-N and slow-P) process corner, 0°C, and 25mV voltage fluctuation; the worst case happens at the FS (Fast-N and slow-P) process corner, 105°C, and -25mV voltage fluctuation. Under one PVT scenario, local variations spread the DRV of the cells, and the tail of the distribution (marked with circle) determines the standby Vmin for this global condition. In contrast, global variations predominantly move the entire DRV distribution around, so the tail point, i.e. the standby Vmin, also shifts with global effects. For this 90nm process, the worst-case Vmin ($Vmin_{wc}$) is about 100mV and 140mV higher than the typical case Vmin ($Vmin_{typ}$) and the best-case Vmin ($Vmin_{bc}$) respectively. For more advanced processes, the variability of global effects might increase and result in a larger difference between $Vmin_{wc}$ and $Vmin_{typ}$/$Vmin_{bc}$. To ensure data safety under all the conditions, we must address this Vmin variability.

The most straight forward method is the worst case approach, which uses a standby V_{DD} based on the worst case at design time and even adds some guard-band for more robustness. For instance, authors of the drowsy cache set the standby V_{DD} 50% higher than the threshold voltage despite the fact that the actual DRV can be much smaller (Kim et al., 2004). A processor with a drowsy mode is also implemented by collapsing the supply voltage well above that required to upset the logic states during standby mode (Clark et al., 2004). Although this open-loop worst-case approach is very robust, it can potentially waste substantial power because of two reasons. First, the worst PVT scenario only occurs in extreme conditions like extremely high temperature, which is very rare for most of the applications. Second, the difference of the Vmin values between the worst case and the non-worst cases can be quite large, and it even becomes larger as CMOS technology continuously scales. We can expect that the conservative worst-case approach would sacrifice more power savings for future CMOS technologies.

In order to gain optimum power reduction for non-worst-case conditions, we propose a closed-loop standby V_{DD} scaling system with online replica cells as monitors for tracking

PVT variations (Wang & Calhoun, 2008). Section 4 will present the details of this new approach.

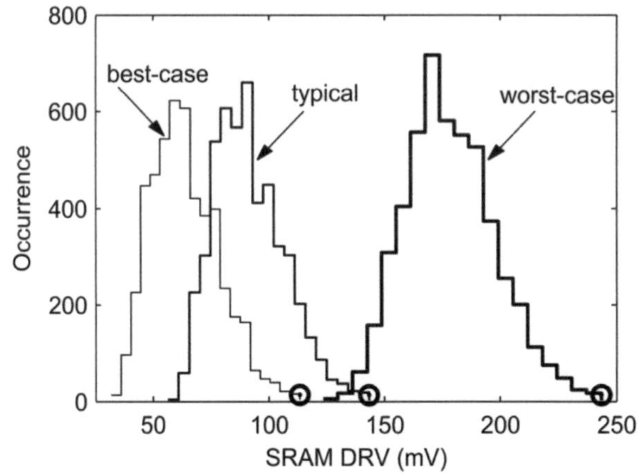

Fig. 5. DRV distribution of a 5Kb SRAM array with global PVT variations and local variations. Three PVT cases (typical, best-case, and worst-case) are shown (© 2008 IEEE)

3. Fast and accurate estimation of standby Vmin

In this section, we propose a fast method to predict standby Vmin, i.e. the tail of the DRV distribution in the presence of random variations. Let us define $P_{cf}(v)$ as the probability that the cell fails when $V_{DD}=v$ during standby. We can compute $P_{cf}(v)$ in two ways. First, in terms of DRV, since DRV is the minimum V_{DD} below which a cell cannot preserve its data we can compute $P_{cf}(v)$ as

$$P_{cf}(v) = P(DRV > v) = 1 - F_{DRV}(v) \tag{2}$$

where F_{DRV} is the cumulative density function (cdf) of DRV. We can also compute $P_{cf}(v)$ in terms of static noise margin (SNM), which is the conventional metric for cell stability. A cell fails at voltage v when its SNM is less than the lowest acceptable noise margin s (e.g. s=0 in a noiseless system), so we can also compute $P_{cf}(v)$ as

$$P_{cf}(v) = P(SNM_v < s) = F_{SNM_v}(s) \tag{3}$$

where SNM_v is the cell's SNM at $V_{DD}=v$ and F_{SNM_v} is the cdf of SNM_v. As we observed in Fig. 4, DRV has a non-Gaussian distribution with a heavy tail on the right side, which makes it hard to directly fit the DRV data into a known distribution. Nevertheless, because of the equivalence of (2) and (3), we can obtain F_{DRV} through the simple transformation of F_{SNM_v} by

$$F_{DRV}(v) = 1 - F_{SNM_v}(s) \tag{4}$$

As we will show in the next section, it is much easier to obtain F_{SNM_v}. Thus we can derive the cdf of DRV from SNM and finally derive the inverse cdf or the quantile function of DRV.

3.1 Statistics of hold static noise margin

The most popular metric for SRAM noise margin is the butterfly curve based SNM, which is the maximum amount of dc voltage noise that a cell can tolerate (Seevinck et al., 1987) and is equivalent to the largest square that can be embedded with the two butterfly curves as shown in Fig. 6. Particularly, the largest square inside the upper-left lobe is defined as SNMH, the SNM for holding '0'; and the largest square inside the lower-right lobe is defined as SNML, the SNM for holding '1'. The true SNM is the minimum of SNMH and SNML. Fig. 6 further shows how SNMH and SNML change with V_{DD} scaling. In the case that the cell is balanced as in Fig. 6(a), both SNMH and SNML decrease to 0 when V_{DD}=65mV. This implies that DRV=DRV0=DRV1=65mV. On the other hand, if the cell is imbalanced by variation as the example in Fig. 6(b), SNMH first drops to 0 while SNML still maintains a positive amount of value when V_{DD}=130mV. Therefore, for this example, DRV=DRV0=130mV. In fact, Fig. 6 uses the same examples as Fig. 3. The same DRV results are obtained by directly simulating the collapse of the internal states as in Fig. 3 and by simulating the decrease of SNM with V_{DD} scaling as in Fig. 6. This verifies that we can use SNM to explore DRV.

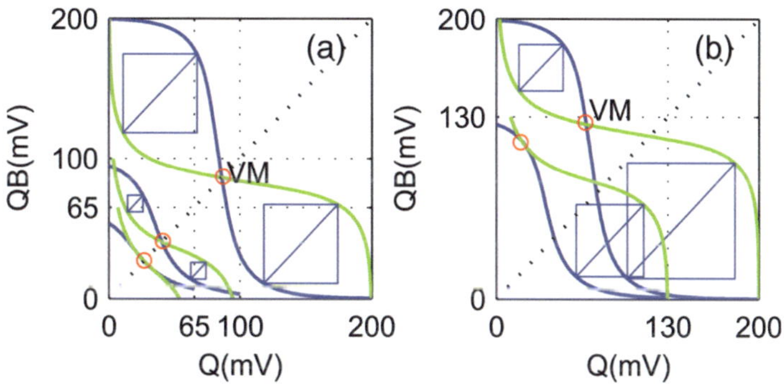

Fig. 6. Butterfly curve based SNM changes with V_{DD} scaling when the cell is (a) balanced and (b) imbalanced by some mismatch (© 2007 IEEE)

The next question we should answer is how local random variations impact SNMH or SNML. Fig. 7 plots the 50,000-point MC simulation results of SNMH and SNML when V_{DD}=300mV. We fit a normal distribution to the data of both SNMH and SNML. The normal distribution closely matches the body of both data. The deviation in the tail points is mainly caused by the error of Monte Carlo simulation, which decreases as we use more Monte Carlo samples. Therefore, it is accurate to approximate the true SNMH and SNML with an identical normal distribution.

Since DRV is the V_{DD} point when SNM is equal to the lowest noise margin (e.g. 0 here), a more important question is how those SNM distributions change with V_{DD} scaling. We further examine the SNMH or SNML distribution at different V_{DD} points. We find that SNMH and SNML remain normally distributed. Moreover, as shown in Fig. 8, the mean (μ) is approximately linear with V_{DD} while the standard deviation (σ) keeps almost constant. If we know that the estimation of the mean and the standard deviation at an initial voltage, v_0, are μ_0 and σ_0, we can quickly obtain the new mean and standard deviation values at any arbitrary V_{DD} point, v, with

$$\mu = \mu_0 + k(v - v_0); \quad \sigma = \sigma_0 \tag{5}$$

where k is the sensitivity of μ to V_{DD} and can be extracted by fitting the mean data in Fig. 8 to the linear curve.

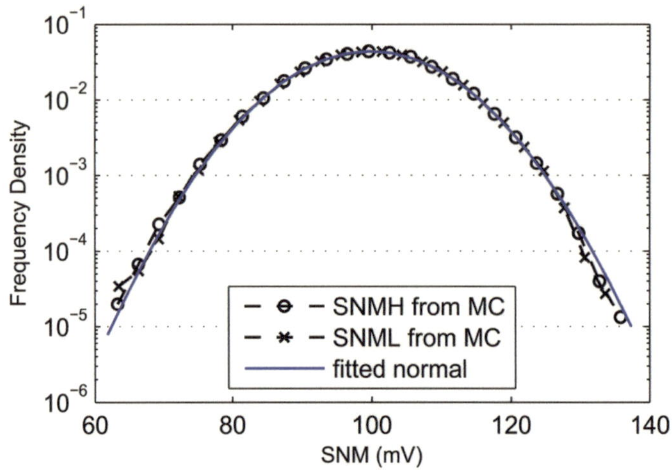

Fig. 7. 50,000-point Monte Carlo results of SNMH and SNML at V_{DD}=300mV and a normal distribution is fitted to both data.

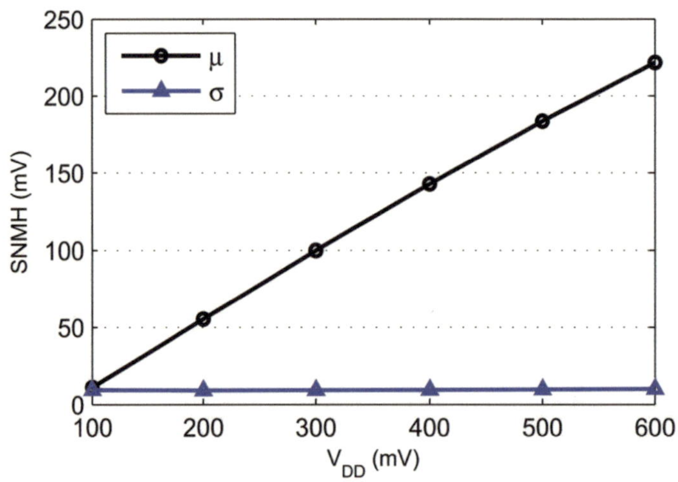

Fig. 8. Estimated mean and standard deviation of SNMH from MC simulations versus V_{DD}.

3.2 DRV and yield model

So far we are able to predict the distribution of SNMH or SNML at any V_{DD} point with (5). The real SNM is the minimum of SNMH and SNML. If we assume SNMH and SNML are independent random variables, according to order statistics, the cumulative density function of the real SNM can be calculated as follows.

$$\begin{aligned} F_{\mathrm{SNM}_v}(s) &= \mathrm{P}(\mathrm{SNM}_v < s) \\ &= \mathrm{P}(\min(\mathrm{SNMH}_v, \mathrm{SNML}_v) < s) \\ &= \mathrm{P}(\mathrm{SNMH}_v < s) + \mathrm{P}(\mathrm{SNML}_v < s) - \mathrm{P}(\mathrm{SNMH}_v < s, \mathrm{SNML}_v < s) \\ &= \mathrm{erfc}(x) - \frac{1}{4}\mathrm{erfc}^2(x), \quad \text{where} \quad x = \frac{\mu_0 + k(v - v_0) - s}{\sqrt{2}\sigma_0}. \end{aligned} \tag{6}$$

Here erfc() is the complementary error function. (6) actually estimates the cell failure probability during standby as expressed in (3). Thus we can quickly estimate the yield of an SRAM array with a given capacity when the standby V_{DD} is equal to v.
Another important estimation is the minimum standby V_{DD} for a given yield or cell failure probability constraint. In other words, we want to estimate the DRV quantile. To derive DRV quantile function, we first obtain the cdf model of the DRV by substituting (6) into (4):

$$F_{DRV}(v) = 1 - \mathrm{erfc}(x) + \frac{1}{4}\mathrm{erfc}^2(x), \quad \text{where} \quad x = \frac{\mu_0 + k(v - v_0) - s}{\sqrt{2}\sigma_0}. \tag{7}$$

Then we obtain the quantile function, i.e. the inverse cdf of DRV, as:

$$v = F_{DRV}^{-1}(p) = \frac{1}{k}\left(\sqrt{2}\sigma_0 \cdot \mathrm{erfc}^{-1}\left(2 - 2\sqrt{p}\right) - \mu_0 + s\right) + v_0, \tag{8}$$

where $\mathrm{erfc}^{-1}()$ is the inverse function of erfc() and p is the probability that DRV≤v.
Both (7) and (8) only require 4 parameters: v_0, μ_0, σ_0, and k. First, we pick m (e.g. m≤6) typical V_{DD} points, say $v_1, \dots, v_m$. Then we run n_{MC} Monte Carlo samples of SNMH at v_i and fit a normal distribution $N(\mu_i, \sigma_i^2)$ to the data. Since we estimate the mean and standard deviation of the distribution body instead of the distribution tail, a small scale of Monte Carlo (e.g. n_{MC}=1,000~5,000) is sufficient. After obtaining μ_i, we extract k by fitting a linear curve to the (v_i, μ_i) data. Finally we pick one V_{DD} point as the initial point v_0, and then μ_0 and σ_0 are chosen accordingly. Therefore, the total number of Monte Carlo samples used in our method is equal to m×n_{MC}, which is 6×5,000 in our test case. To further reduce the run time, we can use a simpler way to approximate k. Instead of running MC simulations on multiple V_{DD} points, we can run a nominal dc simulation of SNM with the sweep of V_{DD}. However, this simplification might cause a slightly larger error.

3.3 Experiment results

We use a 6T cell in a commercial 90nm process to test our DRV model. Without loss of generality, we choose the lowest acceptable noise margin s=0 in the test. Since SRAMs usually contain at least 1,000 cells, we are interested in the DRV quantiles $F_{DRV}^{-1}(p)$ that have the probability p ≥0.999. For the same probability p, the quantile of a theoretical standard normal variable $M \sim N(0,1)$ is m= $\Phi^{-1}(p)$, where Φ^{-1} is the inverse of standard normal cdf. We thereby plot the estimated DRV quantile versus the normal quantile (m) that has the equivalent probability p ≥0.999. Fig. 9 plots the estimates of the DRV quantiles equivalent to m∈ [3,8] from several methods.

1. *Analytical model*: The DRV quantiles estimated from (8) with p = Φ(m) are plotted with the solid curve. We select v_0=100mV. μ_0 and σ_0 are obtained by fitting a normal distribution to the 5,000-point MC result for SNMH at v_0. The parameter k, the sensitivity of the mean of SNMH to V_{DD}, is obtained from linear fitting the curve in Fig. 8.
2. *Standard Monte Carlo or fast Monte Carlo with the Recursive Statistical Blockade*: The DRV quantiles estimated from a 1-million-point Monte Carlo simulation of DRV are plotted with the circles. With 1-million raw MC samples, the maximum DRV quantile we can estimate with a high confidence is equivalent to the normal quantile m≈4. For m>4, we use the fast Monte Carlo method with the recursive statistical blockade tool (Singhee et al., 2008) to reduce run time.

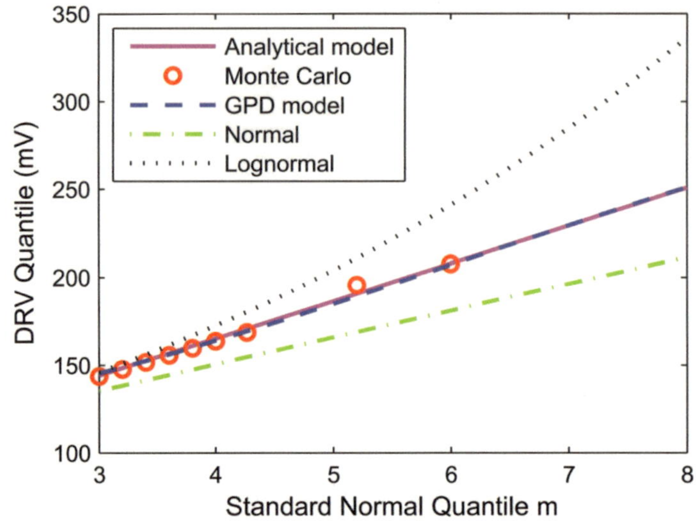

Fig. 9. The DRV quantiles estimated from different methods against the theoretical standard normal quantiles; our new model (8) and the GPD model from the Statistical Blockade tool (Singhee & Rutenbar, 2007) (lines coincident on the plot) closely track Monte Carlo simulation and match farther out in the tail (© 2007 IEEE).

3. *GPD model from the Statistical Blockade (SB)*: The 1,000 tail points from the last recursion stage of the recursive statistical blockade run are used to fit a generalized Pareto distribution (GPD) (Singhee & Rutenbar, 2007). The results estimated from the GPD model are plotted as the dashed curve.
4. *Normal model*: A normal distribution is fit to the DRV data from a 5,000-point MC simulation. The DRV quantiles estimated from the fitting normal distribution are plotted as the dash-dotted curve.
5. *Lognormal model*: A lognormal distribution is fit to the same set of the 5,000 MC points for DRV. The DRV quantiles estimated from the fitting lognormal distribution are plotted as the dotted curve.

Fig. 9 shows that both the results from our model and from the GPD model closely match the MC results up to m=6. In addition, our model matches well with the GPD model at the tail region of m>6, where the tail event has the probability smaller than 9.86e-10. Extrapolation with either normal or lognormal distribution is inaccurate, especially for the points farther out in the tail. The normal model underestimates DRV while the lognormal model overestimates it.

With the comparable accuracy, our method offers a significant speedup over the standard Monte Carlo method because it only requires a small number (e.g. 5,000) of MC simulations for SNMH at a couple of V_{DD} points (totally ≤30,000) to predict any extreme DRV tail values. However, if the probability of the tail event is p_t, the standard MC method requires at least $1/p_t$ samples to obtain one estimate of the quantile. For example, when p_t=9.86e-10 (i.e. m=6), we must run at least 1-billion simulations. Thus, our method provides a speedup of at least 30,000× over standard MC. The recursive statistical blockade requires about 41,700 simulations (Singhee et al., 2008), so our method offers a slight speedup of 1.4× over it. For m>6, standard MC would need thousands of billions of simulations. In this case, the speedup over MC is extremely large.

4. Canary based closed-loop standby V_{DD} scaling

In this section, we deal with the impact of global variations on DRV and present a closed loop V_{DD} scaling system for aggressive leakage power reduction while protecting data by maintaining V_{DD} above the DRV of the worst SRAM cell (Wang & Calhoun, 2007).

4.1 Principle

Fig. 10(a) shows the basic architecture of the system. An on-chip or off-chip voltage regulator supplies V_{DD} to the SRAM cells and to the canary replicas. Multiple canary categories are designed to fail across a range of voltages above the average DRV of the SRAM cells as illustrated in Fig. 10(b). The most important feature of the canary cell is its ability to duplicate the impact of global changes on SRAM stability. With this ability, when the failure voltage of the SRAM cell increases or decreases by some amount due to certain global effect, the failure voltage of each canary category will also change by the same amount. In other words, the DRV of each canary category can maintain a predefined proximity to the DRV of the SRAM cells despite changes in global conditions. Note that, just as SRAM cells, the canary cells are also sensitive to local variations. We employ redundancy

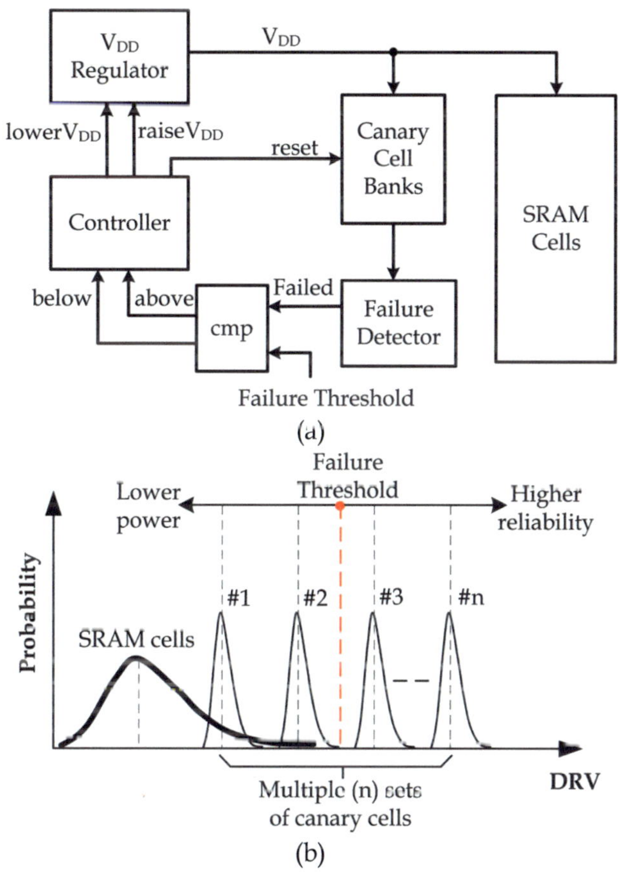

Fig. 10. (a) Architecture and (b) mechanism of the closed loop V_{DD} scaling system (© 2008 IEEE).

and a voting strategy to sharpen the distribution of canary cells within the same category. The failures of the canary categories are monitored by online failure detectors. SRAM data safety is ensured by a programmable failure threshold, which defines the critical failure status of the canary categories and determines the proximity of the standby V_{DD} to the tail of the SRAM DRV distribution. When entering the standby mode, the controller starts lowering V_{DD} until the canary failures meet the failure threshold. Once the global stimuli occur, the canary failures will exceed or drop below the failure threshold, which triggers the controller to raise or lower V_{DD} accordingly.

Besides the improvement of power reduction under variations, this system also allows a trade-off between power savings and data reliability by altering the failure threshold. When the application needs a higher data reliability, a failure threshold that allows less canary sets to fail should be chosen. On the other hand, when the data reliability constraint is lowered or some data errors can be tolerated by redundancy or error correction techniques, we can change the failure threshold to allow more canary sets to fail so that V_{DD} can be reduced for more power savings.

4.2 Major components

4.2.1 Canary cell

The canary cell is the most important component in our system. It must replicate the impact of global variations on SRAM cell stability. Moreover, it must fail before the SRAM cells to prevent the loss of data in SRAM. The canary DRV distribution is not a good indicator of the SRAM cell DRV distribution because there are too few canary cells. Therefore, we must use a design that makes it more sensitive to V_{DD} than it would be simply due to the impact of local variation.

We propose the circuit in Fig. 11(a) and (b) as canary cells for holding '1' and '0', respectively. Each canary cell contains the same 6T transistors (M1~M6) as any SRAM cell, an additional pmos pass transistor (M7) for enhancing the ability of writing a '1' at lower voltage, and a pmos header transistor (M8) for tuning the virtual supply of the cell. The input signal, W, and its inversion, WB, act as the bit lines and word line. During reset mode, W rises, and the pass transistors M5~7 are turned on; '1'/'0' is written into the canary cell '1'/'0'. During standby mode, W switches to low and turns off M5~7. In addition, the bitlines are holding the opposite states with the internal nodes, which creates the worst leakage current through M5~7 and contributes to a higher DRV for the canary cell. The header M8 plays the key role for tuning canary DRV. By tuning the input signal VCTRL at its gate, the virtual supply of the canary cell, VV_{DD}, becomes smaller than V_{DD}, which results in a higher V_{DD} to flip the storage nodes, i.e. a higher DRV for the canary cell. Fig. 12 shows the simulated canary DRV values against the VCTRL values. For comparison, the histogram of the SRAM cell DRV from a 5000-point Monte Carlo simulation is also plotted. Two interesting observations make this tuning knob more appealing. First, there is a nice linearity between canary DRV and VCTRL. Thereby we can create multiple canary categories by simply using regularly increased VCTRL signals, which are easy to implement (e.g. in our test chip, we use a resistor ladder to generate a series of VCTRL signals). Second, the canary DRV can be potentially moved to any point in a wide range. Thus we can always find at least one canary category with its DRV higher than the tail value of SRAM DRV distribution, which could be quite large for big SRAM arrays in scaled technologies.

Now let us further examine the canary cell's capability for tracking PVT variations, which is essential to protecting data in this approach. We use a 1-Kb SRAM and 8 canary sets (#0~#7)

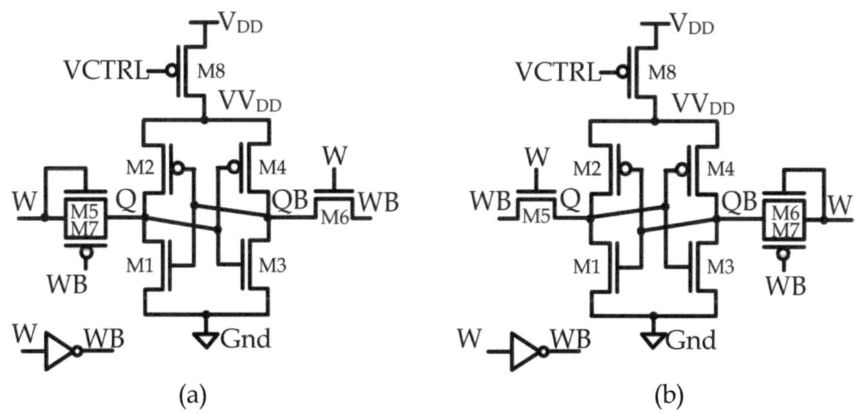

Fig. 11. Schematic of canary cell (a) for holding a '1' and (b) holding a '0' (© 2008 IEEE).

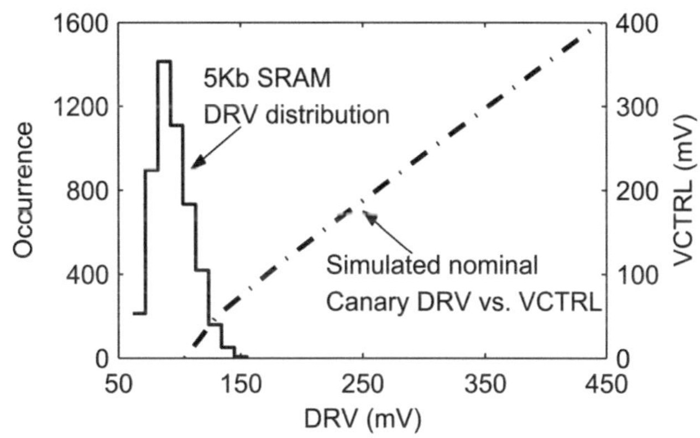

Fig. 12. Simulated nominal canary cell DRV versus VCTRL relative to a 5 Kb SRAM DRV distribution (© 2008 IEEE).

as an example. We first obtain the worst DRV value, i.e. Vmin, of the 1-Kb SRAM with Monte Carlo simulations at normal condition (i.e. at TT process corner & 25°C). Then at the same normal condition, we configure the canary cells by tuning their VCTRL values so that $DRV_{C,7}$ >$DRV_{C,6}$ > … >$DRV_{C,1}$ >Vmin>$DRV_{C,0}$. Here, $DRV_{C,i}$ is the DRV of the canary set #*i*. In order to protect SRAM data, the canary set #1 can be chosen as the first set that should never fail. After configuration, the canary VCTRL values are fixed. Then we change either the temperature or the process corner and rerun the simulations to obtain the new SRAM Vmin and $DRV_{C,i}$ values, which are shown in Fig. 13(a) and (b). The SRAM Vmin is plotted as the curve with circles. $DRV_{C,i}$ is plotted as the curve with triangles. For all the temperature and process changes, the DRV of each canary set moves almost by the same amount as the SRAM Vmin. This indicates that the canaries can successfully track global effects. The only exception here is the SF (Slow-N Fast-P) corner because the technology we use is a strong-N process. At the SF corner, the impact of global variation on the tail of SRAM DRV is overwhelmed by the impact of large local variations. However, the canary DRV is still affected by global variation, so $DRV_{C,1}$ becomes smaller than Vmin at SF corner. To fix this, we can either reconfigure $DRV_{C,1}$ so that $DRV_{C,1}$ >Vmin at this corner or reset the failure threshold to choose the canary set #2 as the first one that does not allow to fail.

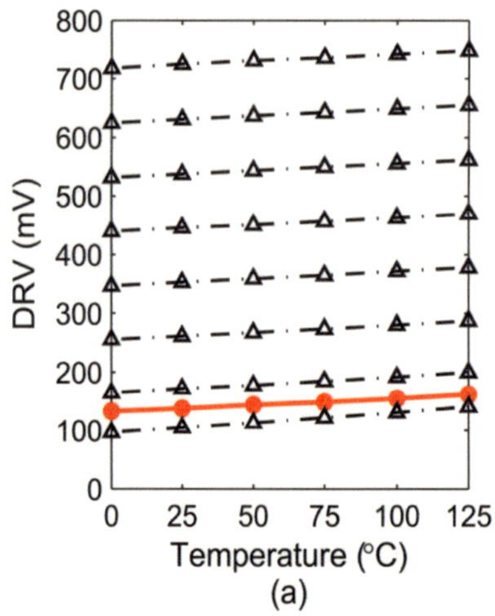

Fig. 13. Simulated DRV of the canary sets (lines with triangles and the upper ones have higher VCTRL) and the worst DRV of a 1 Kb SRAM (the line with circles) change consistently with (a) temperature and (b) process corner for the 90 nm technology (© 2008 IEEE).

4.2.2 Failure detector and canary bank

In our system, the failure of the canary cell is detected online. To enable a quick sensing, the failure detector directly monitors the storage nodes Q and QB of the canary cell. As shown in Fig. 3(a), Q and QB of an SRAM cell might converge if the cell is balanced. However, we set the two bitlines of the canary cell with the complementary values of W and WB (see Fig. 11). This asymmetry makes Q and QB mainly flip when the current V_{DD} is below the cell's DRV. Thus we propose to use a differential sense amplifier shown in Fig. 14 as the failure

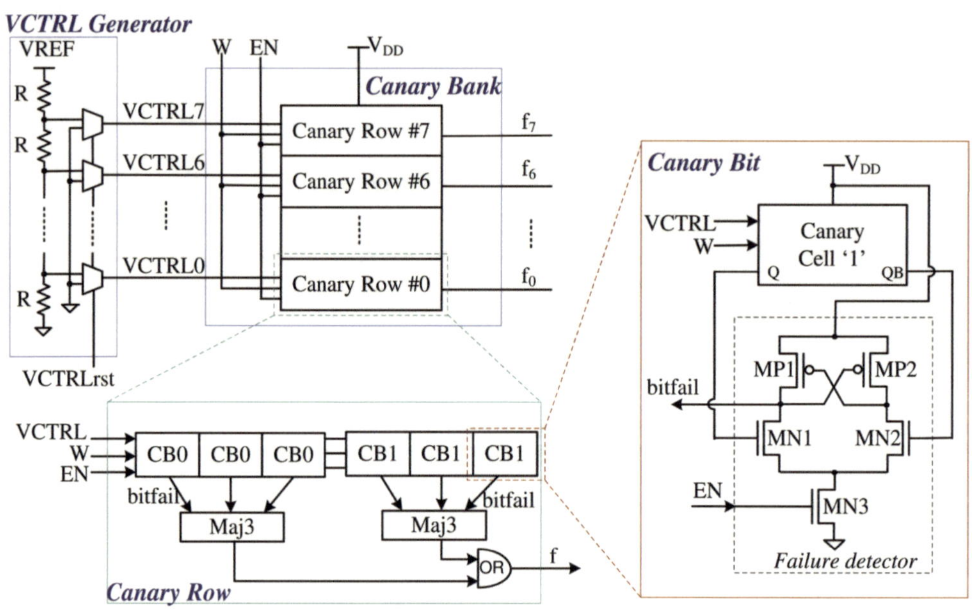

Fig. 14. Canary bank and VCTRL generator.

detector. It shares V_{DD} with the canary cell, and its input differential pair MN1 and MN2 directly connect to Q and QB. One canary cell and its own failure detector compose a canary bit.

The canary sets are deployed as rows in a bank structure as illustrated in Fig. 14. Each canary set occupies one row of the bank. To reduce the variance of the canary DRV, we employ redundancy and majority voting circuits. Thus one canary set (row) consists of n copies of canary bit '1' and n copies of canary bit '0'. Although a larger n can decrease the variance, the area and complexity overhead would dramatically increase. By trading off between the efficiency of variance reduction and the overhead cost, we choose n=3. The failure signals from the three replicas of canary bit '1'/'0' go into the majority-3 gate to generate the voted failure signal. The whole canary set fails when either the majority of the canary bit '1' or the majority of the canary bit '0' fails.

Fig. 14 also shows the VCTRL generator, which is a resistor ladder with a reference voltage VREF and a series of identical resistors. Each canary set (row) is connected to one VCTRL signal from the VCTRL generator.

4.2.3 Feedback controller

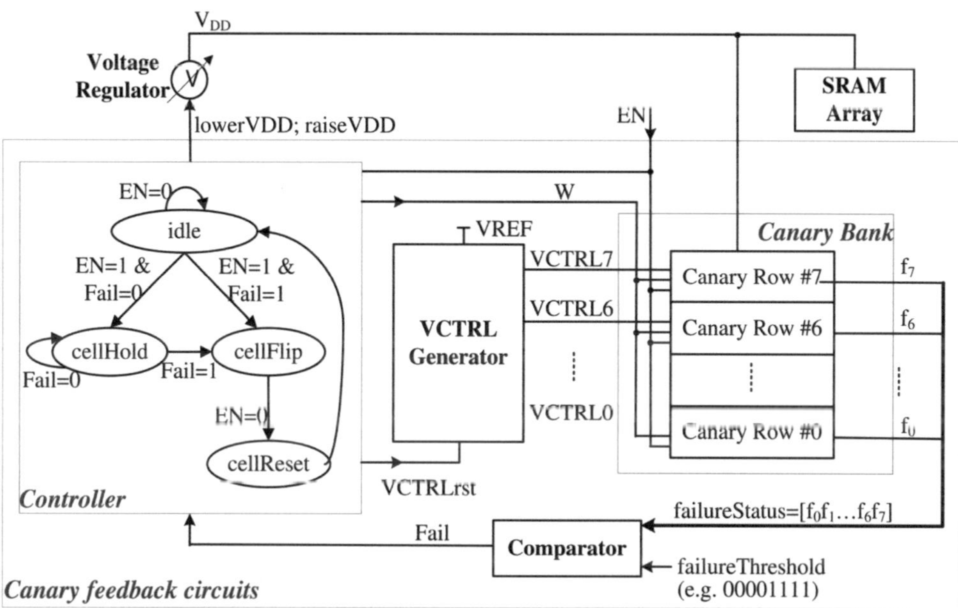

Fig. 15. The feedback controller connects other components in the feedback system.

The feedback controller plays an important role in our system. As shown in Fig. 15, it ties all the other blocks together to form a complete feedback loop. The controller receives the final failure signal 'Fail' from the comparator, which asserts 'Fail' when the failure status of the canary sets ($f_0f_1...f_6f_7$) exceeds the predefined failure threshold. The controller then sends out different control signals to different blocks. The 'lowerVDD' and 'raiseVDD' are sent to the voltage regulator to lower or raise V_{DD} by one step (e.g. 10mV). The 'W' signal is sent to the canary bank for rewriting all the canary cells. The 'VCTRLrst' signal is sent to the VCTRL generator for occasionally resetting all the VCTRL signals to 0.

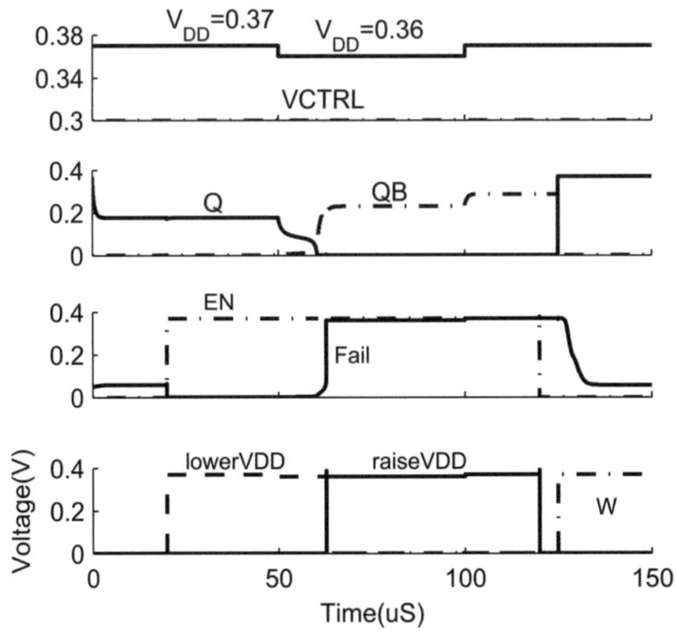

Fig. 16. The timing diagram of the controller (© 2008 IEEE).

Fig. 15 also illustrates the major state transitions in the controller. There are four states: *idle*, *cellHold*, *cellFlip* and *cellReset*. Fig. 16 gives the timing diagram that shows how the states transfer. Suppose the failure threshold is set to 00001111, which implies that the canary set #3 is the first set not allowed to fail. We configure its VCTRL=0.3V. For simplicity, we do not consider redundancy here. After we assert the enable signal 'EN', the failure detector of each canary set evaluates its own Q and QB. When V_{DD}=0.37V, Q and QB of the canary set #4-7 flip, but those of the canary set #0-3 maintain their original values. Thus the failureStatus is 00001111, which is no larger than the failure threshold. Therefore, 'Fail' maintains zero, which causes the controller change from the *idle* state to the *cellHold* state, and the signal 'lowerVDD' rises up to inform the voltage regulator to decrease V_{DD} by 10mV. Once V_{DD} is lowered to 0.36V, Q and QB of the canary set #3 flip to the opposite value, resulting in failureStatus=00011111, which is larger than the failure threshold. Thus 'Fail' rises up and the *cellFlip* state becomes valid. This state asserts 'raiseVDD'. As a result, the regulator increases V_{DD} by one step and V_{DD} returns to the previous value 0.37V, which is actually the DRV of the canary cell #3. After that, 'EN' goes low to disable the failure detection, and the controller enters the *cellReset* state, which asserts the 'W' signal to write the original values into Q and QB for next check.
Since SRAM Vmin can be near or even smaller than the threshold voltage V_T, all the circuits including the failure detector, the comparator, and the controller are designed to function in the sub-threshold region, where $V_{DD}<V_T$ (Wang et al., 2006).

4.3 Model for canary cell tuning

We have observed in Fig. 12 that the canary DRV changes approximately linearly with VCTRL. By analyzing the current through the pmos header (M8 in Fig. 11(a)), we can derive the theoretical model for this linear dependency. We denote DRV_C as the canary DRV. It is equal to the V_{DD} value when the actual supply voltage of the canary cell, VV_{DD}, reaches the cell's true DRV, DRV_t, i.e. the cell DRV without the header. Let us denote I_{min} as the leakage current for holding the cell data when $VV_{DD}=DRV_t$. We assume that the header M8 operates in the sub-threshold region. Since the sub-threshold leakage current is the dominant source of the leakage current, we can compute I_{min} as

$$I_{min} = I_0 \cdot exp\left[\frac{\mathrm{DRV}_C - \mathrm{VCTRL} - V_{T,8} + \eta_8(\mathrm{DRV}_C - \mathrm{DRV}_t)}{n_8 V_{th}}\right] \cdot \left[1 - exp\left(\frac{-\mathrm{DRV}_C + \mathrm{DRV}_t}{V_{th}}\right)\right] \quad (9)$$

where $V_{T,8}$ is the threshold voltage of M8, η_8 is its DIBL coefficient, n_8 is its sub-threshold swing factor, V_{th} is the thermal voltage, and I_0 is its off current. For a given canary cell, we assume that the DRV_t remains the same no matter what VCTRL is. This is reasonable because M1-M7 are not changed. Therefore, I_{min} also remains constant. We further ignore the rolling-off term when $DRV_C - DRV_t > 4V_{th}$ (V_{th}=26mV at 300K). Then we can solve DRV_C as

$$\mathrm{DRV}_C = \frac{\mathrm{VCTRL}}{1+\eta_8} + b, \quad \text{where} \quad b = \frac{V_{T,8} + \eta_8 \cdot \mathrm{DRV}_t}{1+\eta_8} + \frac{n_8 V_{th}}{1+\eta_8} ln(\frac{I_{min}}{I_0}). \quad (10)$$

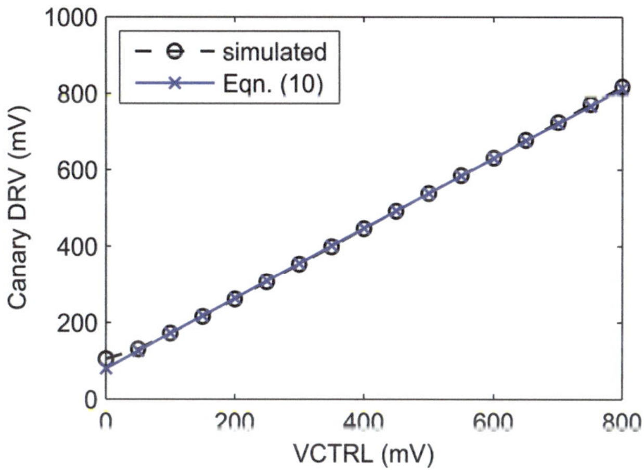

Fig. 17. Estimated canary DRV from (10) versus VCTRL is compared with the simulated result (© 2008 IEEE).

This proves the linear relationship between the canary DRV and VCTRL and implies that the slope can be approximated as $1/(1+\eta_8)$. To verify this model, we first obtain DRV_t and I_{min} from simulation without the header. Then we compute the canary DRV values against VCTRL with (10) and compare them with the simulated results. Fig. 17 shows that our first order linear model provides an excellent approximation for all the VCTRL values across a wide range. A slightly larger error occurs only when VCTRL<50mV. In this region, the canary DRV (DRV_C) is very close to DRV_t, so the rolling-off term cannot be ignored, in which case numerically solving (9) can give a more accurate estimation.

In section 3.2, we proposed the model to predict SRAM DRV quantile and yield in the presence of random variations. Now by combining (8) and (10), we can estimate the VCTRL value y that is needed for a canary cell in order to satisfy a given SRAM cell yield as:

$$y = \frac{1+\eta_8}{k}\left[\sqrt{2}\sigma_0 \cdot erfc^{-1}(2-2\sqrt{p}) - \mu_0 + s\right] + (v_0 - b)\cdot(1+\eta_8). \tag{11}$$

where p=P(DRV_S <$DRV_C(y)$), the probability that the SRAM DRV (DRV_S) is less than $DRV_C(y)$, i.e. the canary DRV when VCTRL is equal to y. All the other parameters are the same as in (8) and (10). Fig. 18 plots the estimated VCTRL values from (11) with the solid curve. In this figure, the point at the coordinates of (x,y) means P(DRV_S <$DRV_C(y)$) is equal to $\Phi(x)$, where Φ is the cdf of a theoretical standard normal variable. For instance, if one application requires 90% yield for a fault-free 100-Kb SRAM, the required failure probability is ~1e-7, which is equivalent to the probability when x=5.2. From Fig. 18, we quickly know that all the canary cells with VCTRL≤120mV should never fail in order to meet this yield. This gives us the guidance to choose the proper VCTRL value for each canary set. Fig. 18 gives an example of the canary configurations. We configure the canary set #2 with VCTRL=120mV. Then we assign 5 points in the region VCTRL>120mV to the canary set #3~7 and assign 2 points in the region VCTRL<120mV to the canary set #0~1. The failure threshold is set to 00011111 so that only the upper 5 canary sets are allowed to fail. This configuration ensures that SRAM can always achieve 90% yield under any PVT variations. If the application changes and needs a different reliability, we can reset the failure threshold or even reconfigure all of the canary sets (by remapping VCTRL values) for better results.

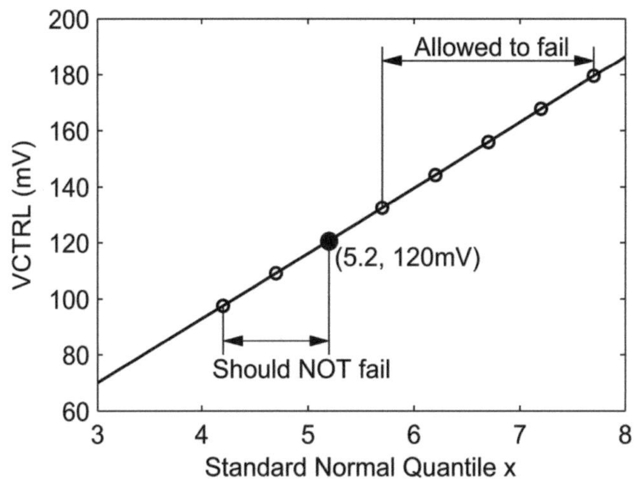

Fig. 18. Estimated VCTRL value y to satisfy that P(DRV_S <$DRV_C(y)$) = $\Phi(x)$, where x is a standard normal quantile. To achieve 90% yield for a fault-free 100Kb SRAM (i.e. x = 5.2), only the canary sets with VCTRL>120mV are allowed to fail (© 2008 IEEE).

4.4 Test chip implementation & measurement

We implement all of the circuits in Fig. 10(a) except the V_{DD} regulator in a 90nm CMOS bulk test chip. In addition to a 16×8Kb SRAM, the test chip contains the canary circuits and test circuits. The area overhead of the canary circuits is about 0.6%. Fig. 19 shows the die photo of the chip. Fig. 20(a) shows the measured average DRV of canary cells versus VCTRL at

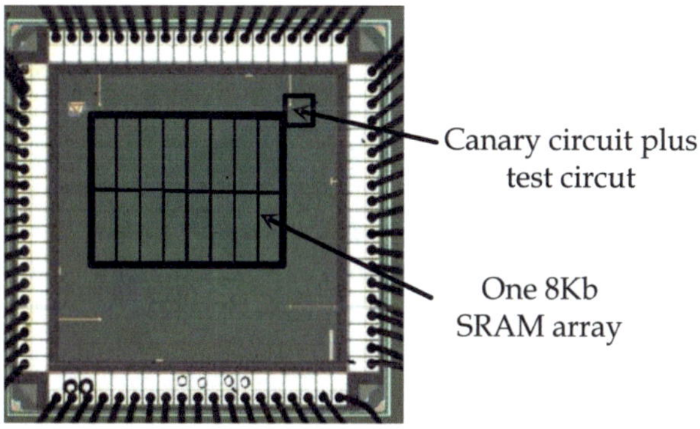

Fig. 19. The die photo of the 90nm test chip (© 2007 IEEE).

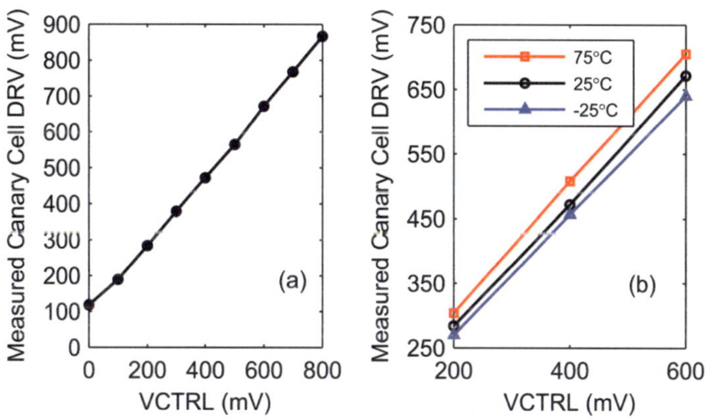

Fig. 20. The measured canary DRV against VCTRL at (a) room temperature and (b) different temperatures (© 2008 IEEE).

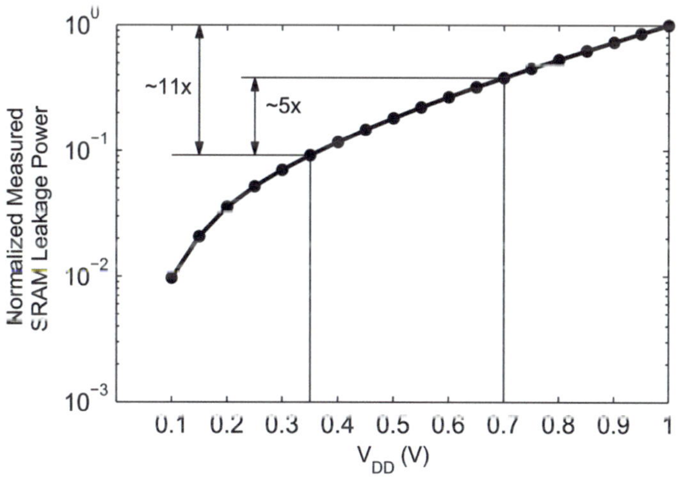

Fig. 21. The normalized measured SRAM leakage power against V_{DD} (© 2008 IEEE).

room temperature. The tuning of VCTRL allows us to provide the desired continuum of failure voltages for the canary cell. It also verifies a good linear relationship between the canary DRV and VCTRL. Fig. 20(b) further shows the measured canary DRV against VCTRL at different temperatures, which verifies that our canary cell can successfully track temperature changes. Fig. 21 plots the normalized measured leakage power of the SRAM array with V_{DD} scaling. Under normal environmental conditions, the measured worst DRV of one 8Kb SRAM array is 0.35V. We estimate that the worst standby Vmin for all the PVT variations plus certain guardband is equal to 0.7V. With the worst case approach, we always set the standby V_{DD} to 0.7V. In contrast, by using our canary-based feedback approach, we can adjust V_{DD} to near the true Vmin value (i.e. 0.35V) at the normal condition. Thereby the canary approach offers ~5× power reduction compared with the conservative worst-case approach and ~11× reduction compared with using the nominal V_{DD}, 1V.

5. Conclusion

Variation has become one of the biggest challenges for circuit design in scaled CMOS technologies. In this chapter, we first investigate the impact of both local and global variations on SRAM data retention voltage (DRV) and then present a method to deal with each type of variation. Local random variations spread the cell DRV across the same array, and the tail of the distribution is the minimum standby V_{DD} (Vmin) that can be applied on the whole SRAM. We propose a fast and accurate method to predict the tail DRV. Our method offers the comparable accuracy with the standard Monte Carlo (MC) method and shows an excellent agreement with another fast method, the Statistical Blockade (SB) tool, for the tails up to 8σ. It offers the speedup of > 10^4× over MC and 1.4× over SB. Global PVT variation results in the shift of Vmin values. The worst-case design approach over-protects non-worst-case scenarios. To enable optimum power savings for any PVT scenario, we propose a closed-loop V_{DD} scaling approach. It uses online canary replica cells and monitors to track global variations, and a feedback circuit to adjust V_{DD} to approach the true Vmin. As device variability continues growing with CMOS technology scaling, SRAM supply voltage scaling requires efficient statistical analysis methods and smart adaptive approaches to maximize power reduction while maintaining correct functionality and acceptable noise immunity.

6. References

Asenov, A. et al. (2003). Simulation of intrinsic parameter fluctuations in decananometer and nanometer-scale MOSFETs, *IEEE Transactions on Electron Devices* 50(9): 1837–1852.

Bhavnagarwala, A. J. et al. (2004). A transregional CMOS SRAM with single, logic V_{DD} and dynamic power rails, *Symposium on VLSI Circuits*, pp. 292–293.

Calhoun, B. & Chandrakasan, A. (2007). A 256-kb 65-nm sub-threshold SRAM design for ultra-low-voltage operation, *IEEE Journal of Solid-State Circuits* 42(3): 680–688.

Clark, L., Morrow,M. & Brown,W. (2004). Reverse-body bias and supply collapse for low effective standby power, *IEEE Transactions on Very Large Scale Integration (VLSI) Systems* 12(9): 947–956.

Ferre, A. & Figueras, J. (2005). *Low-Power Electronics Design, 2nd*, CRC Press, chapter Leakage in CMOS Nanometric Technologies, pp. 3_1-3_19.

Flautner, K. et al. (2002). Drowsy caches: simple techniques for reducing leakage power, *Proceeding of International Symposium on Computer Architecture*, pp. 148–157.

Gu, C. & Roychowdhury, J. (2008). An efficient, fully nonlinear, variability-aware non-montecarlo yield estimation procedure with applications to SRAM cells and ring oscillators, *Proceedings of Asia and South Pacific Design Automation Conference*, pp. 754–761.

Joshi, R. et al. (2007). 6.6+ GHz low Vmin, read and half select disturb-free 1.2 Mb SRAM, *Symposium on VLSI Circuits*, pp. 250–251.

Kanj, R., Joshi, R. & Nassif, S. (2006). Mixture importance sampling and its application to the analysis of SRAM designs in the presence of rare failure events, *Proceedings of Design Automation Conference (DAC)*, pp. 69–72.

Kim, N. S. et al. (2004). Circuit and microarchitectural techniques for reducing cache leakage power, *IEEE Transactions on Very Large Scale Integration (VLSI) Systems* 12(2): 167–184.

Mistry, K. et al. (2007). A 45nm logic technology with high-k+metal gate transistors, strained silicon, 9 Cu interconnect layers, 193nm dry patterning, and 100% pb-free packaging, *Proceedings of IEEE International Electron Devices Meeting (IEDM)*, pp. 247–250.

Morita, Y. et al. (2006). A Vth-variation-tolerant SRAM with 0.3-V minimum operation voltage for memory-rich SoC under DVS environment, *Symposium on VLSI Circuits*, pp. 13– 14.

Nakagome, Y. et al. (2003). Review and future prospects of low-voltage RAM circuits, *IBM Journal of Reseach & Development* 47: 525–552.

Qin, H. et al. (2004). SRAM leakage suppression by minimizing standby supply voltage, *Proceedings of International Symposium on Quality Electronic Design (ISQED)*, pp. 55–60.

Roy, K., Mukhopadhyay, S. & Mahmoodi-Meimand, H. (2003). Leakage current mechanisms and leakage reduction techniques in deep-submicrometer CMOS circuits, *Proceedings of the IEEE* 91(2): 305–327.

Seevinck, E., List, F. & Lohstroh, J. (1987). Static-noise margin analysis of MOS SRAM cells, *IEEE Journal of Solid-State Circuits* 22(5): 748–754.

Singhee, A. & Rutenbar, R. (2007). Statistical blockade: A novel method for very fast monte carlo simulation of rare circuit events, and its application, *Proceedings of Design, Automation & Test in Europe Conference & Exhibition DATE '07*, pp. 1–6.

Singhee, A. et al. (2008). Recursive statistical blockade: An enhanced technique for rare event simulation with application to SRAM circuit design, *Proceedings of 21st International Conference on VLSI Design VLSID 2008*, pp. 131–136.

Verma, N. & Chandrakasan, A. (2008). A 256 kb 65 nm 8T subthreshold SRAM employing sense-amplifier redundancy, *IEEE Journal of Solid-State Circuits* 43(1): 141–149.

Wang, A., Calhoun, B. H. & Chandrakasan, A. P. (2006). *Sub-Threshold Design for Ultra Low-Power Systems*, Springer.

Wang, J. & Calhoun, B. (2007). Canary replica feedback for near-drv standby V_{DD} scaling in a 90nm SRAM, *Proceedings of IEEE Custom Integrated Circuits Conference (CICC)*, pp. 29–32.

Wang, J. & Calhoun, B. H. (2008). Techniques to extend canary-based standby V_{DD} scaling for SRAMs to 45 nm and beyond, *IEEE Journal of Solid-State Circuits* 43(11): 2514–2523.

Wang, J., Singhee, A. et al. (2007). Statistical modeling for the minimum standby supply voltage of a full SRAM array, *Proceedings of European Solid State Circuits Conference (ESSCIRC)*, pp. 400–403.

Wang, Y. et al. (2007). A 1.1GHz 12uA/Mb-leakage SRAM design in 65nm ultra-low-power CMOS with integrated leakage reduction for mobile applications, *IEEE International Solid-State Circuits Conference*, pp. 324–606.

7

Ultralow-power LSI Technology with Silicon on Thin Buried Oxide (SOTB) CMOSFET

Takashi Ishigaki, Ryuta Tsuchiya, Yusuke Morita,
Nobuyuki Sugii and Shin'ichiro Kimura
Central Research Laboratory, Hitachi, Ltd.
Japan

1. Introduction

For a variety of applications from mobile to high-performance computing, the power consumption of very-large-scale-integrated (VLSI) circuits is a serious issue. The scaling rule has been a paradigm for miniaturizing complementary metal-oxide-semiconductor (CMOS) field-effect-transistors (FETs) in VLSI circuits for a long period. In the ideal scaling rule, the supply voltage V_{dd} should decrease in proportion to the miniaturization of the transistor. This V_{dd} reduction has roughly been successful so far. In extremely scaled transistors such as those in the 45-nm logic node and beyond, however, it is very difficult to further decrease V_{dd}. Unless V_{dd} is reduced with the scaling rule, the power consumption of the LSI will increase significantly due to an increase in both operational and standby-leakage power (Sakurai, 2004; Chen, 2006). The primary cause of this difficulty is widely recognized as the increase in threshold voltage (V_{th}) variation of CMOSFETs, because V_{dd} should be set higher considering the margin to the increased V_{th} variation (Takeuchi et al., 1997).
Variation of transistor characteristics, primarily V_{th} variation, is increasing substantially in sub-100-nm technologies. This makes the V_{dd} reduction, required by the scaling rule, difficult, and significantly increases the power consumption of an LSI chip. Here, power consumption P of an inverter, which is the representative LSI unit circuit, is defined as

$$P = CV_{dd}^2 f + I_{leak} V_{dd} \tag{1}$$

where C, f, and I_{leak} are load capacitance, operation frequency, and leakage current, respectively. The first and second terms on the right-hand side represent operational and standby power, respectively. As the scaling proceeds, C and I_{leak} decrease due to the size reduction of transistors, and f increases. Since the miniaturization enables the number of circuits crammed onto a single chip to increase exponentially, it is extremely important to lower V_{dd} to maintain power consumption of an LSI chip (Moore, 1979). In the ITRS 2008 roadmap, the rate of V_{dd} reduction below 1 V is forecasted to be extremely small.
The origin of V_{th} variation is not only due to lithographic variations and layer thicknesses, but also due to line edge roughness (LER) and random dopant fluctuation (RDF) (Asenov et al., 2003; Mizuno et al., 1994). In particular, it has been pointed out that the V_{th} variation caused by the number and special distribution of impurities in the channel of transistors (RDF) becomes serious with the scaling. The magnitude of the V_{th} variation is described by

standard deviation σV_{th}, since the distribution of V_{th} usually shows a normal distribution. The V_{th} variation becomes small with the wider area of the gate because the impurity distribution should be random. This relationship is well known (Pelgrom et al., 1989) and is defined as

$$\sigma V_{th} = A_{Vt} / (L_g W_g)^{1/2} \tag{2}$$

where L_g and W_g are the length and width of a gate, and the gradient A_{Vt} is called the Pelgrom coefficient. Moreover, in conventional bulk CMOSFETs, the following relationship exists

$$A_{Vt} \propto t_{ox} (N_{channel})^{1/4} \tag{3}$$

where t_{ox} and $N_{channel}$ are the thickness of a gate oxide and a channel impurity density. Because $N_{channel}$ has increased to suppress the short channel effect (SCE), A_{Vt} has increased with scaling. Thus, it is understood that the present V_{th} variation problem is inevitably caused by the conventional bulk CMOSFETs' miniaturization.

To solve these problems, it is necessary to first decrease the V_{th} variation due to size variations by suppressing SCE, and secondly to decrease RDF by lowering the impurity densities of the channel. FinFETs, which have strong immunity from SCE without increasing the impurity density of the channel, are reported to have low V_{th} variation (Thean et al., 2006). On the other hand, to continue to both improve the speed and reduce the power from the system-LSI designer's viewpoint, it is necessary to set V_{th} and V_{dd} to the best value in every circuit block or set of transistors in LSI circuits. The multiple V_{th} design, such as that with two or three kinds of V_{th} setting, is already indispensable. Additionally, a technique that controls V_{dd} adaptively according to the state of operation has also been applied (Nakai et al., 2005). A technique to apply substrate bias V_{bb} to control V_{th} flexibly is used in some applications, also (Miyazaki et al., 2000). This V_{bb} control technique is a strong tool that can minimize the performance deviation due to temperature fluctuation as well as the variation of each chip. However, in present scaled bulk CMOSFETs, it is difficult to apply V_{bb} because of the increase in the junction leakage current between the source/drain and the substrate.

To solve the power consumption and V_{th} variation issues, we have proposed a fully depleted silicon-on-insulator (FD-SOI) CMOSFET with an ultrathin buried oxide (BOX), named "silicon on thin BOX (SOTB)". In this chapter, we will describe the features, process and characteristics of the SOTB CMOSFET. Its wide-range back-gate controllability, which enables the optimization of both performance and power after fabrication, will also be described. In addition, to solve some intrinsic problems with SOI technology such as poor electrostatic discharge (ESD) susceptibility and low breakdown voltage, an SOTB/bulk hybrid technology for system-on-chip (SoC) applications will be presented. In the later part of this chapter, we will show the variability reduction and back-gate bias control in the SOTB scheme and demonstrate its impact on the power reduction of VLSI circuits.

2. SOTB CMOSFET

2.1 Features of SOTB CMOSFET

To solve the V_{th} variation problem due to RDF and satisfy the demand from circuit designers, we have proposed the SOTB CMOSFET (Tsuchiya et al., 2004; Ishigaki et al., 2008; Morita et al., 2008). Figure 1 shows a schematic cross-section of the SOTB structure.

Ultrathin SOI and BOX layers make the transistor highly immune from SCE, and its intrinsic channel without halo implant suppresses the V_{th} variation due to RDF. The thin BOX and impurity doping in the substrate just beneath the thin BOX enables a multiple V_{th} design. This thin BOX and the doped region also enable the wide-range back-gate controllability which realizes optimization of both performance and power after fabrication.

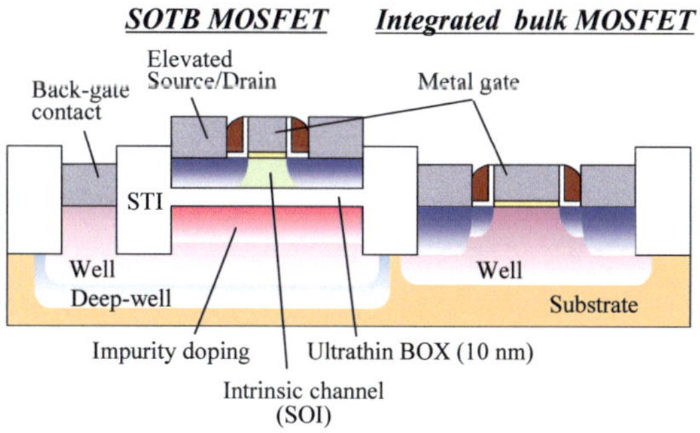

Fig. 1. Schematic cross-sectional view of hybrid SOTB/bulk CMOSFETs.

Some intrinsic problems with (FD-) SOI technology such as poor electrostatic discharge (ESD) susceptibility and low breakdown voltage are well known (Yoshimi et al., 1990). Such adverse effects can be easily avoided by combining bulk technology on the same wafer. The previous approach to integrating SOI and bulk technologies required the use of selective epitaxial growth to compensate for the height difference (Yang et al., 2003), which significantly increases process complexity. In this SOTB structure, bulk CMOSFETs for high-voltage I/O operation, ESD protection, and analog circuits can be easily integrated by removing the thin SOI/BOX layers. This simplified SOTB/bulk hybrid technology is preferable for SoC applications because no design change for the conventional peripheral circuits is required.

Regarding the isolation between transistors and back-gate contacts, this SOTB technology uses a conventional shallow-trench isolation (STI) process similar to the bulk technology. The slight change is that the trench is formed by dry etching three layers: the SOI, BOX, and substrate, as shown in Fig. 1. Consequently, the isolation between devices on the SOI and back-gate contacts is ensured by STI. The well region, acting as a back gate and ground plane, was formed beneath the BOX layer and connected to the region of the back-gate contact through the area underneath the STI. The back-gates for NMOS and PMOS are also isolated by STI. This back-gate contact structure in SOTB technology can be fabricated with the same mask layout as the conventional bulk CMOSFET. A triple-well structure is also adopted to prevent leakage for back gate biasing.

2.2 Device design and fabrication

For low-standby power (LSTP) applications, a single mid-gap metal gate with an intrinsic channel is easily introduced in the fabrication process and is suitable for the desired V_{th} (Fenouillet-Beranger et al., 2008). However, unless the gate-induced drain leakage (GIDL) current is suppressed to less than the subthreshold leakage, off-current I_{off} cannot be reduced at the desired V_{th}. GIDL increases when gate bias V_g becomes negative (for NMOS, vice versa for PMOS). This is because of band-to-band tunneling in and around the drain junction below the gate edges. This leakage current is a source-to-drain current in SOI structures because of the BOX layer, unlike the bulk transistor in which the GIDL current flows to the substrate. In the SOTB structures, the intrinsic channel is also effective to suppress GIDL because of low electric fields around the drain junction below the gate edges. Note that in an SOTB device, it is unnecessary to consider either the junction depth (because it is controlled only by SOI thickness) or the channel-impurity profile (because the SOTB device has an intrinsic channel with no halo implant). Therefore, the control of GIDL in an SOTB is simple. Only the gate overlap length L_{ov}, which is defined as the length of the overlapped region between gate and source/drain extensions, needs to be controlled.

Figure 2 plots the calculated GIDL with various L_{ov} values by using the ATLAS simulation (http://www.silvaco.com). The parameters used in this calculation are as follows: L_g, SOI thickness, BOX thickness, and gate oxide thickness are 65, 15, 10, and 2 nm at V_{dd} = 1.2 V, respectively. The inset shows the potential distribution with L_{ov} of 10 nm when V_g is -0.5 V and V_{dd} is 1.2 V. It is shown that the potential on the drain edge is steep due to the bias difference between the gate and drain. When L_{ov} is 10 nm, I_{off} (at V_g = 0 V) increases because of the subthreshold leakage and the GIDL since the large overlap enhances both the SCE and the GIDL. By decreasing L_{ov} to less than a few nm, GIDL can be reduced sufficiently. At the same time, however, on-state current I_{on} also decreases. Therefore, this indicates that L_{ov} should be carefully optimized for a target specification.

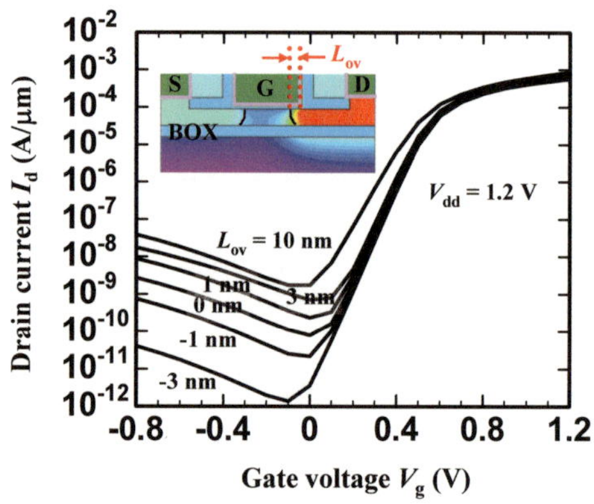

Fig. 2. Simulated NMOS I_d - V_g characteristics as a function of gate overlap length L_{ov} at V_{dd} = 1.2 V. The inset shows the potential distribution in SOTB structure at V_g = -0.5 V.

Figure 3 shows a process flow and a cross-sectional transmission electron microscope (TEM) image of an SOI/bulk hybrid structure. After the STI formation, the SOI layers (t_{SOI} ~ 12 nm) on both the well contact and bulk active regions were removed by dry etching using a BOX layer (t_{BOX} ~ 10 nm) as a stopper, followed by the removal of the BOX layer. Due to the small step height of 22 nm (t_{SOI} + t_{BOX}), gate patterning can easily be performed on both the SOI and bulk regions simultaneously. This process enables the SOTB/bulk hybrid structure to be fabricated without requiring epitaxial growth to compensate for the height difference. The SOI region is on the left side of the TEM image, and the integrated bulk region is on the right. Smooth gate patterning without voids on both the SOI and bulk regions was confirmed. These integrated hybrid bulk CMOSFETs with a 7.5-nm-thick gate dielectric were fabricated on the exposed surface of the silicon support wafer by removing SOI/BOX layers. The quality of the surface after the BOX removal is a concern in this process. Carrier mobilities as high as a universal curve and gate oxide interface trap density (D_{it}) as low as

10^{11} $eV^{-1}cm^{-2}$ were confirmed with no sacrificial oxidation of the surface, indicating that little damage was caused by dry etching during the SOI-layer removal (Ishigaki et al., 2008). A sufficiently long time-dependent dielectric breakdown (TDDB) lifetime is ensured at V_g = 3.3 V, as shown in Fig. 4.

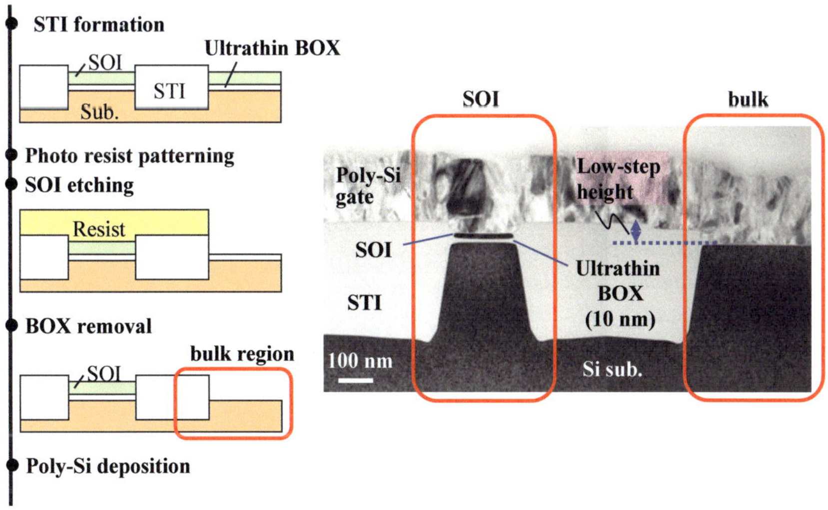

Fig. 3. Process flow of hybrid SOI/bulk fabrication and a cross-sectional TEM image of poly-Si gate on hybrid SOI/bulk regions.

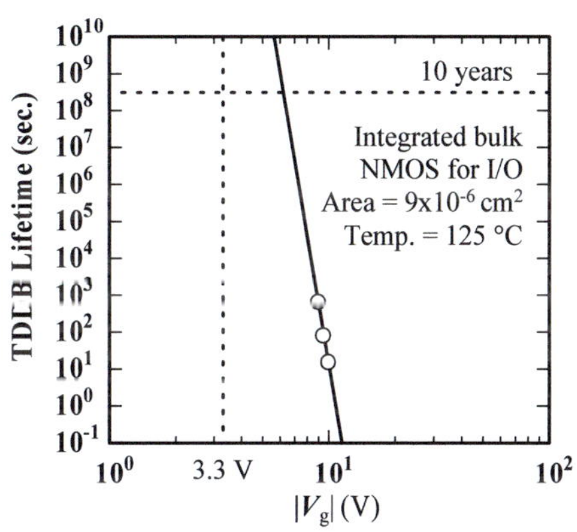

Fig. 4. TDDB lifetime of integrated bulk NMOS for I/O operation.

Figure 5 shows the following process steps as well as a cross-sectional TEM image of a fabricated 50-nm-gate-length SOTB MOSFET. In the dual oxidation process, SiON gate dielectrics were formed at an equivalent oxide thickness (EOT) of 1.9 nm for SOTB core CMOSFETs, and 7.5 nm for bulk I/O CMOSFETs. To precisely control L_{ov}, an additional SiN offset spacer was formed after the first SiO_2 spacer formation and before the source/drain extension implantation. After forming a sidewall, an elevated source/drain structure was formed by selective epitaxial growth to obtain low external resistance. To prevent recesses in the SOI, the conditions for gate- and sidewall- etching and precleaning before epitaxy were carefully optimized, resulting in a low external resistance, as shown in Fig. 6. In low-power FD-SOIs with intrinsic channels, no dual metal technology between NMOS and PMOS gates, such as nickel silicide phase control (Veloso et al., 2006), is required when using nickel silicide as a gate electrode material. The gate poly-Si and the source/drain epitaxial Si were set to their optimal heights before gate-cap removal and fully silicided simultaneously in a single step without using chemical mechanical polishing (CMP). The metal inserted poly-silicon stack (MIPS) metal-gate structure is also applicable for SOTB CMOSFETs.

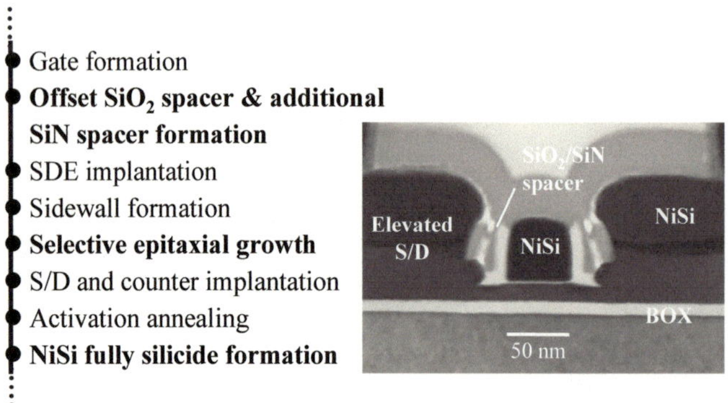

Fig. 5. Process steps and cross-sectional TEM image of 50-nm-gate-length SOTB MOSFET.

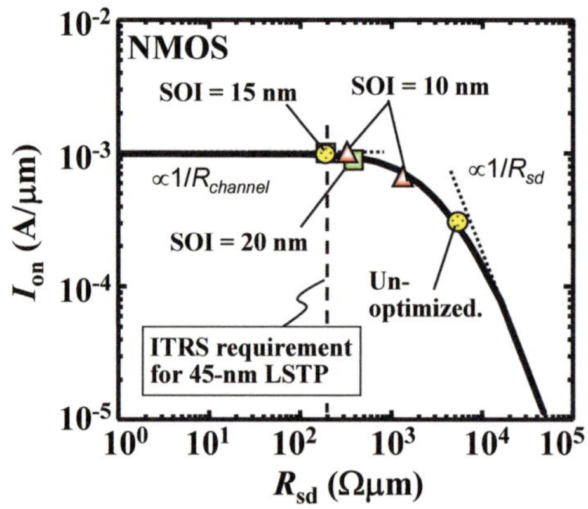

Fig. 6. Relationship between on-current I_{on} and external resistance R_{sd}.

3. Characteristics of SOTB CMOSFET

The typical subthreshold characteristics of the 50-nm-gate-length SOTB CMOSFETs at V_{dd} = 1.2 V are plotted in Fig. 7. The desired symmetrical characteristics were successfully obtained with a single Ni FUSI gate. The off-state drain currents were less than 20 pA/μm due to the reduction of GIDL with properly controlled L_{ov}. At the same time, comparable on-currents were obtained with the conventional bulk CMOSFETs. The I_{off} could be further reduced to 1 pA/μm by reducing L_{ov} (Ishigaki et al., 2009). These I_{on}/I_{off} values are in good agreement with the data based on bulk or FD-SOI technology for LSTP applications (Kimizuka et al., 2005). These SOTB CMOSFETs also suppressed the SCE even with the intrinsic channel because of the thin SOI and BOX layers. Figure 8 demonstrates that the

SOTB CMOSFET is free from the self-heating effect thanks to the thin BOX. That is, negative drain conductance (decreasing I_d with increasing V_d) was not observed.

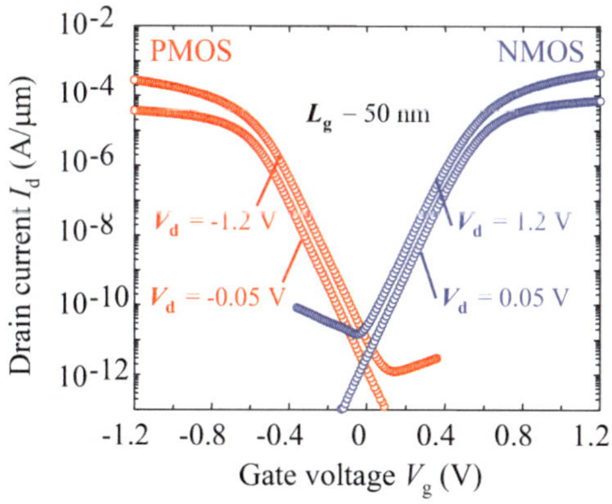

Fig. 7. Typical I_d-V_g characteristics of 50-nm-gate-length SOTB CMOSFET.

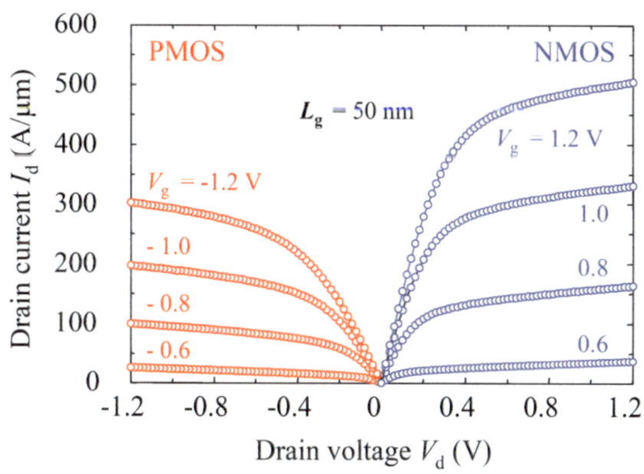

Fig. 8. I_d-V_d characteristics of 50-nm-gate-length SOTB CMOSFET. No self-heating is evident due to the thin BOX.

In the SOTB scheme, time-to-time or area-to-area (including die-to-die or wafer-to-wafer) device characteristics can be widely controlled using the back-gate bias V_{bb}. In particular, forward back-gate bias can be effectively used because there is no substrate leakage. Note that a forward bias higher than 0.6 V can never be applied in conventional bulk CMOSFETs owing to the significant increase in p-n junction leakage current from source to substrate. The dependences of V_{th} and the subthreshold slope (*SS*) of a 50-nm-gate-length SOTB CMOSFET on V_{bb} at V_{dd} = 1.2 V are shown in Fig. 9. By applying a reverse V_{bb}, V_{th} increased to above 0.6 V, and the *SS* decreased to less than 80 mV/decade. In contrast, by applying a forward back-gate bias of 1.2 V, V_{th} can be lowered by more than 0.3 V while keeping the *SS* small. In such a high forward bias, there is no increase in the substrate leakage currents.

As for conventional bulk structures, reverse biasing can be used to reduce the standby leakage after fabrication. However, this is less effective because both V_{th} variation and GIDL increases (Yasuda et al., 2007). In the SOTB scheme, GIDL is sufficiently suppressed by the intrinsic channel and the controlled L_{ov}, and forward V_{bb} is also effective in adjusting the V_{th} variation. The die-to-die compensation for the V_{th} of each chip is demonstrated in Fig. 10.

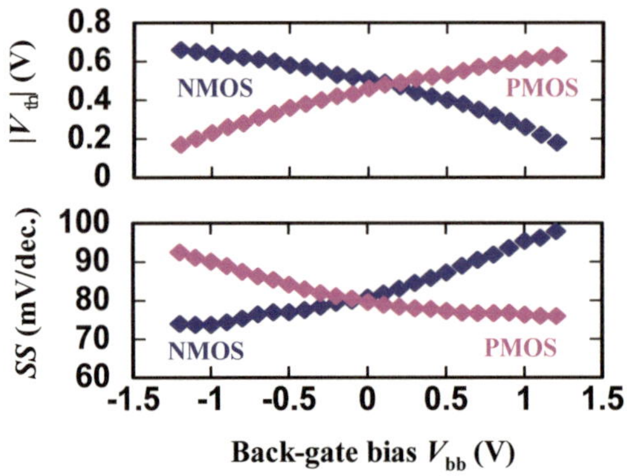

Fig. 9. Dependences of V_{th} and SS for a 50-nm-gate-length SOTB CMOSFET as a function of back-gate bias V_{bb} at V_{dd} = 1.2 V.

Each circle represents a 50-nm-gate-length NMOS of a chip. The open circles indicate the V_{th} distribution without V_{bb} control, and the closed circles indicate the distribution with V_{bb} control, that is, when the V_{bb} was adjusted for each transistor to approach the target V_{th}. The V_{bb} values range from -1.2 to 1.2 V, in 0.2-V increments. Without V_{bb} control, the range of V_{th} distributions is about 0.1 V due to size (gate length or layer thicknesses such as SOI) variations or channel dose fluctuations. The standard deviation σV_{th} with V_{bb} control was suppressed to 1/4 (case A) even with such a wide V_{bb} step. In addition, the typical V_{th} can be set arbitrarily within a range of 0.18 V (cases B and C), which is larger than the 0.1 V of the original V_{th} distribution, while keeping the variation suppressed. It is assumed, for instance, that setting the optimum V_{bb} according to the speed and the power of the chip will improve the yield.

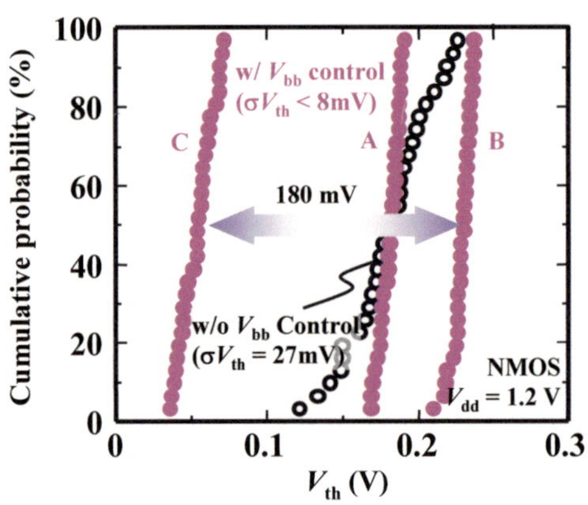

Fig. 10. V_{th} shift and variation reduction of Poly-Si gate SOTB NMOS using back-gate bias V_{bb} (V_{bb}: -1.2 < V_{bb} < 1.2 V, increments = 0.2 V).

4. Reduction of power consumption

The nominal V_{th} cumulative probability plot (not shown) of SOTB CMOSFETs indicates that the distribution is random and SCE is suppressed even down to 50 nm (Morita et al., 2008).

The Pelgrom plot is shown in Fig. 11. The slope of the plot, the Pelgrom coefficient A_{vt}, is 1.8 and 1.5 mVμm for NMOS and PMOS, respectively. These values are about half those of conventional bulk CMOSFETs of the same technology generation due to the intrinsic channel without halo implant. Impurities below the BOX layer have a small impact on the variability. The local component of variation is also plotted. To extract the local variation of V_{th}, the difference between the forward and the reverse measurement by exchanging the source and drain was used (Tanaka et al., 2000). It has already been confirmed that this method simply represents the local variation of adjusting pair transistors (Sugii et al., 2008). These results suggest the SOTB CMOSFET is robust in terms of variability.

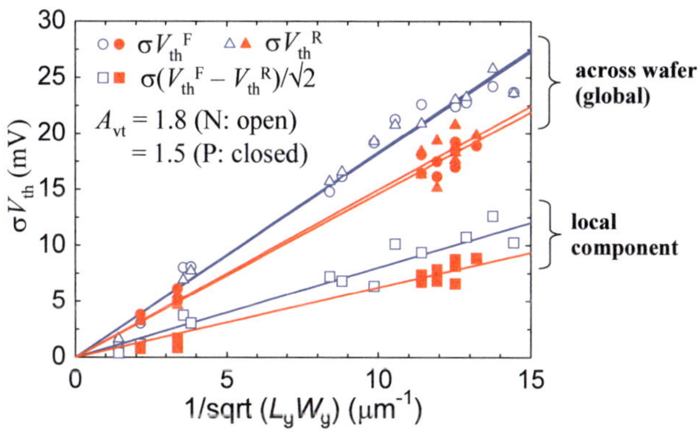

Fig. 11. Pelgrom plot of SOTB CMOSFETs for both global and local components.

Because static random access memory (SRAM) has been integrated in recent VLSI circuits with large capacities occupying large areas, reducing the power consumption of SRAMs is becoming increasingly important. Moreover, since the SRAM circuit is most sensitive to the local V_{th} variation, it is assumed that achieving low V_{dd} is most difficult with SRAM. Figure 12 plots the characteristics of 6T-SOTB SRAM cells 0.99 μm² in size. The static noise margins (SNM) of 0.357 V at V_{dd} = 1.2 V and 0.142 V at V_{dd} = 0.6 V indicate a much more stable operation in comparison with conventional bulk ones. The fail bit count (FBC) analysis indicated that the SOTB-SRAM can operate as low as 0.6 V, whereas the bulk SRAM with the same cell size operates at 1.1 V (Tsuchiya et al., 2009).

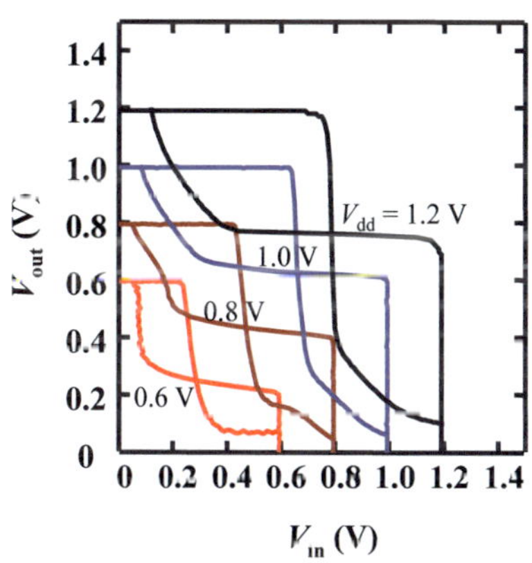

Fig. 12. Measured butterfly curves of 50-nm-gate-length SOTB SRAM cell.

The variability of SOTB CMOSFETs has a significant impact on standby power consumption. The total standby leakage of a conventional 6T-SRAM cell is roughly expressed as

$$I_{standby} = 3I_{off} + 2I_{gate} \tag{4}$$

where I_{gate} is the gate leakage when the gate node of a transistor is high. We calculated the standby leakage of an SRAM between SOTB and bulk devices, taking variability into account. When V_{th} decreases from a typical value (ideally at the minimum I_d point determined both by subthreshold and GIDL currents), I_{off} increases on the subthreshold slope of 80 mV/decade. On the other hand, I_{off} also increases with larger V_{th} because of the GIDL with a slope of 400 mV/decade. The summation of off-currents taking V_{th} distribution into account is expressed as

$$\Sigma I_{off} = \int P\,(\Delta V_{th}) \times I_{off}\,(\Delta V_{th}) \tag{5}$$

where P is a probability of V_{th} in the distribution, and ΔV_{th} is the deviation from the typical V_{th}. The σV_{th} of SOTB and bulk devices is 27 and 54 mV, respectively. I_{gate} is calculated as a constant without taking variability into account because its value is much smaller for LSTP applications, where J_g is 2 x 10^{-3} A/cm2. The integrated standby leakage of 1-Mbit SRAM is shown in Fig. 13. One typical I_{off} of SOTB devices is calculated as 10 pA/μm compared with bulk devices. The other typical I_{off} value of SOTB devices is 1 pA/μm. When the typical I_{off} = 10 pA/μm, the standby leakage of the SOTB device is 44% that of the bulk ones. This result indicates that reducing σV_{th} by half also reduces the standby leakage by half. Moreover, when the typical I_{off} of an SOTB device = 1 pA/μm, the standby leakage can be further reduced to 6%. In the case of I_{off} = 1 pA/μm, which indicates a lower on-state current, the driving performance can be boosted by using back-gate biasing in the SOTB technology.

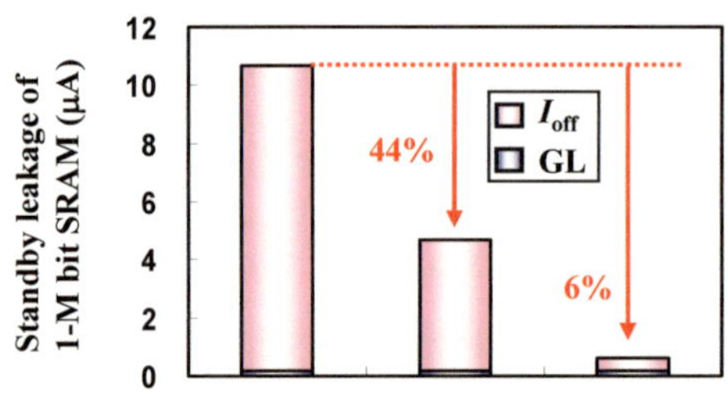

Device	bulk	**SOTB**	**SOTB**
Typ. I_{off}	10 pA/μm	**10 pA/μm**	**1 pA/μm**
σV_{th}	54 mV	**27 mV**	**27 mV**

Fig. 13. Estimated standby leakage of 1-Mbit SRAM taking V_{th} variability into account.

5. Conclusion

Recently, the scalability of CMOSFETs has become a topic of utmost importance. In SOTB technology, scalability can be pursued by reducing the SOI and BOX thicknesses. The minimum SOI thickness is considered to be 6 nm, after which the influence of the quantum effect or mobility degradation appears (Uchida et al., 2001). Given this value and considering the thickness variation, the minimum gate length of the SOTB CMOSFET is expected to be about 20 nm while maintaining a small V_{th} variation (Sugii et al., 2009). In addition, the applicability of back-gate biasing is important. Even if the uniformity of transistors is not maintained, the characteristic variation can be eased by correcting V_{bb}.

In this chapter, it was shown that both operating voltage and standby power can be substantially reduced due to the low variability of the SOTB CMOSFET. This indicates that the power consumption of VLSI circuits can be drastically reduced. Moreover, when combined with V_{bb} control, it is possible to obtain the optimum power efficiency by flexibly changing V_{dd} and V_{th} to the operation situation. It is hoped that these flexible voltage controls will be applied to VLSI circuits in the future to meet the increasingly complex application demands.

6. Acknowledgements

The authors would like to thank the staff of Renesas Technology and the Hitachi Central Research Laboratory (HCRL) for fabricating the devices. We would also like to thank K. Torii of HCRL and T. Onai, M. Odaka and K. Kasai of the Hitachi R&D group for their encouragement. This work was partly supported by the Ministry of Education, Culture, Sports, Science and Technology of Japan.

7. References

Asenov, A. et al. (2003). Intrinsic parameter fluctuations in decananometer MOSFETs introduced by gate line edge roughness. *IEEE Trans Electron Devices*, 50, pp. 1254-1260, 0018-9383

Chen, T. C. (2006). Where CMOS is going : trendy hype vs. real technology, *Plenary of ISSCC*, pp. 1-18, 1-4244-0079-1, San Francisco

Fenouillet-Beranger, C. et al. (2008). FDSOI devices with thin BOX and ground plane integration for 32 nm node and below. *ESSDERC.*, pp. 206-209, 978-1-4244-2363-7

Ishigaki, T. et al. (2008). Wide-range threshold voltage controllable silicon on thin buried oxide integrated with bulk complementary metal oxide semiconductor featuring fully silicided NiSi gate electrode. *Jpn J Appl Phys.*, 47(4), pp. 2585-2588

Ishigaki, T. et al. (2009). Silicon on thin BOX (SOTB) CMOS for ultralow standby power with forward-biasing performance booster. *J. Solid-State Electronics*, 53, pp. 717-722, 0038-1101

Kimizuka, K. et al. (2005). Ultra-Low Standby Power (U-LSTP) 65-nm node CMOS Technology Utilizing HfSiON Dielectric and Body-biasing Scheme. *Proceedings of VLSI technology symp.*, pp. 218-219, 4-900784-00-1, Kyoto

Miyazaki, M. et al. (2000). A 1000-MIPS/W microprocessor using speed adaptive threshold-voltage CMOS with forward bias, *Dig. Tech. Papers IEEE Int. Solid-State Circuits Conf.*, pp. 420-421, 0-7803-5856-2

Mizuno, T. et al. (1994). Experimental study of threshold voltage fluctuation due to statistical variation of channel dopant number in MOSFET's. *IEEE Trans Electron Devices*, 41, pp. 2216-2221, 0018-9383

Moore, G. E. (1979). Are We Really Ready for VLSI?. *Keynote Address of ISSCC*, pp. 54-55

Morita, Y. et al. (2008). Smallest V_{th} variability achieved by intrinsic silicon on thin BOX (SOTB) CMOS with single metal gate, *Proceedings of VLSI technology symp.*, pp. 166-167, 978-1-4244-1802-2, Honolulu

Nakai, M. et al. (2005). Dynamic voltage and frequency management for a low-power embedded microprocessor. *IEEE J. Solid-State Circuits,* 40, pp. 28-35, 0018-9200

Pelgrom, M. J. M. et al. (1989). Matching properties of MOS transistors. *IEEE J. Solid-State Circuits,* 24, pp. 1433-1439, 0018-9200

Sakurai, T. (2004). Perspectives of low-power VLSI's. *IEICE Trans. Electron.,* E87-C, pp. 429-436

Sugii, N. et al. (2008). Comprehensive Study on Vth Variability in Silicon on Thin BOX (SOTB) CMOS with Small Random-Dopant Fluctuation: Finding a Way to Further Reduce Variation. *IEEE Trans Electron Devices,* pp. 249-252, 1-4244-2377-4, San Francisco

Sugii, N. et al. (2009). Local Vth Variability and Scalability in Silicon on Thin BOX (SOTB) CMOS with Small Random-Dopant Fluctuation. *IEEE Trans Electron Devices,* to be published

Takeuchi, K. et al. (1997). Channel Engineering for the Reduction of Random-Dopant-Placement-Induced Threshold Voltage Fluctuation, *Tech. Dig. Int. Electron Devices Meet.,* pp. 841-844, 0-7803-4100-7, Washington, D. C.

Tanaka, T. et al. (2000). Direct Measurement of V_{th} Fluctuation Caused by Impurity Positioning. *Proceedings of VLSI technology symp.,* pp. 136-137, 0-7803-6308-6, Honolulu

Thean, A. et al. (2006). Performance and Variability Comparisons between Multi-Gate FETs and Planar SOI Transistors, *Tech. Dig. Int. Electron Devices Meet.,* pp. 881-884, 1-4244-0439-8

Tsuchiya, R. et al. (2004). Silicon on thin BOX : a new paradigm of the CMOSFET for low-power and high-performance application featuring wide-range back-bias control, *Tech. Dig. Int. Electron Devices Meet.,* pp. 631-634, 0-7803-8684-1, San Francisco

Tsuchiya, R. et al. (2009). Low Voltage (V_{dd}~0.6 V) SRAM Operation Achieved by Reduced Threshold Voltage Variability in SOTB (Silicon on Thin BOX). *Proceedings of VLSI technology symp.,* pp. 150-151, 978-4-86348-009-4, Kyoto

Uchida, K. et al. (2001). Experimental evidences of quantum-mechanical effects on low-field mobility, gate-channel capacitance, and threshold voltage of ultrathin body SOI MOSFETs, *Tech. Dig. Int. Electron Devices Meet.,* pp. 633-636, 0-7803-7050-3, Washington D. C.

Veloso, A. et al. (2006). Dual work function phase controlled Ni-FUSI CMOS (NiSi NMOS, Ni_2Si or $Ni_{32}Si_{12}$ PMOS): Manufacturability, Reliability & Process Window Improvement by Sacrificial SiGe cap. *Proceedings of VLSI technology symp.,* pp. 116-117, 1-4244-0005-8, Honolulu

Yang, M. et al. (2003). High Performance CMOS Fabricated on Hybrid Substrate With Different Crystal Orientations, *Tech. Dig. Int. Electron Devices Meet.,* pp. 453-456, 0-7803-7872-5, Washington D. C.

Yasuda, Y. et al. (2007). Design methodology of body-biasing scheme for low power system LSI with Multi-V_{th} transistors. *IEEE Trans Electron Devices,* 54(11), pp. 2946-2952, 0018-9383

Yoshimi, M. et al. (1990). Analysis of the drain breakdown mechanism in ultra-thin-film SOI MOSFETs. *IEEE Trans Electron Devices,* 37, pp. 2015-2021, 0018-9383

8

The Progress and Challenges of Applying High-k/Metal-Gated Devices to Advanced CMOS Technologies

Hsing-Huang Tseng, Ph.D.
Professor of Electrical Engineering
Ingram School of Engineering
Texas State University
601 University Drive, San Marcos, TX 78666, USA

1. Introduction

1.1 Motivation for implementing high dielectric constant gate dielectric for advanced CMOS scaling

Semiconductor devices need to have good performance, with a low cost and low power dissipation. For decades, research and development of semiconductor processing technology and device integration have focused on enhancing performance and reducing costs using SiO_2 as the gate dielectric and doped polysilicon as the gate electrode. The most effective way to enhance performance and reduce costs is to scale the device gate length and gate oxide. Scaling the gate length results in fabricating more devices per wafer (i.e., increase the device density) and thus reduce the cost per chip, while scaling the gate oxide enhances the drive current and reduces the short channel effects due to gate length scaling. However, as the gate oxide becomes thinner, the power to operate transistors increases because of greater gate oxide leakage current. To resolve this high gate oxide leakage problem, the mechanism of the carriers tunneling through the gate dielectric must be better understood. In an ideal metal-insulator-semiconductor (MIS) device, the current conduction in the insulator should be zero. In a real MIS device, however, current can flow through the insulating film by various conduction mechanisms. The two primary conduction mechanisms for electron tunneling through high quality gate dielectric are discussed below.

1.1.1 Direct tunneling

In a metal-insulator-semiconductor (MIS) stack, when the oxide voltage (V_{ox}) is smaller than the metal-insulator barrier height, the electron tunnels directly from the metal electrode into other semiconductor electrode through the insulator. This is known as the direct tunneling process. The equation for direct tunneling current density (current normalized by device area) is proportional to

$$J_{DT} = A \exp (- m^{1/2} V_b{}^{1/2} T_{ox}) = B \exp (- m^{1/2} V_b{}^{1/2} K) \qquad (1)$$

where m is the effective mass of the electron, V_b is the barrier height, T_{ox} is the physical thickness of the gate oxide SiO_2, K is the dielectric constant of the insulator, and A and B are pre-exponential factors. From this equation, one can observe that the T_{ox} or K is the dominating factor in controlling the direct tunnel current density. The tunneling current density increases dramatically as the SiO_2 becomes thinner. In general, the gate leakage increases 100 times for every 0.5 nm that the SiO_2 is thinned. The gate leakage density is as high as the 10 E^4 Amp/cm^2 range for SiO_2 as thin as 1.1 nm. The high gate leakage increases standby power consumption according to the following equation:

$$P_{STANDBY} = (I_{subth} + I_{GIDL} + I_g).V_{dd} \quad (2)$$

Where: I_{subth}: subthreshold leakage current
I_{GIDL}: gate-induced drain leakage current
I_g: gate leakage
V_{dd}: supply voltage

On the other hand, Eq. [1] shows that increasing the dielectric constant of an insulator dramatically reduces the direct tunneling current. This is because the gate oxide physical thickness (T_{ox}) and the dielectric constant K of an insulator are correlated by the equivalent oxide thickness (EOT) defined as

$$EOT = (K_{SiO2}/K_{insulator})T_{insulator} \quad (3)$$

where K_{SiO2} and $K_{insulator}$ are the dielectric constant of SiO_2 and the insulator, respectively, and $T_{insulator}$ is the physical thickness of the insulator. Based on this definition, an insulator material with a five times greater dielectric constant than SiO_2 would require a five times greater physical thickness than SiO_2 to keep the same EOT as SiO_2. Therefore the tunneling current for a device using this insulator would be orders of magnitude lower than that using SiO_2 because the tunneling leakage current decays exponentially as the insulator becomes thicker.

1.1.2 Fowler nordheim tunneling

When the oxide voltage (V_{ox}) is greater than the metal/insulator barrier height, the electron tunnels from the metal electrode into the insulator conduction band first and then travels toward the other semiconductor electrode. This is known as the Fowler-Nordheim (FN) tunneling process. The equation for FN tunneling current density is proportional to

$$J_{FN} = C \exp(-m^{1/2} V_b^{3/2} T_{ox}) = D \exp(-m^{1/2} V_b^{3/2} K) \quad (4)$$

where C and D are pre-exponential factors. From this equation, one can observe that the barrier height is the dominating factor in controlling the FN tunnel current density.

Table 1 compares the exponent (the product of m, V_b, and K) shown in the equations of direct tunneling (low field) and FN tunneling (high field) for insulator materials with different V_b and K values. The results show the following:

1. Although oxynitride processed by incorporating nitrogen into SiO_2 was developed to increase the dielectric constant, the reduction of leakage current density is not enough to be used for highly scaled devices such as for the 45 nm technology node. Even for pure nitride (Si_3O_4) shown in the table, the exponent shown in the direct tunneling

equation is only about two times greater than that for SiO_2 while the exponent shown in the FN tunneling equation is similar to that of SiO_2.

2. On the other hand, it is clear that an insulator with a *K* value of 25 reduces the exponent significantly in the direct tunneling Eq. [1] and more than two times in the FN tunneling Eq. [2].

Material	V_b (Volts)	K	Low Field	High Field
SiO_2	3.0	3.9	-6.75	-20.3
Si_3N_4	2.0	7.8	-11.0	-22.1
Ta_2O_6, HfO_2, ZrO_2	1.5	25	-30.6	-45.9

Table 1.

1.2 Motivation for implementing metal gates for advanced CMOS scaling

1.2.1 SiO_2/Polysilicon stack

Figure 1 shows a schematic of a MOSFET in which the gate oxide is SiO_2 and the gate electrode is doped polysilicon. Figure 2 shows the equivalent circuit of an MOS capacitor. The total MOS capacitance C_g can be expressed as

$$C_g^{-1} = C_{ox}^{-1} + C_s^{-1} + C_p^{-1} \tag{5}$$

where C_{ox} is the oxide capacitance, C_s is the silicon capacitance, and C_p is the polysilicon gate electrode depletion capacitance.

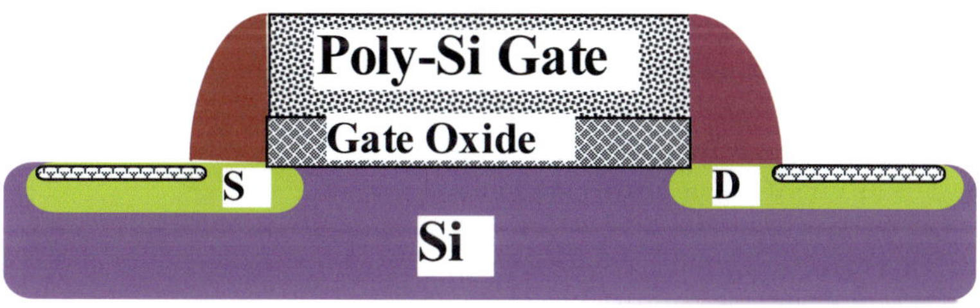

Fig. 1.

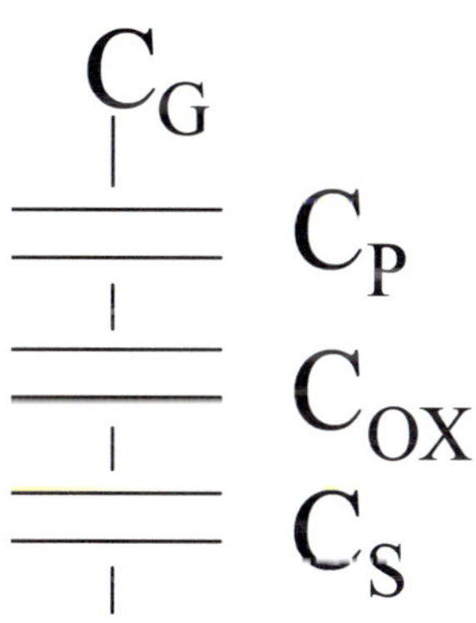

Fig. 2.

When the MOSFET is operated in inversion mode, the doped polysilicon gate energy band bending and charge distribution form a thin space-charge region. This results in a finite, bias-dependant value of C_p and causes polysilicon depletion. The C_p will reduce the value of the C_{ox} for an applied gate voltage (V_g), which, in term, will degrade the MOSFET performance. Although the C_p can be reduced by increasing the dopant concentration in the polysilicon gate electrode, a high dopant concentration would result in dopant penetrating the gate oxide and induce a threshold voltage (V_t) instability problem. On the other hand, using a metal gate as the gate electrode could eliminate the polysilicon gate depletion problem without V_t instability concern because there is no need for dopant to be incorporated into the gate electrode. Another advantage of a metal gate is that the resistance of the metal gate electrode is less than a polysilicon gate.

1.2.2 High dielectric constant insulator compatibility with gate electrode

The compatibility of polysilicon gate electrodes with high dielectric constant (high-k) insulators raises some concern. Most metal oxides with a high dielectric constant used as a gate insulator react with polysilicon and degrade the gate dielectric. Furthermore, this interaction makes it difficult to control the MOSFET threshold voltage. On the other hand, a metal gate is more compatible with high-k metal oxides.

2. Materials screening of high dielectric constant insulators and metal gates

2.1 High-k gate dielectric screening

2.1.1 Issues of high-k gate dielectric

Some fundamental issues with implementing a high-k gate dielectric in MOSFETs are as follows:

- Thermodynamic stability of high-k on silicon
- Trade-off between the dielectric constant (K) and band gap (E_g)
- Film microstructure: crystalline vs. amorphous
- O_2 and dopant diffusion through the grain boundary
- Impact of the amorphous interfacial layer (IL) on the overall dielectric constant of the film, EOT scalability, interfacial roughness, and MOSFET mobility
- Possible mobility degradation and high fixed charge caused by a high-k insulator
- Compatibility with the gate electrode

a. Thermodynamic stability of high-k on silicon

An analysis of the Gibbs free energies governing the following chemical reactions for metal-Si-oxygen ternary systems is important in predicting stability.

To avoid instability with Si to form SiO_2,

$$Si + M_{ox} \rightarrow M + SiO_2$$

To avoid silicide formation,

$$Si + M_{ox} \rightarrow M_{Si} + SiO_2$$

b. Trade-off between the Dielectric Constant (K) and Band Gap (E_g)

From the direct tunneling Eq. [1], it is desirable to find an insulator with a high dielectric constant and high barrier to ensure low gate leakage current density. Since the barrier height values for most high-k metal oxides are not reported, the closest and most readily available indicator for the band offset is the band gap values. Figure 3 shows the plot of band gap versus dielectric constant for various metal oxides.

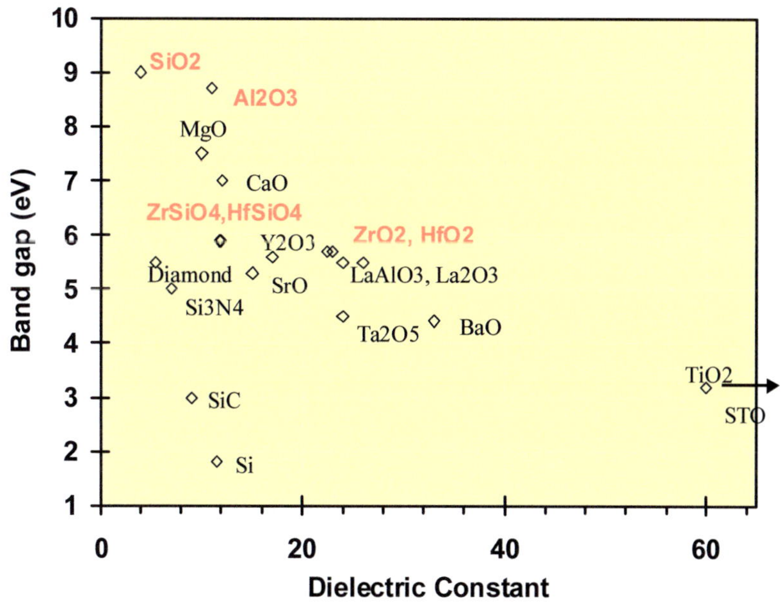

Fig. 3. The trade-off between dielectric constant and band gap limits the choice of metal oxides.

c. Film Microstructure: Crystalline versus Amorphous
For polycrystalline metal oxide, the triple point may generate defects/voids, which will cause a device yield issue. In addition, oxygen, dopant, and impurities diffuse swiftly in the polycrystalline structure primarily through the grain boundary and degrade the electrical properties of the gate stack. Another potential concern is controlling the grain size among small devices and wafers. Amorphous metal oxides can reduce O_2 and dopant diffusion and lower defectivity; however, they usually have a lower dielectric constant than those metal oxides with a polycrystalline structure.

d. Oxygen Diffusion through the Grain Boundary of Metal Oxide
There is a distinct processing difference between the metal oxides and conventional thermal oxide SiO_2. Metal oxides are deposited on the silicon substrate instead of thermally grown like SiO_2. The intrinsic quality of the deposited film is inferior to thermally grown film. A post-deposition anneal under dilute oxygen ambient is necessary for high performance devices. Most metal oxides with a high dielectric constant, however, form a crystalline structure after a relatively high temperature anneal. Therefore the oxygen contained in the ambient of the post-metal oxide deposition anneal diffuses through the grain boundaries of the metal oxides and reacts with silicon substrate, which forms a SiO_2 interfacial layer (IL).

e. Impact of the Amorphous Interfacial Layer (IL) on the Overall Dielectric Constant of the Insulator, EOT Scalability, Interfacial Roughness, and MOSFET Mobility
The SiO_2-like interfacial layer reduces the overall dielectric constant of the bi-layer gate dielectric (high-k on top of a SiO_2 IL). The IL is about 1 nm thick, which would make scaling the EOT to less than 1 nm difficult. The interface between the IL and silicon substrate is rougher than the interface between conventional thermally grown SiO_2 and

silicon, which may degrade channel carrier mobility and generate interface state defects. Figure 4 shows a transmission electron microscopy (TEM) image of the metal oxide MOS structure and the key concerns about the gate stack.

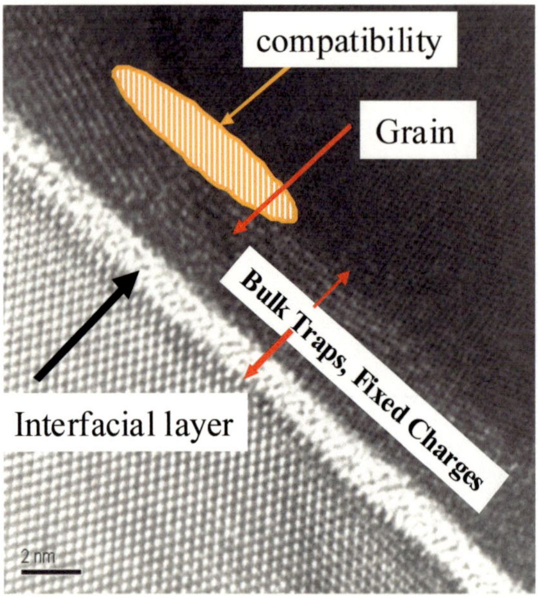

Fig. 4.

f. Possible Mobility Degradation and High Fixed Charge Caused by the High-k Insulator
Soft phonons in the metal oxide bonding structure contribute to the high dielectric constant property of metal oxides. Soft phonons are generated by high atomic number atoms resonating in their bonding structures. These phonons make a lattice contribution to the overall polarizability and therefore the high dielectric constant. There is a concern that the mobility of the channel carriers may be degraded by interactions with these soft phonons.

2.1.2 High-k Gate Dielectric Candidates

a. Metal Oxide
After eliminating the metal oxides that are thermodynamically reactive with silicon substrates or that have a relatively low dielectric constant or small band gap, the remaining candidates fall under group IVB, IIIA, and IIIB of the periodic table.
  i. Metal Oxides Used for Memory Capacitors [1, 2, 3, 4]
  Candidates such as TiO_2 and Ta_2O_5 have the advantage of having a relatively high dielectric constant and a history of processing in the industry. However, the following concerns make them unattractive for logic devices:
  - Small band gap
  - Instability with silicon substrates
  - High density of oxygen vacancies
  - Require an oxygen anneal to improve film quality, which results in oxidation of bottom leading to an undesirable increase in EOT

- Unstable microstructure

ii. Group IIIA and IIIB Metal Oxides: Al_2O_3 and La_2O_3 [5, 6, 7]
Both Al_2O_3 and La_2O_3 are thermodynamically stable with silicon substrates. In addition, Al_2O_3 is amorphous at 1000°C and has a relatively high band gap (8.7 eV ~ SiO_2's value). However, it has a relatively low dielectric constant; high oxygen, B, and P diffusion; and easily absorbs H_2O. La_2O_3 has a relatively high dielectric constant (K ~ 27), but the band gap is small (4.3 eV) and it very easily absorbs moisture from the ambient.

iii. Group IVB Metal Oxides: HfO_2 and ZrO_2 [8-18]
These metal oxides have reasonably high dielectric constants and band gaps (see Figure 3). Both ZrO_2 and HfO_2 devices have demonstrated a many orders of magnitude reduction in gate leakage with an EOT around 1.0 nm and well-behaved transistors. However, they have high O_2 and dopant diffusivity due to their crystalline microstructure. ZrO_2 has a relatively stronger interaction with polysilicon gates than HfO_2. Figure 6 shows the interfacial layer thickness beneath ZrO_2 increases as the post-deposition anneal temperature increases from 550 to 650°C.

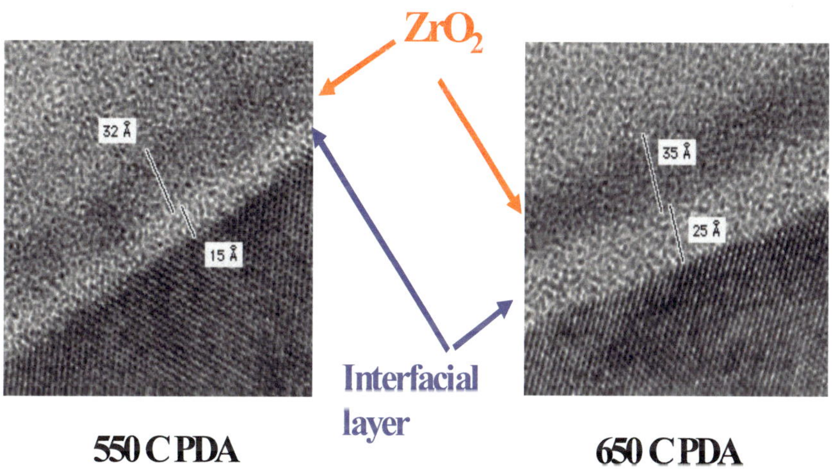

Fig. 6. [13]

Figure 7 shows the x-ray photoelectron spectroscopy (XPS) spectra of the interface between ZrO_2 and the polysilicon gate. The polysilicon gate was deposited in situ on ZrO_2. XPS reveals that ZrO_2 decomposes into a Zr metal compound when the gate stack is annealed in nitrogen at 950°C under ultra-high vacuum leading to a high gate leakage current. It also shows the formation of interfacial SiO_2 between ZrO_2 and the polysilicon gate during polysilicon deposition. These results suggest a strong interaction between ZrO_2 and SiH_4 during polysilicon deposition at 550 to 620°C.

On the other hand, HfO_2 metal oxide is thermodynamically more stable with silicon than ZrO_2. Figure 8 shows the XPS spectra of the interface between HfO_2 and the polysilicon gate. Unlike the ZrO_2 film, the HfO_2 film remains stable after polysilicon deposition and a post-anneal in nitrogen up to 950°C.

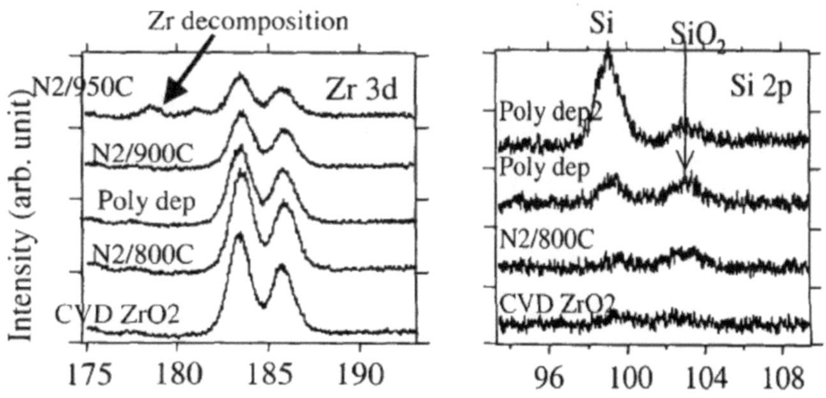

Fig. 7. [14]

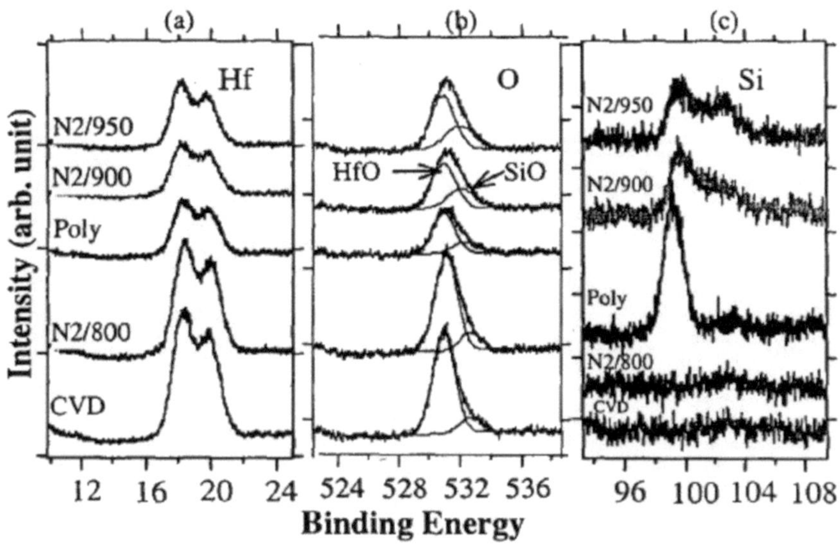

Fig. 8. [9]

b. Metal Silicates (M-Si-O) [19-22]

Adding silicon to metal oxide can maintain the amorphous phase up to a medium temperature such as 800°C depending on the silicon concentration. These metal silicates are thermodynamically stable with the silicon substrate. Figure 9 shows the TEM cross-sections of a gate stack composed of $ZrSi_xO_y$ silicate deposited on a silicon substrate with an aluminum metal electrode. No interfacial layer forms between the zirconium silicate and the silicon substrate after annealing at 800°C for 30 minutes in nitrogen ambient. The interface is atomically sharp, and the film remains amorphous after the anneal. However, there is a possible phase separation with a high temperature anneal and the dielectric constants are lower than metal oxide candidates such as ZrO_2 and HfO_2.

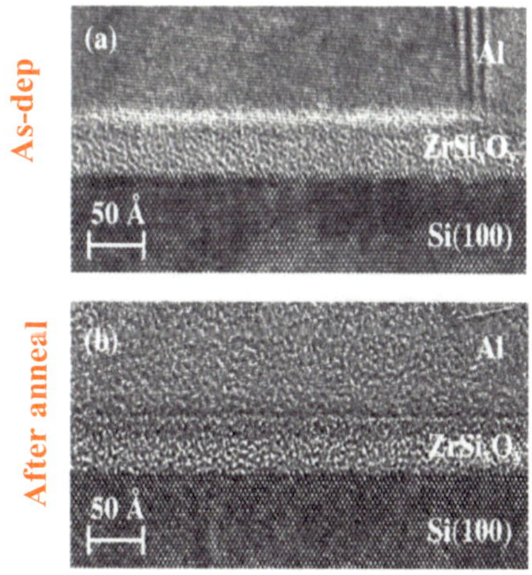

Fig. 9. [21, 22]

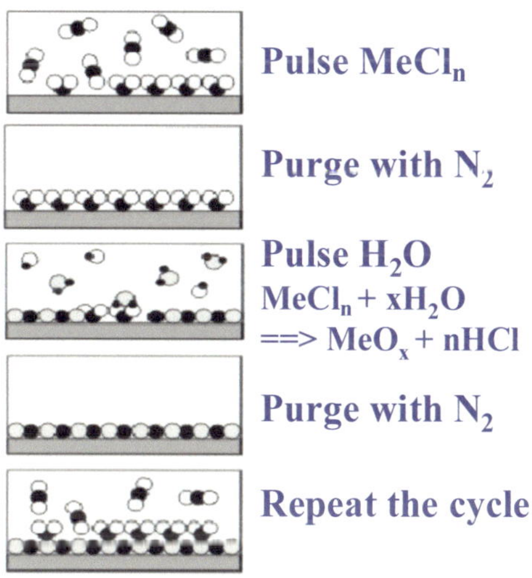

Fig. 10.

2.1.3 High-k dielectric deposition techniques

An important factor in determining the final choice of high-k dielectric is the deposition process, which must be compatible with current CMOS processing, cost, and throughput. In general, there are four major deposition techniques:

- Metal organic chemical vapor deposition (MOCVD) is commonly used, but the C, H, and OH impurities contained in the film are a major concern.
- Physical vapor deposition (PVD) is good for evaluating new materials. However, the purity of the target and plasma-induced damage are common concerns.
- Molecular beam epitaxy (MBE) is good for interfacial control, but the throughput is rather low.
- Atomic layer deposition (ALD) has high uniformity control and good conformality. However, throughput is low and the process is sensitive to surface preparation.

Potential contamination with Cl, C, H, and OH impurities is also a concern. Figure 10 shows a typical ALD process to fabricate metal oxide [23, 24].

In summary, same high-k materials fabricated by different deposition tools, processes, and precursors result in different properties. The final choice of the deposition technique needs to balance cost, throughput, tool reliability, film properties, and device performance and reliability.

2.1.4 High-k gate dielectric device integration issues [17]

a. Device Size Dependence of Gate Current Density

Figure 11 shows that the gate current density of the ZrO_2/polysilicon gate stack has a strong dependence on device size. At 2 V, the leakage current density for the 14 µm NMOS (PMOS) device is 9X (7X) that of a 1.4 µm device. Figure 12 shows the TEM cross-section of the PMOS capacitor with a longer gate length. It is clear that some reactions have occurred at the polysilicon/ZrO_2 interface. The TEM cross-section (Figure 13) shows a nodule extending above and below the polysilicon/ZrO_2 interface. The penetration into ZrO_2 creates a conduction path that results in a high gate leakage current. The longer gate length results in a higher probability that conduction paths will be formed.

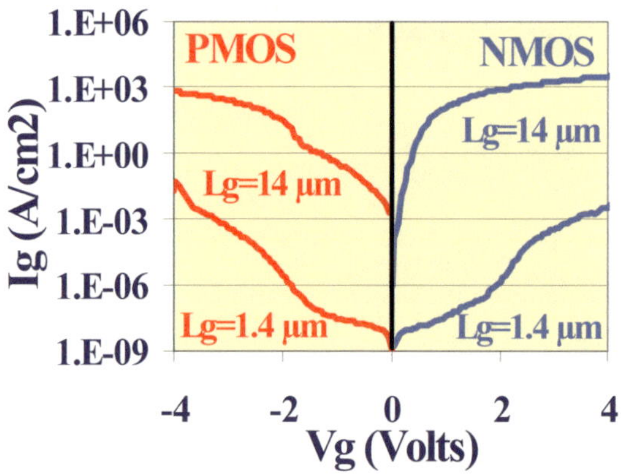

Fig. 11.

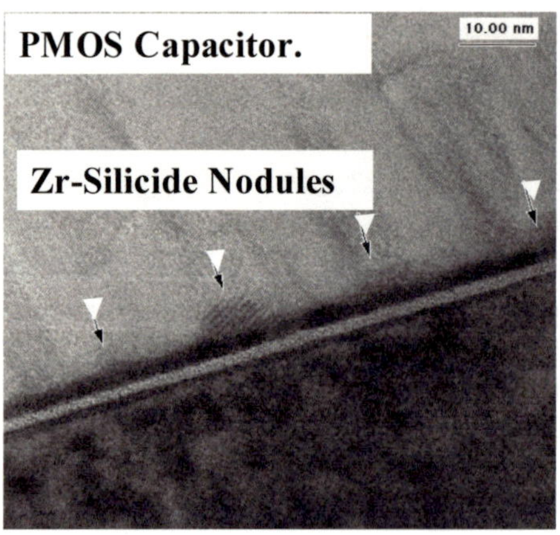

Fig. 12.

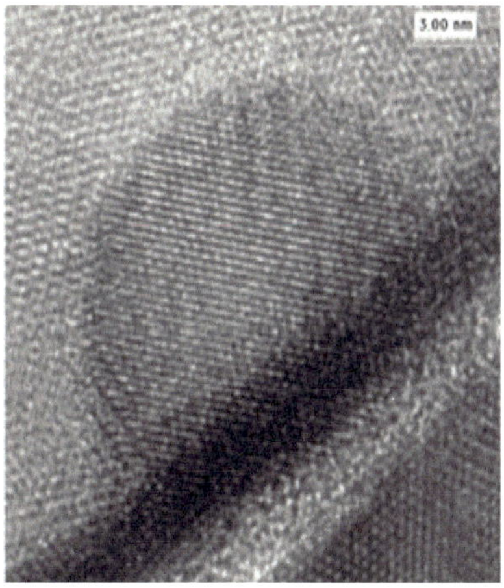

Fig. 13.

b. Lateral Oxidation Model [17]

Figure 14 shows that no nodule is observed at the edge of the devices. A detailed XTEM analysis found the first nodules located ~ 2–3 μm from the edge of the polysilicon gate. This observation suggests that devices with 4–6 μm or smaller gate lengths are nodule-free. It is proposed that active oxygen diffuses through the metal oxide and grows an oxide (SiO_2) at the polysilicon interface (Figure 15). The oxide prevents nodule formation during a high temperature anneal such as a source/drain anneal. Oxidation at the center of large devices is insufficient to prevent the formation of nodules, which cause high leakage current.

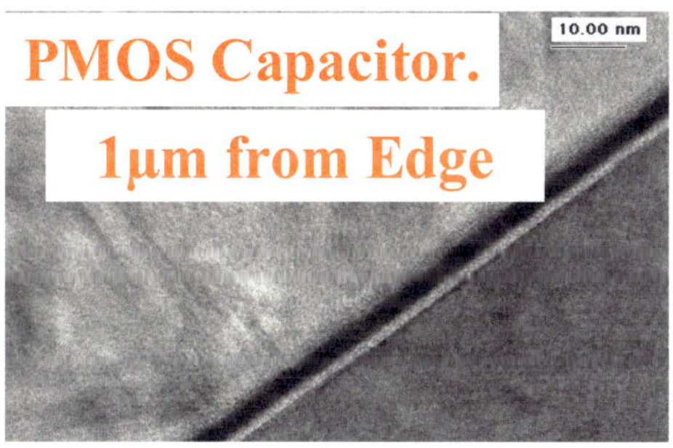

Fig. 14.

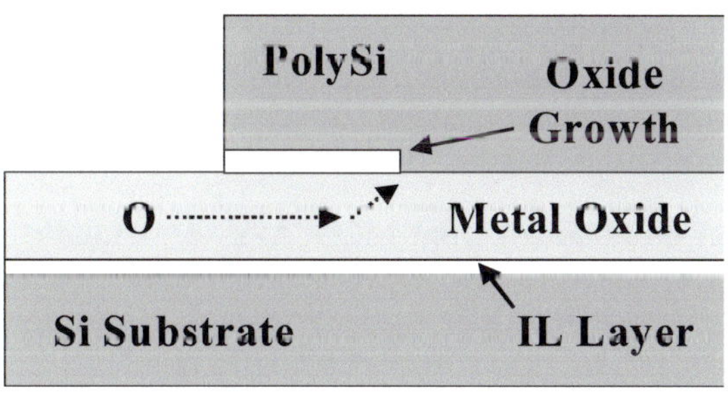

Fig. 15.

2.1.5 High-k dielectric summary

Select high-k metal oxides with thin EOTs have demonstrated an orders of magnitude lower gate leakage than SiO_2. However, it is still a challenge to achieve an EOT thinner than 1.0 nm after transistor fabrication. Degradation in device mobility is observed when using a high-k dielectric. This is more problematic for low EOT applications. Since compatibility with polysilicon gate electrodes is a major concern for some potential metal oxides and metal gate has several advantages, a high-k/metal gate stack is the choice for advanced devices.

2.2 Metal gate electrode materials screening

2.2.1 Materials constraints

A key issue for metal gate materials research is controlling the work function of metal gate electrodes after CMOS processing. There are two choices of metal gate implementation. The first type is a single metal electrode with a work function near the mid-gap (~ 4.6 eV) using additional processing or incorporating additional materials to control the work function of CMOS devices. The second type is a dual metal gate electrode with one metal having a work function (4.1 eV) near the conduction band of the silicon substrate (E_C) for NMOS and the other one having work function (~5.2 eV) near the valence band of the silicon substrate (E_V) for PMOS (Figure 16). The metal electrode materials should have thermal, chemical, and mechanical stability with the high-k gate dielectric and surrounding material during CMOS processing. It is desirable to have a metal gate with a low sheet resistance that is compatible with CMOS process integration, either a conventional "gate first" or replacement "gate last" approach. Metallic elements, compounds (nitrides, silicides, carbides, borides, etc.), and solid solutions are possible candidates.

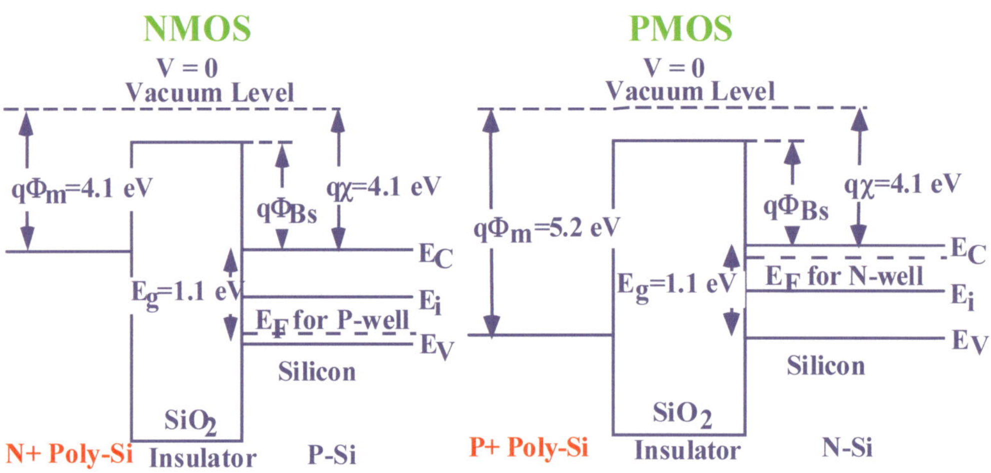

Fig. 16.

Work functions can be obtained from MOSCAPs of various oxide thicknesses using the following equation:

$$V_{fb} = \phi_{MS} + Q_F / C_{OX} \tag{6}$$

An intersect in the Y-axis of the V_{fb} versus EOT plot is the work function (Figure 17). Table 2 shows the work function values for some potential metal gate electrode candidates.

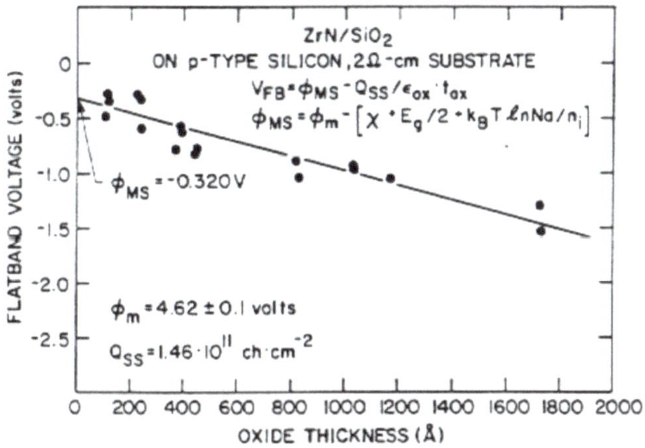

Fig. 17. [25]

Gate Material	Work Function (MOS) (eV)
Ti	4.17, 4.33, 4.6
TiN	4.95
TiSi2	3.67-4.25
Zr	4.05
ZrN	4.6
ZrSi2	-
Ta	4.25, 4.6, 4.15-4.25
TaN	5.41
TaSi2	4.15
Nb	4.3, 4.02-4.3
NbN	-
NbSi2	4.35-4.53
W	4.75, 4.72, 4.55-4.63
WNx	5
WSi2	4.55-4.8
Mo	4.64, 4.53 4.6
MoN	5.33
MoSi2	4.6-4.8, 4.9
Al	4.1

Table 2.

2.2.2 Metal electrode material deposition method and film properties

Metal deposition processing parameters and post-deposition processing affect the metal film properties such as resistivity, microstructure (grain size and orientation), stress, and adhesion. Deposition processes such as CVD and PVD are common methods. In general, CVD films provide better conformality and negligible damage compared to PVD films, but the process may incorporate contaminants. Atomic layer deposition (ALD) shows excellent conformality, but throughput is low. Figures 18 and 19 show the impact of the metal gate deposition process on device performance and gate leakage current, respectively. The device with a SiO_2/PVD TiN metal gate stack results in lower mobility than CVD TiN or polysilicon gated devices. The devices with a SiO_2/PVD TiN gate stack result in a 100X higher gate leakage current than devices fabricated with CVD and ALD.

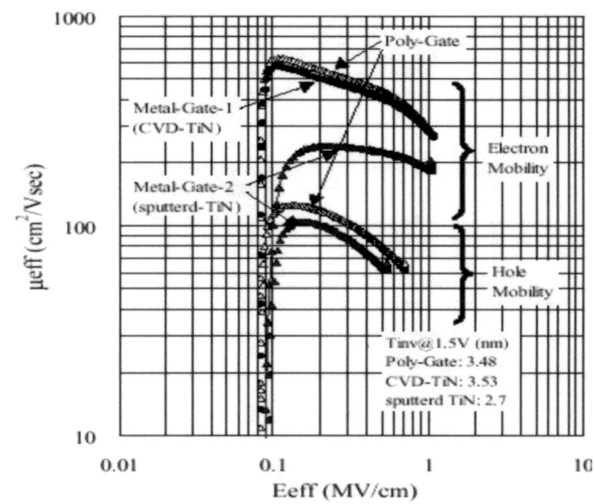

Fig. 18. [26]

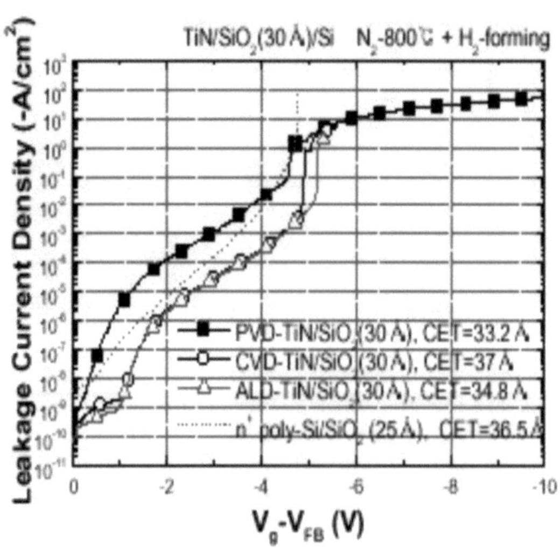

Fig. 19. [27]

2.2.3 Metal gate device integration issues

The dual metal gate approach that needs different metal gate materials for NMOS and PMOS, respectively, increases process complexity dramatically. Contamination (mostly from CVD) and plasma damage (mostly from PVD) affect gate device parameters such as Q_f, D_{it}, V_t stability, and gate oxide integrity. Another concern with metal gates is poor oxidation resistance. The etchability of metal materials, especially selectivity to metal oxides, is a key area for process development.

a. Conventional "Gate First" CMOS Integration

Conventional gate first integration involves the ability to etch the gate material. For example, Figure 20 shows a nitride/W/TiN gate electrode stack on top of gate oxide. The gate electrode materials must be protected from oxidation or attack by wet chemicals. The gate electrode materials must also be stable with their surrounding materials during high temperature steps. These requirements may limit the choices of materials for metal gate materials along with alternative high-k dielectrics.

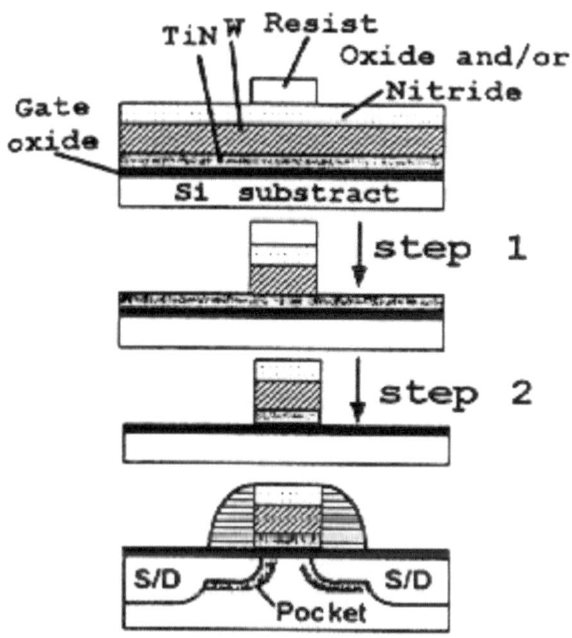

Fig. 20. [28, 29]

b. Replacement "Gate Last" CMOS Integration

This integration eliminates many constraints posed by conventional "gate first" integration such as process-induced damage, the requirement to etch new materials, thermal/chemical stability concerns, stress-induced diffusion issues. After completing conventional MOSFET fabrication with replacement polysilicon, gate oxide is deposited and using chemical mechanical polishing (CMP) technique to flatten the top surface (Figure 20a). A wet etch or dry etch process is used to etch off the polysilicon gate (Figure 20b) followed by a new metal gate electrode material and a new gate dielectric deposition (Figure 20c). Another CMP process or a second patterning of the gate is used to form the final structure.

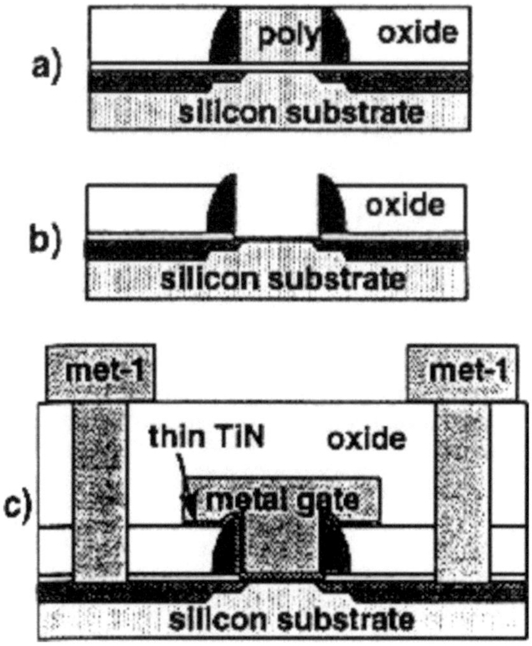

Fig. 20. [30]

c. Summary

A single metal gate with a mid-gap work function may not be able to achieve a low MOSFET threshold voltage and boost performance; however, it has potential for fully depleted silicon-on-insulator (FDSOI) applications. Dual metals with work functions similar to n^+ polysilicon and p^+ polysilicon pose significant integration challenges. Thermodynamic and mechanical stability is an important issue in the choice of metal gate materials. Both the gate first and gate last integration have advantages and disadvantages. The final choice of integration scheme will be decided by performance, yield, and cost.

3. Device characteristics using High-k/Metal Gate (HKMG) stack

3.1 Impact of defects in the HKMG stack on device performance and reliability

As mentioned in section 2.1.1.d, the deposited high-k gate dielectric contains a high defect density in the bulk even after a post-deposition anneal (PDA). The quality of the interface between the silicon substrate and interfacial layer (IL) is not as good as the interface between silicon and conventional thermally grown gate oxide SiO_2. A TEM cross-section of the HKMG gate stack, shown in Figure 21, highlights the different regions of the gate stack. The defects in the bulk of the high-k gate dielectric and at the interface between the silicon substrate and IL have a significant impact on device performance and reliability, such as threshold voltage stability. As discussed in Section 2.1.2.a.iii, the best candidate for a high-k dielectric is Hf-based metal oxide. Therefore the following discussion is based on HfO_2/metal gate stacks.

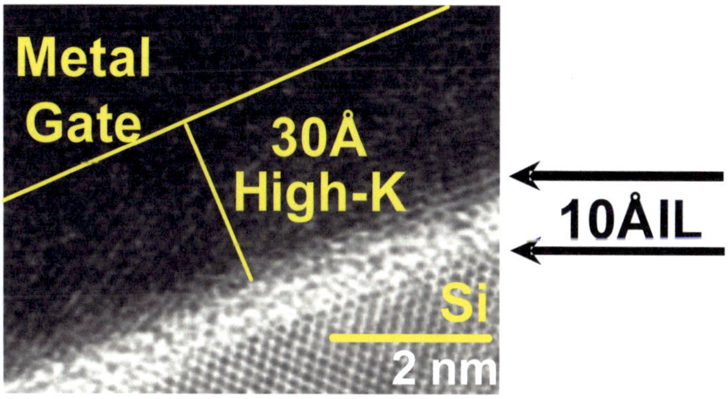

Fig. 21.

3.1.1 Si/IL interface improvement – stressed relaxed pre-oxide [31]

Depositing high-k on top of thick, high quality, thermally grown SiO_2, taking the advantage of its better interface quality, is not desirable because we need a thin gate dielectric to increase transistor speed. To solve this problem, a stress relieved pre-oxide (SRPO) process has been developed to improve the interfacial properties between the high-k dielectric and silicon substrate while maintaining the required thinness to meet the speed enhancement requirement for integrated circuits. The experiment discussed here is as follows. The SRPO is formed by growing a relatively thick thermal oxide with a high temperature anneal (higher than the SiO_2 glass flow temperature of ~ 980°C) for stress relief followed by etching the thermal oxide back to 10 Å using a diluted 700:1 hydrofluoric acid: H_2O solution. The HfO_2 is then deposited by ALD. After the high-k dielectric undergoes a PDA, a TaSiN metal gate is deposited. A commercial CMOS process technology with a source/drain anneal at 1000°C was used to fabricate metal gate/high-k stack nMOSFETs on bulk silicon [32]. Fig. 22 compares the threshold voltage (V_t) shift under constant voltage stress for the control split using the standard process (an RCA clean followed by ALD HfO_2 with a TaSiN metal gate) and the new SRPO process with a TaSiN/HfO_2 gate stack for short channel devices (W/L = 10 µm/0.15 µm). The SRPO with a TaSiN/HfO_2 stack results in a 3X smaller V_t shift than the standard process. These results suggest that devices with the standard process suffer process-induced gate edge damage during transistor fabrication, which increases the V_t shift due to greater trap generation. The process-induced gate edge damage is reduced significantly when using SRPO due to the high quality interfacial layer under the HfO_2, which suppresses interface trap and border trap generation during constant voltage stress. To assess process-induced gate edge damage, the charge pumping current was measured on short channel devices using a pulse string with a fixed base level and varying pulse heights [33] under drain or source bias to detect local charge at the gate edge. The results show less interface and border state trap density for HfO_2/SRPO devices than for HfO_2/RCA devices. The stressed charge pumping results clearly show that the SRPO pre-treatment is much more robust than the RCA pre-treatment under process-induced local charge generation near gate edges. Fig. 23 shows that the normalized transconductance (G_m) of TaSiN-gated short channel devices with SRPO is higher than the standard pre-treatment due to better interface properties with the SRPO process. The fundamental difference between a chemical oxide formed after an RCA clean and SRPO is as follows. The quality of the chemical oxide is poor and has a lower density than thermally grown SiO_2. It is susceptible to process-induced gate damage. On the other hand, for thermally grown SiO_2, a mechanical stress builds up as the SiO_2 becomes thicker during oxidation caused by the molar volume mismatch between Si and SiO_2 and their different expansion coefficients. The stress degrades the Si-O bonding configuration near the interface between SiO_2 and the silicon substrate. The quality of the sub-oxide layer underneath the bulk SiO_2 is poor. It is therefore important to relieve the stress build-up during oxidation to have a high quality sub-oxide. Annealing the oxide at a temperature higher than the SiO_2 glass flow temperature (~980°C) allows the SiO_2 to "flow" and the bonding configuration near the interface to be rearranged, improving the sub-oxide quality. The SRPO process grows a relatively thick thermal oxide during a high temperature (~ 980°C) anneal, resulting in a high quality sub-oxide followed by an etch back to 10 Å using a diluted HF solution. The 10 Å SiO_2 serving as a pre-oxide before high-k deposition thus becomes a high quality film. Excellent cross-wafer inversion T_{ox} uniformity is demonstrated using the SRPO pretreatment for an array of twenty-eight 10 μm × 10 μm (W/L) devices as shown in Fig. 24.

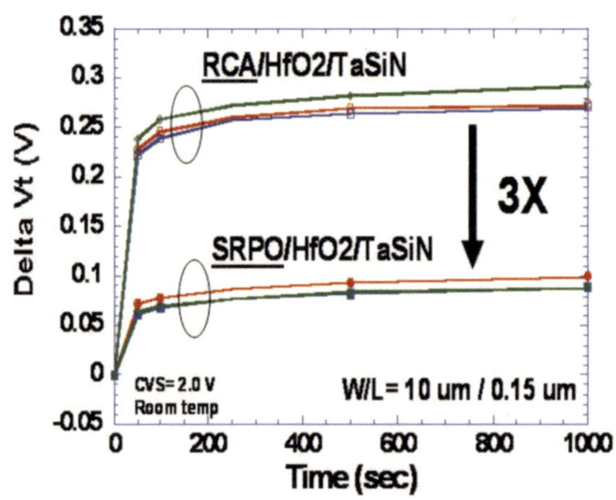

Fig. 22.

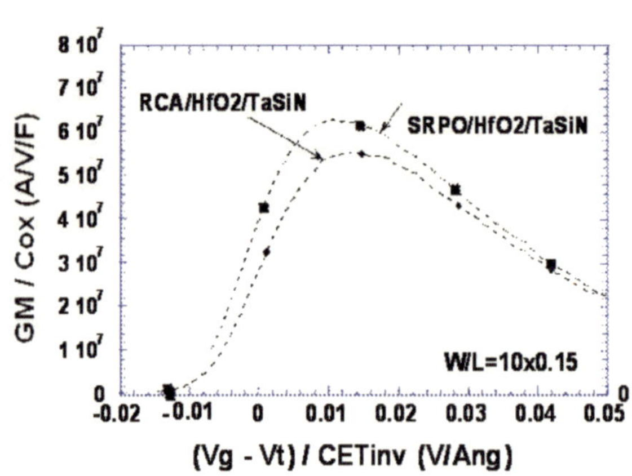

Fig. 23.

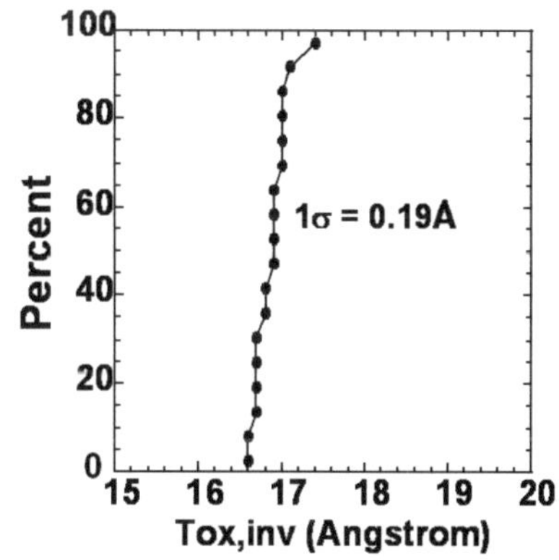

Fig. 24.

3.1.2 Si/IL interface improvement – deuterium incorporation [34]

The other approach to improve interface robustness is to incorporate deuterium into the interface between the silicon substrate and IL because the Si-D bonds are much harder to break than Si-H bonds. One effective way to incorporate deuterium is to use D_2O instead of H_2O as the oxidant precursors during ALD [34]. Introducing deuterium in situ during ALD is an effective way to passivate the Si dangling bonds because it does not require the pre-existing hydrogen to be replaced. Hydrogen can be introduced easily from the processes after high-k gate dielectric deposition. Fig. 25 compares the threshold shift for nMOSFETs under positive bias stressing at 25°C and 125°C. A significant reduction in V_t shift is observed for D_2O-processed HfO_2 devices. The presence of deuterium at the interface is supported by the low energy secondary ion mass spectroscopy (SIMS) analysis (Fig. 26), which reveals a spike of deuterium at the 7E17 at/cm^3 level. To improve the depth resolution of SIMS, we used 100 Å D_2O-processed HfO_2 to prepare the test wafers. The SIMS analysis of another sample prepared by using 100 Å H_2O-processed HfO_2 shows no deuterium at the bottom interface.

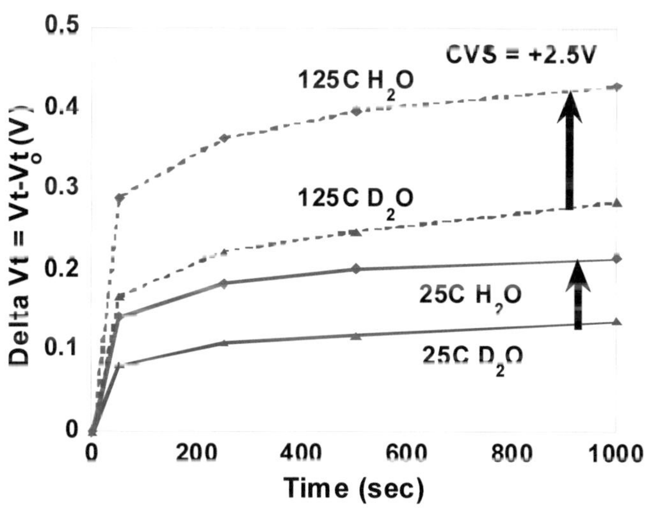

Fig. 25.

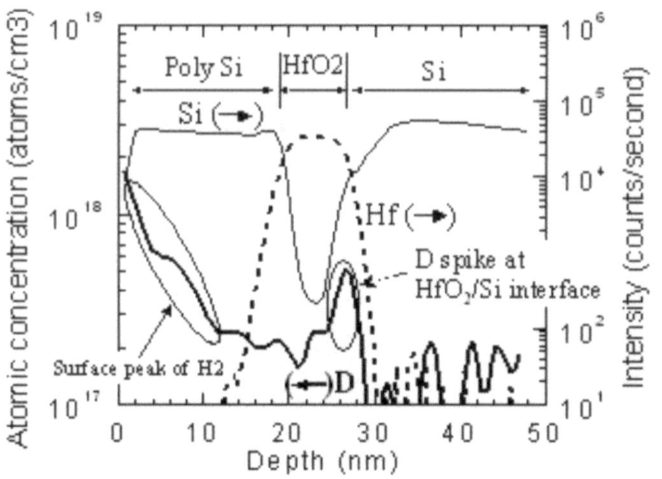

Fig. 26.

3.1.3 Fluorine passivation coupled with SRPO [35]

In addition to the defects at the interface, defects contained in the bulk HfO_2 will degrade device reliability and performance. Incorporating fluorine into the bulk HfO_2 and interfaces can passivate the defects, thus improving the device robustness and speed. Combining SRPO interface engineering and defect passivation with fluorine in a high-k/Ta_xC_y stack showed excellent V_t stability [35]. The positive bias temperature instability (PBTI) time to failure (TTF)-lifetime extraction using stress voltages in the direct tunneling regime is shown in Fig. 27. Fluorinated devices exceed the V_t TTF lifetime target (<30 mV Vt shift in 10 years at 105°C for V_{dd} = 1 V) with a sufficient margin and reveal about a four orders of magnitude longer PBTI lifetime than the control that does not incorporate fluorine. SIMS analysis results confirmed the incorporation of fluorine at the interfaces and bulk high-k [35], which is expected to reduce the defect density and thus reduce charge trapping. It has been shown that oxygen vacancies are the major defects in the bulk HfO_2 [36]. Table 3 shows the results of atomistic calculations of the formation energies of fluorine-passivated oxygen vacancies in the bulk of the Hf-based high-k using density functional theory (DFT) [37, 38] as implemented in the local orbital SIESTA code [39]. The results show that the fluorine passivation of threefold (V3) or fourfold (V4) oxygen vacancies is energetically favorable and can therefore lead to less trapping. Fig. 28 shows the structure used to calculate a fluorine atom at a V3 site. The structure of the fluorine-passivated V4 defect (not shown) is similar. To electrically support the bulk trap density reduction from fluorine incorporation, stress-induced leakage current (SILC) was measured. Fig. 29 compares the SILC of fluorinated devices with the control. Fluorine incorporated in the bulk reduces the SILC significantly. The results are consistent with bulk defect passivation by incorporating fluorine. Finally, Fig. 30 shows the gate leakage for the fluorinated high-k/Ta_xC_y stack devices is over 4 orders of magnitude lower than that of silicon oxynitride. The CETinv is the sum of EOT and the quantum mechanical effect in the silicon substrate, which is around 0.4 nm.

Reaction	Energy Gain (eV)	F Coordination #	F Charge
Hf-V3-Hf + F(interstitial) → Hf-F-Hf	-7.66	3	7.18
Hf-V4-Hf + F(interstitial) → Hf-F-Hf	-7.66	3–4	7.18

Table 3.

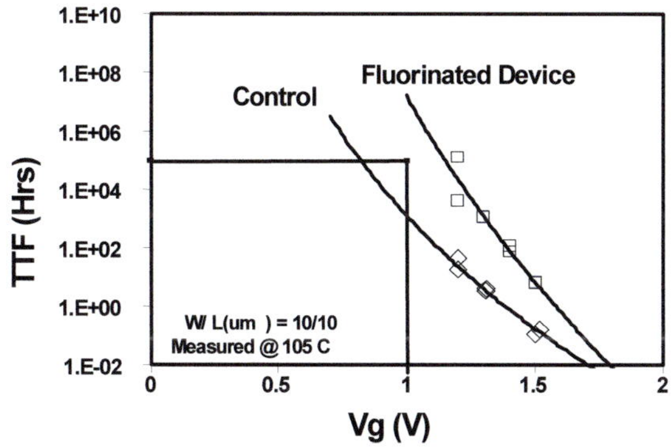

Fig. 27.

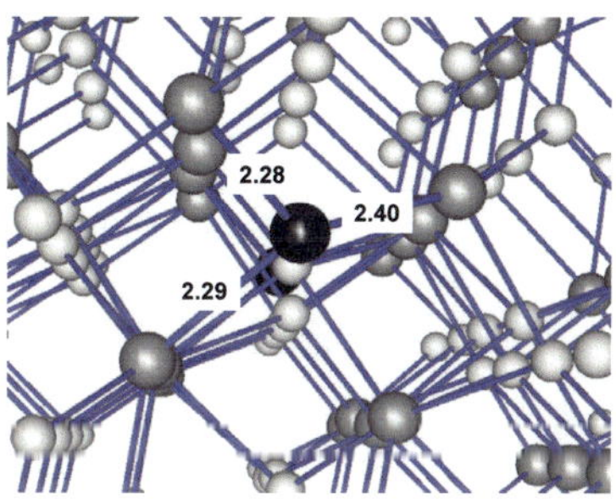

Fig. 28.

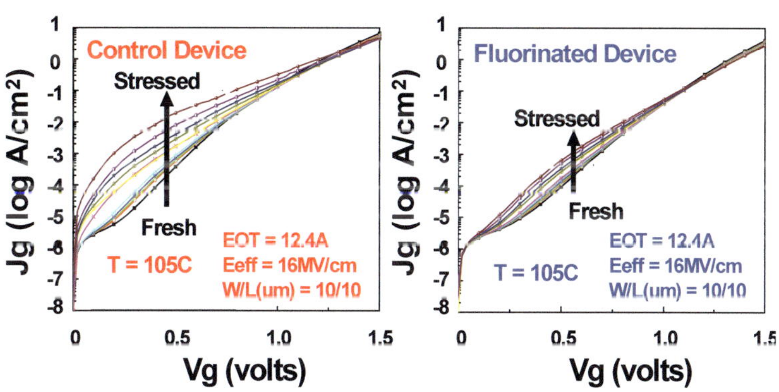

Fig. 29.

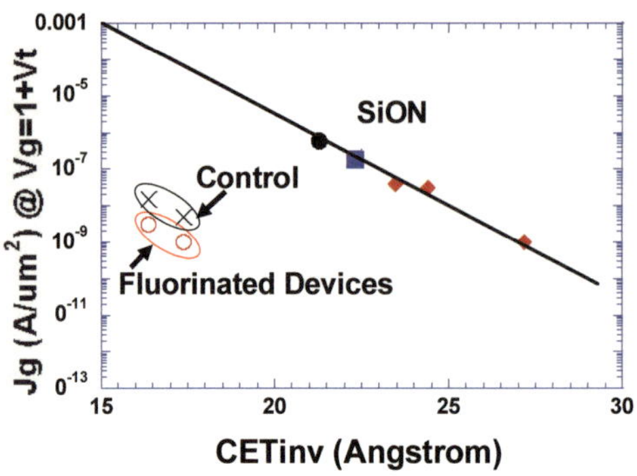

Fig. 30.

3.2 Threshold voltage turning

3.2.1 Effective work function of metal gates

In an nMOSFET device, threshold voltage (V_t) is the voltage at which electrons in the inversion layer formed at the substrate Si/dielectric interface are sufficient to produce a conducting path between the MOSFET source and drain (S/D). In CMOS applications, the effective work function (EWF) of metal gate electrodes is an important parameter as it determines the flatband voltage (V_{fb}) and, subsequently, the V_t of MOSFETs. If no charge is present in the oxide or at the Si/dielectric interface, the V_{fb} equals the work function difference between the gate metal and the semiconductor substrate as shown in Eq. [7]. The work function values shown in Table 2 are based on this simple method. However, the work function of the metal, and thus the V_t obtained after the CMOS processing, is likely different from that obtained from Eq. [7]. The work function obtained after the CMOS processing is called the EWF, which determines the final V_t of CMOS.

A precise measurement of the metal gate EWF is crucial to identifying the optimal electrode material. However, EWF measurements of metal electrodes on high-k dielectrics have shown a dependence on the particular dielectric material and are further complicated by Fermi-level pinning at the high-k/metal interface [40]. Insufficient understanding of metal-electrode systems and their interactions with underlying dielectrics has contributed to inconsistent EWF values. It has further been reported that factors such as specific processing and associated thermal budgets affect the final EWF of high-k/metal electrode systems [41] due to composition variations and the temperature-driven crystalline phase production of metal (metal oxides, metal nitrides, and metal oxynitrides) electrodes.

The EWF of metal electrodes on high-*k* material may be extracted more precisely from MOS capacitors fabricated by depositing and patterning the high-*k* dielectric and metal gate capacitor structures on a "terraced" oxide layer consisting of incrementally thicker layers of thermally grown SiO_2 [42]. The multiple oxide thicknesses (1.0, 1.5, 2.5, 3.5 nm) on a single wafer are achieved by growing a relatively thick oxide and selectively etching regions with diluted hydrofluoric acid. Complete details of this etch process and evaluated work function results have been described previously [42]. Fig. 31 illustrates the bottom SiO_2 terrace step thickness measurements recorded by spectroscopic ellipsometry (SE) in a diameter scan across the wafer. The wafer image insert in Fig 30 illustrates the concentric thickness bands. To evaluate metal gate work functions on high-k films, a fixed amount of high-k (~ 2.0 nm) is deposited on the terraced oxide followed by ALD of a 10 nm TiN (or other metal) gate electrode. The EWFs for terraced oxide stacks were extracted using Eq. [7b] [42], which is a simplified form of the general model given by R. Jha [43] (Eq. [7a]) to enable linear extraction of the V_{fb}-EOT relationship:

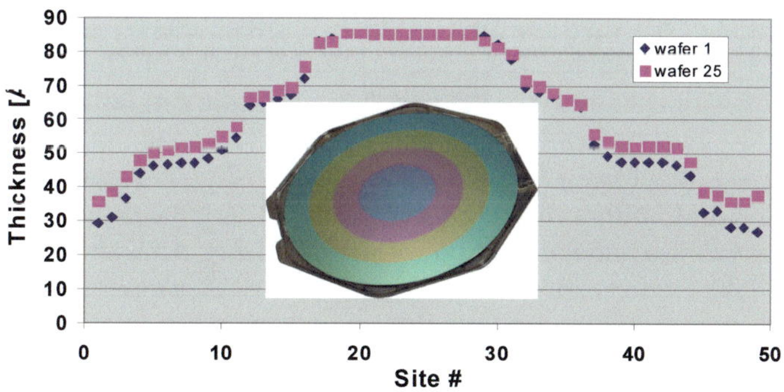

Fig. 31. Illustration of etched oxide thickness as a function of position across the diameter of a 200 mm wafer. [42]

$$a)\; V_{fb} = \Phi'_{ms} - \frac{Q_f * EOT}{\varepsilon_{ox}} - \left(\frac{Q_i * t_h}{\varepsilon_h} - \frac{1}{\varepsilon_h}\int_0^{t_h} x\rho_b(x)dx - \frac{1}{\varepsilon_{ox}}\int_{t_h}^{t-t_h} x\rho_{ox}(x)dx\right),\; t_h = \frac{\varepsilon_h}{\varepsilon_{ox}} EOTh$$

$$b)\; V_{fb} = \left(\Phi'_{ms} - \frac{Q_i * EOTh}{\varepsilon_{ox}} - \frac{\rho_b * \left(\varepsilon_h / \varepsilon_{ox}\right) * EOTh^2}{2 * \varepsilon_{ox}}\right) - \frac{Q_f * EOT}{\varepsilon_{ox}} \quad (7)$$

With the terraced oxide structure, a constant fixed interface charge between the Si substrate and dielectric (Q_f) is maintained across the varying oxide thicknesses, minimizing wafer-to-wafer variation associated with multiple wafer extraction methods. Q_f can be calculated from the slope of the V_{fb}-EOT relationship. The SiO_2 bulk charge term in Ref. [43] can be neglected to simplify the extraction since this bulk charge contribution is far less than that of Q_f [44]. The extracted y-axis ordinate intercept value contains the contribution of the metal WF as well as the high-k/SiO_2 interface charge (Q_I) and high k bulk charge density (ρ_b) terms (t_h: high-k physical thickness, t_{ox}: SiO_2 physical thicknesses, t: total physical thickness of high-k/SiO_2 stack; EOT_h: high-k EOT; EOT: EOT of high-k/SiO_2 stack, ε_h: high-k dielectric constant, ε_{ox}: SiO_2 dielectric constant). The contributions of charges in the high-k on V_{fb} are controlled and can be minimized (~ +50 mV) by using a fixed and thinned high-k film (2–3 nm), thus enabling accurate extraction of the EWF. The resultant terraced oxide capacitors exhibit excellent linear V_{fb}-EOT fits with minimal effects from variations in fixed charge at the Si/SiO_2 interface.

3.2.2 Dielectric capping for work function tuning

Although TaSiN (nMOS), TiN (pMOS), and Ru (pMOS) on HfO_2 have been demonstrated in CMOS integration [45, 46], the use of dual metal gates must address the significant complexity of optimizing two different metal etch processes. A single metal gate approach for CMOS integration provides several advantages over the dual metal electrode process, specifically a more straightforward integration and less demanding gate etching process optimization and control. A single metal gate, however, requires tuning the EWF value to obtain the proper n- and p-type threshold voltage. These tuning efforts have given rise to an extensive study of the impact of capping layer, a thin metal oxide incorporated between the Hf-based high-k dielectrics and metal gates, whereby a single metal gate CMOS process can be implemented with either single or dual cap layer approaches [47]-[52].

Thin cap layers (≤ 1 nm) such as Dy_2O_3 and Al_2O_3 deposited on the high-k gate dielectric followed by thermal annealing drives the metal atoms of the cap layer into the high-k layer (and bottom SiO_x interface layer). Appropriate control of cap layer doping (depth of diffusion) has tuned the EWF of metal electrodes by dipole formation at the interface between the high-k and bottom interfacial SiO_x layer. The EWF of metal electrodes can be shifted toward the valence and conduction bands of Si using AlO_x and LaO_x capping layers, respectively [52]-[55]. Performance data suggests that AlO_x capping is also effective after high temperature processing, indicating that Al atoms need to diffuse to the interface between the HfO_2 and the interfacial SiO_2 layer to shift the EWF [56]. Modulating LaO_x within the HfO_2 has been shown to contribute to this EWF shift as a result of La concentration at the SiO_x/HfO_2 interface. The relative concentration trends with the amount of EWF shift [57]. Coincident Hf and La EELS element profiles and SIMS profiles in Fig. 32 and Fig. 33, respectively, verify that La migrates from a La_2O_3 cap layer into un-annealed HfO_2 [58]. These observations can make the dipole moment formed at the internal interface a more feasible explanation for EWF tuning.

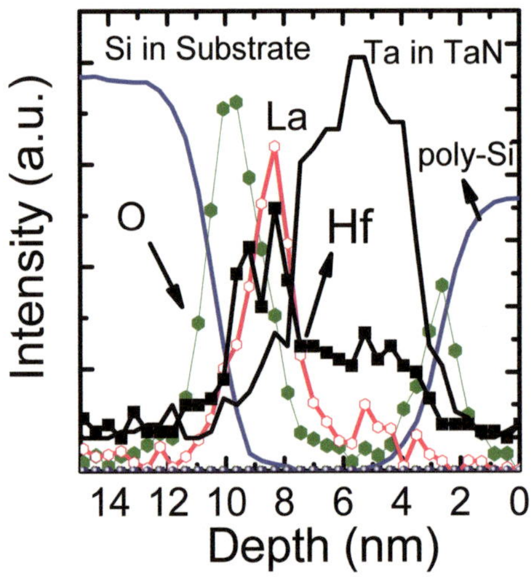

Fig. 32. EELS scan showing intermixing of HfSiO and La_2O_3.

Analysis of the energy band diagram of the multilayer dielectric stack suggests that the EWF of the gate stack can be modified by the dipole at the interface between two dielectric layers, typical of the high-k gate stack (i.e., the $Si/SiOx/HfO_2$/electrode, where a SiO_2-like IL is formed). As indicated above, such a dipole layer can be formed by introducing metal ions into the IL near its interface with the high-k film. The electronegativity of the metallic elements with respect to Hf atoms presents a plausible case for a model explaining EWF tuning [59]. Fig. 34 illustrates the effect of a band offset change between the high-k and interfacial oxide induced by incorporating metal ions into the interfacial layer. A metallic element can be incorporated in the IL by diffusion during thermal processing, specifically with a high temperature (~ 1000°C) S/D dopant activation anneal. Based on their electronegativity relative to Hf, metallic elements can generate either "positive" (i.e., Al (Ti)-O-Hf) or "negative" dipoles resulting in a higher or lower EWF, respectively. Furthermore, ab initio calculations show that the Al-O-Hf dipole results from substituting Al for Si in SiO_2 near its interface with HfO_2, thereby significantly reducing the SiO_2/high-k valence band offset and thus effectively increasing the EWF [60].

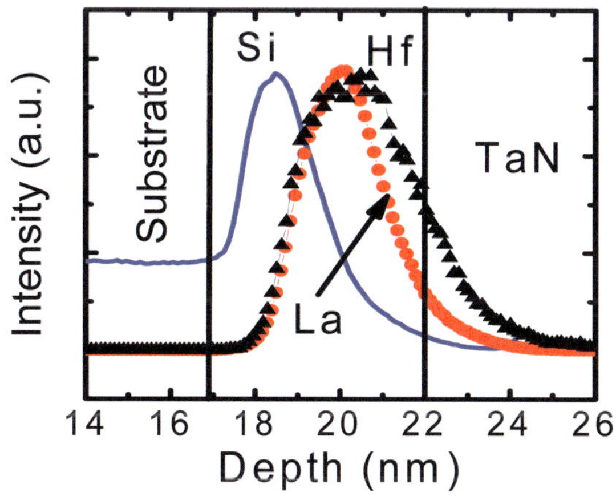

Fig. 33. SIMS profile confirms intermixing of La_2O_3 with the HfSiO layer.

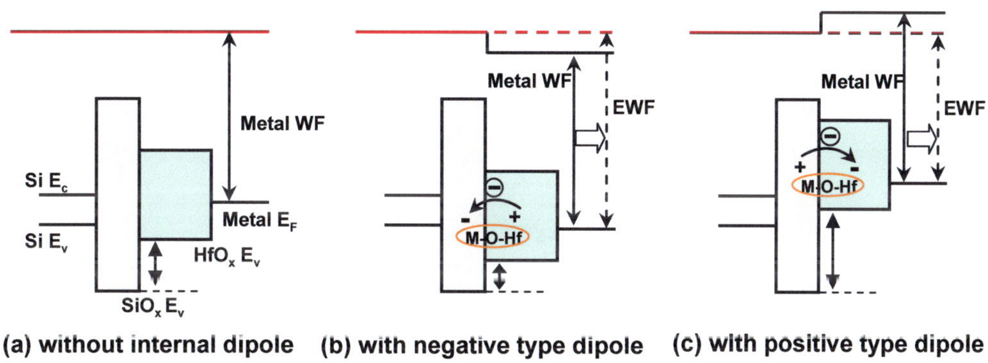

Fig. 34. Band diagram illustrating the EWF change of the metal gate depending on the type of dipole.

In nFETs, La is found to be the most effective dopant based on its overall effect on V_t, EOT scaling, mobility, and reliability [50, 53, 54]. An Al_2O_3 cap has been widely used for pFETs, but it increased the EOT [52] (since it has a relatively lower dielectric constant value) as well as raises reliability concerns due to Al diffusing into the interfacial oxide layer [52]. In addition, the V_{fb}-EOT roll-off phenomenon (see next section) was observed in Al_2O_3-capped gate stacks, making it even more difficult to achieve a proper V_{fL} (V_t) for pMOS.

3.2.3 V_{fb}-EOT Roll-Off

When gate stack film systems consisting of a metal electrode, high-k dielectric, and SiO_2 IL were used in devices of practical interest with scaled EOTs, their V_{fb} values at thin gate stacks were found to be significantly less than those obtained in test structures with thicker gate stacks. This is most remarkable in gate stacks with high EWFs [61], as seen in Fig. 35 showing a grand plot of V_{fb} vs. EOT in pMOSCAPs with various metal gates fabricated on terraced oxide structures. The V_{fb} roll-off starts at a certain minimal thickness of the SiO_2 IL and increases as the SiO_2 becomes thinner. One can clearly see a gradual reduction of V_{fb} as the gate stack EOT scales below a certain value. It is more prominent in gate stacks annealed at higher temperatures and/or for a longer anneal time. This V_{fb}-EOT roll-off phenomenon is a thermally activated process that suggests an intrinsic relation to the EWF values of the gate stack and/or the change in electrical integrity of the physically scaled bottom interface layer.

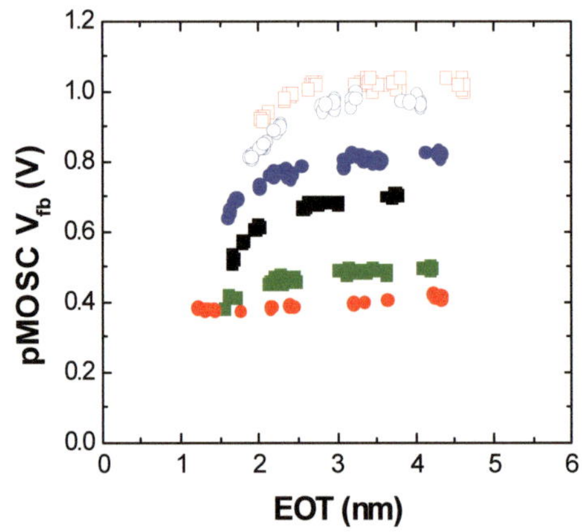

Fig. 35. Dependence of V_{fb} on EOT in high-k terraced oxide capacitors with metal electrodes of different WF values.

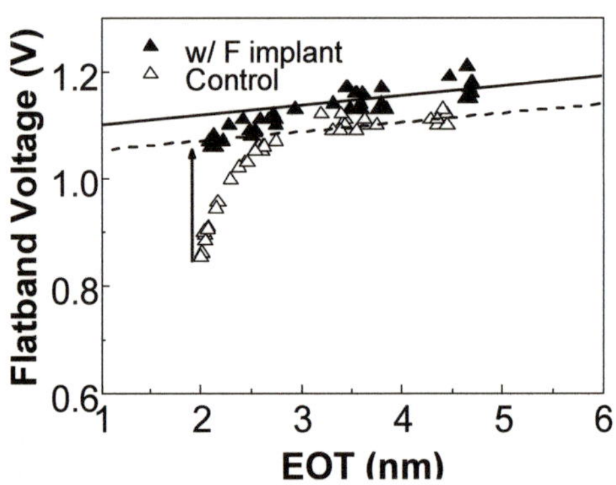

Fig. 36. Advantages of fluorine incorporation for V_{fb}-EOT roll-off reduction

The likely root cause of the V_{fb}-EOT roll-off problem in HKMG devices is the oxygen vacancies in the high-k stack, which trigger the generation of defects at the bottom of the SiO_2/Si interface. Therefore processes that can minimize the oxygen vacancy density in the high-k bulk and/or enhance the robustness of the SiO_2/Si interface are critical to reducing the V_{fb}-EOT roll-off in advanced HKMG devices. From the point of view of interface quality, stress in the transitional region of the SiO_2 interfacial layer enhances the diffusion of oxygen up from the interface, which makes the V_t roll-off problem more severe. Therefore a stress-relieved SiO_2/Si interface is desirable to minimize V_t roll-off. The SRPO process, deuterium incorporation, and the combination of SRPO with defect passivation with fluorine (see section 3.1.3) are effective approaches to minimize the V_{fb}-EOT roll-off problem. Indeed, F+ implanted in the gate stack with 1 nm SiO_2 under 2 nm HfSiOx followed by a 1000°C/10 sec. anneal [62] significantly reduces the roll-off (Fig. 36).

4. Conclusion

The high-k/metal gate stack has been used for high performance and low power semiconductor products replacing the SiO_2/polysilicon stack, which has been used for decades. The motivation for this replacement as well as materials screening for the high-k/metal gate stack have been discussed. Two major advantages of implementing HKMG stacks are that they contribute to reducing the gate leakage current and scaling the EOT of advanced CMOS. Progress in high-k/metal gate device threshold voltage tuning, including an evaluation of capping layers, has also been presented. While nMOS V_t is acceptable for various applications of Si CMOS devices, achieving a targeted pMOS V_t is still challenging due to V_{fb}-EOT roll-off issues. Approaches to minimize the pMOS V_{fb}-EOT roll-off problem have been outlined. Clearly, CMOS scaling will continue to provide opportunities for exciting research in high-k/metal gate modules in the future.

5. References

[1] X. Guo, X. Wang, Z. Luo, T. P. Ma, and T. Tamagawa, "High quality ultra-thin (1.5 nm) TiO2/Si3N4 gate dielectric for deep sub-micron CMOS technology," IEDM Tech. Dig., pp. 137 - 140, December 1999

[2] C. Hobbs, R. Hegde, B. Maiti, H. Tseng, D. Gilmer, P. Tobin, O. Adetutu, F. Huang, D. Weddington, R. Nagabushnam, D. O'Meara, K. Reid, L. La, L. Grove and M. Rossow, "Sub-Quarter Micron CMOS Process for TiN-Gate MOSFETs with TiO2 Gate Dielectric formed by Titanium Oxidation," Symp. on VLSI Tech. Dig., 1999, P. 133-134

[3] Donggun Park; Ya-chin King; Qiang Lu; Tsu-Jae King; Chenming Hu; Kalnitsky, A.; Sing-Pin Tay; Chia-Cheng Cheng, "Transistor characteristics with Ta2O5 gate dielectric," IEEE EDL, P. 441-443, 1998

[4] Campbell, S.A.; Gilmer, D.C.; Xiao-Chuan Wang; Ming-Ta Hsieh; Hyeon-Seag Kim; Gladfelter, W.L.; Jinhua Yan, "MOSFET transistors fabricated with high permitivity TiO2 dielectrics," IEEE, TED, P. 104-109, 1997

[5] D. A. Buchanan, E. P. Gusev, E. Cartier, H. Okorn-Schmidt, K. Rim, M. A. Gribelyuk, A. Mocuta, A. Ajmera, M. Copel, S. Guha, N. Bojarczuk, A. Callegari, C. D'Emic, P. Kozlowski, K. Chan, R. J. Fleming, P. C. Jamison, J. Brown, and R. Arndt, "80 nm Poly-silicon gated n-FETs with ultra-thin Al2O3 gate dielectric for ULSI applications," IEDM Tech. Dig., pp. 223 - 226, December 2000.

[6] J. H. Lee, K. Koh, N. I. Lee, M. H. Cho, Y. K. Kim, J. S. Jeon, K. H. Cho, H. S. Shin, M. H. Kim, K. Fujihara, H. K. Kang, and J. T. Moon, "Effect of polysilicon gate on the flatband voltage shift and mobility degradation for ALD-Al2O3 gate dielectric," IEDM Tech. Dig., pp. 645 - 648, December 2000.

[7] A. Chin, Y-H Wu, S. B. Chen, C. C. Liao, and W. J. Chen, "High quality La_2O_3 and Al_2O_3 gate dielectrics with equivalent oxide thickness 5-10 A," Symp. on VLSI Tech. Dig., 2000, P. 16-19
[8] Katsunori Onishi, Laegu Kang, Rino Choi, Easwar Dharmarajan, Sundar Gopalan, Yongjoo Jeon, Chang Seok Kang, Byoung Hun Lee, Renee Nieh, and Jack C. Lee, "Dopant Penetration Effects on Polysilicon Gate HfO2 MOSFET's," Symp. on VLSI Tech. Dig., 2001, P. 131
[9] S. J. Lee, H. F. Luan, W. P. Bai, C. H. Lee, T. S. Jeon, Y. Senzaki, D. Roberts, and D. L. Kwong, "High quality ultra thin CVD HfO2 gate stack with poly-Si gate electrode," IEDM Tech. Dig., pp. 31 - 34, December 2000.
[10] C. Hobbs, H.-H. Tseng, K. Reid, B. Taylor, L. Dip, L. Mebert, R. Garcia, R. Hegde, J. Grant, D. Gilmer, A. Granke, V. Dhandapani, M. Azrak, L. Prabhu, R. Rai, S. Bagchi, J. Conner, S. Backer, F. Dumbuya, B. Nguyen, and P. Tobin, "80 nm poly-Si gate CMOS with HfO_2 gate dielectric," IEDM Tech. Dig., 2001, P. 651-654
[11] W. Zhu and T. P. Ma, "HfO_2 and HfAlO for CMOS: thermal stability and current transport," IEDM Tech. Dig., 2001, P. 463-466
[12] L. Kang, K. Onishi, Y. Jeon, B. H. Lee, C. Kang, W. Qi, R. Nieh, S. Gopalan, R. Choi, and J. C. Lee, "MOSFET Devices with polysilicon on single-layer HfO2 high-K dielectrics," IEDM Tech. Dig., pp. 35 - 38, December 2000.
[13] W. Qi, R. Nieh, B. H. Lee, L. Kang, Y. Jeon, K. Onishi, T. Ngai, S. Banerjee, and J. C. Lee, "MOSCAP and MOSFET characteristics using ZrO2 gate dielectric deposited directly on Si," IEDM Tech. Dig., pp. 145 - 148, December 1999.
[14] C. H. Lee, H. F. Luan, W. P. Bai, S. J. Lee, T. S. Jeon, Y. Senzaki, D. Roberts, and D. L. Kwong, "MOS Characteristics of ultra thin rapid thermal CVD ZrO2 and Zr silicate gate dielectrics," IEDM Tech. Dig., pp. 27 - 30, December 2000.
[15] Z. J. Luo, T. P. Ma, E. Cartier, M. Copel, T. Tamagawa, and B. Halpern, "Ultra-thin ZrO2 (or Silicate) with High Thermal Stability for CMOS Gate Applications," Symp. on VLSI Tech. Dig., 2001, P. 135-136
[16] W.-J. Qi, R Nieh, B. H. Lee, K. Onishi, L. Kang, Y. Jeon, J. Lee, V. Kaushik, B-Y Neuyen, L. Prabhu, K. Eisenbeiser, and J. Finder, "Performance of MOSFETs with ultra thin ZrO_2 and Zr silicate gate dielectrics," Symp. on VLSI Tech. Dig., 2000, P. 40-41
[17] C. Hobbs, L Dip, K. Reid, D. Gilmer, R. Hegde, T. Ma, B. Taylor, B. Cheng, S. Samavedam, H. Tseng, D. Weddington, F. Huang, D. Farber, M. Schippers, M. Rendon, L. Prabhu, R. Rai, S. Bagchi, J. Conner, S. Backer, F. Dumbura, J. Locke, D. Workman, and P. Tobin, "Sub-Quarter micron Si-gate CMOS with ZrO2 gate dielectric," Symp. on VLSI Technology, Systems, and Applications, Tech. Dig., P. 204-207, 2001, Taipei, Taiwan
[18] H.-H. Tseng, P. J. Tobin, S. Kalpat, J. K. Schaeffer, M. E. Ramón, L. Fonseca, Z. X. Jiang, R. I. Hegde, D. H. Triyoso, and S. Semavedam , "Defect Passivation with Fluorine and Interface Engineering for Hf-based High-K/Metal Gate Stack Device Reliability and Performance Enhancement ," IEDM Tech. Dig., 2005, P. 713-716
[19] R. Beyers, "Thermodynamic considerations in refractory metal-silicon-oxygen systems," J. Appl. Phys. 56, 147 (1984)
[20] S. Q. Wang and J. W. Mayer, "Reactions of Zr thin films with SiO2 substrates," Journal of Applied Physics Volume 64, Issue 9, Nov 1988 Page(s):4711 - 4716

[21] G. D. Wilk and R. M. Wallace, "Stable zirconium silicate gate dielectrics deposited directly on silicon," Applied Physics Letters Volume 76, Issue 1, Jan 2000 Page(s):112 - 114
[22] G. D. Wilk, R. M. Wallace, and J. M. Anthony, "High-κ gate dielectrics: Current status and materials properties considerations," Journal of Applied Physics Volume 89, Issue 10, May 2001 Page(s):5243 - 5275
[23] E. P. Gusev et al., AVS Topical Conf. on Atomic Layer Deposition, May 14, 2001
[24] M. Ritala and M. Leskela, " Atomic Layer Deposition," Handbook of Thin Film Materials, H. S. Nalwa ed., Academic Press, 2001, Vol.1, Chap. 2
[25] L. Krusin Elbaum et al., Mat. Res. Soc. Symp. Proc Vol. 171, P. 351, 1986
[26] S. Matsuda, H. Yamakawa, A. Azuma and Y. Toyoshima, "Performance Improvement of Metal Gate CMOS Technologies," Symp. on VLSI Tech. Dig., 2001, P. 63-64
[27] Dae-Gyu Park, Kwan-Yong Lim, Heung-Jae Cho, Tae-Ho Cha, Joong-Jung Kim, Jung-Kyu Ko, In-Seok Yeo, and Jin Won Park, "Novel Damage-free Direct Metal Gate Process Using Atomic Layer Deposition," Symp. on VLSI Tech. Dig., 2001, P. 65-66
[28] J. C. Hu, H. Yang, R. Kraft, A. L. P. Rotondaro, S. Hattangady, W. W. Lee, R. A. Chapman, C. Chao, A. Chatterjee, M. Hanratty, M. Rodder, and I. Chen, "Feasibility of using W/TiN as metal gate for conventional 0.13μm CMOS technology and beyond," IEDM Tech. Dig., pp. 825 - 828, December 1997.
[29] B. H. Lee, R. Choi, L. Kang, S. Gopalan, R. Nieh, K. Onishi, Y. Jeon, W. Qi, C. Kang, and J. C. Lee, "Characteristics of TaN gate MOSFET with ultrathin hafnium oxide (8 Å-12 Å)," IEDM Tech. Dig., pp. 39 - 42, December 2000.
[30] A. Chatterjee, R. A. Chapman, G. Dixit, J. Kuehne, S. Hattangady, H. Yang, G. A. Brown, R. Aggarwal, U. Erdogan, Q. He, M. Hanratty, D. Rogers, S. Murtaza, S. J. Fang, R. Kraft, A. L. P. Rotondaro, J. C. Hu, M. Terry, W. Lee, C. Fernando, A. Konecni, G. Wells, D. Frystak, C. Bowen, M. Rodder, and I. Chen, "Sub-100nm gate length metal gate NMOS transistors fabricated by a replacement gate process," IEDM Tech. Dig., pp. 821 - 824, December 1997.
[31] H.-H. Tseng, C. C. Capasso, J. K. Schaeffer, E. A. Hebert, P. J. Tobin, D. C. Gilmer, D. Triyoso, M. E. Ramon, S. Kalpat, E. Luckowski, W. J. Taylor, Y. Jeon, O. Adetutu, R. I. Hegde, R. Noble, M. Jahanbani, C. El Chemali, and B. E. White, "Improved short channel device characteristics with stress relieved pre-oxide (SRPO) and a novel tantalum carbon alloy metal gate/HfO2/ stack ," in IEDM Technical Digest, 2004, pp. 821-824
[32] A. H. Perera, B. Smith, N. Cave, M. Sureddin, S. Chheda, R. Islam, J. Chang, S.-C. Song, A. Sultan, S. Crown, V. Kolagunta, S. Shah, M. Celik, D. Wu, K. C. Yu, R. Fox, S. Park, C. Simpson, D. Eades, S. Gonzales, C. Thomas, J. Sturtevant, D. Bonser, N Benavides, M. Thompson, V. Sheth, J. Fretwell, S. Kim, N. Ramani, K. Green, M. Moosa, P. Besser, Y. Solomentsev, D. Denning, M. Friedemann, B. Baker, R. Chowdhury, S. Ufmani, K. Strozewski, R. Carter, J. Reiss, M. Olivares, B. Ho, T. Lii, T. Sparks, T. Stephens, M. Schaller, C. Goldberg, K. Junker, D. Wristers, J. Alvis, B. Melnick, and S. Venkatesan, "A versatile 0.13 μm CMOS platform technology supporting high performance and low power applications," IEDM Technical Digest, 2000, pp571-574

[33] W. Chen and T. P. Ma, "A new technique for measuring lateral distribution of oxide charge and interface traps near MOSFET junctions," IEEE Electron Device Letters, vol. 12 , no. 7, pp. 393 - 395, July, 1991
[34] H.-H. Tseng, M. E. Ramon, L. Hebert, P. J. Tobin, D. Triyoso, J. M. Grant, Z. X. Jiang, D. Roan, S. B. Samavedam, D. C. Gilmer, S. Kalpat, C. Hobbs, W. J. Taylor, O. Adetutu, and B. E. White, "ALD HfO2 using heavy water (D2O) for improved MOSFET stability," IEDM Technical Digest, 2003, pp 83-86
[35] H.-H. Tseng, P. J. Tobin, S. Kalpat, J. K. Schaeffer, M. E. Ramon, L. Fonseca, Z. X. Jiang, R. I. Hegde, D. H. Triyoso, and S. Semavedam, " Defect Passivation with Fluorine and Interface Engineering for Hf-based High-K/Metal Gate Stack Device Reliability and Performance Enhancement," IEEE Transactions on Electron Devices, Volume 54, Number 12, December 2007, pp. 3267-3275
[36] X. Xiong and J. Robertson, "Defect energy levels in HfO2 high-dielectric-constant gate oxide," Applied Physics Letters, vol. 87, 183505, 2005
[37] P. Hohenberg and W. Kohn, "Inhomogeneous Electron Gas," Phys Rev. 136, B864-B871, 1964
[38] W. Kohn and L. J. Sham, "Self-Consistent Equations Including Exchange and Correlation Effects," Phys. Rev. 140, A1133-A1138, 1965
[39] J. M. Soler, E. Artacho, J. D Gale, A. García, J. Junquera, P. Ordejón, and D. Sánchez-Portal, "The SIESTA method for ab initio order-N materials simulation," J. Phys.: Condens. Matter 14, pp. 2745-2779, 2002
[40] Y.-C. Yeo, P. Ranade, T.-J. King, and C. Hu, "Effects of high-κ gate dielectric materials on metal and silicon gate workfunctions," IEEE Electron Device Lett., vol. 23, no. 6, pp. 342–344, Jun. 2002.
[41] H. Y. Yu, C. Ren, Y.-C. Yeo, J. F. Kang, X. P. Wang, H. H. H. Ma, M.-F. Li, D. S. H. Chan, D.-L. Kwong, "Fermi Pinning Induced Thermal Instability of Metal Gate Work Functions," IEEE Electron Device Letters, vol. 25, p337, May 2004.
[42] G. A. Brown, G. Smith, J. Saulters, K. Matthews, H.-C. Wen, H. AlShareef, P. Majhi, and B. H. Lee, "An improved methodology for gate electrode work function extraction in SiO2 and high-k gate stack systems using terraced oxide structures," in Proc. SISC, 2004, p. 15. 2004.
[43] R. Jha, J. Gurganos, Y. H. Kim, R. Choi, J. Lee, and V. Misra, "A capacitance-based methodology for work function extraction for metals on high-K," IEEE Electron Device Letter, vol. 25, pp. 420-423, 200
[44] G. D. Wilk, M. L. Green, M.-Y. Ho, B. W. Busch, T. W. Sorsch, F. P. Klemens, B. Brijs, R. B. van Dover, A. Kornblit, T. Gustafsson, E. Garfunkel, S. Hillenius, D. Monroe, P. Kalavade, and J. M. Hergenrother, " Improved Film Growth and Flatband Voltage Control of ALD HfO2 and Hf-Al-O with n+ poly-Si Gates using Chemical Oxides and Optimized Post-Annealing," in VLSI Symp. Tech. Dig., 2002, pp. 88-89.
[45] S. B. Samavedam, L. B. La, J. Smith, S. Dakshina-Murthy, E. Luckowski, J. Schaeffer, M. Zavala, R. Martin, V. Dhandapani, D. Triyoso, H. H. Tseng, P. J. Tobin, D. C. Gilmer, C. Hobbs, W. J. Taylor, J. M. Grant, R. I. Hegde, J. Mogab, C. Thomas, P. Abramowitz, M. Moosa, J. Conner, J. Jiang, V. Arunachalarn, M. Sadd, B.-Y. Nguyen, B. White, "Dual-metal gate CMOS with HfO2 gate dielectric," in IEDM Tech. Dig., 2002, pp. 433-436.

[46] Z. B. Zhang, S. C. Song, C. Huffman1, J. Barnett, N. Moumen, H. Alshareef, P. Majhi, M. Hussain, M. S. Akbar, J. H. Sim, S. H. Bae, B. Sassman, and B. H. Lee, "Integration of Dual Metal Gate CMOS with TaSiN (NMOS) and Ru (PMOS) Gate Electrodes on HfO2 Gate Dielectric," in VLSI Symp. Tech. Dig., 2005, pp.50-51.
[47] S. Kubicek, T. Schram, V. Paraschiv, R. Vos, M. Demand, C. Adelmann, T. Witters, L. Nyns, L.-Å. Ragnarsson, H.Yu, A. Veloso, R. Singanamalla, T. Kauerauf, E.Rohr, S. Brus, C. Vrancken, V. S. Chang, R. Mitsuhashi, A. Akheyar, H.-J. Cho, J. C. Hooker, B. J. O'Sullivan, T. Chiarella, C. Kerner, A. Delabie, S. Van Elshocht, K. De Meyer, S. De Gendt, P. Absil, T. Hoffmann and S. Biesemans, "Low V_T CMOS using doped Hf-based oxides, TaC-based Metals and Laser-only Anneal," in IEDM Tech. Dig., 2007, pp. 49-52.
[48] T. Schram, S. Kubicek, E. Rohr, S. Brus, C. Vrancken, S.-Z. Chang, V.S. Chang, R. Mitsuhashi, Y. Okuno, A. Akheyar, H.-J. Cho, J.C. Hooker, V. Paraschiv, R. Vos, F. Sebai, M. Ercken, P. Kelkar, A. Delabie, C. Adelmann, T. Witters, L-A. Ragnarsson, C. Kerner, T. Chiarella, M. Aoulaiche, Moon-Ju Cho, T. Kauerauf, K.De Meyer, A. Lauwers, T. Hoffmann, P. P. Absil and S. Biesemans, "Novel Process To Pattern Selectively Dual Dielectric Capping Layers Using Soft-Mask Only," in VLSI Symp. Tech. Dig., 2008, pp.44-45.
[49] X. Chen, S. Samavedam, V. Narayanan, K. Stein, C. Hobbs, C. Baiocco, W. Li, D. Jaeger, M. Zaleski, H. S. Yang, N. Kim, Y. Lee, D. Zhang, L. Kang, J. Chen H. Zhuang, A. Sheikh, J. Wallner, M. Aquilino, J. Han, Z. Jin, J. Li, G. Massey, S. Kalpat, R. Jha, N. Moumen, R. Mo, S. Kirshnan, X. Wang, M. Chudzik, M. Chowdhury, D. Nair, C. Reddy, Y. W. Teh, C. Kothandaraman, D. Coolbaugh, S. Pandey, D. Tekleab, A. Thean, M. Sherony, C. Lage, J. Sudijono, R. Lindsay, J. H. Ku, M. Khare, A. Steegen, "A Cost Effective 32nm High-K/ Metal Gate CMOS Technology for Low Power Applications with Single-Metal/Gate-First Process," in VLSI Symp. Tech. Dig., 2008, pp. 88-89.
[50] P. Kirsch, M. A. Quevedo-Lopez, S. A. Krishnan, C. Krug, H. AlShareef, C. S. Park, R. Harris, N. Moumen, A. Neugroschel, G. Bersuker, B.H. Lee, J.G. Wang, G. Pant, B. E. Gnade, M. J. Kim, "Band Edge n-MOSFETs with High-k/Metal Gate Stacks Scaled to EOT=0.9nm with Excellent Carrier Mobility and High Temperature Stability," in IEDM Tech. Dig., 2006, p.639-642.
[51] J. Huang, P. D. Kirsch, D. Heh, C. Y. Kang, G. Bersuker, M. Hussain, P. Majhi, P. Sivasubramani, D. C. Gilmer, N. Goel, M.A. Quevedo-Lopez, C. Young, C. S. Park, C. Park, P. Y. Hung, J. Price, H. R. Harris, B.H. Lee, H.-H. Tseng and R. Jammy, "Device and Reliability Improvement of HfSiON+LaOx/Metal Gate Stacks for 22nm," Node Application in IEDM Tech. Dig., 2008, pp. 45-48.
[52] S. C. Song, Z. B. Zhang, M. M. Hussain, C. Huffman, J. Barnett, S. H. Bae, H. J. Li, P. Majhi, C. S. Park, B. S. Ju, H. K. Park, C. Y. Kang, R. Choi, P. Zeitzoff, H. H. Tseng, B. H. Lee, and R. Jammy, "Highly manufacturable 45 nm LSTP CMOSFETs using novel dual high-κ and dual metal gate CMOS integration," in VLSI Symp. Tech. Dig., 2006, pp. 16–17.
[53] V. Narayanan, V. K. Paruchuri, N. A. Bojarczuk, B. P. Linder, B. Doris, Y. H. Kim, S. Zafar, J. Stathis, S. Brpwn, J. Arnold, M. Copel, M. Steen, E. Cartier, A. Callegari, P. Jamison, J.-P. Locquet, D. L. Lacey, Y. Wang, P. E. Batson, P. Ronsheim, P. Jammy, M. P. Chudzik, M. Ieong, S. Guha, G. Shahidi, and T. C. Chen , "Band-edge high-

performance High-k/metal gate n-MOSFETs using cap-layers containing group IIA and IIB elements with gate-first processing for 45 nm and beyond," in VLSI Symp. Tech. Dig., 2006, pp. 224–225.
[54] H. Alshareef, M. Quevedo-Lopez, H. Wen, R. Harris, P. Kirsch, P. Majhi, B. Lee, R. Jammy, D. Lichtenwalner, J. Jur, and A. Kingon, "Work function engineering using lanthanum oxide interfacial layers," Appl. Phys. Lett., vol. 89, no. 23, pp. 232 103-1–232 103-3, Dec. 2006.
[55] L.-Å. Ragnarsson, V. S. Chang, H.Y. Yu, H.-J. Cho, T. Conard, K. M. Yin, A. Delabie, J. Swerts, T. Schram, S. De Gendt, and S. Biesemans, "Achieving conduction band-edge effective work functions by La_2O_3 capping of hafnium silicates," IEEE Electron Device Lett., vol. 28, no. 8, pp. 486–488, Jun. 2007.
[56] H.-C. Wen, S. C. Song, C. S. Park, C. Burhamn, G. Bersuker, O. Sharia, A. Demkov, B. S. Ju, M. A. Quevedo-Lopez, H. Niimi, K. Choi, H. B. Park, P. S. Lysaght, P. Majhi, B. H. Lee, and R. Jammy, "Highly manufacturable MoAlN PMOS electrode for 32 nm low standby power applications," in VLSI Symp. Tech. Dig., 2007, pp. 160–161.
[57] Y. Yamamoto, K. Kita, K. Kyuno, and A. Toriumi, "Study of La concentration dependent VFB shift in metal/$HfLaO_x$/Si capacitors," in Proc. Ext. Abs. of SSDM, 2006, pp. 212–213.
[58] C. S. Park, P. Lysaght, M. M. Hussain, J. Huang, G. Bersuker, P. Majhi, P. D. Kirsch, H. H. Tseng, and R. Jammy, "Advanced High-k/Metal Gate Stack Progress and Challenges - A Materials and Process Integration Perspective," accepted to be published in International Journal of Materials Research, 2010
[59] P. Sivasubramani, T. S. Böscke, J. Huang, C. D. Young, P. D. Kirsch, S. A. Krishnan, M. A. Quevedo-Lopez, S. Govindarajan, B. S. Ju, H. R. Harris, D. J. Lichtenwalner, J. S. Jur, A. I. Kingon, J. Kim, B. E. Gnade, R. M. Wallace, G. Bersuker, B. H. Lee, and R. Jammy, "Dipole moment model explaining nFET V_t tuning utilizing La, Sc, Er, and Sr doped HfSiON dielectrics," in VLSI Symp. Tech. Dig., 2007, pp. 68–69.
[60] O. Sharia, A. A. Demkov, G. Bersuker, B. H. Lee, "Theoretical study of the insulator/insulator interface: Band alignment at the SiO_2 /HfO_2 junction," Phys. Rev. B, v. 75, p. 035306, 2007.
[61] B. H. Lee, J. Oh, H. H. Tseng, R. Jammy and H. Huff, "Gate stack technology for nanoscale devices," Materials Today, Vol. No. 9, p. 36, 2006.
[62] K. Choi, T. Lee, S. Kweon, C.D. Young, H. Harris, R. Choi, S.C. Song, B.H. Lee, and R. Jammy, "Impact of the bottom interfaceial layer on the threshold voltage and device reliability of fluorine incorporated pMOSFETs," in Proc. IEEE IRPS, 2007, p. 374.

9

Metal Gate Electrode and High-κ Dielectrics for Sub-32nm Bulk CMOS Technology: Integrating Lanthanum Oxide Capping Layer for Low Threshold-Voltage Devices Application

HongYu Yu
School of EEE / Nanyang Technological University
Singapore
Also with IMEC / Leuven
Belgium

1. Introduction

Metal gate electrode together with high dielectric constant or high-κ insulator is considered as one of the critical technology enablers to scale the CMOS devices into sub-45nm region (ITRS, 2007; Mistry et al., 2007), due to the following concerns on the conventional poly-Si electrode and Si oxynitride dielectrics stack:

1. Poly-depletion effect to add an equivalent oxide thickness or EOT up to ~0.5 nm to the gate stack, which is a significant portion for the overall targeted EOT requirement of ~1 nm;
2. Excess gate leakage when the EOT of the gate stack is reduced to sub-1nm;
3. High resistance for the poly electrode.

Additional benefit of using metal gate / high-κ dielectrics is on the improvement of the device variability as no poly-Si doping is needed. Integration of metal gate /high-κ dielectrics using a conventional gate-first route (i.e. the gate stack undergoes a source/drain activation annealing) is attractive as compared to a gate-last route, as the gate first approach is more compatible with the conventional poly-Si/SiON flow, and hence low-cost fabrication is feasible. In addition, in the gate-first flow, the gate stack can afford a high thermal budget process, which is required for embedded application (e.g. DRAM). In this chapter, Lanthanum Oxide, (LaO_x, with κ ~20 and an E_g ~ 5.5eV) dielectric capping incorporation into the Hf-based host high-κ dielectrics is firstly demonstrated as a practical solution to achieve low threshold-voltage or V_T metal-gated uni-channel nMOSFETs fabricated using a gate-first flow (Kubicek et al., 2007; Narayanan et al., 2006). Further, a comprehensive study is presented on the integration of LaO_x capping layer for sub-32nm metal gated CMOS devices with Hf-based high-K dielectrics in a gate first manner. Two different integration routes, i.e. Dual Metal Dual Dielectric flow or DMDD (hard-masks to pattern selectively nMOS and pMOS) and Single Metal Dual Dielectric flow or SMDD (soft-mask processes), are presented and compared. The device reliability study is also provided.

2. Experimental

Hf-based high-κ dielectrics, e.g. 1.8nm $HfSiO_x$ with 60% of Hf by metal-oxide chemical vapor deposition, or 1.5nm HfO_2 by atomic layer deposition, were used as host dielectrics. An interfacial layer of ~1nm thermal SiO_2 was formed before high-κ dielectrics deposition. LaO_x capping layer with various thickness was deposited via atomic layer deposition, and incorporated immediately below and above Hf-based high-κ layer. A 10nm Ta_2C electrode by physical vapor deposition or TaCNO electrode by metal-oxide chemical vapor deposition with a 100nm Poly-Si cap layer was then deposited as metal gate. Considering the ultra-shallow junction requirement, source/drain was activated with various thermal budgets: i.e. via Low (1150°C), Medium (1250°C), and High (1350°C) Laser Power anneals (LLP, MLP and HLP), or spike anneals (1035°C). CMOS transistors were fabricated via either DMDD or SMDD approach. Note that Al_2O_3 by atomic layer deposition was used as the dielectrics capping incorporated in Hf-based host dielectrics in pFETs to tune the V_T.

3. Results and discussion

3.1 NFETs V_T dependence on LaO_x capping layer thickness, post-annealing condition, and location

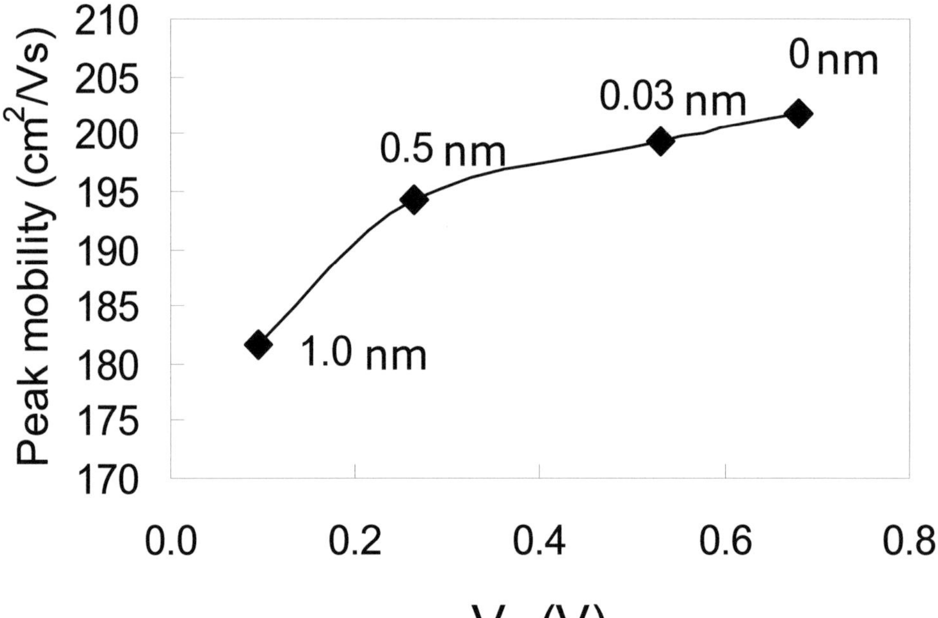

Fig. 1. Relation between nMOS peak mobility and V_T for different La_2O_3 cap thicknesses on HfSiON with Ta_2C metal gate electrode.

In Fig.1, it is seen the nFETs V_T (L_g = 1μm) is effectively reduced up to 600mV when increasing the La_2O_3 cap thicknesses. However there is a penalty of considerable mobility degradation for the case of using 1nm think La_2O_3 cap. Thus 0.5nm thickness is considered as the optimum La_2O_3 cap thickness for the device integration described in the following part of this paper.

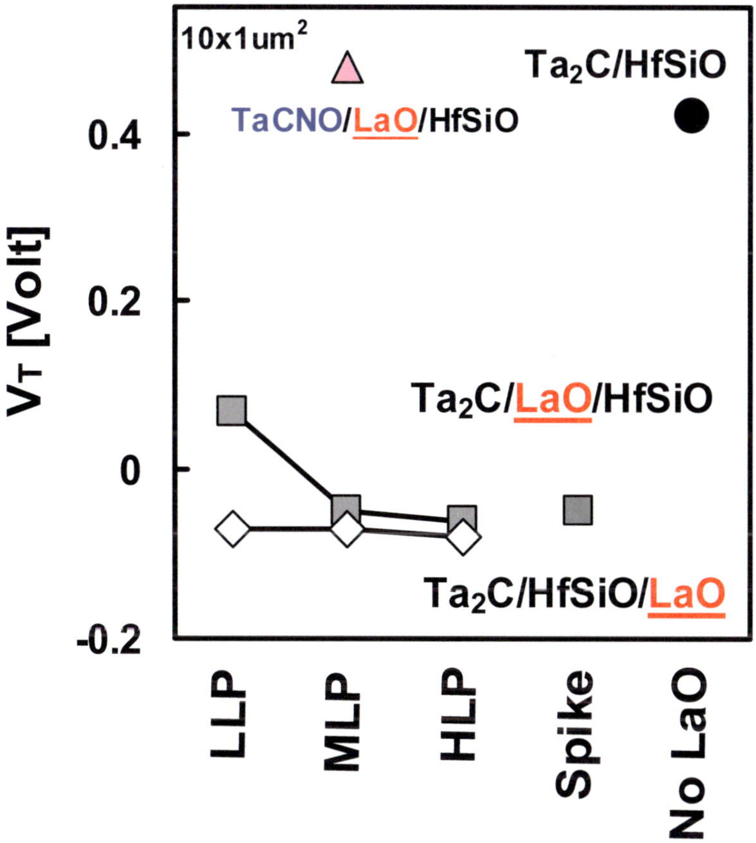

Fig. 2. Ta_2C metal gated NFETs V_T dependence on various thermal budgets applied for source/drain activation when positioning LaO_x capping layer above or immediate below HfSiO host dielectrics.

In Fig. 2, the impact of laser annealing conditions (low, medium and high laser power) on V_T of nFETs with LaO capping layer positioning above or immediately below HfSiO is shown and compared to the spike- rapid thermal annealed reference. When LaO_x is on top of HfSiO, it is seen that only when applying high laser power, V_T lowering is comparable to the reference sample. On the other hand, device V_T can be effectively reduced regardless the thermal budget applied when positioning LaO_x immediately below HfSiO. It is naturally concluded that the La at the interface between HfSiO and SiO_x interfacial layer plays a critical role to modulate the nFETs V_T. In case of LaOx is on top of HfSiO, when applying high thermal budget (i.e. the high power laser annealing or the spike annealing in this work), La can be driven to diffuse to reach the interface between HfSiO and interfacial layer, effectively driving down the V_T. It is worth mentioning that the gate leakage vs. EOT would not be degraded with the adding of LaO_x capping layer into the HfSiO host dielectrics, as shown in Fig. 3, partially due to the excellent κ and E_g value of La_2O_3. Further from Fig. 3, it is noted that Ta_2C gated devices exhibit better EOT scalability than the TaCNO case, and the reason shall be discussed in part 3.3 of the paper.

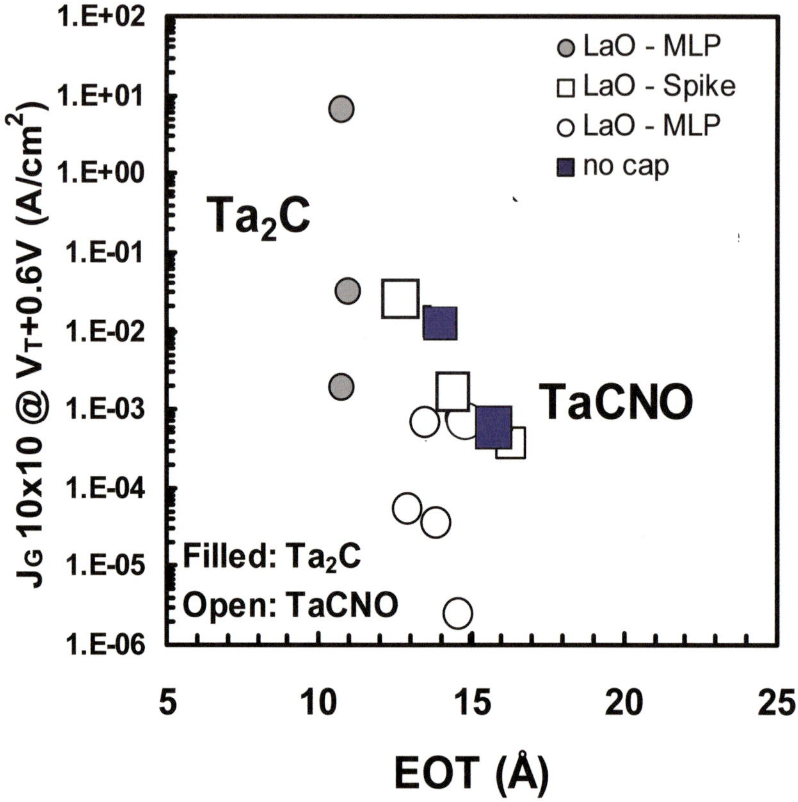

Fig. 3. J_G *vs.* EOT for Ta_2C/ TaCNO gated devices with LaO capping incorporated HfSiO dielectrics.

3.2 Integration LaO_x capping layer into CMOS devices

CMOS transistors were fabricated using both DMDD and SMDD approaches. Fig.4 outlines the schematic DMDD integration flow. The first gate stack (Ta_2C/ LaO cap/ HfSiO) is deposited (Fig.4a) and selectively removed from the complementary side using a Si hard mask (Fig.4b). The second gate (TaCNO/ AlO cap/ HfSiO) is formed again using a Si hard mask (Fig.4c). Next, the poly-Si is deposited (Fig.4d) and gate patterning is done by immersion lithography and dry etch (Fig.4e). The remainder of the flow follows conventional CMOS processing. Cross-sectional high resolution transmission electron microscopy of n- & p- MOSFETs fabricated using DMDD approach with gate lengths of 45nm are shown in Fig.5 along with a detailed view of the gate stack interfaces after gate etch. The n- & p- MOSFETs boundaries on an inverter circuit can be seen as inset of Fig.6 (after silicidation). Symmetric low V_T values of ±0.25V can be obtained for both n- and p-MOSFETs (Fig.6).

In Fig. 7(a), both short-channel n- and p- FETs (L_g = 55nm) fabricated using DMDD approach exhibit well-behaved I_d-V_g characteristics. As shown in Fig. 7(b), the unstrained I_{DSAT} of 1035/500 μA/μm for n- / p- MOSFETs at I_{OFF}=100nA/μm and an operating voltage or $|V_{DD}|$=1.1V are demonstrated on a single wafer.

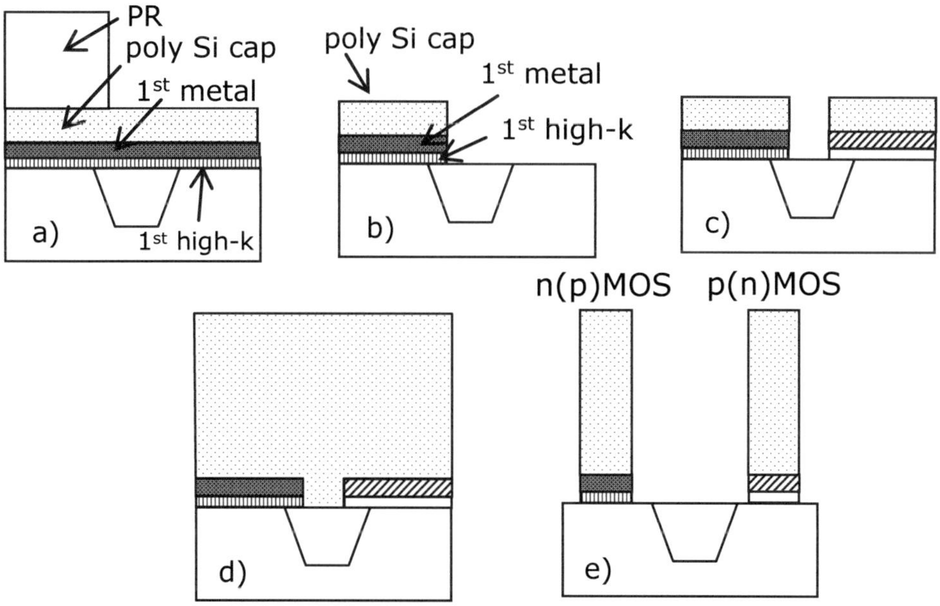

Fig. 4. Dual Metal Dual Dielectrics (DMDD) CMOS integration scheme

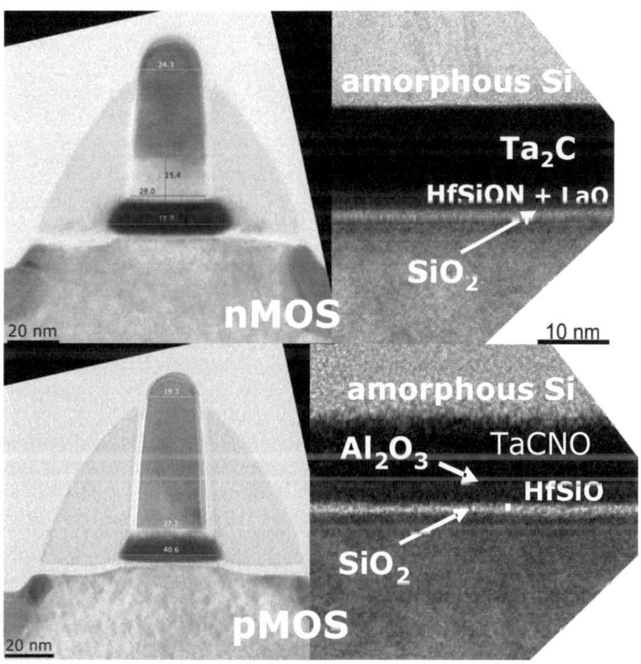

Fig. 5. XTEM of the n-& pFETs fabricated using DMDD.

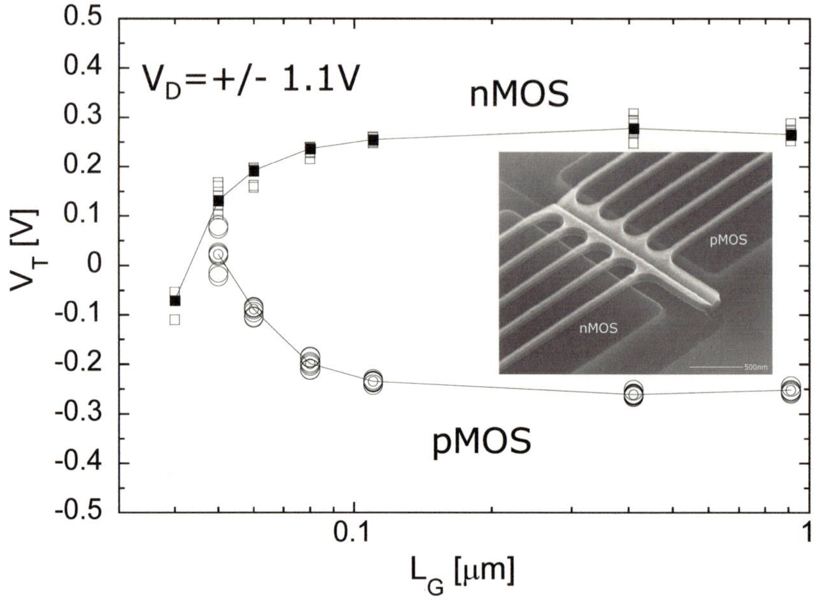

Fig. 6. V_T roll-off for both n- & p- FETs fabricated using DMDD. Inset: SEM views of the nMOS and pMOS boundaries.

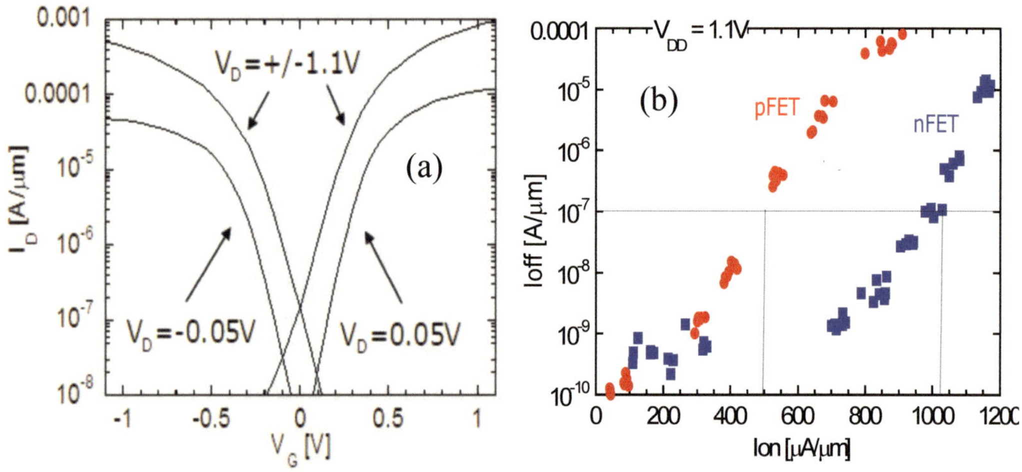

Fig. 7. (a): I_d-V_g of both n- & p- MOSFETs with a L_g ~55nm; (b) I_{ON}-I_{OFF} curves of both n- & pMOS fabricated using DMDD.

Next, we explain the SMDD process flow. As schematically shown in Fig.8, SMDD involves a simple resist-based selective high-κ dielectric capping removal process (in this work: La_2O_3 or Al_2O_3 over both HfSiO and SiO_2). Several key process modules development in this SMDD route is discussed in this section. 1) For the sake of a simple patterning strategy, a wet developable Bottom-Anti-Reflection-Coating or BARC layer is developed to be patterned directly on the dielectric capping and to be selectively removed from the complementary areas (La_2O_3 from pMOS and Al_2O_3 from nMOS). This wet BARC layer

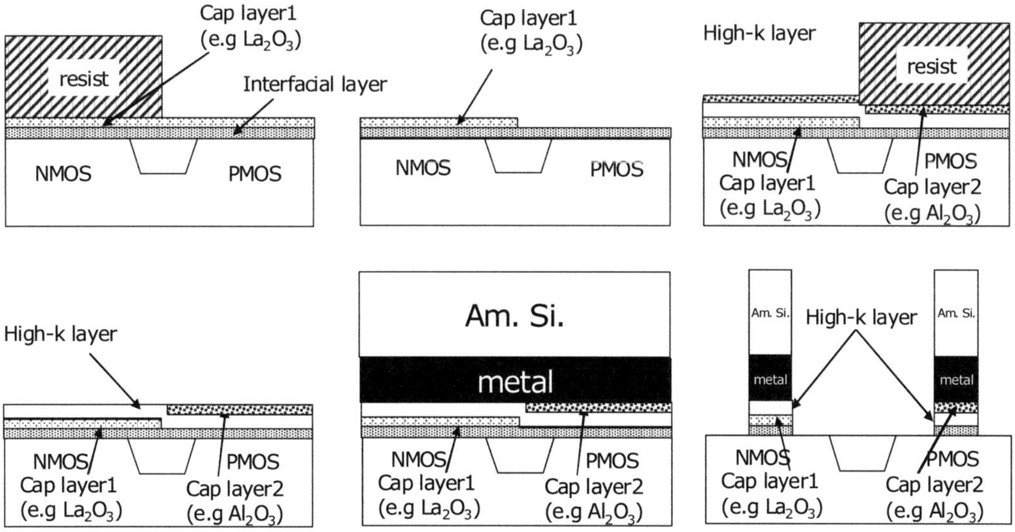

Fig. 8. Single Metal Dual Dielectrics or SMDD CMOS integration scheme

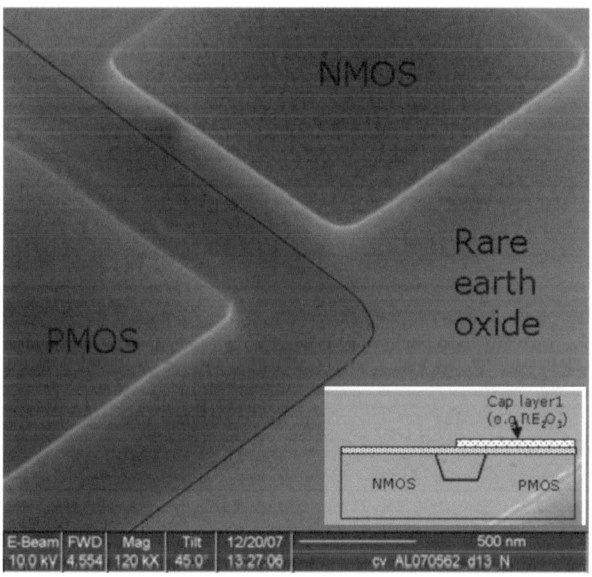

Fig. 9. N-P MOSFETs boundary after etching and resist removal using the wet bottom-anti-reflection-coating or BARC based process (developed for SMDD).

could guarantee an excellent adhesion towards the dielectrics layer, which can not be achieved via 248nm photo-resist only. Fig.9 illustrates the superior adhesion and sharp patterning achieved with wet BARC. 2) The high-κ wet capping removal required for the proposed process flow must be resist-compatible, highly selective (>100) to the underlying layer (SiO_2 or HfSiON). As summarized in Table 1, diluted HCl is the chemistry of choice for La_2O_3, and TMAH for Al_2O_3. 3) Once the high-κ capping has been selectively removed, the photo resist must be stripped without damaging the exposed materials. The resist strip (NMP- based) and post-cleans (APM- based) process details are provided in Table 1. It's worthy noting that during SMDD process, both selective high-κ removal and resist strip processes have been characterized physically and electrically indicating no major impact on V_T, EOT, gate leakage, mobility and gate dielectric integrity.

	Resist	Cap removal	Resist strip	Post-clean
La_2O_3	*Wet BARC+ 248 nm*	Dilute HCl	Wet organic strip (NMP/AEE)	APM at 65C
Al_2O_3	248 nm	During litho development step (1 min ~3.5 % TMAH) (ER 2.4 nm/min)	Wet organic strip (NMP/AEE)	APM at RT (with some sacrificial Al_2O_3 removal)

Table 1. Processes used to selectively remove the cap layers (La_2O_3 or Al_2O_3) to high-k dielectrics and subsequent strips.

In Table-2, a comparison is made between SMDD and DMDD. The key advantage of SMDD is that the number of process step can be significantly reduced by 40%, which means much lower manufacturing cost. It also allows relatively easier and simpler gate etch profile control since the same metal is used for both n- and p-MOS areas. On the other hand, the V_T tuning flexibility is scarified for SMDD process, as only dielectrics capping layer can be utilized for such a purpose. In contrast, in DMDD process, the combination of dielectrics capping layer and metal gate itself allows a wider V_T tuning capability. In addition, for for SMDD approach, attention needs to be paid to avoid the potential impact of capping layer removal process to the gate dielectrics integrity.

	DMDD	**SMDD**
# of extra process steps compared to conventional SiO_2/Si case	15	**9**
# of extra mask steps	2	2
Gate etch aspects	Different metals etched at the same time	**Only one metal etched**
V_T tuning flexibility	cap layer and metal	Cap layer only
Gate dielectric integrity	Gate dielectric not toughed be removal processes	Gate dielectric only exposed to wet chemistries

Table 2. A comparison between DMDD and SMDD.

3.3 Positive Bias Temperature Instability (PBTI) study of n- MOSFETs with LaO_x capping layer

The PBTI of nFETs using LaO_x capping layer is measured at 110°C by using sense-and-measure technique. V_T relaxation with a 100s recovery time after each stress cycle is also measured for dielectric trapping/de-trapping investigation. The measurement set-up is depicted in Fig. 10.

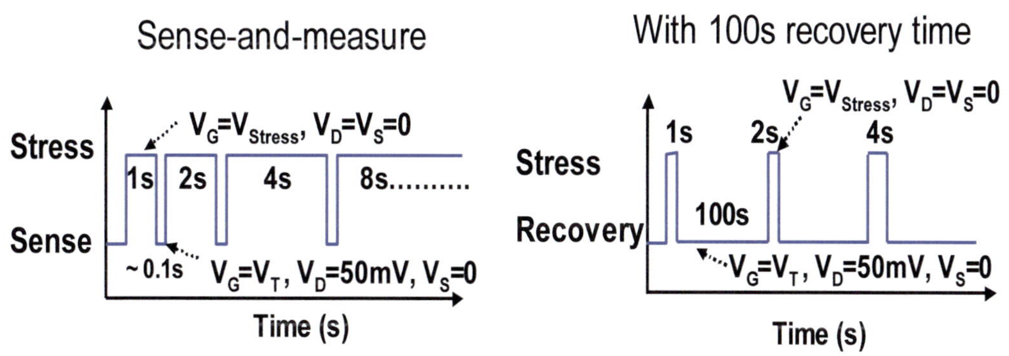

Fig. 10. PBTI measurement set-up: sense-and-measure method and the V_T relax with 100s recovery time after each stress cycle

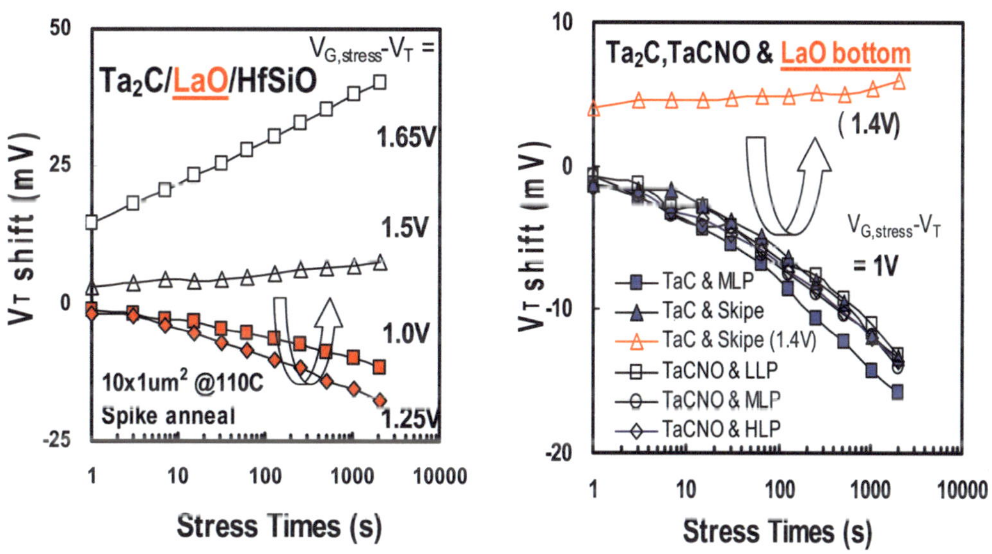

Fig. 11. Stress-field dependent polarity-change PBTI V_T shift is observed in the Ta_2C gated n MOSFETs when incorporating LaO_x capping layer, regardless the position (i.e. either on top or immediately below HfSiO). Both laser and spike annealing were applied to the device under study.

Fig. 11 plots the PBTI induced V_T shifts vs. stress times for the Ta_2C gated n- MOSFETs. A stress-field dependent two polarities V_T shift is observed, regardless the LaO_x capping layer position (i.e. either on top or immediately below HfSiO). This phenomenon was also reported in the Dysprosium silicate gate stack (Yu et al., 2008), and can be explained by the competition between electron de-trapping (dominate at low-stress field) and electron trapping / defect generation (dominate at high-stress field).
The V_T relaxation on these devices with various source/drain activation processes (i.e. spike or laser annealing) during PBTI recovery periods (100s) is also examined, as shown in Fig. 12. It is observed that the LLP annealed device exhibits a different relaxation behavior as compared to MLP/HLP case, when positioning LaO_x- cap either on top or below HfSiO: V_T follows $HfSiO_x$-like (i.e. no La) recovery behavior initially and then changes to the La silicate-like gradually as the stress time increases. It is believed that the relaxation behaviors can be explained by the electron de-trap from bulk traps, which are generated by LaO/ HfSiO (or SiO) intermixing during PBTI stress, and trap back during the recovery period. Insufficient intermixing is expected for the devices under low power anneal, which not only reduces V_T relaxation amplitude (less trap generation) but also makes relaxation of both Hf-host dielectrics and La-silicate seen simultaneously.

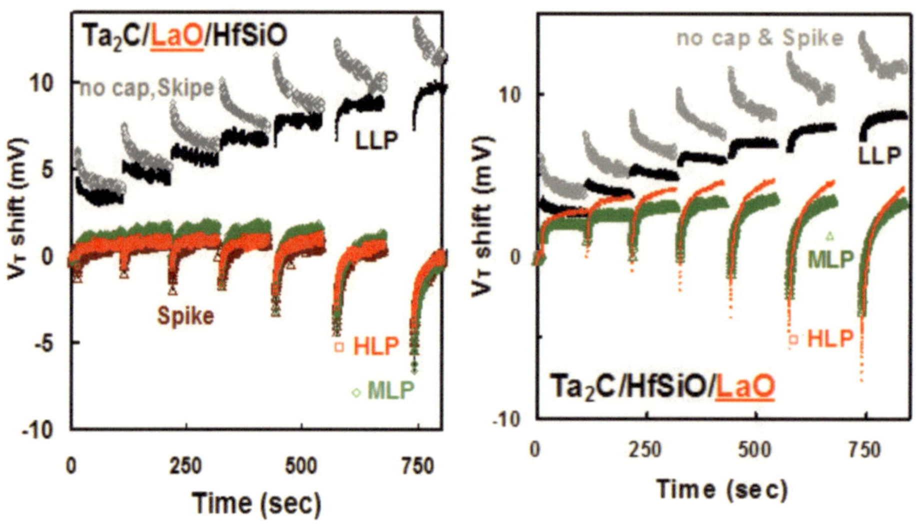

Fig. 12. V_T relaxation vs. time during PBTI stress for the Ta_2C gated n- MOSFETs when incorporating LaO_x capping layer either on top or immediately below HfSiO. $V_G - V_{Stress}$ = 1.25V in this case.

In the case of TaCNO gated n- MOSFETs (Fig.13), normal PBTI and pure HfSiO-like V_T relaxation (see Fig.12) are observed. Further, cross-sectional TEM images together with electron-energy loss spectroscopy or EELS study (Fig. 14) suggest the LaO / $HfSiO_x$ intermixing, and also interactions between dielectrics and electrodes (Ta_2C or TaCNO). Interestingly, both image contrast and EELS analysis identifies an oxygen-less region (~1nm) at the bottom of TaCNO electrode. Likely there, the oxygen is incorporated from TaCNO into dielectrics during the intermixing process, and this also links to the worse EOT scalability of TaCNO than Ta_2C (see Fig.3 also). Considering these, we thus believe the trapping / de-trapping defects generated from dielectric intermixing are probably related to the oxygen vacancies incorporation (Shen et al., 2004): TaCNO can provide oxygen, suppressing bulk trapping generation in La / Dy based silicates. It is more evident when placing cap layers above high κ layer. Schematic diagrams illustrating these phenomena are provided in Fig. 15.

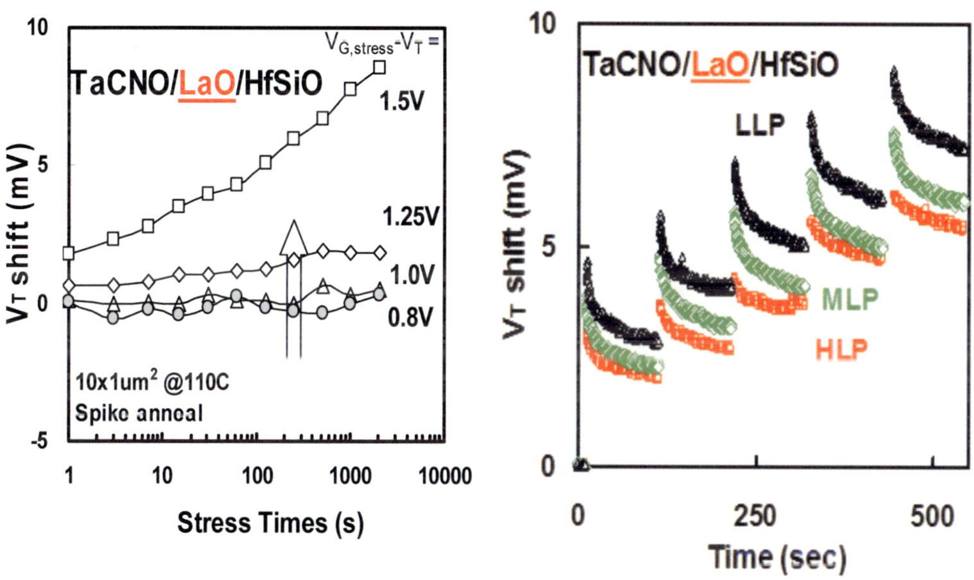

Fig. 13. Normal PBTI V_T vs. stress and V_T relaxation curves *vs.* time for TaCNO gated n FETs. Positive V_T and HfSiO-like relaxation behaviors (Fig. 12) are observed.

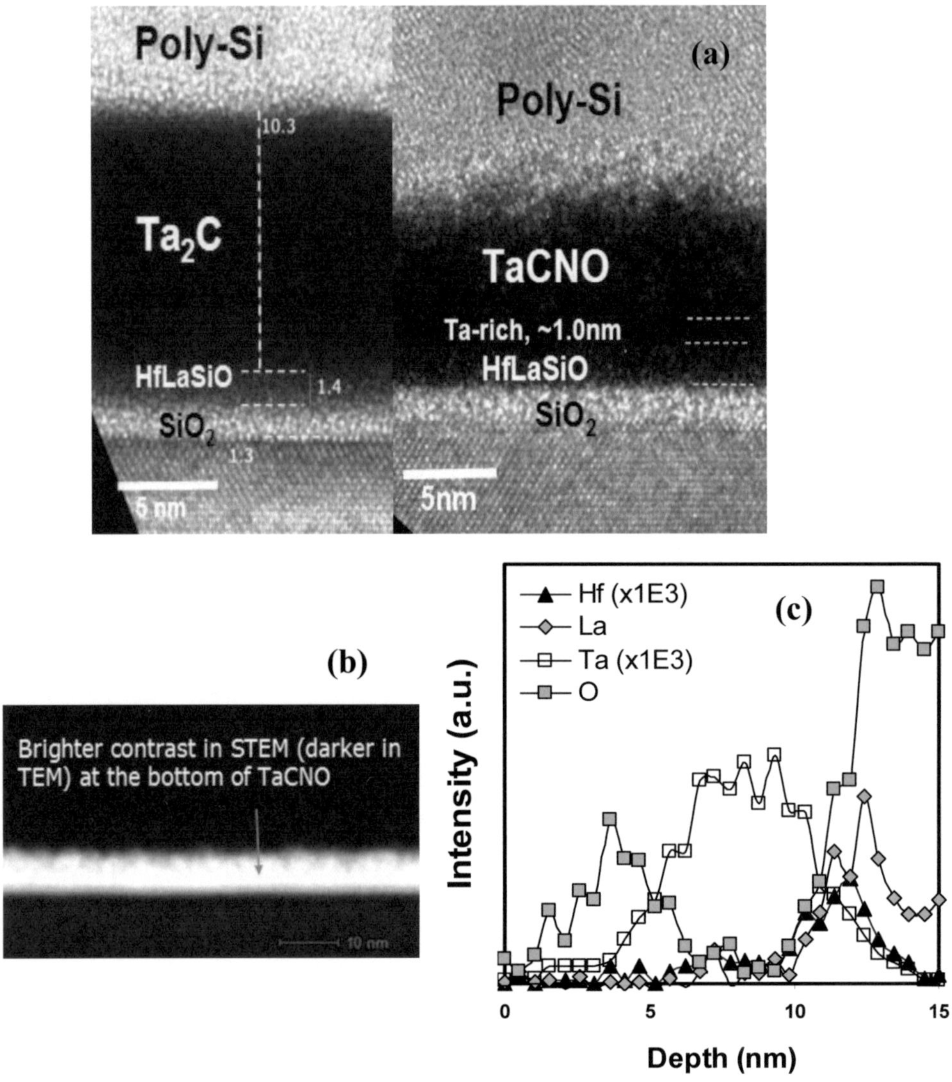

Fig. 14. **(a)** Cross sectional TEM shows LaO / HfSiO intermixing after annealing, with both Ta_2C and TaCNO electrodes. A less-oxygen layer (or Ta rich) at the bottom of TaCNO electrode is observed from **(b)** image contrast, and **(c)** electron-energy loss spectroscopy.

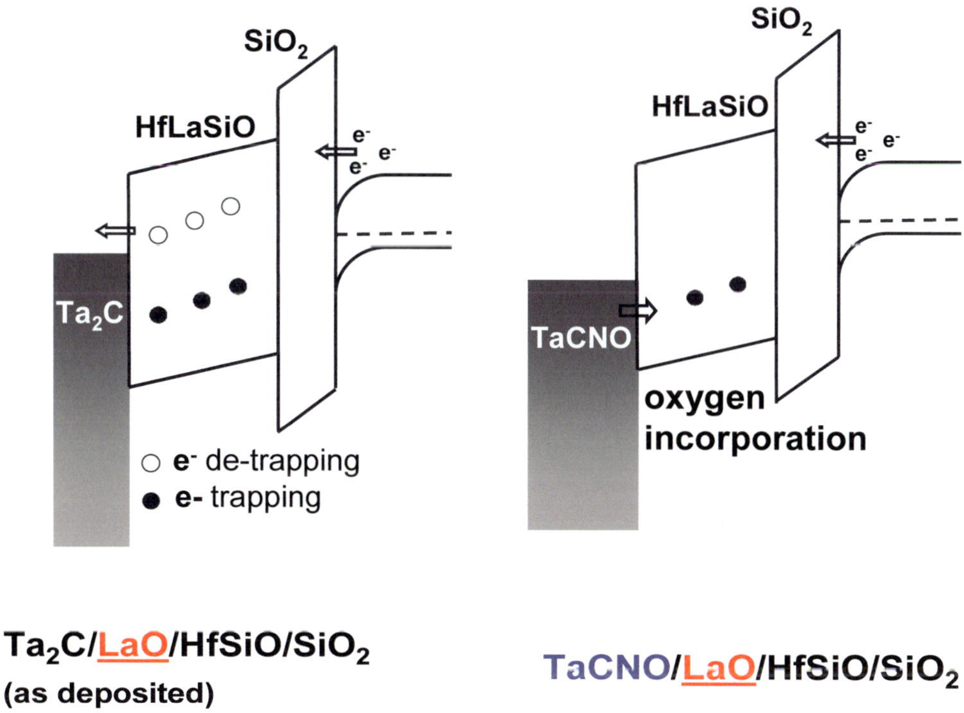

Fig. 15. Schematic diagrams (after thermal anneals) illustrate the negatively charged traps (●) and electron de-trapping (○) during the PBTI stress. Oxygen incorporation from TaCNO can result in less trap generation in the gate stack.

4. Conclusion

A comprehensive study is presented on the integration of LaO_x capping layer for sub-45nm metal gated CMOS devices with Hf-based high-κ dielectrics in a gate first manner. Two different integration routes, i.e. DMDD and SMDD flow, are reported and compared. The device PBTI study is also provided.

5. References

ITRS (2007). *International Technology Roadmap for Semiconductors*, www.itrs.net

Mistry K; Allen C; Auth C; Beattie B; Bergstrom D; Bost M; Brazier M; Buehler M; Cappellani A; Chau R; Choi C; Ding G; Fischer K; Ghani T; Grover R; Han W; Hanken D; Hattendorf M; He J; Hicks J; Huessner R; Ingerly D; Jain P; James R; Jong L; Joshi S; Kenyon C; Kuhn K; Lee K; Liu H; Maiz J; McIntyre B; Moon B; Neirynck J; Pae S; Parker S; Parsons D; Prasad S; Pipes L; Prince M; Ranade P; Reynolds T; Sandford J; Shifren L; Sebastian J; Seiple J; Simon D; Sivakumar S; Smith P; Thomas, T; Roeger T; Vandervoorn P; Williams S. & K. Zawadzki. (2007). A 45nm Logic Technology with High-k+Metal Gate Transistors, Strained Silicon, 9 Cu Interconnect Layers, 193nm Dry Patterning, and 100% Pb-free Packaging,

Proceedings of Internation Electron Device Meeting, pp. 247-250, Washington D.C. USA, December 2007, IEEE EDS

Kubicek S; Schram T; Paraschiv V; Vos R; Demand M; Adelmann C; Witters T; Nyns L; Ragnarsson L.-Å., Yu H.Y; Veloso A; Singanamalla R; Kauerauf T: Rohr E; Brus S; Vrancken C; Chang V; Mitsuhashi R; Akheyar A; Cho H; Hooker J; O'Sullivan B; Chiarella T; Kerner C; Delabie A, Van Elshocht S; De Meyer K; De Gendt S; Absil P; Hoffmann T. & Biesemans S. (2007). Low V_T CMOS using doped Hf-based oxides, TaC-based Metals and Laser-only Anneal, *Proceedings of Internation Electron Device Meeting*, pp. 49-52, Washington D.C. USA, December 2007, IEEE EDS

Narayanan V; Paruchuri V; Bojarczuk N; Linder B; Doris B; Kim Y; Zafar S; Stathis J; Brown S; Arnold J; Copel M; Steen M; Cartier E; Callegari A; Jamison P; Locquet J; Lacey D; Wang Y; Batson P; Ronsheim P; Jammy R; Chudzik M; Ieong M; Guha S; Shahidi G. & Chen T.C. (2006). Band-Edge High-Performance High-κ /Metal Gate n-MOSFETs using Cap Layers Containing Group IIA and IIIB Elements with Gate-First Processing for 45 nm and Beyond, *Digest of Technical Papers, 2009 Symposium on VLSI Technology*, pp. 22.2-1 - 22.2-2, Hawaii, USA, June 2006, IEEE EDS

Yu H.Y; Chang S.Z.; Aoulaiche M; Adelmann C; Wang X.P; Kaczer B; Absil P; Lauwers A. & Biesemans S. (2008). Transistors threshold voltage modulation by DyO rare-earth oxide capping: the role of bulk dielectrics charge. *Applied Physics Letters,* Vol. 93, (December 2008) pp.263502, ISSN 0003-6951

Shen C; Li M.F; Wang X.P; Yu H.Y.; Feng Y.P; Lim A.T.L; Yeo Y.C.; Chan D.S.H. & Kwong D.L. (2004). Negative U Traps in HfO_2 Gate Dielectrics and Frequency Dependence of Dynamic BTI in MOSFETs, *Proceedings of Internation Electron Device Meeting*, pp. 733-736, San Francisco, USA, December 2004, IEEE EDS

10

Computational Study of the Effects of Channel Materials & Channel Orientations and Dimensional Effects on the Performance of Nanowire FETs

Chee Shin Koong and Gengchiau Liang
National University of Singapore, Singapore, 117576, Republic of Singapore

1. Introduction

Silicon has been extensively studied for decades due to its successful applications in semiconducting devices such as metal-oxide-semiconductor field-effect-transistors (MOSFETs). With the demand for high performance devices and packing density, scaling of Si based MOSFETs was drastically driven into nano-scale regime. However, quantum tunneling starts play an important role in degrading the device performance of a conventional Si MOSFET, such as drain-induced barrier-lowering (DIBL) in nano-scale regime. Furthermore, silicon based devices will face its own physical limitation in near future (ITRS, 2007) due to this. Therefore, in order to overcome the challenges of scaling limitation, search for other potential channel materials, such as high carrier mobility material and structure modification have been the heart of research. Among the various proposed materials and device structures, gate-all-around (GGA) Si nanowire (NW) field-effect-transistors (FETs) stand out because their perfect surrounding gates enhance the ability of gate control to suppress the problem of DIBL and fully compatible with Si based technology integration. With the successful fabrication of Si nanowires in the different laboratories (Singh, N. *et. al.*, 2006), nanowires (NWs) have been extensively studied as they are promising for building blocks as nanowire MOSFETs (Cui, Y. *et. al.*, 2003; Pecchia, A. *et. al.*, 2007; Kumar, M. Jagadesh *et. al.*, 2008; Wei, Lu & Lieber, C.M., 2006; Wei, Lu. *et. al.*, 2008), nanophotonic systems (Greytak, A.B. *et. al.*, 2005; Agarwal, R. & Lieber, C.M, (2006); Tian, B. *et. al.*, 2007; McAlpine, M.C. *et. al.*, 2004) and as biochemical sensors (Patolsky, F. & Lieber, C. M., 2005; Hahm, J. & Lieber, C. M., 2004; Cui, Y. *et. al.*, 2001; Gengchiau, L. *et. al.*, 2007). Recent advanced development reveals that physical properties of nanowires could be modified depending on the NW growth direction and diameter. This suggests that material structure such as channel orientations play an important role in device performance optimization. Coupled with the fact that besides silicon, other semiconductor materials such as germanium (Ge) demonstrates promising results (Wang, J. *et. al.*, 2005; Rahman, A. *et. al.*, 2005), a new chapter of study on alternate high mobility channels in nano world has been opened.

Previous theoretical study on this topic using the effective mass model has been carried out (Wang J. *et. al.*, 2004) with the lack of detailed information portraying the electronic structures in the nano-scale regime. Recently, tight-binding (TB) method has been used (Neophytou, N. *et. al.*, 2008; Rahman, A. *et. al.*, 2003) as an alternative procedures to evaluate device performance. However, the former focuses their analysis and simulations on cylindrical cross-section nanowires (Wang, J., *et. al.*, 2004) and the latter discussed on ultra-thin body dual gate (UTB DG) MOSFET (Rahman, A. et. al., 2003). Although TB and non-equilibrium Green's Functions (NEGF) has been developed and implemented to study nanowire MOSFETs (Luisier, M., et. al., 2008) as it provides better transport results compared to top-of-barrier approach, it is time-consuming especially in simulating large nanowire size and long channel. As the objective of this work is to investigate the effects of nanowire orientations to the ultimate performance of nanowire, it is suffice to apply top-of-barrier approach in our work to do the comparison. Therefore, in this work, we explore and compare the ultimate performance of a set of cylindrical nanowire devices with different semiconductor materials (Si and Ge) and channel orientations using TB model and top-of-barrier model (Neophytou, N. *et. al.*, 2008) approaches. TB approach is employed to investigate the electronic properties of NWs in terms of E-k dispersion in order to accurately capture the orientation as well as quantum effects in a nano-scale system. Based on the calculated E-k dispersion, we engaged a semi-classical top-of-barrier MOSFET model to evaluate the ballistic I-V characteristics of NW FETs by self-consistently solving Poisson equation in order to evaluate the ultimate performance of these semiconductor NW FETs with various channel orientations. The schematic of the device structure is shown in Fig. 1. The simulation is conducted in two parts: a) simulate I-V characteristics of circular nanowire (CW) with diameter of 3nm for Si and Ge with different orientations to study the effects of materials and orientations on the device performance and b) extend this simulations to explore performance of CW with different diameters. In this work, we explore four different diameters: 3nm, 5nm, 8nm and 10nm.

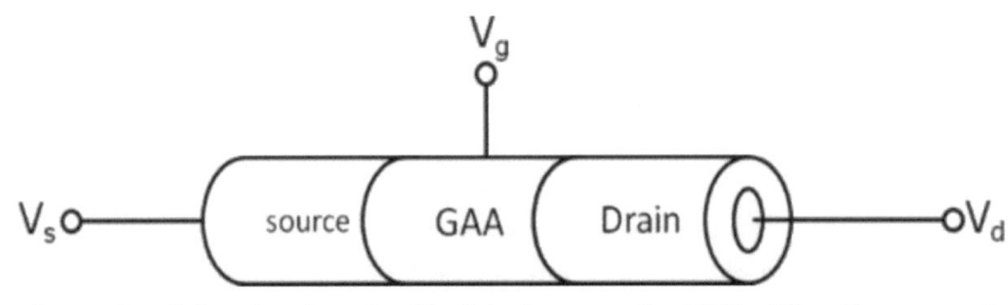

Fig. 1. A schematic of the simulated cylindrical nanowire FETs. The diameter of the circular cross-section (D) varies from 3nm, 5nm, 8nm and 10nm. The oxide thicknesses (tox) for this simulation are set at 1.6nm and 0.5nm for comparison.

2. Methodology

To investigate the ultimate performance of NW MOSFETs based on structural effects, we follow a two-step approach. Firstly, we assume a NW with certain size and orientation, and then, a $sp^3d^5s^*$ tight-binding model is implemented to investigate the electronic properties of NWs in terms of *E-k* dispersion relations. Next, we use the simulated dispersion relations obtained from TB model to calculate the ballistic current-voltage (*I-V*) characteristics of both p-channel and n-channel NW MOSFETs using a semi-classical "top-of-the-barrier" MOSFET model. The results shown in (Neophytou, N. *et. al.*, 2008) with and without including self-consistency of bandstructure do not differ significantly to render the results invalid. Therefore, in this work, we do not calculate self-consistency of bandstructure.

A. Electronic Bandstructure Calculations:

To obtain the bandstructure of the Si NWs of different orientations, we assume an unrelaxed nanowire atomic geometry and construct the Hamiltonian of the NW unit cell using the orthogonal-basis $sp^3d^5s^*$ tight-binding method developed for bulk electronic structure. (Boykin, T.B. et. al., 2004) In this approach, we model each atom using 10 orbitals per atom in total. The different energy parameters in this TB model were obtained by fitting a genetic algorithm to reproduce the bandgap and the electron/hole effective masses in different valleys. The simulated NW is assumed to be infinitely long, and the nanowire surface is passivated by hydrogen atoms, which in this case is treated numerically using a hydrogen termination model of the sp3 hybridized interface atoms. This technique is reported to be successfully removing all the interface states from the bandgap. (Lee, S. et; al., 2004) This model has shown good agreement with the measured bandgap vs. diameter of silicon nanowires despite the exclusion of relaxation or strain effects. Once the transport direction is specified, the size t_{Si} of the NW and unit cell can be defined, cf. Fig. 1. The Hamiltonian of atoms within the unit cell and atoms within neighboring unit cells are obtained as H_l, where l is the unit cell index with $l=0$ for the center unit cell and $l\neq 0$ for the *l-th* nearest-neighboring cell. In this model, we only consider the nearest neighbors i.e., $l=1$ and -1. Then, the Hamiltonian in *1D k*-space is calculated by taking the Fourier transform:

$$H_{k_m} = \sum_l H_l \exp[-j\bullet k_m(z_l - z_0)] \quad (1)$$

where $k_m=m\pi/L$ is the wavevector within the 1st Brillouin zone, m is the real number between 0 and 1, and L is the periodicity of the 1D lattice, $z_l=lL$, referring to the unit cell position. Due to the orthogonality of the TB basis sets, a simple eigenvalue problem, $H_{k_m}\Psi = E_{k_m}\Psi$, can be solved for each k_m within the first Brillouin zone. The electronic structures of the nanowires such as bandgap, bandstructure, etc., can be provided and used for the transport calculations. The conduction and valance bands for Si and Ge are shown in Fig. 2(a) to 2(d). From Fig. 2, it can be seen that valley splitting occurs and the degeneracy is lifted, giving rise to two subbands. Figs. 3a and 3b show the valley splitting as a function of nanowire diameters for n-type Si and Ge and p-type Si and Ge for [110] orientation, respectively. Fig. 3a shows that valley splitting is more evident when the diameter falls beyond 5nm. In general, valley splitting for Ge is always larger than Si, as shown in Fig. 3a due to Ge having lighter mass. However, a great distinction in valley splitting can be observed when the NW diameter is smaller than 5nm for Si and 8nm for Ge, respectively.

B. Transport Calculation:

Next, we use a semi-classical top-of-the-barrier MOSFET model (Rahman, A. *et. al.*, 2003) to simulate the ballistic I-V characteristics. In this model, a simplified 3-dimensional self-consistent electrostatic model with ballistic treatment of carrier transport is used. In addition, quantum capacitance effects are included into this approach. Three-dimensional electrostatics is described by a simple capacitance model (Rahman, A. *et. al.*, 2003). The capacitors represent the electrostatic coupling of the gate (C_G), drain (C_D), and source terminals (C_S) to the top of the potential barrier at the source end of the channel. These

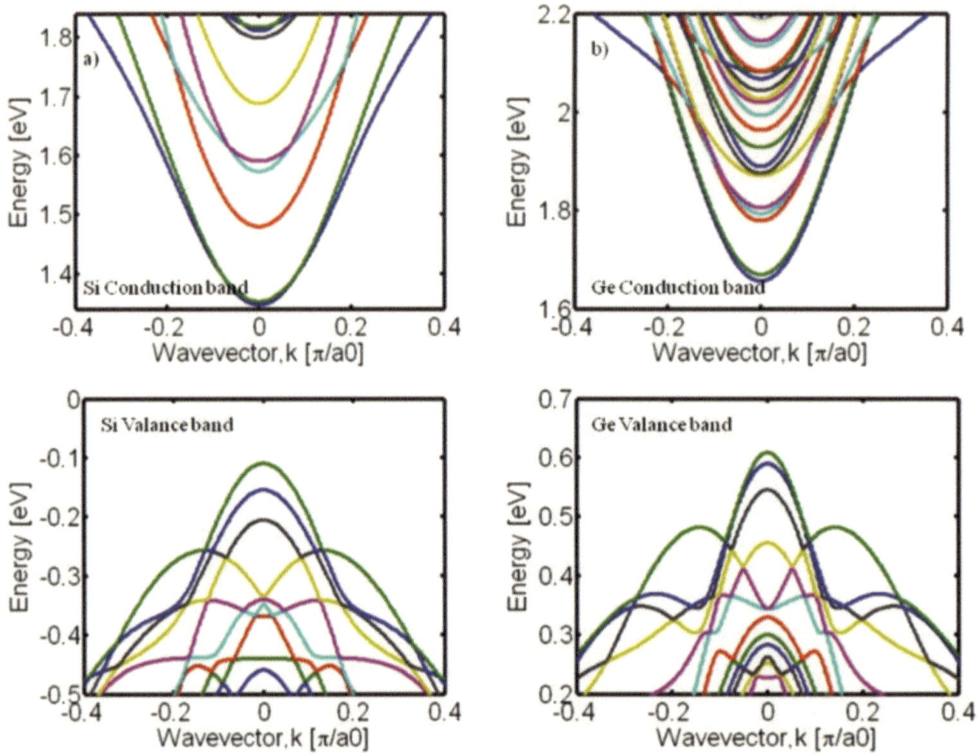

Fig. 2. Conduction and valence band bandstructure for (a) 3nm Si with [110] orientation and (b) 3nm Ge with [110] orientation.

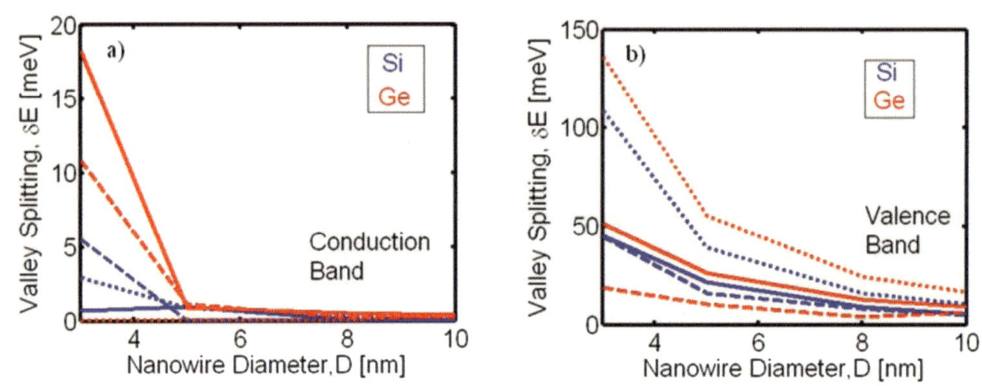

Fig. 3. (a) and (b) show the valley splitting as a function of nanowire diameters for n-type and p-type respectively. It can be observed that valley splitting is a strong function of quantum confinement. In terms of semiconducting material, valley splitting is more evident in Ge compare to Si. This is due to Ge having lighter mass compared to Si. The solid line, dashed line, and dotted line represnt NW's orientation along [100], [110], and [111], respectively.

capacitors control the subthreshold swing, S, of the transistors and the drain-induced barrier lowering ($DIBL$) according to the following equations:

$$\frac{C_G}{C_\Sigma} = \frac{2.3k_B T / q}{S} \tag{2a}$$

$$\frac{C_D}{C_\Sigma} = \frac{2.3k_B T / q}{S} \times DIBL \tag{2b}$$

$$C_\Sigma = C_G + C_D + C_S \tag{2c}$$

Then, the Poisson's equation is solved using a simplified capacitance model. The Poisson's potential (U_P) is equal to $U_0 \cdot (N - N_0)$, where $U_0 = q/C_\Sigma$ is the single electron charging energy, N_0 and N are the number of mobile carriers at the top of the barrier at equilibrium and under applied bias, respectively, while C_Σ is the total capacitance. The carrier density N, can be directly computed from E-k relations determined earlier by applying the below equation,

$$N = \int_{-\infty}^{\infty} \frac{dk}{\pi} [f(E - E_{fs} + U_{scf}) + f(E - E_{fs} + qV_D + U_{scf})] \tag{3}$$

where $f(E)$ is the Fermi function and E_{fs} is the chemical potential in the source region. Iteration between N and U_{scf} is repeated until the self-consistency converges. The NW MOSFET current is then evaluated using the semi-classical transport equation in the ballistic limit, which is given by:

$$I = \frac{2q}{h} \int_{U_{scf}}^{\infty} dE[f(E - E_{fs}) - f(E - E_{fs} + qV_D)] \tag{4}$$

3. Simulation results and discussions

Firstly, using top-of-barrier model, the I-V characteristics of Si and Ge for different orientations of CW NW transistors with the effective gate oxide thickness (EOT) of 1.6nm and 0.5nm are investigated. The off-state currents of all cases are set to be $0.2\mu A / \mu m \cdot (2D)$ in our simulation. Solid lines represent [100] orientation while dash lines represent [110] orientation and dotted lines represent [111] orientation. Red lines represent p-type and blue lines represent n-type devices. Fig. 4(a) and 4(b), respectively, show the Ids-Vds curves for 3nm Si and Ge at Vgs=0.6V for EOT of 1.6nm while Fig. 4(c) and 4(d), respectively, show the Ion/Ioff ratio as a function of nanowire diameter for Si and Ge with EOT of 1.6nm. For N-type NW FETs, as shown in Fig. 4(a) and (b), Si and Ge of [110] orientation give the highest on-currents, which is about 45% and 146%, respectively, compared to the current along [100] orientation. Comparing best orientation with the highest ON-state currents for different materials, Ge [110] outperforms Si [110] by 1.18 times due to Ge [110] having lighter effective mass compared to Si [110]. Similarly for p-type NW FETs, Si of [110] orientation and Ge of [111] orientation give the highest ON-state currents, with Ge [111] outperforms Si [110] by 1.78 times. Moreover, for Si NW FETs, n-type device has similar performance as p-type device. This is due to lifting of degeneracy of the dispersion as an effect of quantum confinement. This resulted in a decrease in effective mass (Neophytou, N. *et. al.*, 2008). From Fig. 4(c) and 4(d), n-type Si and Ge NW FETs with [110] orientation has the highest Ion/Ioff ratio while for p-type NW FETs, Si and Ge with [111] orientation has highest Ion/Ioff ratio. Furthermore, it is also observed that Ion/Ioff decreases as nanowire diameter increases because small nanowire has better gate control due to larger capacitance.

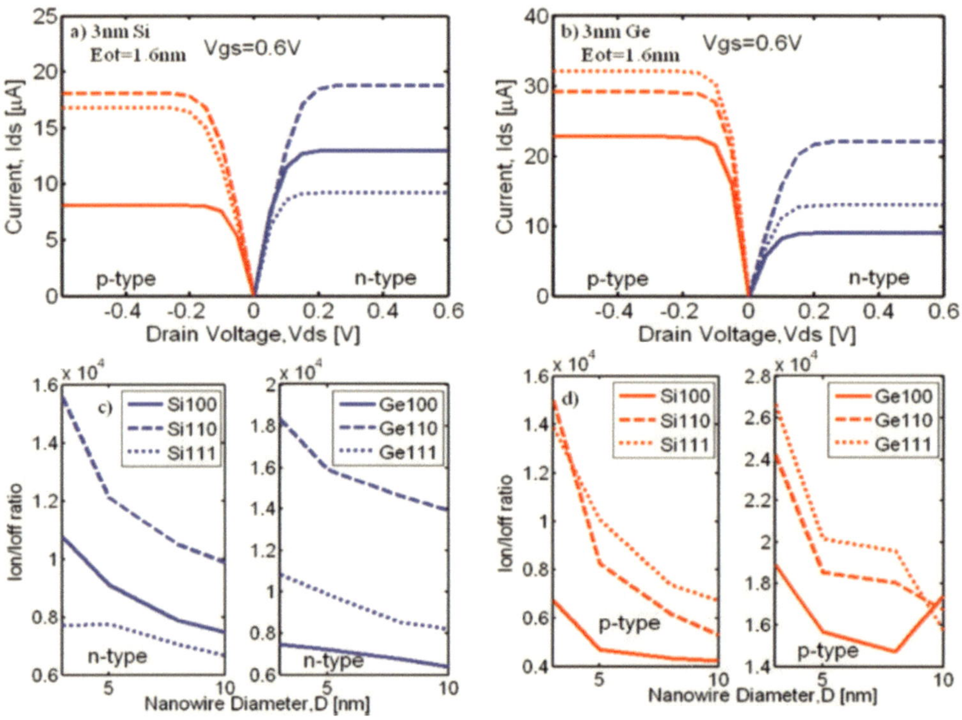

Fig. 4. (a) and (b) shows the I-V characteristics for 3nm n-type and p-type Si and Ge , respectively while (c) and (d) respectively show the Ion/Ioff ratio for n-type and p-type devices, respectively. In general for n-type devices, Si and Ge with [110] orientation give highest on-current while p-type devices, [110] Si and [111] Ge give highest on-current. This is due to these orientations having lightest effective mass.

Next, the current density of n-type and p-type Si and Ge for different orientations with EOT of 1.6nm and 0.5nm were investigated, as shown in Fig. 5. As expected, the current density for EOT of 0.5nm is higher, about doubled comparing to Si and Ge with EOT of 1.6nm for all orientations due to better gate control as a result of larger gate capacitance. In terms of semiconducting materials, Ge always outperforms Si regardless of nanowire diameters for p-type NW FETs (Fig. 5b and Fig. 5d). However, for n-type NW FETs (Fig. 5a and Fig. 5c), the current density for Si does not differ much from Ge. This phenomenon could be explained by the effective mass of Si and Ge. From calculation of hole effective mass, it could be deduced that the effective mass of Ge is far apart compared to Si while for the electron effective mass, the differences of Si and Ge is not significantly far apart. Furthermore, two important points could be obtained for both Si and Ge electronic bandstructures: a) It is shown, from calculation, that [110] orientation has the lightest electron effective mass compared with the other two orientations and b) Ge has lighter mass compared to Si in terms of semiconducting material, regardless of NW diameter. As a result, Ge with [110] orientation outperforms Si with [110] orientation, giving rise to better performance, in terms of the higher ON-state current.

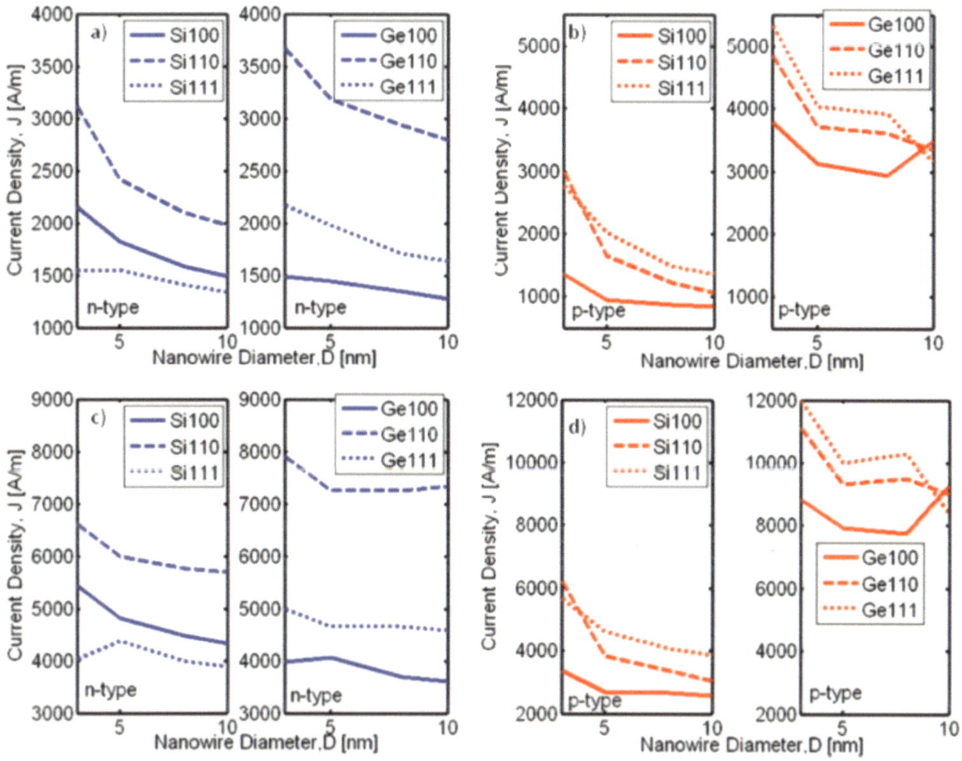

Fig. 5. (a) and (b) shows the current density as a function of nanowire size for n type and p type Si and Ge with EOT=1.6nm while (c) and (d) shows the current density as a function of nanowire size for n-type and p-type Si and Ge with EOT=0.5nm. In general, the current density with EOT of 0.5nm is twice larger than that of 1.6nm due to better gate control.

In addition, we explore the capacitance effect on the device as the device current tightly depends on the capacitance, gate capacitance in particular. The C_g/C_{ox} ratio as a function of nanowire diameter for n-type and p-type Si and Ge NW GAA FETs with oxide thickness of 1.6nm and 0.5nm are shown in Fig. 6(a), (b), (c), and (d), respectively. It can be observed that the capacitance value degraded from the gate oxide capacitance for both Si and Ge regardless of oxide thickness. Detailed calculation with capacitance value given by $C_g = \frac{C_{ox}C_s}{C_{ox}+C_s}$ shows that [110] orientation for Si and Ge encounter greatest degradation from oxide capacitance by 31.6% while [100] orientation encounters degradation of 15.4% and [111] Si and [111] Ge both encounter degradation of 7.14% and 25%, respectively. All these translate to an effective increase in gate oxide thickness, which in general reduces the gate control to the device. However, in all cases, we found $C_g<C_{ox}$ and as gate oxide

thickness, t_{ox} decreases, the difference increases. It suggests that C_{OX} is not much larger than C_S and dominating C_G, under the approximation of a conventional Si planar MOSFET.
As shown in Fig. 6, in general, we could see that as the diameter of cylindrical NW is large, given t_{ox} is thick, $C_g \approx C_{ox}$, where $C_g = \left.\frac{dQ}{dV_g}\right|_{V_{ds}=0.05V}$. It indicates that $C_s \gg C_{ox}$ which is in agreement with the classical approximation. However, as the diameter decreases, we found that in all cases, C_g/C_{ox} ratio is less than 1, indicating that the approximation $C_s \gg C_{ox}$ does not hold. For example, using $C_{ox} \approx 0.6448nF/m$ and C_g obtained from simple calculation for 8nm cylindrical NW with tox=1.6nm, C_g/C_{ox} for n-type Si and Ge with [110] orientations is 0.8375 and 0.7444, respectively, and C_g/C_{ox} for p-type [110] Si and [111] Ge is 0.8685 and 0.8995, respectively. Similarly, for 3nm cylindrical NW with tox=1.6nm, C_g/C_{ox} for n-type Si and Ge with [110] orientations is 0.6993 and 0.7467, respectively, and for p-type [110] Si and [111] Ge, C_g/C_{ox} for both cases is about 0.76. For comparison, we did a calculation for 3nm cylindrical NW with t_{ox} of 0.5nm and the C_g/C_{ox} ratio falls to below 0.7 for n-type [110] Si and Ge as well as for p-type [110] Si. This is due to the larger gate capacitance caused by the thinner gate oxide.

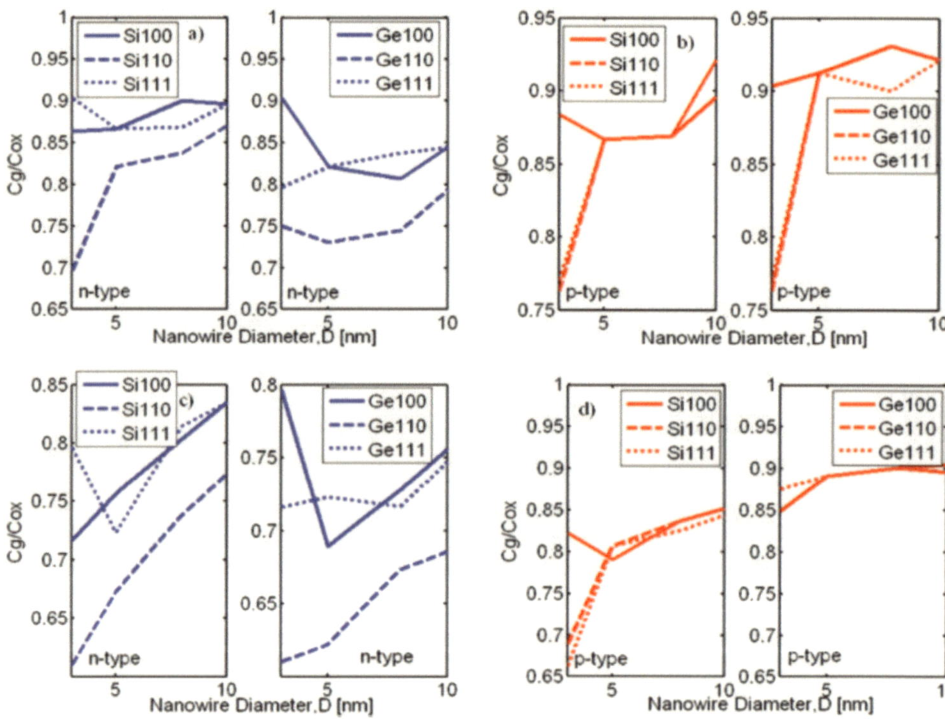

Fig. 6. Cg/Cox as a function of nanowire diameter for n-type and p-type Si and Ge with oxide thickness of 1.6nm (6a and 6b) and oxide thickness of 0.5nm (6c and 6d). The capacitance value is degraded from the gate oxide capacitance for both Si and Ge regardless of oxide thickness.

On the other extreme region, when the oxide thickness is reduced further due to shrinking of devices, $C_s \ll C_{ox}$, and $C_g \approx C_S$. At this point, the current does not depend on the effective mass of the material, and channel orientations. As a result, all the I-V curves of all cases with different materials and different orientations will overlap (Liang, G.C., *et. al.*, 2007).
To further explore device performance from a different perspective, the transconductance $\left(g_m = \left.\frac{\partial I_{ds}}{\partial V_{gs}}\right|_{V_{ds}=0.05V} \right)$, of FETs is investigated. For an ideal nanowire FET at low bias, the transconductance is given by $g_m = \frac{W \mu_{eff} C_g}{L} v_{ds}$, where C_g is the gate capacitance per unit length. In this simulation, we have chosen gate length, L to be 16nm. This value is obtained from IRTS 2007 PIDS table, to be consistent with the production year in which off-current for dual gate is chosen. Fig. 7(a) and 7(b) show the transconductance of n-type Si and Ge and p-type Si and Ge NW FETs for EOT of 1.6nm at low drain to source bias of 0.05V for n-type and -0.05V for p-type, respectively, as a function of different diameters. Similarly, Fig. 7(c) and 7(d) show the transconductance of n-type Si and Ge and p-type Si and Ge NW FETs for

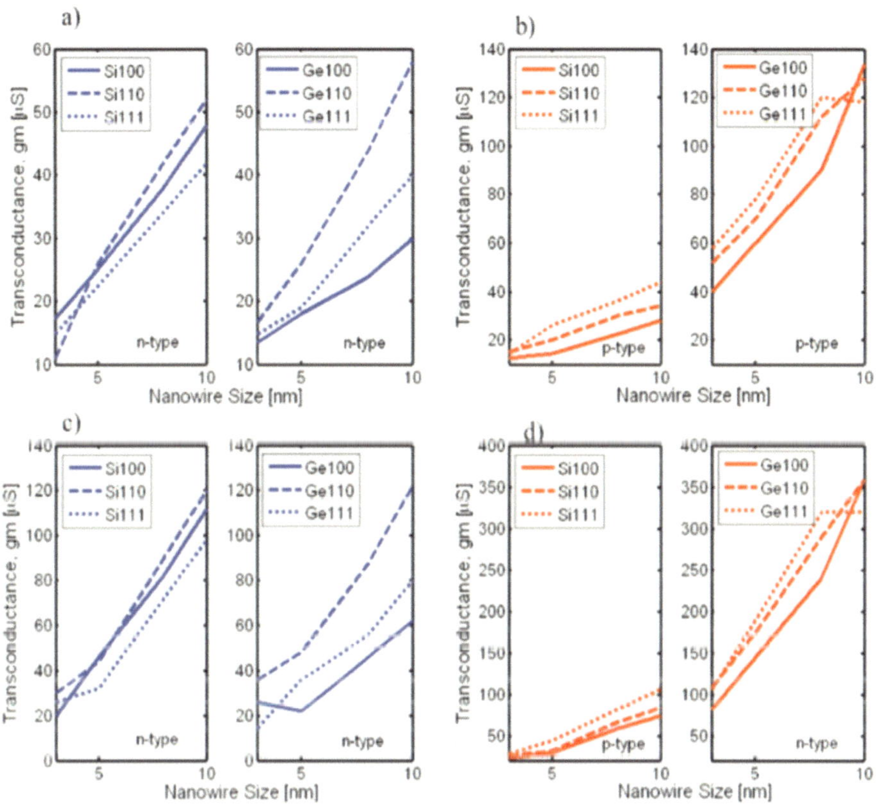

Fig. 7. (a) and (b) show the transconductance for Si and Ge with EOT of 1.6nm while (c) and (d) show the transconductance for Si and Ge with EOT of 0.5nm. In general, for n-type devices, [110] orientation has highest transconductance regardless of channel material and diameter size while for p-type devices, [110] Si and [111] Ge give highest transconductance.

EOT of 0.5nm at drain to source bias of 0.05V for n-type and -0.05V for p-type, respectively, as a function of NW's diameter. In general, the transconductance decreases as the diameter decrease due to lower ON-currents of the smaller diameter NWs. As the difference in gate voltage is a constant of 0.05V, the transconductance, g_m is proportional to on-current.
For n-type devices, Si and Ge of [110] orientation, in general, have the highest transconductance compared to the other two orientations regardless of nanowire area, as shown in Fig. 7(a). The transconductance of Ge [110] is about 1.15 times the value of Si [110] at 3nm and slowly widens to 1.5 times at 10nm. For p-type devices, in general, Si and Ge with [111] orientation have larger transconductance. Fig. 7(c) and 7(d) show the transconductance for EOT of 0.5nm for n-type Si and Ge and p-type Si and Ge. As expected, the transconductance for EOT of 0.5nm in general is more than twice the value with that of EOT of 1.6nm. This is in agreement with the trend of on-current simulation results for drain voltage of ±0.05V. For amplifiers, we require the gain to be large so as to amplify the output by a few magnitudes. Transconductance is a measurement of gain of amplifier. As such, if NW FETs were to be used as amplifiers, large transconductance is required. From the above observation, Ge is the best candidate for both n-type and p-type NW FETs.

4. Conclusion

In this work, we present the device performance of Si and Ge and channel orientations in NW FETs and extend the discussion for different NW diameters. We show that in terms of channel orientation, for n-type devices, Si [110] and Ge [110] give the highest on-current compared to other orientations while in terms of channel material, Ge outperforms Si by between 1.17 to 1.42 times due to the lighter effective mass. Moreover, it is also observed that valley splitting is a strong function of quantum confinement, and it is more significant for NW diameter smaller than 5nm. We also explore the effect of different oxide thickness on the performance of devices as the oxide thickness determines the device capacitance. In investigating the effects of gate capacitance on devices of different NW sizes, we conclude that gate capacitance degrades as the device shrinks into sub-nanometer regime. Therefore, conventional approximation to calculate transport property does not apply. As we examine the gate oxide capacitance further, we found that it has not reached the other extreme where $C_s \ll C_{ox}$ as at this extreme, the on-current for same material will overlap as the on-current only depends on effective mass. This phenomenon is not observed in the Ids-Vds curves even at EOT of 0.5nm. For transconductance, n-type Si and Ge of [110] orientation gives best performance while in terms of semiconducting material, both Si and Ge does not differ much. For p-type devices, Ge NW FETs show the better transconductance than Si NW FETs.

5. Acknowledgements

This work was supported by A*STAR 082-101-0023 and computational support by the Ministry of Education of Singapore under Grant Nos. R-263-000-416-112 and R-263-000-416-133.

6. References

Agarwal, R. & Lieber, C.M. (2006). Semiconductor nanowires: optics and optoelectronics. *Applied Physics a-Materials Science & Processing*, Vol. 85, No. 3 (November 2006), 209-215, 0947-8396

Boykin, T.B., Klimeck, G. and Oyafuso F.(2004). Valence band effective-mass expressions in the sp(3)d(5)s(*) empirical tight-binding model applied to a Si and Ge parametrization, *Physical Review B*, Vol. 69, No. 11 (March 2004), 1152011-11520110, 1098-0121

Cui, Y.; Wei, Q.; Park, H.; Lieber, C.M. (2001). Nanowire nanosensors for highly sensitive and selective detection of biological and chemical species, *Science*, Vol. 293, No. 5533 (August 2001), 1289-1292, 0036-8075

Cui, Y.; Zhaohui Zhong; Deli Wang; Wang, W.U.; Lieber, C.M. (2003). High performance silicon nanowire field effect transistors, *Nano Letters*, Vol. 3, No. 2 (February 2003), 149-52, 1530-6984

Greytak, A.B.; Barrelet, C.J.; Yat Li; Lieber, C.M. (2005). Semiconductor nanowire laser and nanowire waveguide electro-optic modulators, *Applied Physics Letters*, Vol. 87, No. 15 (October 2005), 151103-1-3, 0003-6951

Hahm, J. & Lieber, C. M. (2004). Direct ultrasensitive electrical detection of DNA and DNA sequence variations using nanowire nanosensors, *Nano Letters*, Vol. 4, No. 1 (January 2004), 51-4, 1530-6984

International Roadmap for Semiconductors. 2005; Available from: http://public.itrs.net/.

Kumar, M. Jagadesh; Reed, Mark A.; Amaratunga, Gehan A. J.; Cohen, Guy M.; Janes, David B.; Lieber, Charles M.; Meyyappan, M.; Wernersson, Lars-Erik; Wang, Kang L.; Chau, Robert S.; Kamins, Theodore I.; Lundstrom, Mark; Yu, Bin; Zhou, Chongwu.(2008). Guest Editorial Special Issue on Nanowire Transistors: Modeling, Device Design, and Technology, *IEEE Transactions on Nanotechnology*, Vol. 7, No. 6 (November 2008), 643-650, 1536-125X

Lee, S.; Oyafuso, F.; Allmen, P.; Klimeck, G. Boundary conditions for the electronic structure of finite-extent embedded semiconductor nanostructures, Physical Review B, Vol. 69, No. 4 (January 2004), 0453161-0453168.

Liang, GC.; Xiang, J.; Kharche, N.; Klimeck, G.; Lieber, C. M.; Lundstrom, M. (2007). Performance analysis of a Ge/Si core/shell nanowire field-effect transistor, *Nano Letters*, Vol. 7, No. 3 (March 2007), 642-646, 1530-6984

Liang, GC.,; Kienle, D.; Patil, S.K.R.; Wang, J.; Ghosh, A.W.; Khare, S.V. (2007). Impact of structure relaxation on the ultimate performance of a small diameter, n-type <110> Si-Nanowire MOSFET, *IEEE Transactions on Nanotechnology*, Vol. 6, No. 2 (March 2007), 225-229, 1536-125X

Luisier, M.(2008). Full-band quantum transport in nanowire transistors, Journal of *Computational Electronics*, Vol. 7, No. 3 (September 2008), 309-314, 1572-8137

McAlpine, M.C.; Friedman, R.S.; Jin S.; Lin, K.; Wang, W.U.; Lieber, C.M. (2003). High-performance nanowire electronics and photonics on glass and plastic substrates, *Nano Letters*, Vol. 3, No. 11 (November 2003), 1531-5, 0065-7727

Neophytou, N; Paul, A.; Lundstrom, M.; Klimeck, G. (2008). Bandstructure effects in silicon nanowire electron transport, *IEEE Transactions on Electron Devices*, Vol, 55, No. 6 (June 2008), 1286-1297, 0018-9383

Patolsky, F. & Lieber, C. M. (2005). Nanowire Nanosensors, *Materials Today*, Vol.8 No. 4 (March 2005) 20-28, 1369-7021

Paul, A.; Mehrotra, S.; Klimeck, G.; Luisier, M. (2009). On the validity of the top of the barrier quantum transport model for ballistic nanowire MOSFETs, *2009 13th International Workshop on Computational Electronics (IWCE 2009)*, 1-4, 0-85261-704-6

Pecchia, A.; Salamandra, L.; Latessa, L.; Aradi, B.; Frauenheim, T.; Di Carlo, A. (2007). Atomistic modeling of gate-all-around Si-nanowire field-effect transistors, *IEEE Transactions on Electron Devices*, Vol., 54, No., 12 (December 2007), 3159-3167, 0018-9383

Rahman, A.; Klimeck, G. & Lundstrom, M. (2005). Novel channel materials for ballistic nanoscale MOSFETs-bandstructure effects, *IEEE International Electron Devices Meeting*, 601-604, 0163-1918, Washington DC, December 2005, Electron Devices Society, California

Rahman, A.; Jing Guo; Datta, S.; Lundstrom, M. (2003). Theory of ballistic nanotransistors, *IEEE Transactions on Electron Devices*, Vol. 50, No. 9 (September 2003), 1853-1864, 0018-9383

Singh, N.; Agarwal, A.; Bera, L.K.; Liow, T.Y.; Yang, R.; Rustagi, S.C.; Tung, C.H.; Kumar, R.; Lo, G.Q.; Balasubramanian, N.; Kwong, D.-L. (2006). High-performance fully depleted silicon-nanowire (diameter <= 5 nm) gate-all-around CMOS devices, *IEEE Electron Device Letters*, Vol. 27, No. 5 (May 2006), 383-386, 0741-3106.

Tian, B.; Zheng, X.; Kempa, T.J.; Fang, Yi; Yu, N.; Yu, G.; Huang, J.; Lieber, C.M. (2007). Coaxial silicon nanowires as solar cells and nanoelectronic power sources, *Nature*, Vol. 449, No. 7164 (October 2007), 885-889, 0028-0836

Wang, J.; Polizzi, E.; Lundstrom, M. (2004). A three-dimensional quantum simulation of silicon nanowire transistors with the effective-mass approximation, *Journal of Applied Physics*, Vol. 96, No. 4 (August 2003), 2192-2203, 0021-8979

Wang J.; Klimeck, G.; Lundstrom, M. and Rahman, A. (2005). Bandstructure and orientation effects in ballistic Si and Ge nanowire FETs, *IEEE International Electron Devices Meeting*, 530-533, 0163-1918, Washington DC, December 2005, Electron Devices Society, California

Wei, L. & Lieber, C.M. (2006). Semiconductor Nanowires, *Journal of Physics D: Applied Physics*, Vol. 39, No. 21 (November 2006), R387-R406, 1361-6463

Wei, L.; Ping Xie; Lieber, C.M. (2008). Nanowire Transistor Performance Limits and Applications, *IEEE Transactions on Electron Devices*, Vol. 55, No. 11 (November 2008), 2859-2876, 0018-9383

11

Integration of Carbon Nanotubes in Microelectronics

Stanislav A. Moshkalev[1], Carla Veríssimo[1], Rogério V. Gelamo[1], Leonardo R. C. Fonseca[2], Ettore Baldini-Neto[2] and Jacobus W. Swart[3]
[1]*State University of Campinas-UNICAMP, Campinas, SP,*
[2]*Center of Advanced Research W. von Braun, Campinas, SP,*
[3]*Center for Information Technology Renato Archer– CTI, Campinas, SP,*
Brazil

1. Introduction

Carbon nanotubes (CNTs) has received much attention since their discovery in 1991 due to unique combination of interesting electrical, mechanical, thermal and other properties, and numerous potential applications (Meyyappan, 2005, Sharma, 2008). Single-wall carbon nanotubes (SWCNTs) can be metallic or semiconducting, while multi-wall nanotubes (MWCNTs) are basically metallic. Semiconducting SWCNTs can be used in nanotubes based field effect transistors (FET-CNT), while metallic SWCNTs and MWCNTs can be employed for electrical and thermal interconnections, in sensors and other micro- and nanodevices. However, the integration of nanotubes into microelectronic circuitry is a very challenging task which requires development of reliable and compatible technologies for controlled synthesis, accurate positioning and contacting of nanotubes and their arrays in new devices. Many difficult issues associated with these technologies have to be addressed. In particular, mechanisms of nucleation and growth of high quality nanotubes still are not well understood. Mechanisms of electrical and thermal conductivity in individual nanotubes and ropes, the role of defects, formation of contacts with metals have to be investigated thoroughly.

2. Electrical properties of carbon nanotubes

SWCNT can be viewed as a single graphite (or graphene) sheet rolled into a cylinder of a nanometer size diameter, and MWCNT as a coaxial array of several single-wall nanotubes separated by approximately 0.34 nm. In graphene layers, sp^2 hybridyzation results in formation of three strong in-plane σ bonds between carbon atoms and one π bond, the latter corresponding to loosely bound π electrons of high mobility that are responsible for a very high conductivity along the graphene plane.
As SWCNTs are essentially one-dimensional structures, at the absence of defects they are characterized by a ballistic transport of electrons (without scattering) at moderate current densities for nanotube lengths up to a few micrometers (Graham et al, 2004). This is in striking contrast to metal (copper) wires where the mean free pass (MFP), determined by the mean grain size, is in the range of a few tens of nanometers.

Another advantage of nanotubes is that small diameter copper vias are subject to failure due to electromigration at high current densities (>10^6 A/cm^2) while CNTs of the same diameter can sustain current densities as high as 10^9 A/cm^2 (Graham et al, 2004). This makes CNTs very attractive for electrical interconnect applications (especially, vias) instead of currently used copper.

The resistance of a SWCNT (or a shell in a MWCNT) has three components (Tan et al, 2007; Matsuda et al, 2007): (i) a contact resistance associated with one-dimensional systems, given by a quantum resistance $G_0^{-1} = h/2e^2 = 12.9$ kΩ corresponding to one conducting state (a factor of two is added due to two possible spin states), (ii) an intrinscic resistance due to scattering which is length dependent, and (iii) an additional contact resistance associated with imperfect contacts between a metal electrode and CNTs. Metals that form carbides (e.g., Ti) are believed to be an optimal electrode material as it is expected that carbides ensure better electrical coupling with nanotubes (Tan et al, 2007).

It is very challenging to evaluate precisely the contribution of contact resistances due to evident experimental difficulties. Comparison between different experiments is also not straightforward because of wide variation of conditions and particular geometries used for studies (for example, side- and end- contacts, presence of surfactants and other contaminants, metal grain sizes and degree of metal annealing after its deposition over nanotubes, etc.). SWCNTs can be metallic or semiconducting, depending on chirality. The energy gap for semiconducting nanotubes is given aproximately by $E_g(eV) = 1/d(nm)$, reducing rapidly with the graphene shell diameter *d*. Usually, the number of metallic SWCNTs in as-grown samples is close to 1/3, the rest are semiconducting. In contrast, MWCNTs of larger diameters are basically metallic and thus are especially appropriate for interconnects.

The number of conducting states per graphene shell depends on its chirality, it can be 0 or 2 for small diameter semiconducting and metallic SWCNTs, respectively, and it increases linearly with the shell diameter for larger nanotubes (Naeemi and Meindl, 2007). MFP also was shown to increase with MWCNT diameter. The theoretical limit for resistance of high-quality MWCNTs (diameter of 15 nm, 10 walls) in a ballistic regime can be lower than 0.1 kΩ (for lengths smaller than MFP), compared with the resistance of about 1 kΩ for a 150 nm long, 10 nm diameter copper damascene wire (Graham et al, 2004). The model of MWCNTs as electrical conductors developed by Naeemi and Meindl, 2007, predicts that they can outperform copper wires for lengths exceeding 5-10 μm, depending on diameter. Bundles of SWCNTs were shown to have potentially superior performance at smaller leghths (<1 μm), however for this, dense nanotube packing is essencial which is a very difficult practical task.

Recently, it has been reported by Jun et al, 2007, that the AC conductance of high quality PECVD (plasma enhanced chemical vapor deposition) grown MWCNTs decreases gradually with increasing frequency (frequencies up to 50 GHz were studied), indicating that nanotubes can be used not only for DC but also in a microwave range.

3. Synthesis of carbon nanotubes

It is important to emphasize that properties and quality of carbon nanotubes depend strongly on the fabrication method. There are two main groups of CNTs synthesis methods: (i) high-temperature processes like arc discharge and laser evaporation where the process temperature can reach T = 2000-4000 °C, and (ii) chemical vapor deposition (CVD) processes performed at much lower temperatures: in the range of 500-1000 °C for thermal CVD and even lower for plasma-enhanced CVD. In high-temperature processes, higher quality nanotubes can be obtained, however the process output is a CNT containing soot which needs to be further processed (dispersed, purified and in some cases functionalized) before applications. The low-temperature CVD methods can be compatible with microelectronic technologies and therefore attract most attention. Note that activation by plasmas in PE-CVD processes can promote formation of higher quality nanotubes at lower temperatures, and thus PE-CVD is a promissing technology for microelectronics applications. Electric fields built-up in the plasma, can also be used to provide directional nanotube growth. Studies of nucleation mechanisms (Moshkalev & Veríssimo, 2007) and search for new methods of synthesis, compatible with microelectronics technologies, must continue to provide better control on the properties, location, growth direction and quality of nanotubes.

4. Measurements of CNT resistances

Experimental measurements of individual nanotube resistances can be performed using 2- or 4-points methods. To deposit nanotubes over pre-fabricated metal electrodes, an AC dielectrophoresis (DEP) method (Krupke et al, 2007; Vaz et al, 2008) can be applied, see Fig. 1a. As contact resistances obtained after DEP are frequently very high, further improvement of CNT-electrode contacts using metal (e.g., Ni or Pd) deposition by electroless methods, usually followed by annealing, is required (Vaz et al, 2008; Liebau et al, 2003). For 4-points measurements, additional electrodes can be made using platinum (Pt) deposition induced by focused ion or electron beams, see example in Fig. 1b.

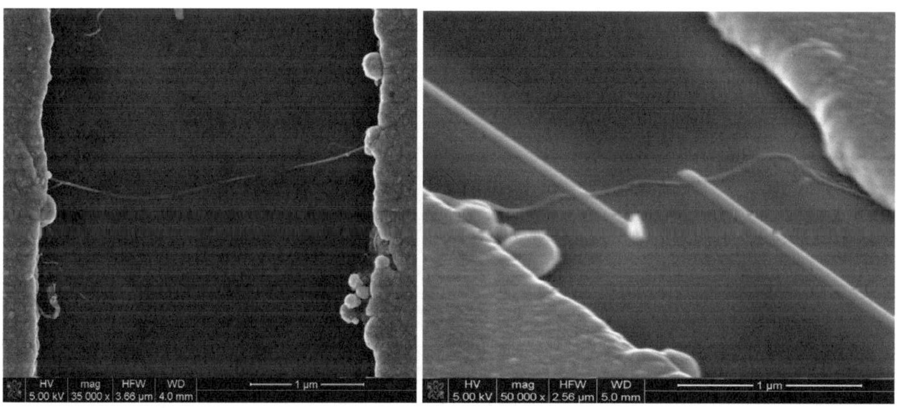

Fig. 1. (a) Individual MWCNT deposited by AC dielectrophoresis over Pd electrodes and then cover by Ni electroless process. (b) For 4-points measurements, 2 intermediate Pt electrodes are fabricated by electron beam induced deposition.

For MWCNTs grown by CVD (15 nm diameter), typical resistances of ~40 kΩ/µm were measured, while resistances of electroless deposited Ni contacts were estimated to be ~10 kΩ per contact (Liebau et al, 2003). In another work, for PE-CVD grown MWCNTs (25 nm diameter, 5µm long) considerably lower resistances (< 10 kΩ/µm) were measured using a 2-point method and Nb electrodes deposited by evaporation (contact resistances were not estimated) (Jun et al, 2007). In our studies (Moshkalev et al, 2008), similar values were obtained for low-bias contact resistances using CVD grown MWCNTs (30 nm mean diameter): ~20 kΩ per Pd or Ni electroless contacts. It should be noted that distinctly different values of MWCNT resistances were obtained for relatively short MWCNTs (nanotube lengths < 1 μm) using 2 and 4 points methods: ~30 kΩ/μm and 100 kΩ/μm, respectively. This was attributed to different contact geometries: in the former, all-around contacts were formed during electroless metal deposition over nanotubes, while in the latter, nanotubes were only side-contacted thus the contribution of internal shells to the measured conductance was considerably smaller. This finding emphasizes the importance of careful evaluation of the measurement conditions, particularly in terms of nanotube/metal contacting.

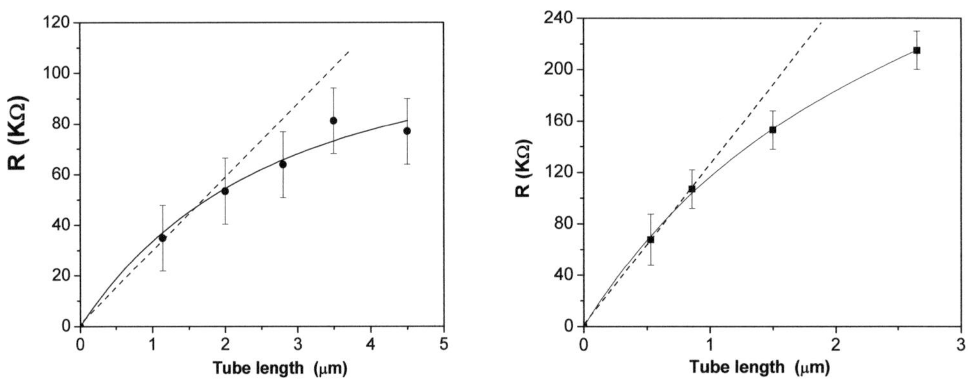

Fig. 2. Left: Two-points resistance vs. nanotube length (the contact resistance subtracted), solid line - fitting to the model, dashed line - linear approximation. Right: Four-points resistance vs. nanotube length, solid line - fitting to the model, dashed line - linear approximation.

More detailed studies of the MWCNT resistance as function of nanotube length (Moshkalev et al, 2008) have shown a non-linear behavior for tubes longer than 1-2 μm, in both 2 and 4 points measurements (Fig. 2). This is likely due to increasing conduction to internal walls as tube length grows. The data can be interpreted using the model of a nanotube as a resistive transmission line consisting of two parallel linear conductors (Bourlon et al, 2004). From the model, one can evaluate the resistance of an external shell ρ_1, of internal shell ρ_2 (only two outermost shells are considered in the model) and the intershell conductance g. For MWCNTs produced by arc discharge method, characteristic values ρ_1 ~ 10 kΩ/μm, ρ_2 ~ 0.1 x ρ_1 and g = (10 kΩ)$^{-1}$/μm were obtained by fitting using the model (Bourlon et al, 2004).
In our study for CVD grown MWCNTs, the following data were obtained, using 2 and 4 points configurations:

i. 2-points, Pd all-around contacted nanotubes (Fig. 2, left): ρ_1 ~ 37 kΩ/μm, ρ_2 ~ 4 kΩ/μm, g ~ (100 kΩ)$^{-1}$/μm;
ii. 4-points, side-contacted nanotubes (Fig. 2, right): ρ_1 ~ 100 kΩ/μm, ρ_2 ~ 22 kΩ/μm, g ~ (100 kΩ)$^{-1}$/μm.

As discussed above, the difference in measured resistances can be explaned by different contact geometries.
The data presented show that resistances of MWCNTs produced by different methods are still far from theoretical limits and thus are not yet suitable for interconnect apllications in

microelectronics. Better quality (lower resistance) is characteristic of nanotubes produced by high-temperature (arc, laser) methods compared with a conventional thermal CVD. Further optimization of growth and contacting tecnologies aiming to obtain lower nanotube resistances and better contacts (in particular, direct contact to internal walls) is strongly required.

5. Contacts of nanotubes with metals: theoretical and experimental approaches

At the most fundamental level, the resistance of a metal contact to a nanotube requires a calculation of the quantum mechanical transmission between the two objects (Lan et al, 2008). Such theories usually assume an ideal interface between a nanotube and the metal contact, which in practice is frequently contaminated with different impurities.
For calculations, usually the interface between graphene (flat graphitic monolayer) and metal is considered. Graphene, the building block for other graphitic materials such as the 3D graphite (stacked graphene planes), 1D carbon nanotubes (rolled graphene sheets), and 0D carbon buckyballs (wrapped graphene specks), consists of a flat monolayer of carbon atoms tightly packed in a two-dimensional honeycomb lattice. It is important to note that despite long known in the literature, with the electronic structure of graphite well established since the 40's, it was the groundbreaking article of Novoselov et al., 2004, which brought up interest in graphene as a potential material for a number of applications. One of important applications is in microelectronics, where graphenes (in a form of nanotubes and, more recently, of few-layer graphites or FLG) with their highly unusual properties deriving from its two-dimensional geometry open new opportunities. Since then graphene has been intensively investigated both theoretically and experimentally as reviewed by Geim & Novoselov, 2007. Several experimental groups have focused on new field effect transistors where silicon is replaced by graphene as the channel material (Novoselov et al, 2004; Wang et al, 2008). These devices take advantage of graphene's high and nearly temperature independent mobility of carriers leading to ballistic transport in the submicrometer scale, its linear I x V characteristics, and its unusually large sustainable currents (> $10^8 A/cm^2$). To create a graphene-based transistor, graphene is typically deposited over some substrate, usually SiO_2 (Ishigami et al, 2007), or grown on top of some carbon-based substrate such as SiC (Berger et al, 2004). Because the substrate may alter graphenes electronic properties, these groups have investigated if and how this interaction happens, and the impact it causes (Akcoltekin et al, 2009).
Theoretical studies of single- and multi-layer graphene have employed the tight binding model which describes their band structures through the Dirac formalism (Castro Neto et al, 2009). In the presence of other chemical species such as dopants or adsorbates, or for graphene on substrates, under gate dielectrics, or on/under metal contacts, graphene may be structurally and/or electronically affected depending on the nature of the species involved. In this context, many-body effects such as electronic exchange and correlations may play an important role in describing correctly the band structure, requiring ab initio techniques for the simulation of such systems.
Density functional theory (DFT) is particularly suitable to simulate large systems, which is typical of graphene/substrate interface models (Zhou et al, 2007). DFT has been employed to investigate the interaction of graphene with metal contacts (Chan et al, 2008; Giovannetti et al, 2008; Ran et al, 2009) and other substrates. In the specific case of metal-graphene

contacts it is important to understand how the two materials interact at the interface, since this information will help to optimize device operation. For example, a large electrical contact resistance degrades device performance. Because good contacts are usually formed under chemical interaction, knowing the bond strength at these interfaces is crucial to the comprehension of the device transport characteristics. Here we mention the results of Chan et al., 2008, obtained with first principles DFT within the generalized gradient approximation (GGA), which show that ionic bonds are formed between graphene and metals from groups I-III, while covalent bonds are formed between graphene and transitional, noble, and group IV metals. Another study by Giovanetti et al, 2008, found that metal/graphene contacts can be divided in two groups (p and n) by observing the Fermi level change with respect to the Dirac point in the band structure. In a more recent study, Ran et al., 2009, claim, also based on first principles DFT calculations, that there exist two groups of metal/graphene contacts depending on the strength of the interaction between d-orbitals in metals and p_z orbitals in graphene.

In the first group (typical example: Ti) strong chemically bonded contacts are formed through the attractive interactions between the 3d electrons of the metal and the p_z states in graphene, while the second group (example: Au) comprises of weak physically bonded contacts. Both situations are essentially determined by the electronic configurations of the metals. The authors also perform transport simulations and their results suggest that metals which form chemical contact with graphene might be best as electrode materials in graphene-based electronics.

Resuming, theoretical studies have indicated that Ti contacts have lower resistances followed by Pd, Pt, Cu and Au (Matsuda et al, 2007), basically confirming the experimentally observed trends. However, it should be noted that the high reactivity of Ti may lead to its oxidation and distortion of a nanotube structure in the contact region.

Finally, theoretical simulations involving single- and multi-layer graphene in contact with different materials are an important tool to investigate these systems, helping to pave the way for the next generation of electronic devices.

In practice, determining the contact resistance is usually a very difficult task, and requires a great number of experiments to give statistically averaged results. An interesting approach to measure the contact properties between an individual multi-wall nanotube and thin metal layer has been recently developed by Lan et al, 2008. For this, sequential cuts by a focused ion beam (beam diameter of ~10 nm) in the area of contact (reducing the contact length) were utilized. Then, from the measured dependence of 2-terminal resistance on the contact length, both specific nanotube resistance and contact resistance can be evaluated. For PE-CVD grown MWCNTs with diameters in the range of 50-60 nm and thin Ag metal film deposited by evaporation, following parameters were obtained: 1) nanotube resistances ~ 4.5 kΩ/μm, 2) specific contact resistances r_c were shown to depend strongly on the thickness of the Ag film, being of 38 kΩ μm and 1.6 kΩ μm (i.e., 6.4 kΩ for 2 contacts of 0.5 μm length each) for Ag layers of 23 and 63 nm, respectively. From the relation $r_c=\rho_c/(\pi d/2)$, where ρ_c is the specific contact resistivity for a nanotube of diameter d, ρ_c values were determined: 35 and 1.3 μΩ cm^2, for 23 and 63 nm thick layers of Ag, respectively. In this case, the contribution of contact resistances (inversely proportional to its length) to a total 2 terminal resistance becomes insignificant for contact lengths exceeding 1 μm. Much higher values for thinner metal films are due to non-complete coverage of the nanotubes.

Note that in the measurements using the transfer length method (TLM) by Jackson & Graham, 2009, the specific contact resistance between a thin film single wall carbon nanotube electrode and a deposited silver contact were found to be considerably higher: 20 mΩ cm^2. The same method was used by Liu et al, 2008, but the test structures for TLM were produced using densified carbon nanotube strips formed from vertically-aligned CNT forests and various metal films. Contact resistances of Ti/CNT, Pd/CNT, Ta/CNT, and W/CNT contacts with the same nominal contact area were extracted to be 40, 49, 108, and 160 Ω, respectively. This corresponds to even higher specific contact resistivity values for the nominal contact area ~0.144 mm^2. The high resistivity is explained by the geometry of the experiments, where intertube tunneling is the main mechanism of lateral conduction. It is also argued that actual metal/CNT contact area can be much smaller than the nominal contcat area, so that ρ_c values cannot be accurately calculated. These results show that much care should be taken while comparing data obtained using different methods and specific experimental conditions. Speaking more generally, considerable contributions still should be developed in the area of metrology of measurements involving nanostructured materials, in particular nanotubes.

6. MWCNTs for sensing applications

Another interesting appilcation of carbon nanotubes is for gas sensing (Star et al, 2006; Zhao, et al, 2007). However, bare nanotubes do not show appreciable sensitivity to some gases, and recently demonstrated decoration of CNTs by nanoparticles (NPs) (Kong et al, 2001) sensitive to the gases of interest (electron-donating or electron-withdrawing) opened the way to CNT/NP based gas sensors with improved performance and wider area of applications. CNT/NP hybrid nanostructures can be selectively sensitive towards various species in a gas or vapor. Nanoparticles of metals like Pd, Al, Pt, Sn, Pd and Rh have been used to decorate CNTs, allowing selective detection of gases like H_2, NH_3, NO_2 (Kim et al, 2006), CH_4 (Lu et al, 2004), H_2S and CO (Star et al, 2006). CNT/NP based gas sensors in different configurations (e.g., CNT-FET, chemical resistors) can have extremely high selectivilty due to high aspect ratio, fast time response and extremely low power consumption (μW range). Currently, considerable research efforts are concentrated on development of technologies (among them: electroless, sputtering, reflux, hydrolysis, super-critical CO_2 and others) capable to decorate both SWCNTs and MWCNTs with different metals and their oxides, selectively sensitive to different gases.

Other nanostructured materials, metal oxide nanowires (NWs) have been recently implemented as gas sensing elements with high surface-to-volume ratios that allow for considerable improvement of sensitivity and reduction of response/recovery times and power consumption (to ~10^{-5} W, at typical bias of 5-10 V) (Kolmakov and Moskovits, 2004). Furthemore, in experiments with SnO_2 NWs, self-heating by Joule effect has been shown to provide local NW temperatures high enough (~200 300 °C, Prades et al, 2008) to avoid the use of external heating in gas sensing experiments. Note that an external heating (and high power consumption) is usually required for conventional sensors based on metal oxide thin films. However, two functions, sensing and self heating, are coupled in the same element: the NW resistance can be changed significantly under exposure to the gas, in turn this will change the power dissipated on a NW and thus its temperature. This may result in non-linearities in the sensor response and requires appropriate calibration procedures.

CNT/NP hybrid structures represent other alternative of nano-scaled gas sensors that can operate at low voltage and power consumption (Gelamo et al, 2009). Depending on the type of nanotube, basically two different sensor configurations are currently under intensive studies: field effect transistors (FETs) and chemiresistors (CRs). FETs using semiconducting single-wall carbon nanotubes, have shown to be very sensitive to various gases however fabrication of these devices is technologically more challenging than those based on CRs. Thin films of mixed metallic and semiconducting SWCNTs deposited between arrays of interdigitated electrodes in a CR configuration, were shown to be very sensitive to gases like NO_2 and CH_4 (Lu et al, 2006). Room temperature methane detection was demonstrated for SWCNTs decorated with Pd clusters even at room temperarure and a few mW power consumption (Lu et al, 2004).

For multi-wall carbon nanotubes, a CR configuration has been studied (Meyyappan, 2005). In principle, the MWCNT based sensors must be less sensitive than those based on SWCNTs, as the measured current (and the associated noise) is supposed to pass through the whole volume of a MWCNT including all internal walls, whereas the reaction with gases should affect mainly the current fraction through the outermost wall. However, as discussed above, for distances shorter than ~1 μm, current redistribution between graphitic shells is small (Moshkalev et al, 2008), i.e., for side contacted MWCNTs and short gaps between electrodes the major fraction of current passes through the outermost wall. In terms of sensing configuration, this effectively transforms a short side-contacted MWCNT in a big-diameter "single-wall" metallic CNT, providing higher signal-to-noise ratio in gas sensing. Self-heating by Joule effect has been observed in nanotubes also, and so can be successfully employed in the case of CNTs based sensors, increasing sensitivity of hybrid CNT/NP systems to gases under interest, see below. Figure 3 shows some examples of MWCNTs decorated with SnO_2 nanoparticles for sensing applications, and Fig. 4 presents an individual MWCNT and MWCNT film deposited over metal electrodes by DEP and decorated by Ni (electroless) and SnO_2 (hydrolysis) nanoparticles, respectively.

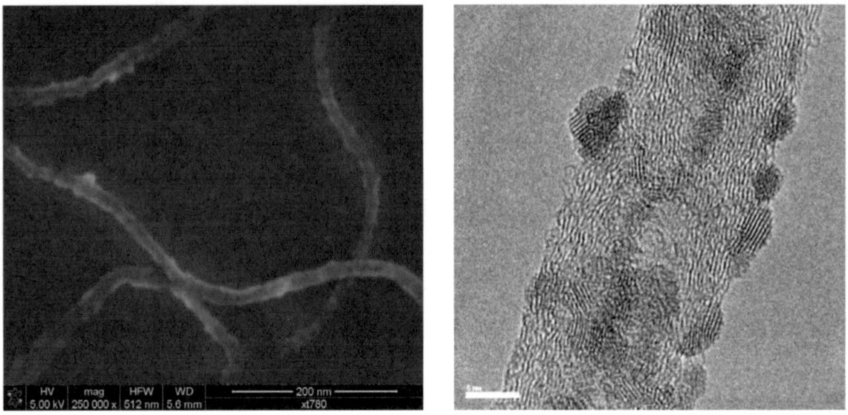

Fig. 3. SEM (left) and TEM (right) images of MWCNTs decorated by SnO_2 nanoparticles.

Multi-wall carbon nanotubes decorated by Ti nanoparticles were used for gas (N_2, Ar, O_2) and pressure sensing at low temperatures (Gelamo et al, 2009). Chemiresistor sensor configurations with supported and suspended nanotubes were tested. For the latter, cuts between electrodes were produced by a focused ion beam before deposition of nanotubes by dielectrophoreris.

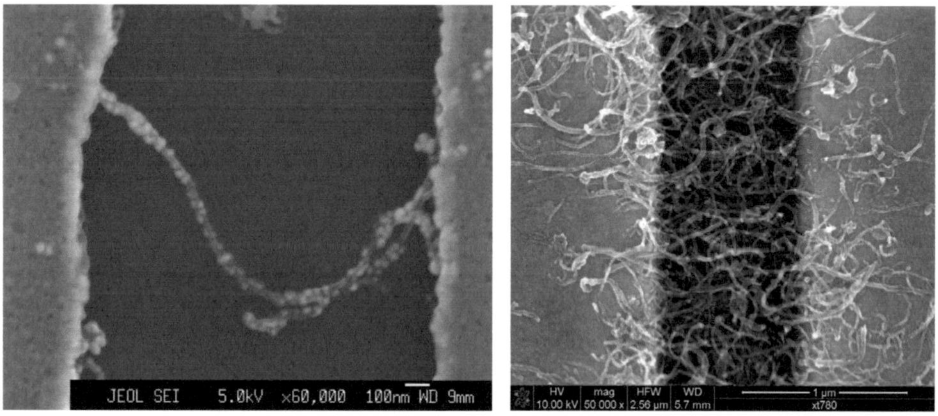

Fig. 4. Individual MWCNT (left) and MWCNT film (right) deposited over metal electrodes and decorated by Ni and SnO_2 nanoparticles, respectively.

As can be seen in Fig. 5, two gas sensing mechanisms (chemical, for O_2, and electrothermal, for chemically inert Ar and N_2) were demonstrated. For the former, current decreases, and for the latter, increases during pulsed gas injection. The contributions of these mechanisms were shown to depend strongly on the CNT heat balance. The electrothermal mechanism is due to changes of the CNT electrical resistance (Kuo et al, 2007). Metallic MWCNTs can be self-heated considerably by current (in the way similar to NWs), and this leads to a rise of resistivity, with the temperature coefficient of resistivity (TCR) ~ 0.1% $^oC^{-1}$ (Kawano et al, 2007). Further, when a gas is injected in the vacuum chamber, fast CNTs cooling by the gas may result in a measurable current increase. This effect was first observed by Kawano et al, 2007. For suspended nanotubes (and attached nanoparticles), heating by Joule effect is much stronger, resulting in strong enhancement of chemical sensitivity to gas (oxygen).

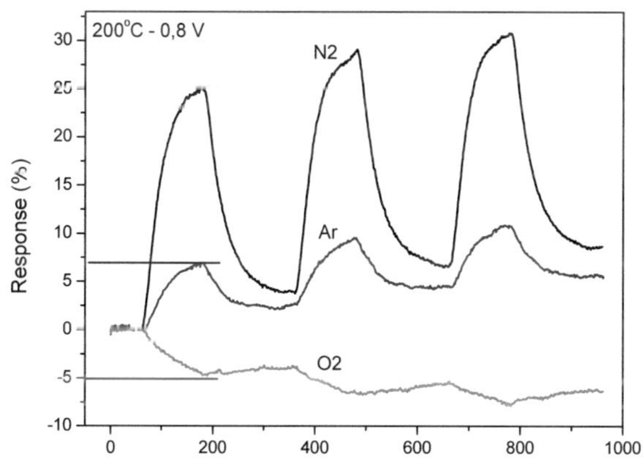

Fig. 5. Sensor response to pulses of gases N_2, Ar and O_2, peak pressures of 150, 30 and 4 mTorr, respectively (Gelamo et al, 2009).

Finally, a CNT/NP hybrid material has been successfully applied for low-pressure gas sensing applications in chemical resistor configuration. In this configuration, multi-wall carbon nanotubes serve as a conductive channel (for electrical signal acquisition), a heating element (for local heating of attached nanoparticles), and a substrate for NPs deposition (for selective gas sensitivity), whereas nanoparticles are employed to provide selective sensitivity to specific gases.

7. Conclusion

Many potential applications of carbon nanotubes in microelectronics are now being investigated extensively in many laboratories. Just a few specific applications are considered in the present work in more detail, showing some current problems and achievements. Some earlier expectations have failed, but many new opportunities arise constantly and, in many cases, unexpectedly. Recent introduction of new related nanocarbon material, graphene, is just one such example. For successful large-scale integration in new microdevices, development of reliable and compatible technologies that provide well controlled synthesis, positioning, characterization, manipulation and modification of nanotubes properties is still a great challenge.

8. Acknowledgements

The authors greatly acknowledge the contributions from Drs. J. Leon, A. R. Vaz, F. P. Rouxinol, A. Abbaspourad, and Mr. A. Flacker (CCS-UNICAMP), and financial support from CNPq, INCT-NAMITEC and FAPESP.

9. References

Akcoltekin, S.; Kharrazi, M. El ; Khler, B.; Lorke A. & Schleberger, M. (2009). Graphene on insulating crystalline substrates. *Nanotechnol.*, 20, (2009), 155601-6, ISSN 1361-6528

Berger, C., Song, Z., Li, T., Li, X., Ogbazghi, A.Y., Feng, R., Dai, Z., Alexei, N., Conrad, M.E.H., First, P.N., De Heer, W.A (2004). Ultrathin epitaxial graphite: 2D electron gas properties and a route toward graphene-based nanoelectronics, J. Phys. Chem B 108, 19912-19916 (2004). ISSN: 1520-6106

Bourlon, B.; Miko, C.; Forró, L.; Glatti, D.C. & Bachtold, A. (2004). Determination of the intershell conductance in multiwalled carbon nanotubes. *Phys. Rev. Lett.*, 93, (2004) 176806-4, ISSN 0031-9007

Castro Neto, A. H. ; Guinea, F.; Peres, N. M. R.; Novoselov, K. S. & Geim, A. K. (2009). The electronic properties of graphene. *Rev. Mod. Phys.*, 81, (2009), 109-162, ISSN 0034-6861, and references therein.

Chan, K. T.; Neaton, J. B. & Cohen, M. L. (2008). First-principles study of metal adatom adsorption on graphene, *Phys. Rev. B* 77, (2008), 235430-12p., ISSN 0163-1829

Geim A. K. & Novoselov, K. S. (2007). The rise of graphene. *Nature Materials*, 6, (2007), 183-191, ISSN 1476-4660

Gelamo, R. V.; Rouxinol, F. P.; Veríssimo, C.; Vaz, A.R.; Bica de Moraes, M. A.; & Moshkalev, S.A. (2009). Low-temperature gas and pressure sensor based on multi-wall carbon nanotubes decorated with Ti nanoparticles. *Chem. Phys. Lett.*, 482, 302–306, ISSN 0009-2614

Giovannetti, G.; Khomyakov, P. A.; Brocks, G.; Karplan, V. M. ; Brink, J. van den & Kelly, P. J. (2008). Doping Graphene with Metal Contacts. *Phys. Rev. Lett.*, 101, (2008), 26803-4, ISSN 0031-9007

Graham, A.P.; Duesberg, G.S.; Seidel, R.; Liebau, M.; Unger, E.; Kreupl, F. & Honlein, W. (2004). Towards the integration of carbon nanotubes in microelectronics, *Diam. Relat. Mater.*, 13, (2004) 1296-1300, ISSN 0925-9635

Ishigami, M.; Chen, J. H.; Cullen, W. G.; Fuhrer, M. S. & Williams, E. D. (2007). Atomic structure of graphene on SiO_2. *Nano Lett.*, 7, (2007) 1643-1648, ISSN 1530-6992

Jackson, R. & Graham, S. (2009). Specific contact resistance at metal/carbon nanotube interfaces, *Appl. Phys. Lett.*, 94, (2009) 012109-3, ISSN 0003-6951

Jun, S.C., Choi, J.H., Cha, S.N., Baik, C.W., Lee, S., Kim, H.J., Hone, J. & Kim, J.M. (2007). Radio-frequency transmission characteristics of a multi-walled carbon nanotube, *Nanotechnol.*, 18, (2007) 255701-5, ISSN 1361-6528

Kawano, T.; Chiamori, H. C.; Suter, M.; Zhou, Q.; Sosnowchik, B. D. & Lin, L. (2007). An electrothermal carbon nanotube gas sensor, Nano Lett., 7, (2007) 3686-3690, ISSN 1530-6992

Kim, B.-K. ; Park, N.; Na, P. S.; So, H.-M.; Kim, J.-J.; Kim, H.; Kong, K.-J.; Chang, H.; Ryu, B.-H. ; Choi, Y. & Lee, J.-O. (2006). The effect of metal cluster coatings on carbon nanotubes. *Nanotechnol.*, 17, (2006) 496-500, ISSN 1361-6528

Kolmakov A. & Moskovits, M. (2004). Chemical sensing and catalysis by one-dimensional metal-oxide nanostructures, *Annu. Rev. Mater. Res.*, 34, (2004) 151-180, ISSN 0084-6600

Kong, J.; Chapline, M. G. & Dai, H. (2001). Capillary force lithography. *Adv. Mater.*, 13, (2001) 1386-1389, ISSN 0935-9648

Krupke, R. ; Hennrich, F.; Weber, H.B.; Beckmann, D.; Hampe, O.; Malik, S.; Kappes, M.M. & Lohneysen, H.V. (2003). Contacting single bundles of carbon nanotubes with alternating electric fields, *Appl. Phys. A*, 76 (2003), 397-400, ISSN 1432-0630

Kuo, C. Y.; C. L. Chan, Gau, C.; Liu, C. W.; Shiau, S. H.; & Ting, J. H. (2007). Nano temperature sensor using selective lateral growth of carbon nanotube between electrodes, *IEEE Trans. Nanotech.*, 6, 1, (2007) 63-69, ISSN 1536-125X

Lan, C. ; Zakharov, D. N.; & Reifenberger, R. G. (2008). Determining the optimal contact length for a metal/multiwalled carbon nanotube interconnect, *Appl. Phys. Lett.*, 92, (2008), 213112-3p., ISSN 0003-6951

Liebau, M., Unger, E., Duesberg, G.S., Graham, A.P., Seidel, R., Kreupl, F. & Hoenlein, W. (2003). Contact improvement of carbon nanotubes via electroless nickel deposition. *Appl. Phys. A*, 77, (2003) 731-734, ISSN 1432-0630

Liu, Z., Ci, L. ; Bajwa, N. ; Ajayan, P. M. ; & Lu J.-Q. (2008). Benchmarking of Metal-to-Carbon Nanotube Side Contact Resistance, *Proc. of International Interconnect Technology Conference*, IITC 2008, pp. 144-146, ISBN 978-1-4244-1911-1, June 2008, Burlingame, CA, USA, IEEE,

Lu, Y.; J. Li, Han, J.; Ng, H.-T.; Binder, C.; Partridge, C. & Meyyappan, M. (2004). Room temperature methane detection using palladium loaded single-walled carbon nanotube sensors. *Chem. Phys. Lett.*, 391, (2004) 344-348, ISSN 0009-2614

Lu, Y.; Partridge, C.; Meyyappan, M. & Li, J. (2006). A carbon nanotube sensor array for sensitive gas discrimination using principal component analysis. *J. Electroanalyt. Chem.*, 593, (2006) 105-110, ISSN 0018-8646

Matsuda, Y.; Deng, W.-Q. & Goddard III, W.A. (2007). Contact resistance properties between nanotubes and various metals from quantum mechanics. *J. Phys. Chem C*, 111, (2007) 11113-11116, ISSN 1932-7447

Meyyappan, M. (2005). Carbon nanotubes: science and applications. (CRC Press LLC, Florida, 2005). ISBN 978-0849321115

Moshkalev S. A. & Veríssimo, C. (2007). Nucleation and growth of carbon nanotubes in catalytic chemical vapor deposition. *J. Appl. Phys.*, 102, (2007) 044303-5.

Moshkalev, S.A. ; Leon, J. ; Verissimo, C. ; Vaz, A.R. ; Flacker, A. ; Moraes, M.B. de & Swart, J. W. (2008). Controlled Deposition and Electrical Characterization of Multi-Wall Carbon Nanotubes, *J. Nano Res.*, 3, (2008) 25-32, ISSN 1662-5260

Naeemi, A. & Meindl, J.D. (2007). Carbon Nanotube Interconnects. Proc. of the 2007 Internat. Symp. Physical Design, ISPD'07., pp. 77-84, March 2007, Austin, Texas, USA. ISBN:978-1-59593-613-4

Novoselov, K. S. ; Geim, A. K. ; Morosov, S. V. ; Jiang, D. ; Zhang, Y. ; Dubonos, S. V. ; Grigorieva, I. V. & Firsov, A. A. (2004). Electric field in atomically thin carbon films. *Science*, 306, (2004) 666-669, ISSN 1095-9203, and supporting on-line material.

Prades, J. D. ; Jimenez-Diaz, R. ; Hernandez-Ramirez, F. ; Barth, S. ; Cirera, A. ; Romano-Rodriguez, A. ; Mathur, S. & Morante, J. R. (2008). Ultralow power consumption gas sensors based on self-heated individual nanowires. *Appl. Phys. Lett.*, 93, (2008) 123110-3, ISSN 0003-6951

Ran, Q. ; Gao, M. ; Guan, X. ; Wang, Y. & Yu, Z. (2009). First-principles investigation on bonding formation and electronic structure of metal-graphene contacts. *Appl. Phys. Lett.*, 94, (2009), 103511-3, ISSN 0003-6951

Sharma, P. & Ahuja, P. (2008). Recent advances in carbon nanotube-based electronics. *Mater. Res. Bull.*, 43, (2008) 2517–2526, ISSN 0025-5408

Star, A. ; Joshi, V. ; Skarupo, S. ; Thomas, D. & Gabriel, J.C. (2006). Gas sensor array based on metal-decorated carbon nanotubes, *J. Phys. Chem. B*, 110, (2006) 21014-9, ISSN 1520-6106

Tan, C.W. & J. Miao (2007). Transmission Line Characteristics of a CNT-based Vertical Interconnect Scheme. *Proc. IEEE 57th Electronic Components & Technology Conference*, 1936-1941. ISBN: 9781424409846, , Sparks, Nevada, May 2007, IEEE

Vaz, A.R. ; Macchi, M. ; Leon, J. ; Moshkalev, S.A. ; & Swart, J.W. (2008). Platinum thin films deposited on silicon oxide by focused ion beam: characterization and application, *J. Mater. Sci.*, 43, (2008) 3429-3434, ISSN 1573-4803

Wang, X. ; Ouyang, Y. ; Li, X. ; Wang, H. ; Guo, J. & Dai, H. (2008). Room-temperature all-semiconducting sub-10-nm graphene nanoribbon field-effect transistors. *Phys. Rev. Lett.* 100, (2008) 206803-4, ISSN 0031-9007

Zhao, L. ; Choi, M. ; Kim, H.-S. & Hong, S.-H. (2007). The effect of multiwalled carbon nanotube doping on the CO gas sensitivity of SnO_2-based nanomaterials. *Nanotechnol.*, 18, (2007) 445501-5, ISSN 1361-6528

Zhou, S. Y. ; Gweon, G. -H. ; Fedorov, A. V. ; First, P. N. ; de Heer, W. A. ; Lee, D. -H. ; Guinea, F. ; Castro Neto, A. H. & Lanzara, A. (2007). Substrate-induced bandgap opening in epitaxial graphene. *Nature Materials*, 6, (2007), 770-775, ISSN 1476-4660

12

Carbon Nanotube Interconnect Technologies for Future LSIs

Mizuhisa Nihei, Akio Kawabata, Motonobu Sato, Tatsuhiro Nozue, Takashi Hyakushima, Daiyu Kondo, Mari Ohfuti, Shintaro Sato and Yuji Awano
MIRAI-Selete
Japan

1. Introduction

Carbon nanotubes (CNTs) are attractive as nanosize structural elements from which devices can be constructed by bottom-up fabrication. A CNT is a macromolecule of carbon and is made by rolling a sheet of graphite into a cylindrical shape. CNTs exhibit excellent electrical properties that include current densities exceeding 10^9 A/cm^2 and ballistic transport along the tube. Because of these factors, with their large electro-migration tolerance and low electrical resistance, CNTs can be used as nano-size wiring materials, and are thus becoming potential candidates for future LSI interconnects. Much effort has been made to produce CNT vias, which use bundles of MWNTs (multi-walled carbon nanotubes), as vertical wiring materials as shown in Figure 1. Sato et al. demonstrated low-resistance CNT vias employing a novel metallization technology, which used preformed catalyst metal particles, to grow dense MWNT-bundles by thermal chemical vapor deposition (CVD).

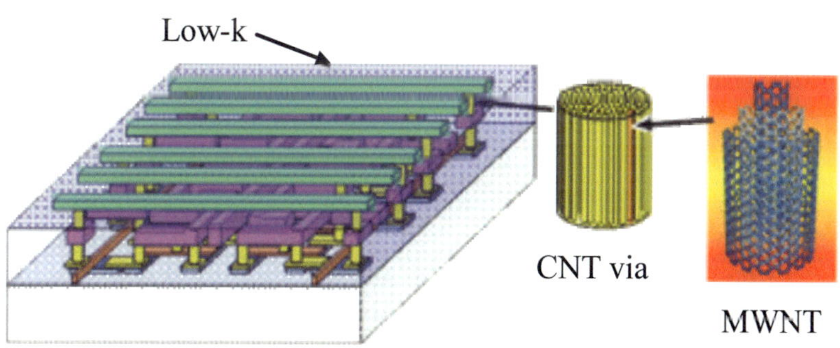

Fig. 1. Schematic of future LSI interconnects consisting of CNT vias and low-k materials.

The advantage of CNT-bundles is their low resistance, which may be the solution to the problem of high resistance in scaled-down vias. As shown in Fig. 2, we estimated the resistance of a 50-nm-diameter via depending on the filling rate of CNTs in the via area. In this estimation we assumed that CNTs have the quantum resistance $R_Q = h/4e^2 = 6.45$ kΩ (conductance $G_Q = 2G_{Q0} = 4e^2/h$, which reaches the maximum conductance limit for ballistic transport in two channels of a CNT), that current flows through each shell of MWNTs, and that there is no dependence of ballistic transport on CNT length. In order to lower the resistance of CNT vias, it is necessary to increase the nanotube's density, by decreasing its diameter. Regarding the electrical properties, CNTs consist of semiconductive CNTs as well as metallic types. Since the energy gap of a semiconductive CNT is inversely proportional to its diameter, smaller-diameter SWNTs may adversely influence the current conduction property. On the other hand, larger-diameter MWNTs seem to have a vanishing energy gap at room temperature. So, we are aiming at using metallic MWNTs with their ballistic transport properties as vias.

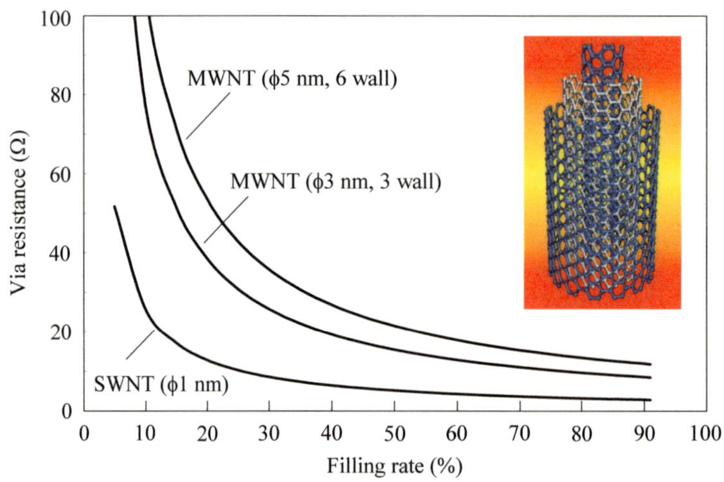

Fig. 2. Estimated resistance of 50-nm-diameter vias dependent on the filling rate of CNTs in a via hole for 1-nm-diameter SWNT, 3-nm-diameter 3-walled MWNT, and 5-nm-diameter 6-walled MWNT.

In this study, we demonstrated vertically scaled-down CNT via interconnects to clarify the current conduction properties of MWNT-bundles grown using thermal CVD. Based on our investigation, the carrier transport is expected to be ballistic for scaled-down vias. The excellent tolerance of CNT vias to a high current density was also demonstrated.

2. Experimental

As schematically shown in Fig. 3, we proposed CNT damascene processes to integrate scaled-down CNT vias with Cu interconnects. The processes were mostly compatible with conventional Cu interconnects. Briefly, a substrate with a Cu interconnect covered by a dielectric layer was first prepared. The dielectric layer was SiOC with k = 3.0 or k = 2.6. Via holes with a diameter of 160 nm were made using conventional photolithography followed by dry etching. A TaN/Ta barrier layer and a TiN contact layer were deposited by physical vapor deposition (PVD). Because CNTs do not need barrier layers, it is favourable to deposit these metals except the sidewall of the via hole. Size-controlled Co particles with an average diameter of about 4 nm were then deposited using a catalyst nano-particles deposition system. Previously we grew CNTs selectively in via holes, but all over the substrate in our new damascene process. For MWNT growth using the thermal CVD system, a mixture of C_2H_2 and Ar at 1 kPa was used as the source gas. The substrate temperature ranged from 400 °C to 450 °C. The chemical mechanical polishing (CMP) process we used is as follows:

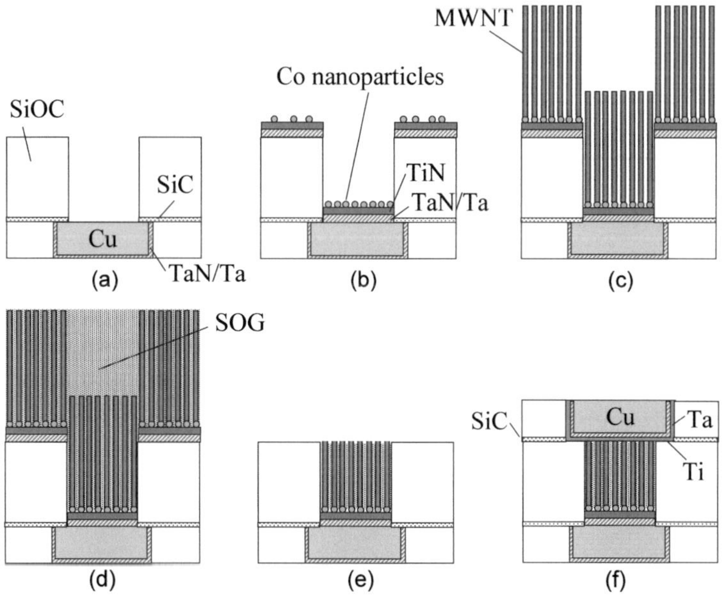

Fig. 3. CNT damascene via process: (a) Via hole formation on bottom Cu interconnect, (b) TaN/Ta barrier layer, TiN contact layer and Co catalyst nanoparticle formation, (c) MWNT growth, (d) SOG coating, (e) CMP Planarization, and (f) Top Cu interconnect formation.

the samples were coated with the spin-on glass (SOG) in order to hold the CNTs during the CMP process. CNTs were polished on the TiN layer on a SiOC layer with a conventional IC1000 pad and silica slurry under pressures of 2 psi (13.8 kPa) for 240 sec. Then, the TiN and TaN/Ta layers were polished with conventional barrier-metal CMP slurry. After polishing, the substrate was slightly wet-etched using buffered HF solution. Finally, the Ti top contact layer, Ta barrier layer and Cu wire were connected to CNT vias by PVD without subsequent annealing.

3. Results and discussion

Figures 4(a) and (b) are the cross-sectional scanning electron microscopy (SEM) images of CNT vias fabricated with growth temperatures of 450 °C and 400 °C. We can see in the images that CNTs grown at 400 °C are a little less straight than those at 450 °C, suggesting CNTs at 400 °C are a little more defective.

To further investigate the quality of CNTs, we performed transmission electron microscopy (TEM) analyses, whose results are shown in Fig. 5. The TEM images indicate that CNTs grown at either temperature are of high quality. However, CNTs at 400 °C appear to be a little more defective.

Figure 6(a) shows a cross-sectional SEM image of CNTs formed all over the substrate, having 160-nm diameter via holes, at the growth temperature of 450 °C. We succeeded in growing vertically-aligned MWNTs with a diameter of 10 nm, a shell number of 7 and a density of 3×10^{11} cm^{-2}. Figure 6(b) shows a cross-sectional SEM image of CNT vias after CMP planarization. MWNT-bundles were successfully polished under pressures as low as those in the conventional Cu/low-k CMP process. Although SOG is filled well with MWNTs inside the 160-nm-diameter via hole, the filling factor of CNT in via is still low in this study.

Fig. 4. Cross-sectional SEM image of the 160-nm-diameter CNT growth temperature (a) 450 °C and (b) 400 °C.

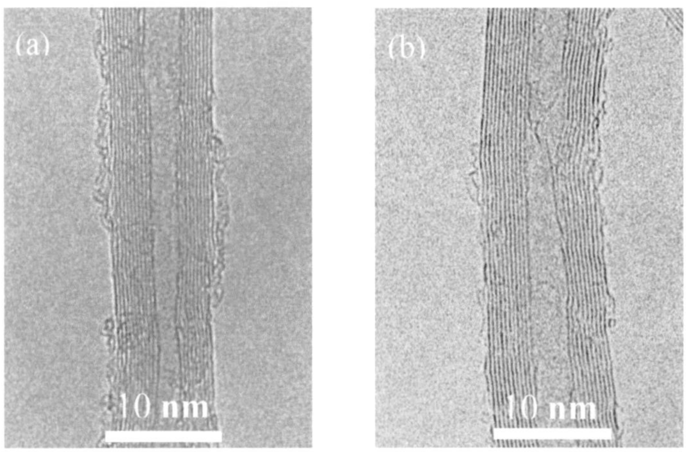

Fig. 5. TEM image of the CNT growth temperature (a) 450 °C and (b) 400 °C.

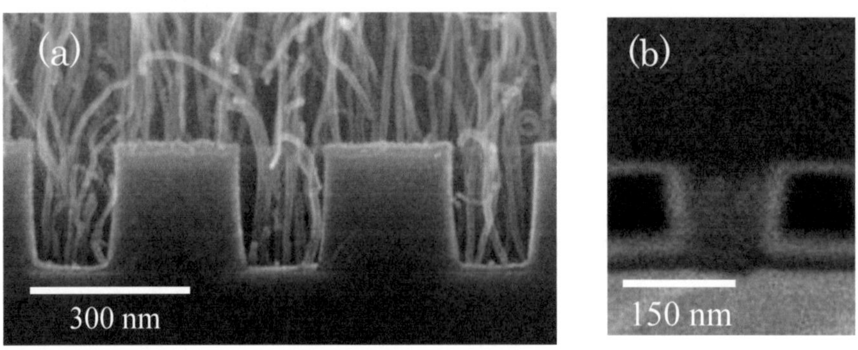

Fig. 6. Cross-sectional SEM image of (a) vertically aligned MWNTs formed all over the substrate, having 160-nm-diameter via holes, and (b) 160-nm-diameter CNT vias after CMP planarization.

We measured the via resistance of 2800-nm-diameter CNT vias with a four-point probe using Kelvin patterns. Figures 7(a) and (b) show the current-voltage characteristic on the low-bias region for the via height of 60 nm and 520 nm, respectively. For both cases, the current increased linearly depending on the voltage, and good ohmic contacts were achieved between the MWNT-bundle and the TiN contact layer. We summarized the electrical properties of 2800-nm-diameter CNT vias for a via height of 60 nm and 520 nm in Table I. The obtained resistance of 0.05 Ω for 60-nm-height 2800-nm-diameter vias is the lowest value ever reported. The most important point of the result is that the via resistance decreased by about 84% as the via height decreased by about 89%.

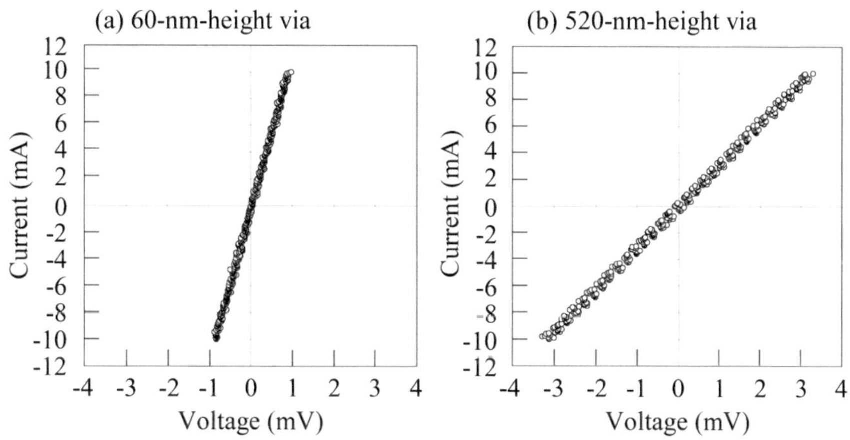

Fig. 7. Current-voltage characteristic of the 2800-nm-diameter CNT vias with a via height of (a) 60 nm and (b) 520 nm.

Sample	Diameter (nm)	Height (nm)	Resistance (Ω)	Resistivity (μΩcm)	Transport property
#1	2800	60	0.05	-	Ballistic
#2	2800	520	0.32	379	Ohmic

Table 1. Summary of electrical properties for CNT vias. The CNT density of $3x10^{11}$ cm^{-2} corresponds to the filling rate of 24%. The diameter and the shell number are 10 nm and 7, respectively. The shell number which contributes to the current conduction was estimated from the assumption of the quantum resistance.

Figure 8 shows via resistance distributions of the 2000-nm-diameter CNT vias with and without CMP planarization. The average via resistance of the sample with CMP decreased by about 25% compared with that without CMP. The scattering for the distribution of the sample with CMP is also smaller than that without CMP. We speculated that cutting the CNT bundles short by CMP could increase the number of electrical contacts between MWNT tips and the top metal electrode, because as-grown CNT bundles have an unfavorable worse uniformity in length.

We also measured the resistance of 160-nm-diameter CNT vias with a four-point probe using Kelvin patterns. Figure 9 shows the current-voltage characteristics on the low-bias region. It was found that the resistance depended on the growth temperature. The via

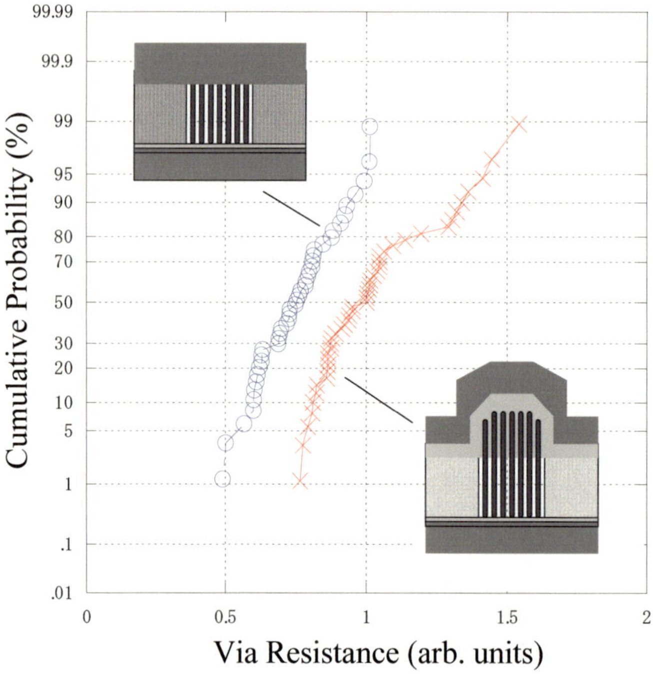

Fig. 8. Via resistance depending on the top metal contacts with and without CMP planarization.

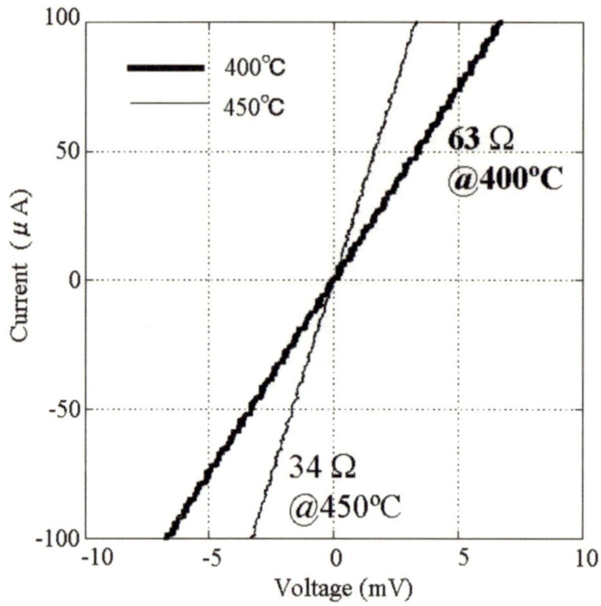

Fig. 9. Current-Voltage characteristics of the 160-nm-diameter CNT via grown at (a) 450 °C and (b) 400 °C.

resistance was 34 Ω for a growth temperature of 450 °C, and 64 Ω for 400 °C. Since the site density of the CNTs was similar for both temperatures, we speculate that the difference in resistance may have been caused by the difference in the CNT quality.

To investigate the transport mechanism, we measured the temperature dependence of the via resistance as shown in Fig. 10. The 520-nm-height vias shows the linear decrease of the resistance by decreasing the temperature. This characteristic is ohmic, which has been attributed to electron-phonon scattering. The corresponding resistivity of 379 μΩcm was obtained for 520-nm-height CNT vias, which are of the same order of magnitude as the value of CVD-tungsten (W) plugs (100-210 μΩcm). On the other hand, the resistance of 60-nm-height vias was independent of temperatures as high as 423 K, which suggests that the carrier transport is ballistic.

In order to estimate the electron mean free path λ_{CNT} of ballistic transport, we assumed the quantum resistance R_Q. The CNT via resistance R_{Via} is given by (1), where R_C is the imperfect metal-CNT contact resistance, n_{CNT} is the number of shells which contributed to the current conduction and H is the via height

$$R_{Via} = \frac{R_C + R_{CNT}}{n_{CNT}} \tag{1}$$

where

$$R_{CNT} = R_Q = \frac{h}{4e^2} \quad \ldots \text{ if } H \ll \lambda_{CNT}$$

$$= H \cdot \frac{R_Q}{\lambda_{CNT}} = H \cdot \left(\frac{h}{4e^2}\right) \cdot \frac{1}{\lambda_{CNT}} \quad \ldots \text{ if } H > \lambda_{CNT}$$

Assuming the imperfect contact resistance R_C is as low as 0.5 kΩ, we estimated that the shell number of 7 contributed as a current conduction channel.

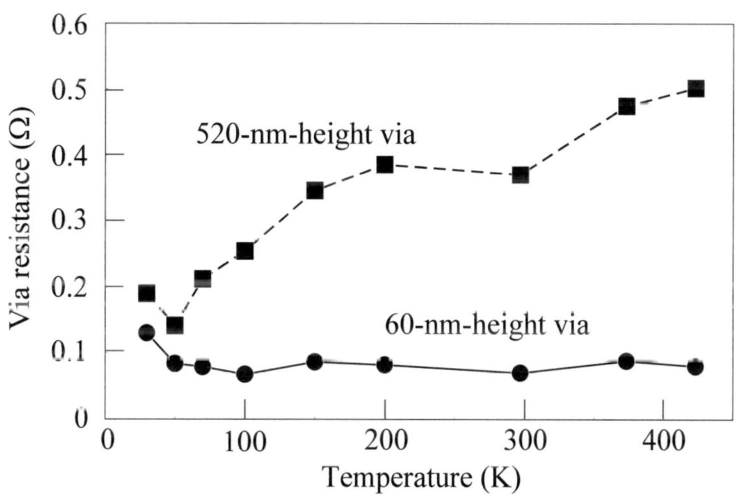

Fig. 10. Temperature dependence of the via resistance for the 60-nm and 520-nm-height CNT via.

Figure 11 shows the via resistance as a function of the via height. The filled circles show the previous results for 2800-nm-diameter vias with a growth temperature of 450 °C. The solid lines indicate the via resistance calculated assuming various electron mean free paths. An solid rectangle or triangle indicates the current result normalized to a diameter of 2800 nm. As can be seen in the figure, the current result for 450 °C falls on the line for an electron mean free path of 80 nm, the same as the previous data. This seems reasonable considering the growth temperature for the previous data was also 450 °C. On the other hand, the resistance for 400 °C falls on the line for an electron mean free path of 40 nm, which suggests the quality of CNTs grown at 400 °C is not as high as that at 450 °C, as also speculated from the SEM and TEM results. We therefore currently work on synthesizing higher-quality CNTs at 400 °C or lower.

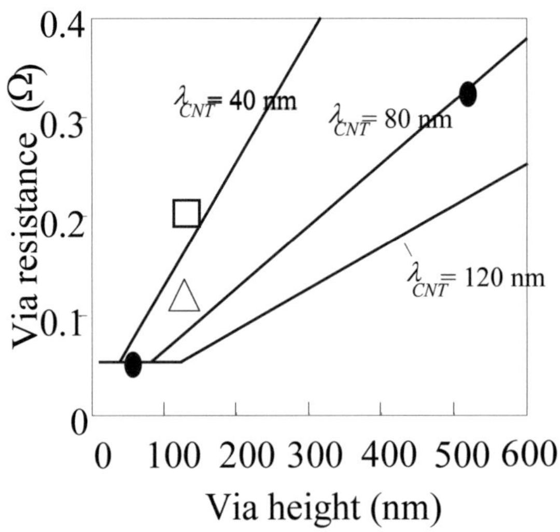

Fig. 11. Via resistance dependence as a function of the via height.
Solid line: the via resistance calculated assuming various electron mean free paths.
•: 2800-nm-diameter via 450 °C growth, △: 160-nm-diameter via 450 °C growth, □: 160-nm-diameter via 400 °C growth.

The stability of the via resistance under an electric current with a density of 5.0×10^6 A/cm^2 is shown in Fig. 12(a). The via diameter and growth temperature were 160 nm and 400 °C, respectively. The dielectric layer was made of SiOC with k = 2.6. The measurement was performed at 105 °C in a vacuum. The resistance remained stable even after running the electric current for 100 hrs. This indicates that the CNT via is robust over a high-density current as we expect. The cross-sectional TEM image of the via is shown in Fig. 12(b). The via shape looks deformed, but this was caused by high-energy electrons during the TEM observation.

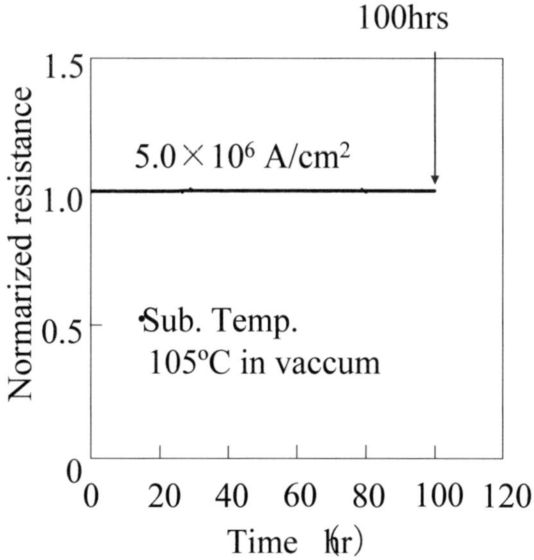

(a)

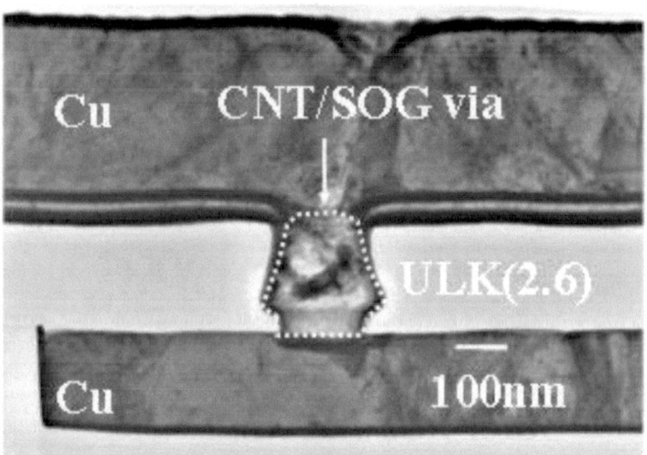

(b)

Fig. 12. (a) EM characteristics at 105 °C in a vacuum and (b) cross-sectional TEM image of the CNT via.

4. Conclusion

In this chapter, we report our trials of using bundles of CNTs with their ballistic transport properties as via interconnects of LSIs. We proposed CNT damascene processes to integrate scaled-down CNT vias with Cu interconnects. Moreover, we demonstrated vertically scaled-down MWNTs via interconnects to clarify the current conduction properties of MWNTs-bundles.

We fabricated a CNT via interconnect and evaluated its electrical properties and robustness over a high-density current. We found that the CNT via resistance was independent of temperatures, which suggests that the carrier transport is ballistic. From the via height dependence of the resistance, the electron mean free path was estimated to be about 80 nm, which is similar to the via height predicted for hp32-nm technology node. This indicates that it will be possible to realize CNT vias with ballistic conduction for hp32-nm technology node and beyond. It was also found that a CNT via was able to sustain a current density as high as 5.0×10^6 A/cm^2 at 105 °C for 100 hours without any deterioration.

5. Acknowledgments

We would like to thank Prof. M. Hirose and Dr. H. Watanabe of MIRAI-Selete, and Dr. N. Yokoyama at Fujitsu Laboratories Ltd. for their support and useful suggestions. This work was completed as part of the MIRAI Project supported by NEDO.

6. References

Awano, Y.; Sato, S.; Kondo, D.; Ohfuti, M.; Kawabata, A.; Nihei, M.; Yokoyama, N. (2006) *phys. stat. sol.*, (a) 203, pp. 3611

Banerjee, K.; Im, S.; Srivastava, N. (2006) *Proceedings of 1st International Conference on Nano-Networks*

Coiffic, J. C. ; Fayolle, M.; Maitrejean, S. ; Foa Torres, L. E. F. ; and Le Poche, H. (2007) *Appl. Phys. Lett.*, vol. 91, pp. 252107

Coiffic, J. C.;, Fayolle, M.; Le Poche, H.; Maitrejean, S. ; Olivier, S. (2008) *Proceedings of IEEE International Interconnect Technology Conference*, pp. 153

Cho, H.; Koo, K. -H.; Kapur, P.; Saraswat, K. C. (2007) *Proceedings of IEEE International Interconnect Technology Conference*, pp. 135

Hoenlein, W. (2001) *Proceedings of International Microprocesses & Nanotechnology Conference*, p. 76

Horibe, M.; Nihei, M.; Kondo, D.; Kawabata, A.; Awano, Y. (2004) *Jpn. J. Appl. Phys.*, Vol. 43, pp. 6499

Horibe, M.; Nihei, M.; Kondo, D.; Kawabata, A.; Awano, Y. (2004) *Jpn. J. Appl. Phys.*, Vol. 43, pp. 7337

Horibe, M.; Nihei, M.; Kondo, D.; Kawabata, A.; Awano, Y. (2005) *Jpn. J. Appl. Phys.*, Vol. 44, pp. 5309

Iijima, S. (1991) *Nature*, Vol. 354, pp. 56

Katagiri, M.; Sakuma, N.; Suzuki, M.; Sakai, T.; Sato, S.; Hyakushima, T.; Nihei, M.; and Awano, Y. (2008) *Jpn. J. Appl. Phys.*, vol. 47, pp. 2024

Katagiri, M.; Yamazaki, Y.; Sakuma, N.; Suzuki, M.; Sakai, T.; Wada, M.; Nakamura, N.; Matsunaga, N.; Sato, S.; Nihei, M.; and Awano, Y. (2009) *Proceedings of IEEE International Interconnect Technology Conference*, pp. 44

Kawabata, A.; Sato, S.; Nozue, T.; Hyakushima, T.; Norimatsu, M.; Mishima, M.; Murakami, T.; Kondo, D.; Asano, K.; Ohfuti, M.; Kawarada, H.; Sakai, T.; Nihei, M.; Awano, Y. (2008) *Proceedings of IEEE International Interconnect Technology Conference*, pp. 237

Kitsuki, H.; Saito, T.; Yamada, T.; Fabris, D.; Jameson, J. R.; Wilhite, P.; Suzuki, M, Yang, C. Y. (2008) *Proceedings of IEEE International Interconnect Technology Conference*, pp. 43

Kong, J.; Yenilmez, E.; Tombler, T. W.; Kim, W.; Dai, H. (2001) *Phys. Rev. Lett.*, Vol. 87, pp. 106801

Kreupl, F.; Graham, A. P.; Liebau, M.; Duesberg, G. S.; Seidel, R.; Unger, E. (2004) *Proceedings of IEEE International Electron Device Meeting*, pp.683

Li, J.; Ye, Q.; Cassell, A.; Koehne, J.; Hg, H. T.; Han, J.; and Meyyappan, M. (2003) *Proceedings of IEEE International Interconnect Conference*, pp.271

Liu, K.; Avouris, Ph.; Martel, R.; Hsu, W. K. (2001) *Phys. Rev. B*, Vol. 63, pp. 161404

Milne, W. I.; Wang, X.; Zhang, Y.; Haque, S.; Kim, S. M.; Udrea, F.; Robertson, J.; Teo, K. B. K. (2008) *Proceedings of IEEE International Interconnect Conference*, pp. 105

Naeemi, A.; Sarvari, R.; and Meindl, J. D. (2004) *Proceedings of IEEE International Electron Devices Meeting*, pp. 699

Naeemi, A.; Meindl, J. D. (2008) *Proceedings of IEEE International Interconnect Conference*, pp. 183

Nihei, M.; Kawabata, A.; and Awano, Y. (2003) *Jpn. J. Appl. Phys.*, Vol. 42, pp. L721

Nihei, M.; Kawabata, A.; Awano, Y. (2004) *Jpn. J. Appl. Phys.*, Vol. 43, pp. 1856

Nihei, M.; Kondo, D.; Kawabata, A.; Sato, S.; Shioya, H.; Sakaue, M.; Iwai, T.; Ohfuti, M.; Awano, Y. (2005) *Proceedings of IEEE International Interconnect Technology Conference*, pp. 234

Nihei, M.; Kawabata, A.; Kondo, D.; Horibe, M.; Sato, S.; Awano, Y. (2005) *Jpn. J. Appl. Phys.*, Vol. 44, pp. 1626

Nihei, M.; Kawabata, A.; Horibe, M.; Kondo, D.; Sato, S.; Awano, Y. (2005) *Materials for Information Technology*, Springer publisher, 978-1-85233-941-8, Germany, pp. 315

Nihei, M.; Kawabata, A.; Hyakushima, T.; Sato, S.; Nozue, T.; Kondo, D.; Shioya, H.; Iwai, T.; Ohfuti, M.; Awano, Y. (2006) *Proceedings of International Conference on Solid State Devices and Materials*, pp. 140

Nihei, M.; Hyakushima, T.; Sato, S.; Nozue, T.; Norimatsu, M.; Mishima, M.; Murakami, T.; Kondo, D.; Kawabata, A.; Ohfuti, M.; Awano, Y. (2007) *Proceedings of IEEE International Interconnect Technology Conference*, pp. 204

Ngo, Q.; Cassell, A.M.; Austin, A.J.; Jun Li; Krishnan, S.; Meyyappan, M.; Yang, C.Y. (2006) *IEEE Electron Device Lett.*, Vol. 27, pp. 221

Sato, S.; Nihei, M.; Mimura, A.; Kawabata, A.; Kondo, D.; Shioya, H.; Iwai, T.; Mishima, M.; Ohfuti, M.; Awano, Y. (2006) *Proceedings of IEEE International Interconnect Technology Conference*, pp. 230

Sato, S.; Kawabata, A.; Kondo, D.; Nihei, M.; Awano, Y. (2005) *Chem. Phys. Lett.*, Vol. 402, pp. 149

Srivastava, N.; Joshi, R. V.; Banerjee, K. (2005) *Proceedings of IEEE International Electron Devices Meeting*, pp. 257

Yao, Z.; Kane, C. L.; Dekker, C. (2000) *Phys. Rev. Lett.*, Vol. 84, pp. 2941

Yamazaki, Y.; Sakuma, N.; Katagiri, M.; Suzuki, M.; Sakai, T.; Sato, S.; Nihei, M.; Awano, Y. (2008) *Appl. Phys. Express*, vol. 1, pp. 034004

Yokoyama, D.; Iwasaki, T.; Ishimaru, K.; Sato, S.; Hyakushima, T.; Nihei, M.; Awano, Y.; Kawarada, H. (2008) *Jpn. J. Appl. Phys.*, vol. 47, pp. 1985
Wei, B. Q.; Vajtai, R.; and P. M. Ajayan, P. M. (2001) *Appl. Phys. Lett.*, vol. 79, pp. 1172

13

On-Chip Interconnects of RFICs

Xiaomeng Shi and Kiat Seng Yeo
Nanyang Technological University
Singapore

1. Introduction

Boosted by the demands of the rapidly growing wireless communication market, there is an increasing interest in the development of the radio frequency integrated circuits (RFICs). As highlighted by the International Technology Roadmap for Semiconductors (ITRS) annually, interconnect has become one of the most critical factors affecting the performance of ICs (ITRS, 2008). Thereafter, incorporating interconnect effects into the RFIC design flow becomes increasingly essential.

Because of the mature technology, low fabrication cost and high packing density, CMOS technology is deemed as a strong contender compared with other available technologies (Shi et al., 2005). Therefore, this chapter will mainly focus on the analysis of interconnects using conventional CMOS technology. Nevertheless, the authors would also like to shed some lights on some emerging interconnect concepts and technologies in the last part of the chapter.

1.1 Physical background

When an electric field, *E*, is applied, free electrons of the conductor begin to accelerate in the opposite direction to the applied *E*. Thus the average electron movement is in one direction. The movement of the charges and the established electric and magnetic fields are the basis for information transfer in interconnects. In order to understand interconnect behaviours in the RF ranges, several physical phenomena must be taken into consideration.

1.1.1 Inductive effect

The movement of the charges results in a magnetic field and hence the storage of the magnetic energy. The ability of a conductor to store the magnetic energy is described by its inductance.

At low frequencies, the impact of the magnetic field is often neglected, and interconnects are usually characterized by the conventional RC model (Kleveland et al., 2002). However, when the frequency increases beyond multi-Gigahertz, the inductive reactance of the interconnects becomes comparable to or dominant over the resistance. Therefore, the inductance and the magnetic field must be considered (Gala et al., 2002) in the Gigahertz frequency range. Hence, it becomes a major concern of the current interconnect modelling.

1.1.2 Skin effect

At low frequencies, current flow is uniformly distributed over the cross section of the conductor. The resistance of an interconnect with length l (m), width W (m) and thickness t (m) is given by (Plett & Rogers, 2003):

$$R = \rho \frac{l}{tW} = R_s \frac{l}{W} \ (\Omega) \tag{1}$$

where ρ ($\Omega \cdot$m) is the resistivity of the interconnect material and R_s (Ω) is the sheet resistance based on DC measurements.

However, at high frequencies, say above 5 GHz, the EM fields attenuate substantially when they pass through the conductor. The current crowds to the surface of the conductor, as shown in Fig. 1. This is known as skin effect.

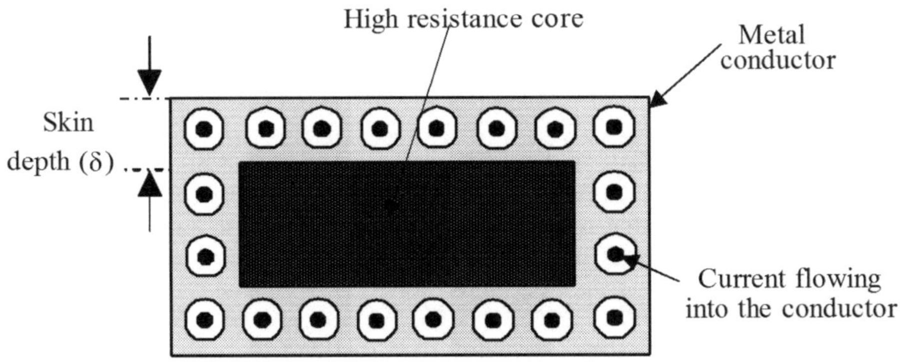

Fig. 1. Illustration of skin effect

The mechanism of skin effect can be explained either from an electrical circuit perspective or an electromagnetic perspective. From the circuit perspective, the currents in the conductor always flow in a way, which has the least impedance, i.e., $R+j\omega L$. For direct current, the imaginary part of the impedance is zero. The currents are distributed uniformly. This way of distribution has the least resistance or impedance. As the frequency increases, the imaginary part becomes more and more significant. While the current crowds to the surface of the conductor, the average distance between the currents is more than that of the currents which are distributed uniformly. Consequently, the magnetic coupling and the inductance are minimal, so is the impedance. From electromagnetic perspective, the electromagnetic waves are attenuated when they pass through the conductor. At a sufficient depth, all electric and magnetic fields are negligible and there is no current flow. The high-frequency voltage between the two terminals of the conductor creates a high-frequency electric field and a high-frequency current in the conductor and thus creates a magnetic field. This is equivalent to the situation where electromagnetic waves penetrate the conductor. Those fields are attenuated as they passing into the conductor. The currents inside the conductor weaken with the attenuation of the electric field.

At a sufficient depth, all the fields are negligible and there is no current. Hence, the effective cross section of the conductor shrinks with the increase of the frequency. Skin depth δ is defined in Eq. 2 in (Plett & Rogers, 2003). It refers to the depth from the surface of a conductor, where the currents are confined to flow.

$$\delta = \sqrt{\frac{2}{\omega \mu \sigma}} = \frac{1}{\sqrt{f \pi \mu \sigma}} \ (\text{m}) \tag{2}$$

where μ (H/m) and σ (S/m) are the permeability and the conductivity of the conductor respectively. ω (rad/s) represents the angular frequency, which is the product of 2π and the operating frequency f (Hz).

We now need to modify the conventional calculation of the resistance in Eq. 1 by replacing the geometrical cross-sectional area with the effective one. When $\delta << W, t$, the resistance formula could be approximated as Eq. 3 (Plett & Rogers, 2003):

$$R = \frac{\rho l}{Wt - (W - 2\delta)(t - 2\delta)} \ (\Omega) \tag{3}$$

As the skin depth decreases with the increasing frequency, the resistance of the conductor becomes frequency-variant. It increases along with the frequency. On the contrary, the inductance reduces. The reason is that at low frequencies, the magnetic energy is stored inside as well as outside the conductor. However, as frequency increases, the current flow is mostly concentrated near the surface of the conductor. Hence, the magnetic field becomes confined to the region outside the conductor.

1.1.3 Substrate effect

In current CMOS technologies, low-resistivity (1 to 20 Ω /cm (Marsh, 2006)) substrate is commonly used to improve yields and suppress the latchup. However, in RF ranges, the low-resistivity substrate causes significant high frequency losses. The silicon substrate therefore appears to be a major concern of the use of CMOS in multi-Gigahertz applications. Therein, its mechanism must be studied thoroughly and its effect must be considered.

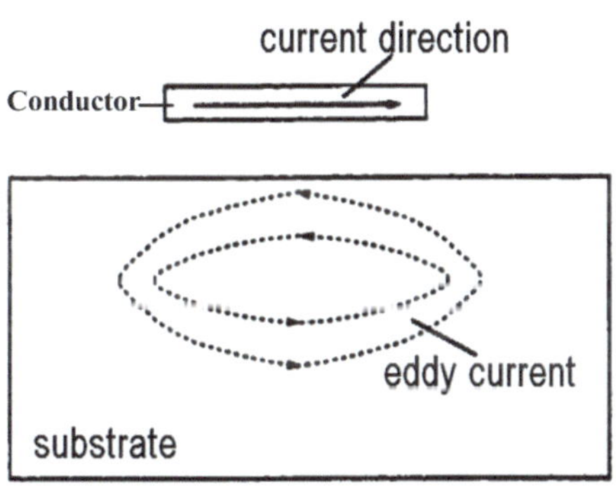

Fig. 2. Eddy currents in the substrate (Zheng, 2003)

The substrate affects interconnects in two ways: eddy current losses and substrate losses induced by the displacement currents injecting into the substrate (Chiprout, 1998). Fig. 2. illustrates the eddy currents in the substrate which are induced by the current flowing through the conductor. The eddy-current, in turn, will change the magnetic field and the inductance of the conductor. Particularly, if a high conductivity substrate is used at high frequencies, the eddy currents are strong and crowded near the surface of the substrate, the inductance is reduced and there are significant eddy current losses (Zheng, 2003). The impact of the eddy current is frequency dependent. For direct current, no eddy current is induced. The inductance is equivalent to that in the free space. As the frequency increases, the eddy current becomes stronger and more crowded to the surface.

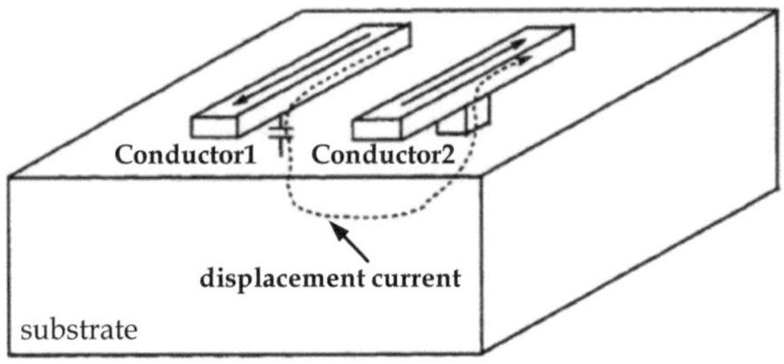

Fig. 3. Displacement current injected into the substrate (Zheng, 2003)

Fig. 3 illustrates the procedure of substrate losses derived from the injection of the displacement currents. The displacement currents flowing through the capacitance terminating on the substrate result in additional resistive losses. The capacitance to the substrate is also frequency-dependent. It is larger at higher frequencies because of skin effect of both the conductor and the substrate, as well as the frequency dependence of the effective permittivity (Zheng, 2003).

1.1.4 Corner effect

In most cases, straight-line interconnects are not adequate for on-chip interconnections. Interconnects with bends are often required. These bends are usually with angles of 90° or 45°. As mentioned in Section 1.1.2, the currents tend to flow in a path with the least impedance. Hence, in consequence of the appearance of the bends, the current distribution is different from that in straight-line interconnects. Fig. 4 illustrates the current distribution in the corners. This difference is known as the corner effect (Edwards & Steer, 2000).

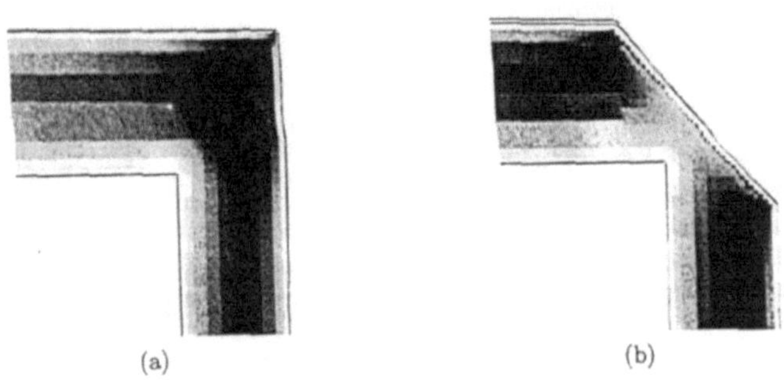

Fig. 4. Magnitude of the current densities at 10 GHz (a) right-angled bend; (b) an optimally mitred bend (Edwards & Steer, 2000).

1.1.5 Distributed effect

When the length of the interconnect is less than $\frac{1}{20}$ of the wavelength λ, the signal can be deemed as reasonably constant along the entire length of the interconnect. Hence it can be characterized with lumped components. However, when the length of the conductor is longer than $\frac{1}{10}$ of λ, the capacitance and inductance are distributed throughout the interconnect. They cannot be confined to a lumped element. This effect is called distributed effects (Edwards & Steer, 2000).

1.2 Model development

Due to increased circuit complexity and higher operating frequency, the circuit performance becomes more and more subjected to interconnect behaviors. Inappropriate decision of interconnects in the design stage may lead to either over-design or excessive design iterations after tapeout. Therefore, there is an increasing need of adequate electronic design automation (EDA) tools for interconnect models from the industry. SPICE (simulation program with integrated circuit emphasis), developed by the University of California, Berkeley has become the industry standard simulation tool. With accurate models and precise model parameters, useful simulation results can be achieved to aid the IC design and significantly shorten the product-to-market time.

Besides SPICE-like circuit simulators, there are also electromagnetic (EM) simulators based on numerical solutions of Maxwell's equations that describe the EM behaviors of physical structures. EM simulators are capable of precisely analyzing the high frequency effects of the devices. However, they take up extremely high computing power and are very time consuming. Moreover, in-depth EM knowledge is required for using those EM simulators (Azadpour & Kalkur, 2002). Therefore, SPICE-compatible circuit models represented in capacitance, resistance and inductance, for instance, which are much easier to handle, are preferred by circuit designers.

In order to develop a desired equivalent circuit model for on-chip interconnects, there are mainly three stages to follow, namely, model construction, parameter extraction and model verification (Shi et al., 2008).

In the first stage, the model structure is established. The constructed interconnect model should be capable of characterizing the high frequency effects as well as incorporable with conventional EDA tools. The main challenge in this stage is that the interconnect behavior becomes frequency-variant at high frequencies. Although behavioral models, which can characterize the frequency-dependent characteristics, can be used in SPICE-like simulators, it is much slower than those only involve frequency-independent components. Therefore, characterizing the frequency dependent characteristics with frequency independent components would be more desirable.

In the second stage, model parameters are extracted. Essentially, the problem in parameter extraction is a multi-parameter and multi-target optimization. The accuracy, convergency and efficiency of the extracted data strongly depend on the chosen algorithm. Therefore, the algorithm should be selected, developed and applied appropriately.

Finally, the proposed model is verified with on-wafer measurements to ensure its accuracy.

2. Interconnect models

2.1 RC model

In many EDA tools, the interconnects are modelled as resistance and capacitance (RC) components (Celik et al., 2002; Shin et al., 2004), as shown in Fig. 5.

The calculation of the resistance for this model is straightforward. For a uniform structure with a rectangle cross-section the resistance can be calculated by Eq. 3. For the nonuniform or nonrectagle structures, the resistance calculation is more difficult. One aproach is to split the conductor into simple regions so that Eq. 3 can be applied to each region. Another approach is to formulate and solve the problem in terms of Laplace equations (Celik et al., 2002).

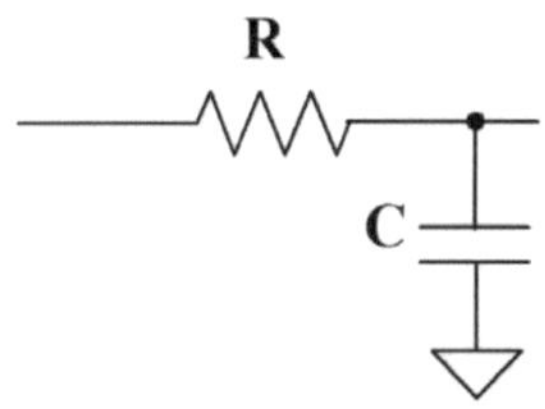

Fig. 5. RC model

For capacitance extraction, many techniques can be used, varying from simple 2-D analytical models to 3-D EM solvers (Celik et al., 2002).
This RC model is simple and straight forward. However, it becomes inadequate in the RF ranges.

2.2 Transmission line model

As stated in Section 1.1, when the operating frequency reaches multi-Gigahertz, inductive effect and distributed effect must be considered. Therefore, transmission line models are mostly studied and employed. The transmission line characteristics of an interconnect line can be mathematically formulated with the Telegrapher's equations (Pozar 1998) as listed below,

$$\begin{aligned} \frac{dV(x,t)}{dx} &= -(R + j\omega L)I(x,t) \\ \frac{dI(x,t)}{dx} &= -(G + j\omega C)V(x,t) \end{aligned} \tag{4}$$

where the voltage V and the current I along the line are both functions of position x and time t. R is per-unit-length (PUL) resistance, L is PUL inductance, G is PUL conductance and C is PUL capacitance. The RLGC model of the classical transmission line is shown in Fig. 6.

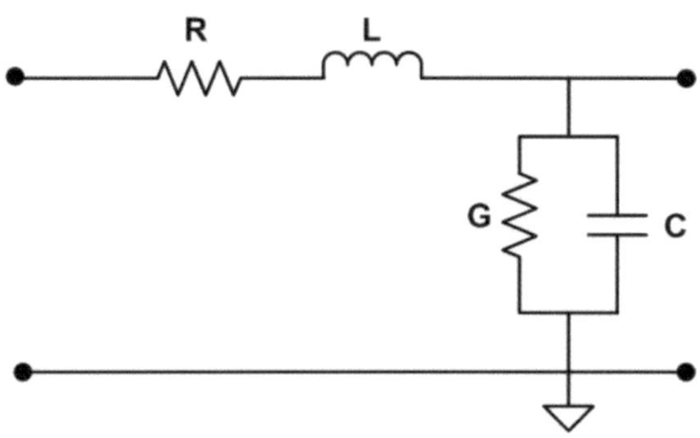

Fig. 6. Classical transmission line RLGC model

The standard solution to the Telegrapher's equations is

$$\begin{cases} V = V^+ e^{-\gamma x} + V^- e^{+\gamma x} \\ I = \dfrac{1}{Z}(V^+ e^{-\gamma x} + V^- e^{+\gamma x}) \end{cases} \tag{5}$$

where

$$\gamma = \sqrt{(R + j\omega L)(G + j\omega C)} \quad (6)$$

is the complex propagation constant and

$$Z = \sqrt{\frac{R + j\omega L}{G + j\omega C}} \ (\Omega) \quad (7)$$

is the characteristic impedance of the interconnect.
The line parameters (γ, Z, R, L, G and C) can be extracted from S-parameter measurements (Eisenstant & Eo 1992).

$$e^{-\gamma x} = \left\{ \frac{1 - S_{11}^2 + S_{21}^2}{2S_{21}} \pm K \right\}^{-1} \quad (8)$$

where

$$K = \sqrt{\frac{(1 + S_{11}^2 - S_{21}^2)^2 - (2S_{11})^2}{(2S_{21})^2}} \quad (9)$$

$$Z^2 = Z_o^{\ 2} \frac{(1 + S_{11})^2 - S_{21}^2}{(1 - S_{11})^2 - S_{21}^2} \quad (10)$$

where Z_o denotes the reference impedance of the S-parameter measurement system, which is usually 50 Ω. During the extraction of γ and Z from $e^{-\gamma x}$ and Z^2, extracted parameters with values that are not physically real, such as negative attenuation constants are ignored (Eisenstant & Eo 1992).
The line parameters R, L, G and C are extracted from S-parameter measurements as follows:

$$R = \mathrm{Re}\{\gamma Z\} \ (\Omega) \quad (11)$$

$$L = \frac{\mathrm{Im}\{\gamma Z\}}{\omega} \ (\mathrm{H}) \quad (12)$$

$$G = \mathrm{Re}\left\{\frac{\gamma}{Z}\right\} \ (\mathrm{S}) \quad (13)$$

$$C = \frac{\mathrm{Im}\left\{\frac{\gamma}{Z}\right\}}{\omega} \ (\mathrm{F}) \quad (14)$$

Since the characteristics of interconnects are frequency-variant, the extracted parameters are also frequency dependent. In order to fully describe the behaviour of high frequency interconnects with frequency independent components, the classical transmission line model is modified. Several model structures could be found in the literature, as shown in Fig. 7 to Fig. 10.

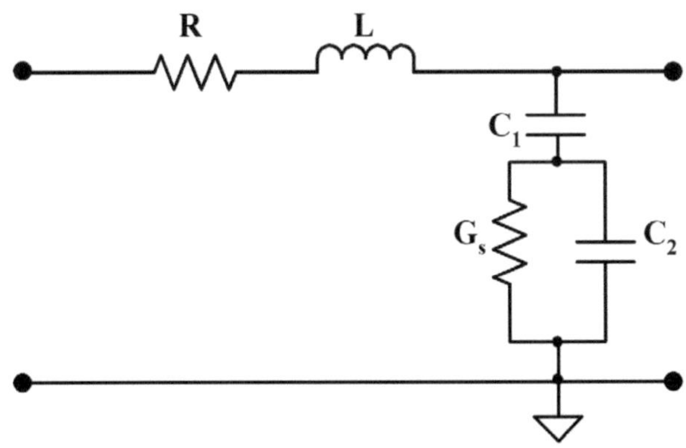

Fig. 7. Improved transmission line model 1 (Eo & Eisenstadt, 1993)

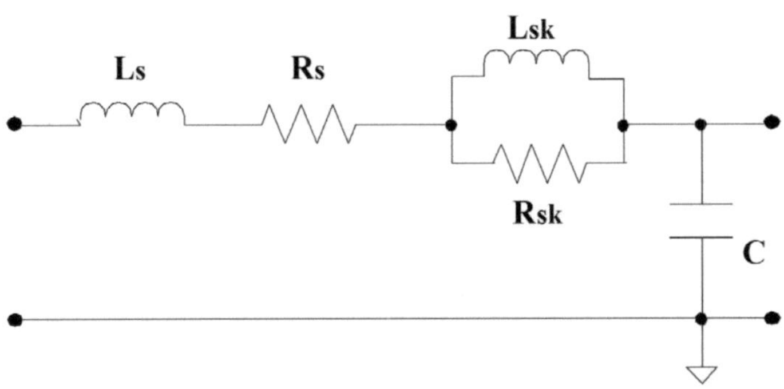

Fig. 8. Improved transmission line model 2 (Deutsch et al., 2001)

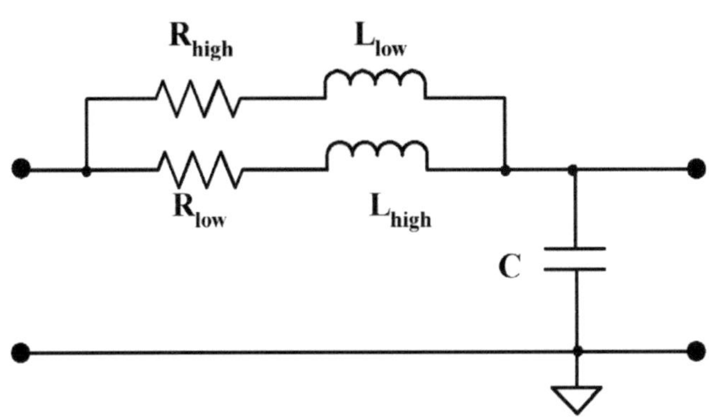

Fig. 9. Improved transmission line model 3 (Kleveland et al., 2002)

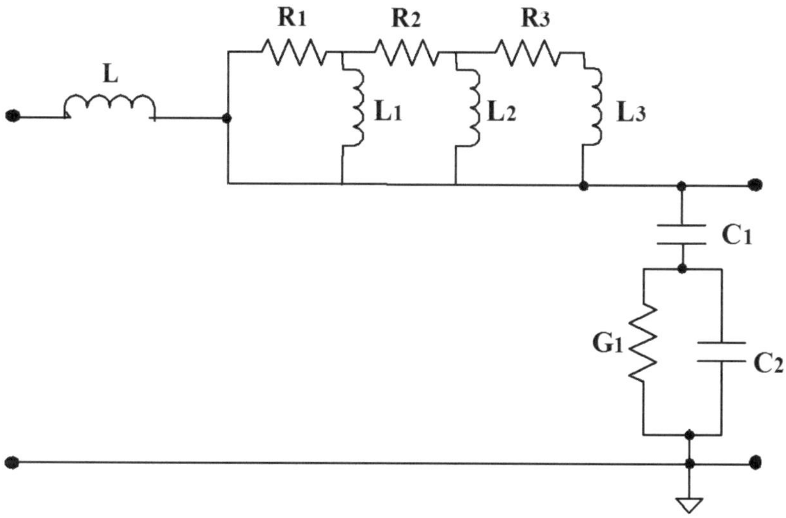

Fig. 10. Improved transmission line model 4 (Zheng et al., 2000)

2.3 Lumped element model

The RLGC parameters of the transmission line model characterize the PUL property. Therefore, the model complexity is proportional to the physical dimension of the interconnects. On the other hand, the on-chip RF interconnects can also be characterized by deliberately proposed lumped element models.

2.3.1 Straight-line interconnects

The function of interconnects is to connect different devices or blocks together. In the low frequency ranges, interconnects can be characterized by frequency-independent resistors (R) and capacitors (C). However, this RC model is not applicable at high frequencies. The reason is that as the frequency increases, the inductive effect, skin effect, substrate effect and distributed effect begin to have significant influences on the characteristics of the interconnects. All these effects are dependent on the frequency. In other words, the characteristics of RF interconnects are frequency-variant. Ideally, frequency-variant models should be used in the simulation. However, behavioural models which can characterize the frequency-dependent elements are much slower than models only involve frequency-independent components.

According to the notion described by Edwards and Steer in (Edwards & Steer, 2000), when the length of the interconnect is less than $\frac{1}{20}$ of the wavelength λ, the signal can be deemed to be reasonably constant along the entire length. Hence a lumped one-Π model shown in Fig. 11 is adequate. This one-Π model topology is widely used in the modelling of on-chip inductors.

With the increase in the length of the interconnect, the distributed effect begins to show its impact. When the length is longer than $\frac{1}{10}$ of λ, the transmission line model should be used (Edwards & Steer, 2000). λ can be calculated using Eq. 15.

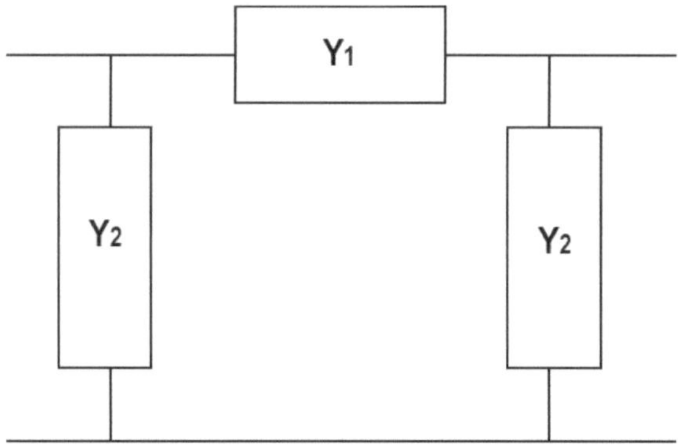

Fig. 11. Schematic block diagram of one-Π model

$$\lambda = \frac{c}{\sqrt{\mu_r \varepsilon_r} f} \quad \text{(m)} \tag{15}$$

where c is the speed of light in free space (3×10^8 m/s), f is the frequency under consideration, μ_r and ε_r are the relative permeability and permittivity of the material in which the signal propagates.

The transmission mode in the on-chip interconnect is not a pure transverse-electromagnetic (TEM) mode but a hybrid of transverse electric (TE) and transverse magnetic (TM) mode, known as a quasi-TEM mode (Marsh, 2006). Therefore, in order to apply Eq. 15, "effective" relative permittivity, which has a value between those of the substrate, the dielectric layer and the air, should be used. Here μ_r =1 and ε_r= (11.9+4.5+1)/3=5.8, where 11.9 is the relative permittivity of the silicon substrate, 4.5 is that of silicon dioxide and 1 is that of air, are used as a rough estimation of the CMOS process.

The criteria for choosing the model topology at various operating frequencies are summarized in Table 1.

Frequency (GHz)	0.3	5	15	30
lumped element model (μm)	20764.1	1245.9	415.3	207.6
Transmission line model (μm)	41528.2	2491.7	830.6	415.3

Table1 Critical Length of various frequencies

From Table 1, it reveals that for the intended frequency range, i.e., from 300 MHz to 30 GHz, the selection of the model topology is complicated. For example, at 30 GHz, the one-Π model is suitable only when the length of the on-chip interconnect is less than 207.6 μm, otherwise the validity of the model cannot be guaranteed. At 300 MHz, the transmission line model is appropriate only when the length is longer than 41528.2 μm; otherwise, it is not necessary to employ this topology. For typical RF circuit sub-blocks, such as low noise amplifier (LNA), voltage controlled oscillators (VCO) and mixer, the total die size is always smaller than 800 μm by 800 μm. Therefore, 800 μm is considered as the maximum length for on-chip interconnects of RFICs. Thus, both the lumped one-Π model and the transmission line model are not viable. The optimal model should be capable of characterizing high frequency behaviours of interconnects while keeping the model simple.

In order to maintain the simplicity, a two-Π model is developed based on the one-Π model. The problem of the one-Π model is that it cannot characterize the distributed effect which is significant at high frequencies. By strategically cascading two-Π lumped blocks together, as illustrated in Fig. 12, the distributed effects can be represented. In order to simplify the model construction and parameter extraction, the series blocks and shunt blocks are made to be identical of the two Πs. This optimization is physically acceptable due to the symmetrical structure of the straight-line interconnect.

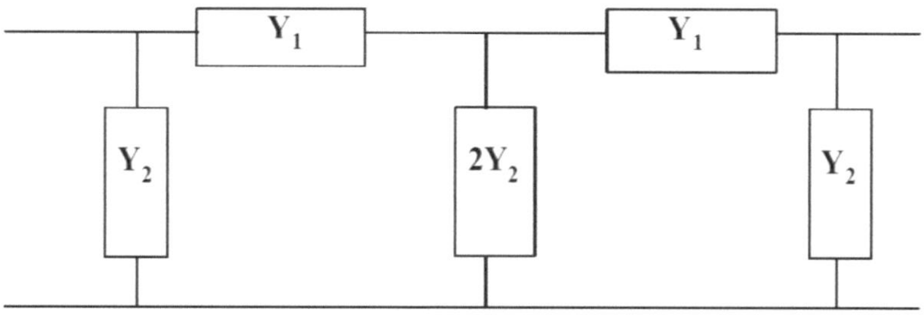

Fig. 12. Schematic block diagram of two-Π model

As shown in Fig. 13, based on the schematic block model, the two-Π equivalent circuit model is proposed from a physical point of view (Shi et al., 2005).

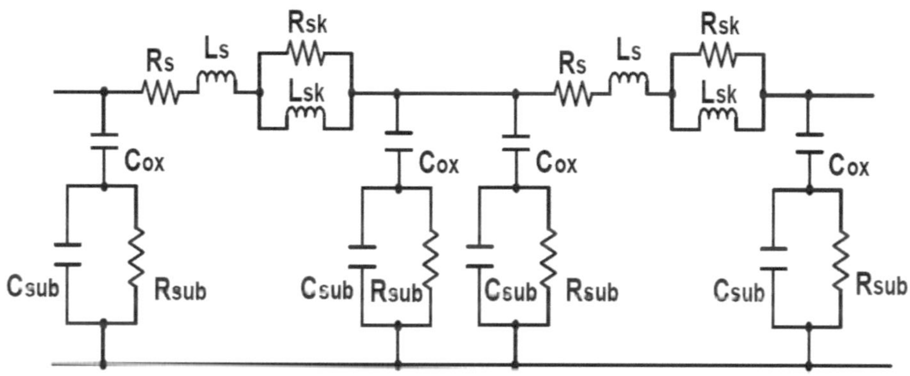

Fig. 13. Equivalent circuit model for straight-line interconnects (Shi et al., 2005)

With the significant increase of the operating frequency, the impact of the magnetic field and the magnetic coupling becomes one of the most emergent concerns of the RFIC design. In the two-Π model, the inductance is introduced by L_s, which represents the ideal series inductance. R_s represents the ideal series resistance. In RFICs, as the operating frequency approaches multi-Gigahertz, the skin effect becomes very significant. Although it must be included in the simulation, frequency-variant components are not supported by conventional circuit simulators. Hence, mimicking the frequency-variant skin effect with frequency-independent components becomes the straightforward solution.

In Fig. 13, the series components R_{sk} and L_{sk} connected in parallel are used to characterize the skin effect. Due to the skin effect, the behaviour of the interconnect becomes more resistive rather than inductive at high frequencies. In this parallel branch at low frequencies, most of the currents pass through L_{sk}. When the operating frequency rises, more currents

shift to the path of R_{sk}. With these two frequency-independent components, the frequency-variant skin effect characteristics are thus well captured.
Besides the skin effect, at Gigahertz frequencies the substrate losses are also substantial. In current CMOS RF technologies, high frequency losses are caused by the low-resistivity substrate (Chiprout, 1998; Zheng et al., 2000). As stated in 1.1.3 the substrate affects interconnects in two ways: eddy current losses and displacement current losses. The eddy currents in the substrate are induced by the current flowing through the conductor. The eddy currents, in turn, change the magnetic field and the inductance of the conductor. Particularly, if a high conductivity substrate is used at high frequencies, strong eddy currents will crowd near the surface of the substrate. As a result, the inductance is reduced and significant eddy current losses occur. This effect is characterized by L_{sk} and R_{sk} as well. As the frequency increases, the flow of the current shifts from L_{sk} to R_{sk}. Hence, the equivalent inductance reduces and the loss increases.
Another part of the substrate losses is derived from the substrate injection of the displacement currents. The displacement currents flow through the capacitance which terminates on the substrate. This results in additional resistive losses. The capacitance in the substrate is frequency-variant as well. It is larger at higher frequencies because of skin effect of both the conductor and the substrate, as well as the frequency dependence of the effective permittivity (Edwards & Steer, 2000). This effect is modelled by the resistor and capacitors in the shunt block. As shown in Fig. 13, C_{ox} represents the oxide layer capacitance, R_{sub} represents the substrate resistance and C_{sub} represents the capacitance of the substrate.
In the parameter extraction stage, an objective function is formulated to which an optimization algorithm is applied. Essentially, it is a multi-parameter and multi-target optimization. Optimizations can be made based on on-wafer measurements of the test structures to ensure the silicon verified accuracy.
At very high frequencies, measuring the voltages and currents is difficult in practice, since direct measurements usually involve the magnitude and phase of wave travelling in a given direction, or of a standing wave. Thus equivalent voltages, currents, related impedance and admittance matrices become somewhat of an abstraction (Pozar, 1998). Therefore, S-parameter is generally employed at radio frequencies.
The parameter extraction process is summarized as follows. Firstly, the admittance of each sub-block in Fig. 13 is derived as a function of the circuit components, as illustrated in Eq. 16 and Eq. 17.

$$Y_1 = \frac{1}{j\omega L_s + R_s + \dfrac{j\omega L_{sk} R_{sk}}{j\omega L_{sk} + R_{sk}}} \tag{16}$$

$$Y_2 = \frac{1}{\dfrac{1}{j\omega C_{ox}} + \dfrac{R_{sub}}{j\omega C_{sub} R_{sub} + 1}} \tag{17}$$

The Y-parameters are presented as functions of the admittance of each sub-block Y_1 and Y_2, as illustrated in Eq. 18 to Eq. 21:

$$Y_{11} = Y_2 + \frac{Y_1(Y_1 + 2Y_2)}{2Y_1 Y_2} \tag{18}$$

$$Y_{12} = -\frac{1}{\frac{Y_1 + 2Y_2}{Y_1^2} + \frac{1}{Y_1}} \tag{19}$$

$$Y_{21} = -\frac{1}{\frac{Y_1 + 2Y_2}{Y_1^2} + \frac{1}{Y_1}} \tag{20}$$

$$Y_{22} = Y_2 + \frac{Y_1(Y_1 + 2Y_2)}{2Y_1Y_2} \tag{21}$$

On the other hand, the measured S-parameters are converted into Y-parameters, based on the equations from Eq. 22 to Eq. 25 (Pozar, 1998) as follows:

$$Y_{11} = \frac{1}{Z_o} \times \frac{(1 - S_{11})(1 + S_{22}) + S_{12}S_{21}}{(1 + S_{11})(1 + S_{22}) - S_{12}S_{21}} \tag{22}$$

$$Y_{12} = \frac{1}{Z_o} \times \frac{-2S_{12}}{(1 + S_{11})(1 + S_{22}) - S_{12}S_{21}} \tag{23}$$

$$Y_{21} = \frac{1}{Z_o} \times \frac{-2S_{21}}{(1 + S_{11})(1 + S_{22}) - S_{12}S_{21}} \tag{24}$$

$$Y_{22} = \frac{1}{Z_o} \times \frac{(1 + S_{11})(1 - S_{22}) + S_{12}S_{21}}{(1 + S_{11})(1 + S_{22}) - S_{12}S_{21}} \tag{25}$$

where Z_0 is the reference impedance of the S-parameter measurement system, which is usually 50 Ω.
By combining Eq. 18 Eq. 21 with Eq. 22 - Eq. 25, Eq. 26 - Eq. 29 are obtained. By solving Eq. 26 - Eq. 29, the values of Y_1 and Y_2 can be obtained from the measurement results.

$$\frac{1}{Z_o} \times \frac{(1 - S_{11})(1 + S_{22}) + S_{12}S_{21}}{(1 + S_{11})(1 + S_{22}) - S_{12}S_{21}} = Y_2 + \frac{Y_1(Y_1 + 2Y_2)}{2Y_1Y_2} \tag{26}$$

$$\frac{1}{Z_o} \times \frac{-2S_{12}}{(1 + S_{11})(1 + S_{22}) - S_{12}S_{21}} = \frac{1}{\frac{Y_1 + 2Y_2}{Y_1^2} + \frac{1}{Y_1}} \tag{27}$$

$$\frac{1}{Z_o} \times \frac{-2S_{21}}{(1 + S_{11})(1 + S_{22}) - S_{12}S_{21}} = -\frac{1}{\frac{Y_1 + 2Y_2}{Y_1^2} + \frac{1}{Y_1}} \tag{28}$$

$$\frac{1}{Z_o} \times \frac{(1 + S_{11})(1 - S_{22}) + S_{12}S_{21}}{(1 + S_{11})(1 + S_{22}) - S_{12}S_{21}} = Y_2 + \frac{Y_1(Y_1 + 2Y_2)}{2Y_1Y_2} \tag{29}$$

Therefore, the model parameter extraction becomes an optimization problem. The objective function $F_0(X)$ (Shi et al., 2005) of the optimization in Eq. 30 can be divided into two parts by the plus sign. The first part is the average error between the derived admittances and those obtained from the measurements. The second part is the variance of the error.

$$F_0(X)\big|_{X=(X_1,X_2,\ldots,X_n)} = \sum_{i=1}^{m}\left\{f_i(X)^2 + [f_i(X) - F_{mean}]\right\}^2 \tag{30}$$

In Eq. (30), the vector $X = (X_1, X_2, \ldots, X_n)$ represents the component values to be extracted, i.e., L_s, R_s, L_{sk} and R_{sk} of sub-block Y_1 and C_{ox}, C_{sub} and R_{sub} of sub-block block Y_2. n is the total number of parameters in each sub-block. m is the total number of frequency points under consideration. $f_i(X)$ is the error between the simulated admittance and the ones obtained from measurement results at each frequency point. The definition of $f_i(X)$ is given in Eq. 31. F_{mean} as defined in Eq. 32 is the mean error of the whole frequency range under consideration.

$$f_i(X) = \left|\frac{Y_{simulated(i)} - Y_{measured(i)}}{Y_{measured(i)}}\right| \tag{31}$$

$$F_{mean} = \frac{\sum_{i=1}^{m} f_i(X)}{m} \tag{32}$$

The values of L_s, R_s, L_{sk} and R_{sk} of sub-block Y_1 and C_{ox}, C_{sub} and R_{sub} of sub-block Y_2 can be determined by searching for the minimum values of $F_0(X)$, starting from the reasonable initial guess.

2.3.2 Interconnects with bends

The interconnect shapes on a real chip are very complicated. Interconnect models which handle straight lines only are far from sufficient. Interconnects with bends are often required. These bends are usually with angles of 90° or 45°.

According to the physical configuration, the entire trace of the interconnects with bends can be divided into different sub-segments, i.e., straight-line segments and corner segments. The structural analysis and nomenclatures are illustrated in Fig 14.

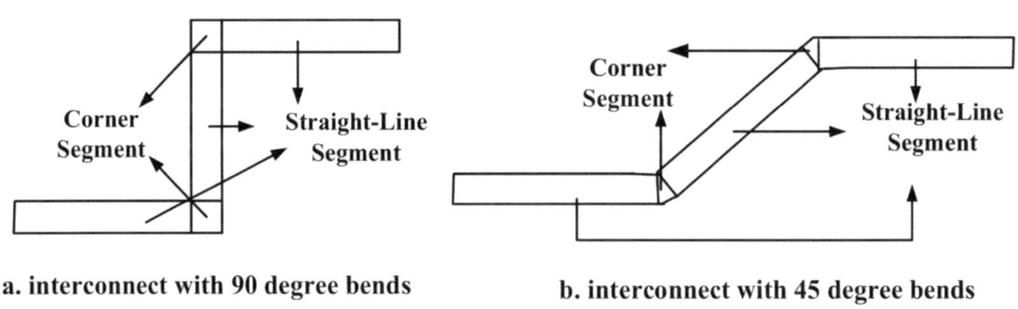

Fig. 14. Structural analysis of interconnect with bends (Shi et al., 2008)

Henceforth, the model development methodology can be proposed. Firstly, a complex-shaped interconnect is decomposed into sub-segments as shown in Fig. 15. Secondly, equivalent circuit models are developed for these sub-segments. Lastly, the sub-segments are cascaded to form the model of the entire interconnect.

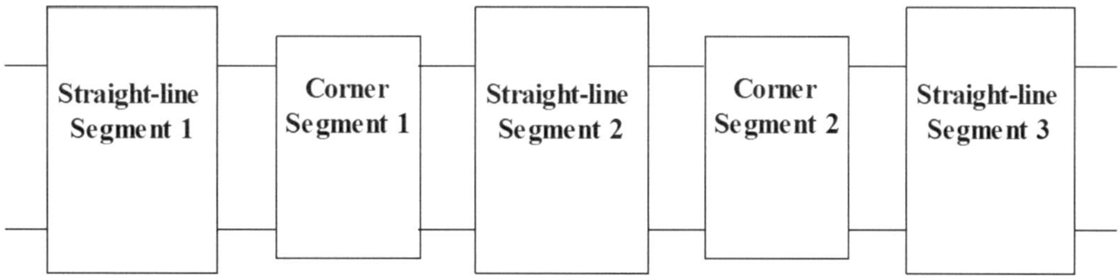

Fig. 15. Schematic block model of interconnect with bends (Shi et al., 2008)

A T-network as shown in Fig. 16 is used to characterize the interconnect bends of the CMOS process.

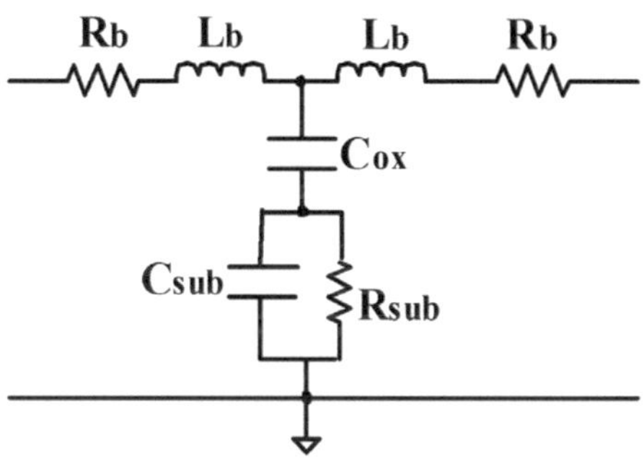

Fig. 16. Equivalent circuit model of the corner segment (Shi et al., 2008)

It is known that currents flowing round the corners distribute unevenly, such that most of the flows crowd around the inner edge (Edwards & Steer, 2000). Given in (Baker et al., 1997), the sheet resistance of straight lines is R_{square}, and the sheet resistance of corners is approximately $0.6 \times R_{square}$. Additionally, considering the relatively smaller physical size of the corner compared to the straight line, R_b can be removed from the series branch. Moreover, a further simplification of the shunt block does not reduce the precision of the model significantly. Thus, the model can be further simplified as shown in Fig. 17.

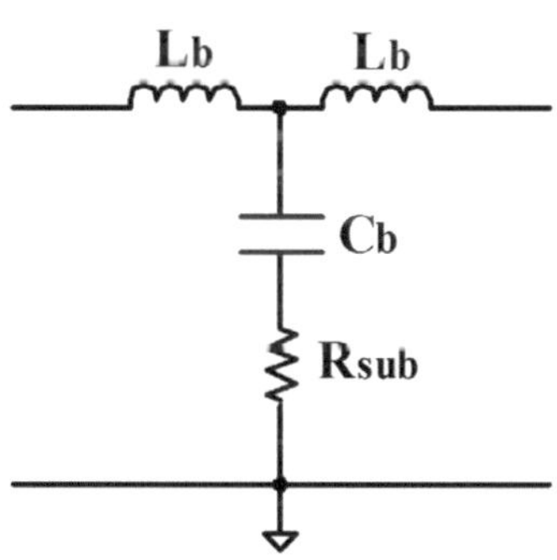

Fig. 17. Simplified equivalent circuit model of the corner (Shi et al., 2008)

The construction of the complex-shaped interconnect model seems quit straightforward. However, simply connecting the sub-segments together will not lead to a precise model. The reason is that the inductance of a curved interconnect does not equal to the sum of the inductances of the straight-line sub-segments. Due to the mutual inductance cancellation of different sub-segments, the general trend is that the larger curvature an interconnect has, the smaller is the inductance. In order to characterize this effect, the series inductance in the Π-network of the straight-line segments which is connected to the corner segment should be modified. An additional parameter α (0< α <1), i.e., the multiplication factor, is introduced to represent this variation of the inductance. As illustrated in Fig. 18, inductance L_s in the second Π-network of *Straight-line Segment 1*, L_s in both of the two Π-networks of *Straight-line Segment 2*, and L_s in the first Π-network of *Straight-line Segment 3* are multiplied with the multiplication factor α. The influences of the corner can be omitted in the shunt blocks of the straight-line segments, so that the parameters are kept unchanged.

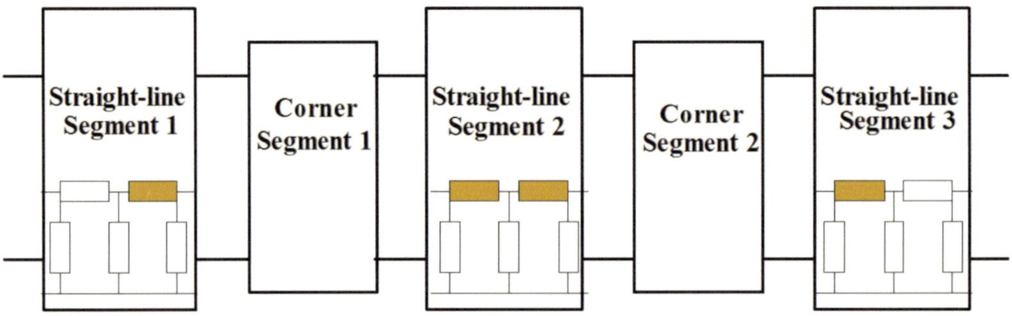

Fig. 18. Illustration of the application of factor α (Shi et al., 2008)

The parameter extraction of the interconnect with bends can also be formulated as an objective function. As shown in Fig. 18, three straight-line segments and two corner segments are cascaded in a sequence. Therefore, in order to get the ABCD matrix of the whole trace, five corresponding ABCD matrixes of each segment are multiplied (Eq. 33).

$$T_{wire} = T_{\Pi} T_{\Pi corner} T_{corner} T_{\Pi corner} T_{\Pi corner} T_{corner} T_{\Pi corner} T_{\Pi} \tag{33}$$

where T_{Π} denotes the ABCD matrix of each Π-network of the equivalent circuit model in Fig. 13; $T_{\Pi corner}$ denotes the ABCD matrix of the Π-network, which is influenced by the corner; and T_{corner} denotes the ABCD matrix of the corner segment.

T_{corner}, as shown in Eq. 34, can be derived as a function matrix of the circuit components L_b, R_b and C_b, based on the model shown in Fig. 17. The elements of T_{corner} are presented in Eq. 35 - Eq. 38.

$$T_{corner} = \begin{bmatrix} A & B \\ C & D \end{bmatrix} \tag{34}$$

$$A = 1 + \frac{j\omega L_b}{\frac{1}{j\omega C_b} + R_b} \tag{35}$$

$$B = 2j\omega L_b - \frac{\omega^2 L_b^2}{\frac{1}{j\omega C_b} + R_b} \tag{36}$$

$$C = \frac{1}{\frac{1}{j\omega C_b} + R_b} \tag{37}$$

$$D = 1 + \frac{j\omega L_b}{\frac{1}{j\omega C_b} + R_b} \tag{38}$$

T_Π is defined based on Eq. 39 - Eq. 43. The derived ABCD matrix elements are functions of the equivalent circuit components L_s, R_s, L_{sk}, R_{sk}, C_{ox}, C_{sub} and R_{sub}.

$$T_\Pi = \begin{bmatrix} A & B \\ C & D \end{bmatrix} \tag{39}$$

where

$$A = 1 + \frac{j\omega C_{ox}(j\omega C_{sub} + \frac{1}{R_{sub}})(j\omega L_s + R_s + \frac{j\omega L_{sk} R_{sk}}{j\omega L_{sk} + R_{sk}})}{j\omega C_{cox} + j\omega C_{sub} + \frac{1}{R_{sub}}} \tag{40}$$

$$B = j\omega L_s + R_s + \frac{j\omega L_{sk} R_{sk}}{j\omega L_{sk} + R_{sk}} \tag{41}$$

$$C = \frac{2j\omega C_{ox}(j\omega C_{sub} + \frac{1}{R_{sub}})}{j\omega C_{ox} + j\omega C_{sub} + \frac{1}{R_{sub}}} + (j\omega L_s + R_s + \frac{j\omega L_{sk} R_{sk}}{j\omega L_{sk} + R_{sk}}) \left[\frac{j\omega C_{ox}(j\omega C_{sub} + \frac{1}{R_{sub}})}{j\omega C_{ox} + j\omega C_{sub} + \frac{1}{R_{sub}}} \right]^2 \tag{42}$$

$$D = 1 + \frac{j\omega C_{ox}(j\omega C_{sub} + \frac{1}{R_{sub}})(j\omega\alpha L_s + R_s + \frac{j\omega L_{sk} R_{sk}}{j\omega L_{sk} + R_{sk}})}{j\omega C_{cox} + j\omega C_{sub} + \frac{1}{R_{sub}}} \tag{43}$$

The matrix elements presented in Eq. 40 to Eq. 43 are for the Π-networks without corner influence. For the corner-influenced Π-networks, $T_{\Pi corner}$ is similar to T_Π. We just have to replace the item L_s with αL_s to account for the corner effect as illustrated in Eq. 44 - Eq. 48.

$$T_{\Pi corner} = \begin{bmatrix} A & B \\ C & D \end{bmatrix} \tag{44}$$

where

$$A = 1 + \frac{j\omega C_{ox}(j\omega C_{sub} + \frac{1}{R_{sub}})(j\omega\alpha L_s + R_s + \frac{j\omega L_{sk} R_{sk}}{j\omega L_{sk} + R_{sk}})}{j\omega C_{cox} + j\omega C_{sub} + \frac{1}{R_{sub}}} \quad (45)$$

$$B = j\omega\alpha L_s + R_s + \frac{j\omega L_{sk} R_{sk}}{j\omega L_{sk} + R_{sk}} \quad (46)$$

$$C = \frac{2 j\omega C_{ox}(j\omega C_{sub} + \frac{1}{R_{sub}})}{j\omega C_{ox} + j\omega C_{sub} + \frac{1}{R_{sub}}} + (j\omega\alpha L_s + R_s + \frac{j\omega L_{sk} R_{sk}}{j\omega L_{sk} + R_{sk}})\left[\frac{j\omega C_{ox}(j\omega C_{sub} + \frac{1}{R_{sub}})}{j\omega C_{ox} + j\omega C_{sub} + \frac{1}{R_{sub}}}\right]^2 \quad (47)$$

$$D = 1 + \frac{j\omega C_{ox}(j\omega C_{sub} + \frac{1}{R_{sub}})(j\omega\alpha L_s + R_s + \frac{j\omega L_{sk} R_{sk}}{j\omega L_{sk} + R_{sk}})}{j\omega C_{cox} + j\omega C_{sub} + \frac{1}{R_{sub}}} \quad (48)$$

The values of all the components in the straight-line segment, namely L_s, R_s, L_{sk}, R_{sk}, C_{ox}, C_{sub} and R_{sub} are obtained from the previous section. Thereafter, the ABCD matrix of the entire wire are interpreted as functions of L_b, C_b, R_b and α. Corresponding S-parameters are expressed as functions of these variables under the following transformation formulas from Eq. 49 to Eq. 52.

$$S_{11} = \frac{A + \frac{B}{Z_o} - CZ_o - D}{A + \frac{B}{Z_o} + CZ_o + D} \quad (49)$$

$$S_{12} = \frac{2(AD - BC)}{A + \frac{B}{Z_o} + CZ_o + D} \quad (50)$$

$$S_{21} = \frac{2}{A + \frac{B}{Z_o} + CZ_o + D} \quad (51)$$

$$S_{22} = \frac{-A + \frac{B}{Z_o} - CZ_o + D}{A + \frac{B}{Z_o} + CZ_o + D} \quad (52)$$

The objective function $F_0(X)$ can then be obtained as Eq. 53.

$$F_o(X)|_{X=(\alpha,L_b,C_b,R_b)} = \sum_{i=1}^{m}\left\{f_{1i}^2(X)+\left[f_{1i}(X)-F_{1mean}\right]^2+f_{2i}^2(X)+\left[f_{2i}(X)-F_{2mean}\right]^2\right\} \quad (53)$$

where m is the total number of the frequency points under consideration. f_{1i} is the error between the simulated S_{11} and those acquired from the measurement results at each frequency point i, which is stated in Eq. 54. f_{2i} is the error between the simulated S_{21} and the measurement results at each frequency point i, which is stated in Eq. 55. F_{1mean} and F_{2mean} defined in Eq. 56 and Eq. 57 are the mean errors of S_{11} and S_{21} at each frequency point i. Given the symmetry of the interconnect test structures as shown in Fig. 14, it is known that $S_{ij} = S_{ji}$ and $S_{ii} = S_{jj}$. We apply the average values of the measured S_{11} and S_{22} and S_{12} and S_{21} as follows. They are denoted by $S'_{11}$and S'_{21}, respectively.

$$f_{1i}(X) = \left|\frac{S_{11-i\_simulated} - S'_{11-i\_measured}}{S'_{11-i\_measured}}\right| \quad (54)$$

$$f_{2i}(X) = \left|\frac{S_{21-i\_simulated} - S'_{21-i\_measured}}{S'_{21-i\_measured}}\right| \quad (55)$$

$$F_{1mean} = \frac{\sum_{i=1}^{m} f_{1i}(X)}{m} \quad (56)$$

$$F_{2mean} = \frac{\sum_{i=1}^{m} f_{2i}(X)}{m} \quad (57)$$

The values of α, L_b, C_b and R_b of the corner segment can be determined by searching for the minimum values of $F_0(X)$ as shown in Eq. 53, starting from the reasonable initial guess. Therefore, the model of the complex shaped interconnect can be constructed.

3. Emerging on-chip interconnect concepts and technologies

The previous sections have emphasized on interconnects in the conventional metal/dielectric system. In this section, the authors would like to shed some lights on some emerging interconnect concepts and technologies. According to ITRS, these interconnect renovations are going to play the key role in satisfying the requirements of performance, reliablility and power consumption of the IC designs in the long run.

3.1 Optical interconnects

Optical interconnects (OIs) have been proposed to overcome the communication bottleneck by replacing electrical wires with optical waveguides (Haurylau et al., 2006). The major advantages of the OIs are speed-of-light signal propagation, large bandwidth and minimum crosstalk between signal transmission paths (ITRS, 2008).

While board--to-board and chip-to-chip of OIs have been actively under development, the feasibility of on-chip OIs is still an open question (Haurylau et al., 2006). The compatibility with CMOS technology is the biggest challenge for on-chip OIs and therefore gaining ever increasing interest from both academia and industry. The block diagram of an on-chip OI system is illustrated in Fig. 19. It consists of the following components (ITRS, 2008; Haurylau et al., 2006):

- Light sources: From the modulation perspective, light source can be either directly modulated or non-modulated. For the non-modulated case, the light source must be used with modulators that can be controlled by electrical signals. From the location point of view, lasers can be either off-die or on-die. Key parameters of the light sources are output power, efficiency, cost, thermal stability, cooling requirements and speed for directly modulated sources. Up to date, high speed, electrically driven, on-chip monolithic light sources are still far from reality.
- Modulators: Modulators are used together with a non-modulated light source, typically off-die. The light provided by the laser is fed into the modulator. The main function of a modulator is to transducer electrical data supplied from the electrical driver into a modulated optical signal. The key parameters are coupling efficiency, operation voltage, switching time, waveguide loss, overall power, modulation depth/extinction ratio and area.
- Waveguides: waveguides are the paths through which light is propagated on-chip with minimum losses. Key parameters of waveguides include loss per unit length, refractive index contrast and pitch.
- Photo detectors: photo detectors are used to converts the incoming optical signal to small output current proportional to the input optical power. Key parameters of photo detectors are responsivity, bandwidth, switching speed and noise performance (Dagli, 2006). The two most widely used semiconductor photo detectors are P-Insulator-N (PIN) photodiodes and avalanche photodiodes (APDs).
- Transimpedance amplifiers (TIAs): TIAs acts as the electrical front-end of an optical receiver. It converts the small current signal generated by the photo detector to voltage signal. Key parameters of TIAs are the input referred noise, the overload current, the transimpedance gain, the bandwidth and the group delay.

The primary challenge for optical interconnects is to develop low-cost, low-power and CMOS-compatible components.

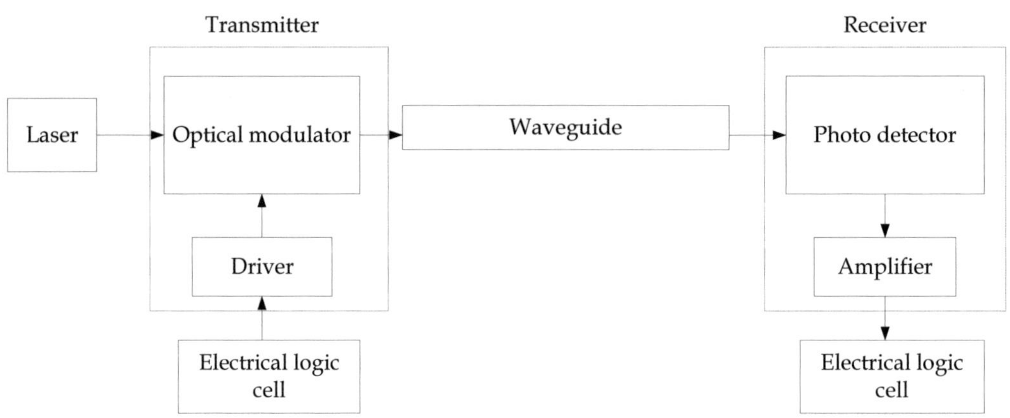

Fig. 19. Block diagram of OI system (Haurylau et al., 2006)

3.2 RF/Microwave interconnects

The basic idea of RF/microwave interconnect is to replace on-chip wires with integrated on-chip antennas to realize communication from one part of a chip to another part via RF or microwaves. RF/microwave signals are transferred either through free space or guided mediums (Chang et al, 2001). In the first case, it essentially takes the form of an on-chip LAN (local-area network), with transmitters, receivers, antennas and appropriate signal generation and signal detection circuitry (ITRS, 2008). The biggest challenge comes from the antenna design. Free space transmission and reception of RF/Microwave signals requires the antenna size comparable to its wavelength. Even at near 100 GHz operating frequency and cut-off frequency, the optimal aperture size of the antenna is of the order of one mm^2, which is too large to be implemented on-chip. In the later case, the RF/microwave signals are transmitted in guided mediums, such as the microstrip transmission line (MTL) or coplanar waveguide (CPW). Microwave transmission in MTLs and CPWs has much lower attenuation as compared with traditional wires. Moreover, since the communication distance is relatively short (several centimetres), the conventional large "far-field" antenna can be replaced by much smaller "near-field" capacitive couplers (Chang et al, 2001).

RF/Microwave interconnect technology is still in the early state of development. According to ITRS, there are four most critical questions must be solved (ITRS, 2008). First of all, in order to compare it with alternative interconnect solutions, characterization of the RF/Microwave interconnect system in terms of cost and performance must be completed. Secondly, full design rules for the electrical and electromagnetic portions of RF/microwave interconnect must be set up. Thirdly, the associated power and design complexity trade-offs must be fully understood. Last but not least, appropriate IC substrate and packaging materials for optimized transmission of RF and microwaves must be identified.

3.3 Carbon nanotubes

As the physical dimension of on-chip interconnects keeps on shrinking with the scaling of CMOS, the increased resistivity and electromigration issues of the conventional metal interconnects have caused serious concern. Carbon nanotubes (CNTs) have been proposed as a replacement for metal interconnects for their high mechanical and thermal stability, high thermal conductivity and large current carrying capacity (Naeemi et al, 2005; Raychowdhury& Roy, 2006).

CNTs are sheets of graphite rolled into cylinders with diameter of the order of one nanometer. Depending on the direction in which CNTs are rolled up (chirality), they demonstrate either metallic or semiconducting properties (Srivastava & Banerjee, 2005).

There are mainly two categories of CNTs, i.e., single-wall (SWCNT) or multi-wall (MWCNT). SWCNTs consist of only one graphene shell, while MWCNTs consist of several concentric graphene cylinders. MWCNTs are predominantly metallic. However, it is more difficult to achieve ballistic transport over long distance as compared to SWCNTs (Srivastava & Banerjee, 2005). Therefore, metallic SWCNTs are determined as preferred candidates as interconnects.

Research has shown a promising outlook for CNTs as a possible alternative to traditional metal interconnects. However, there are still numerous technical challenges to be addressed (ITRS, 2008), such as achieving a high-density integration with CNTs, selective growth of metallic SWCNTs, directional growth in CNTs, achieving low-resistance metal-CNT contacts, achieving defect free CNTs and back-end-of-the-line compatible CNT growth.

4. Conclusion

Boosted by the great demand from the wireless telecommunication market, RFICs are gaining more and more attention. As circuit performance is getting increasingly dependent on interconnects, the RFIC design faces a big challenge that is the interconnect. In this chapter, the physical background of on-chip interconnects and the basic ideas of model development have been introduced. Various existing interconnect models have been presented. Extractions of model parameters have been discussed. Three of the most promising on-chip interconnection technologies, i.e., optical interconnects, RF/microwave interconnects and carbon nanotubes have also been introduced.

5. References

Azadpour, M. A. & Kalkur, T. S. (2002). Interconnect model at multi-GHz frequencies incorporating inductance effect, *Proceedings of ICSE 2002*, pp. 82-86, ISBN 0780375785, Penang Malaysia, Dec. 2002, IEEE, Piscataway, New Jersey.

Baker, R. J., Li, H. W. & Boyce, D. E. (1997). *CMOS: circuit design, layout and simulation.* Wiley-IEEE, ISBN 0-7803-3416-7, New York, United States of America.

Celik, M.; Pileggi, L. & Odabasioglu A. (2002). *IC interconnect analysis*, Kluwer Academic Publishers, ISBN 14020-7075-6, Boston. Dordrecht. London.

Chang, M.F.; Roychowdhury, V.P.; Liyang Zhang; Hyunchol Shin & Yongxi Qian (2001). RF/wireless interconnect for inter- and intra-chip communications, *Proceedings of the IEEE*, Vol. 89, No. 4, Apr. 2001, pp456-466, ISSN 0018-9219.

Chiprout, E. (1998). Interconnect and substrate modeling and analysis: an overview, *IEEE J. Solid-State Circuits*, Vol. 33, No. 9, Sep., 1998, pp. 1445 - 1452, ISSN 0018-9200.

Dagli, N. (2006). *High-speed photonic devices*, CRC Press, Boca Raton Florida, ISBN 0750308893.

Deutsch, A.; Coteus, P.W.; Kopcsay, G.V.; Smith, H.H.; Surovic, C.W.; Krauter, B.L.; Edelstein, D.C. & Restle, P.L. (2001). On-chip wiring design challenges for gigahertz operation, *Proceedings of the IEEE*, Vol. 89, No. 4, Apr., 2001, pp. 529-555, ISSN 0018-9219.

Edwards, T. C. & Steer, M. B. (2000). *Foundations of interconnect and microstrip design*, John Wiley & Sons, ISBN 0-471-60701-0, Chichester, England.

Eisenstant, W. R. & Eo, Y. (1992). S-parameter-based IC interconnect transmission line characterization, *IEEE Trans. Comp., Hybrids, Manufact. Technol.*, Vol. 15, No. 4, Aug., 1992, pp. 483-490, ISSN 1070-9894.

Eo, Y. & Eisenstadt, W. R. (1993). High-speed VLSI interconnect modeling based on S-parameter measurements, *IEEE Trans. Comp., Hybrids, Manufact. Technol.*, Vol. 16, No. 5, Aug., 1993, pp. 555-562, ISSN 1070-9894.

Gala, K.; Blaauw, D.; Zolotov, V.; Vaidya P.M. & Joshi, A. (2002). Inductance model and analysis methodology for high-speed on-chip interconnect. *IEEE Trans. VLSI Syst.*, Vol. 10, No. 6, Dec., 2002, pp. 730-745, ISSN 1063-8210.

Haurylau, M.; Guoqing Chen; Hui Chen; Jidong Zhang; Nelson, N.A.; Albonesi, D.H.; Friedman, E.G. & Fauchet, P.M. (2006). On-Chip Optical Interconnect Roadmap: Challenges and Critical Directions. *IEEE J. Sel. Topics Quantum Electron.*, Vol. 12, No. 6, Nov./Dec., 2006, pp. 1699-1705, ISSN1077-260X

ITRS 2008 (2008) http://www.itrs.net/

Kleveland, B.; Qi, X.; Madden, L.; Furusawa, T.; Dutton, R. W.; Horowitz, M. A. & Wong, S. S. (2002). High-frequency characterization of on-chip digital interconnects. *IEEE J. Solid-state Circuits*, Vol. 37, No. 6, Jun., 2002, pp. 716-725, ISSN 0018-9200.

Marsh, S. (2006). *Practical MMIC Design*, Artech House, ISBN 1-59693-036-5,

Naeemi, A.; Sarvari, R.& Meindl, J.D. (2005). Performance comparison between carbon nanotube and copper interconnects for gigascale integration (GSI). *IEEE Electron Device Lett.*, Vol. 26, No. 2, Feb. 2005, pp. 84-86, ISSN 0741-3106, Boston. London.

Plett, C. & Rogers, J. (2003). *Radio frequency integrated circuit design*, Artech House, ISBN 1-58053-502-x, Boston .London.

Pozar, D. M. (1998). *Microwave Engineering*, John Wiley & Sons Inc., ISBN 0-471-17096-8, New York. Chichester. Weinheim. Brisbane. Singapore. Toronto

Raychowdhury, A.& Roy, K. (2006). Modeling of metallic carbon-nanotube interconnects for circuit simulations and a comparison with Cu interconnects for scaled technologies. *IEEE Trans. Comput.-Aided Des. Integr. Circuits Syst.*, Vol. 25, No. 1, Jan. 2006, pp. 58-65, ISSN 0278-0070.

Shi, X.; Ma, J.-G.; Yeo, K. S.; Do, M. A. & Li, E. (2005). Equivalent circuit model of on-wafer CMOS interconnects for RFICs. *IEEE Trans. VLSI Syst.*, Vol. 13, No. 9, Sep., 2005, pp. 1060-1071, ISSN 1063-8210.

Shi, X.; Yeo, K. S.; Ma, J.-G.; Do, M. A. & Li, E. (2008). Complex shaped on-wafer interconnects modeling for CMOS RFICs. *IEEE Trans. VLSI Syst.*, Vol. 16, No. 7, Jul., 2008, pp. 922-926, ISSN 1063-8210.

Shin, S.; Eo, Y., Eisenstadt, W. R. & Shim, J. (2004), Analytical models and algorithms for the efficient signal integrity verification of inductance-effect-prominent multicoupled VLSI circuit interconnects. *IEEE Trans. VLSI syst.*, Vol. 12, No. 4, Apr., 2004, pp. 395 - 407, ISSN 1063-8210.

Srivastava, N.& Banerjee, K. (2005). Performance analysis of carbon nanotube interconnects for VLSI applications. *Proceedings of IEEE/ACM ICCAD 2005,* ISBN 078039254X, San Jose California, Nov. 2005, IEEE, Piscataway, New Jersey.

Wang, G. ; Qi, X. & Yu, Z. (2001). Device Level modeling of metal-insulator-semiconductor interconnects. *IEEE Trans. Electron Devices*, Vol. 48, No. 8, Aug. 2001, pp. 1672-1682, ISSN 0018-9383.

Zheng, J. ; Hahm, Y.-C. & Tripathi, V. K. & Weisshaar, A. (2000). CAD-oriented equivalent-circuit modeling of on-chip interconnects on lossy silicon substrate. *IEEE*

Trans. Microwave Theory Tech., Vol. 48, No. 9, Sep. 2000, pp. 1443-1451, ISSN 0018-9480.

Zheng, Y. (2003). High-frequency on-chip interconnect characterization and measurment. *PhD Dissertation,* Columbia University.

14

Highly Energy-Efficient On-Chip Pulsed-Current-Mode Transmission Line Interconnect

Tomoaki Maekawa, Shuhei Amakawa, Hiroyuki Ito,
Noboru Ishihara, and Kazuya Masu
Tokyo Institute of Technology
Japan

1. Introduction

System-on-a-chip (SoC) has become possible since a great number of circuit elements can be integrated into a single chip by the miniaturization technologies for Si CMOS. Network-on-Chip (NoC) has been investigated actively, and it is expected to be a new approach for designing the communication subsystems of SoC (Lee et al., 2008). Enormous circuit blocks are loaded onto the NoC, and on-chip networks like local area networks (LANs) in the NoC communicate among these circuit blocks. Since the performance of the NoC is strongly affected by on-chip networks, the construction of efficient on-chip communications infrastructures will be increasingly significant.

Some of the important characteristics for on-chip interconnects are bandwidth, latency, and power. In particular, power saving technologies are very important in realizing Green IT (in- formation technology). Power dissipation in on-chip networks mainly occurs at interconnects due to the increase of wiring resistance and capacitance. A significant issue is that power consumption of conventional on-chip interconnects, i.e. so-called RC lines, is proportional to the signal frequency; hence, it is very difficult to reduce energy dissipation per bit. Given the recent trend of high speed signaling, we have to solve this problem and offer some good solutions. One solution is the use of copper lines and low-k dielectric, and these techniques have been widely applied and reduce power consumption for transmitting signals. However long interconnects still consume large power as in the case of RC lines.

Another solution is the introduction of on-chip transmission line interconnects (TLIs). The applications of TLIs have been widely demonstrated. Modulation (Chang et al., 2003), pulsed-current-mode (Jose et al., 2006), current-mode-logic (Ito et al., 2004, 2005; Ishii et al., 2006; Gomi et al., 2004), low voltage differential signaling (Ito et al., 2007) and multi-drop (Ito et al., 2008) techniques are proposed, and these techniques enable the improvement of bandwidth, latency and extensibility of on-chip networks. Figure 1 is an image of on-chip networks with TLIs.

It is also reported that TLIs have a better power efficiency than the conventional on-chip lines as the line length and signal frequency increase (Ito et al., 2004; Gomi et al., 2004; Ito et al., 2005; Ishii et al., 2006; Ito et al., 2007). Further improvement of the power efficiency at low frequencies is the design challenge in the case of on-chip TLIs. Since current-mode differential amplifiers are usually used for transmitters (Txs) and receivers (Rxs) in TLIs, Tx and Rx consume static power regardless ardless of the signal frequency. This means TLIs waste power if they are applied to paths with a low activity factor or to transmit low bit-rate signals.

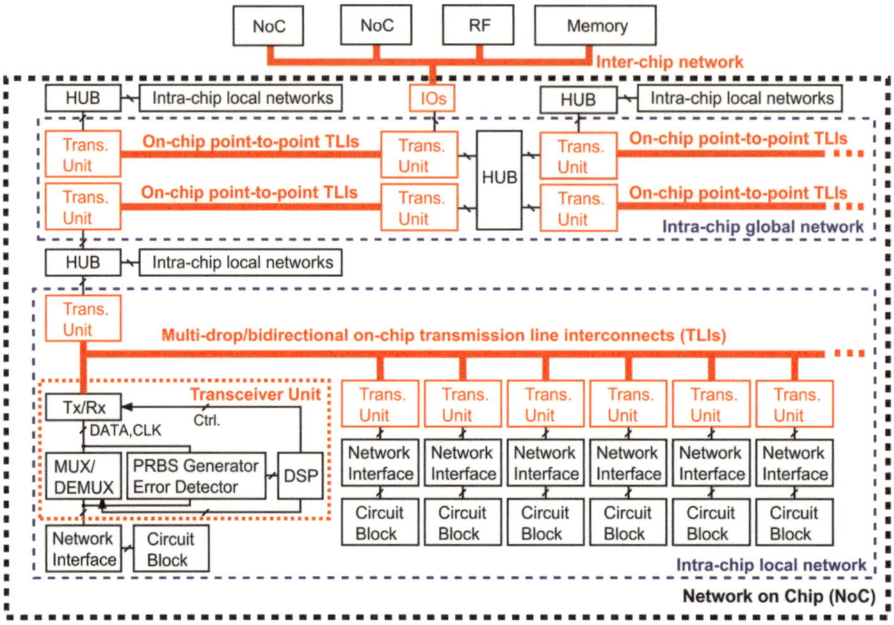

Fig. 1. On-chip networks with transmission line interconnects (TLIs).

Another challenge is the signal amplitude attenuation of on-chip transmission lines. Tx has to output large amplitude signals in order to compensate for the loss of the transmission lines; therefore it usually consumes more power than Rx. Thus, the improvement of Tx power efficiency is crucial for further power saving of on-chip TLIs, and also an important challenge for applying transmission line technologies to on-chip networks.

This paper proposes an on-chip pulsed-current-mode transmission line interconnect (PTLI) with a stacked-switch Tx that does not consume static power and generates return-to-zero (RZ) codes. PTLIs using RZ signals have been proposed in an earlier report (Jose et al., 2006). The features of our interconnects are as follows:

1. Our interconnect mainly consists of transistors and does not have inductors and capacitors that generally occupy a large area.
2. Pulse width, which should be optimized by considering spectral efficiency and power consumption, is adjustable.

Our Tx outputs pulse-shaped RZ signals and consumes power only during signal transitions. Tx has high output impedance in the standby states, and our Tx and Rx can be applied to bidirectional and multi-drop signaling; this can save the area occupied by the TLIs and would improve the extensibility of on-chip networks (Ito et al., 2008).

This paper is organized as follows. The design of on-chip transmission lines is discussed in Section 2. The circuit details of the proposed PTLI are presented in Section 3. The point-to-point and multi-drop PTLIs fabricated by 90nm CMOS process and their measurement results are introduced in Section 4. The concluding remarks are presented in Section 5.

2. Design of on-chip interconnects

Generally, it is often a serious challenge to design on-chip interconnects while taking into account the large resistive losses involved. Since inductance effects ωL become significant as signal frequency increases, ωL must be considered carefully at high frequency. Otherwise on-chip interconnects act as RC-dominant lines.

Inductance effects on on-chip signaling are widely investigated. The line length ℓ for which inductance effects become apparent can be calculated using the following equation (Ismail et al., 1998).

$$\frac{t_r}{2\sqrt{LC}} < \ell < \frac{2}{R}\sqrt{\frac{L}{C}}, \tag{1}$$

where t_r is the signal rise time. In the case of 10 Gb/s signaling with a rise time of 20 ps, the inductive behavior of the line becomes significant at line lengths of is 1mm to 14mm.

Figure 2 shows an image of signal transmissions on an on-chip wire. Tx turns on at time 0, and the voltage rises at the near-end of the line. The voltage wave propagates toward Rx at the electromagnetic wave speed v as shown in Figure.2 (a). L and C are dominant in signal transmissions before the electromagnetic wave reaches the far-end of the line. At the time of l/v, the voltage of the line increases as capacitance of the line is charged as shown in Figure.2 (b). R and C are dominant in this region. Transmission line interconnects use the LC dominant portion for signal transmission by choosing suitable resistive loss, characteristic impedance and termination. Thus, the on-chip transmission line interconnects can achieve a lower latency than the conventional RC lines.

The line width required for on-chip transmission lines is greater compared to that required for conventional RC lines. Figure 3 shows the structure of our on-chip transmission line. Differential and small-amplitude signaling is applied for achieving small rise-time of signals. A differential transmission line that consists of two signal lines and does not have ground lines for area saving is applied to on-chip wiring. Since transmission lines should have wide line width, it is preferable to use thick metal layers to implement transmission lines. In this work, transmission lines are built on the top layer. The line width is 6 μm and the space between signal lines is 4.6 μm. The transmission line is made of aluminum, and the dielectric used is silicon dioxide.

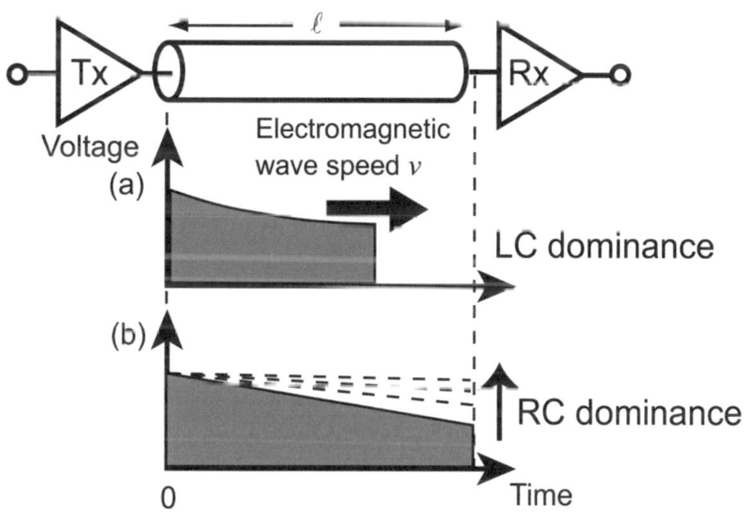

Fig. 2. Signal transmission on an on-chip wire.

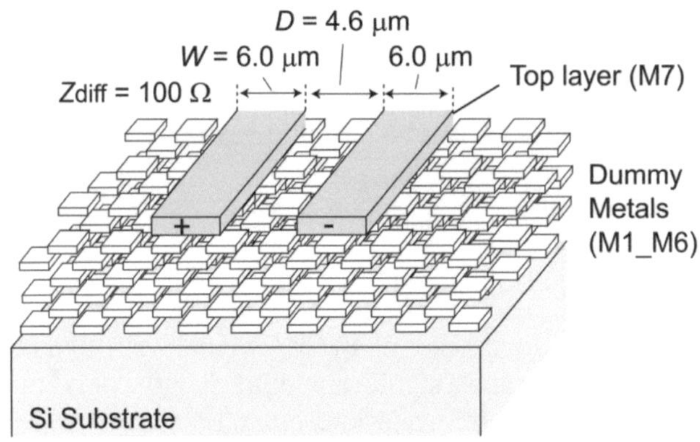

Fig. 3. On-chip transmission line structure.

Characteristic impedance Z_0 affects losses, crosstalk noises, area required for the transmission lines, and power consumption. Let us focus on the losses of transmission line. The attenuation constant α of the transmission line can be approximated as

$$\alpha \approx \frac{1}{2}\left(\frac{R}{Z_0} + GZ_0\right). \tag{2}$$

In multilevel interconnects, the resistive loss is dominant because the underlying metals shield the transmission lines from Si substrate and reduce dielectric loss. High Z_0 helps reduce the losses, thereby reducing the power dissipation at the interconnects. In terms of energy dissipation, Z_0 directly affects the power of Tx. The output signal amplitude v_{out} of Tx is calculated by the formula $v_{out} = Z_0 i$, where i is a current which flows into the transmission lines. i can be reduced by using high Z_0 transmission lines after determining v_{out}. Thus, high Z_0 is acceptable if we focus only on power saving. On the other hand, low Z_0 is better for area and crosstalk robustness. When the line width W is determined by DC resistance, Z_0 can be adjusted by varying the line space D between the signal lines shown in Figure 3. Lower Z_0 lines have smaller D than higher Z_0 lines. Coupling between the signal lines in a differential pair becomes strong as D reduces. Lines with smaller D have higher crosstalk robustness. Thus, it appears to be preferable to choose low Z_0 while building transmission lines for multilevel interconnect. Since this work focuses on power saving, we choose a differential impedance of 100Ω.

Figure 4 shows the frequency response of a 5-mm-long on-chip interconnect. The frequency response of an RC line is provided for comparison with that of the transmission line. The characteristics of the transmission line are obtained by measurement, while those of the RC line are obtained by 2D analysis of the electromagnetic field. The attenuations of the transmission line and the RC line are 2 dB and 14 dB, respectively, and the cutoff frequency of the transmission line is higher than that of the RC line. Thus, it is apparent that high-speed and small-amplitude signaling can be achieved by using on-chip transmission lines.

3. On-chip pulsed-current mode transmission line interconnect (PTLI)

3.1 Level diagram

LSI (large-scale integration) designers who design on-chip transmission line interconnects have to take into account the characteristics of the transmission lines. Figure 5 shows a level diagram based on the frequency response shown in Figure 4.

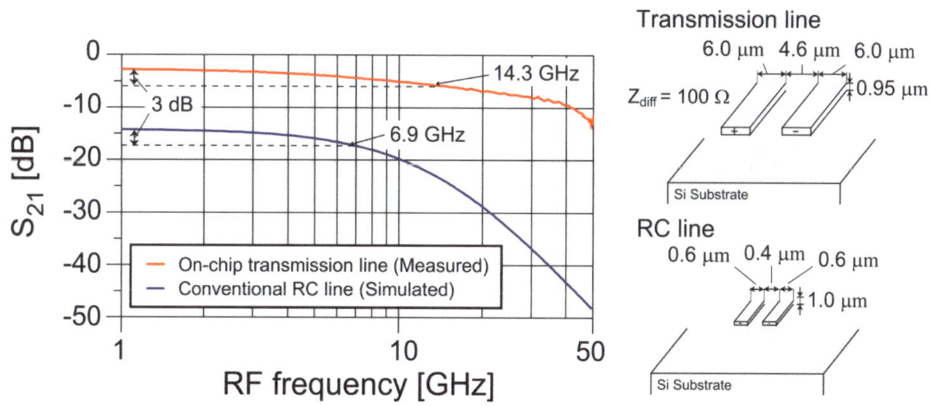

Fig. 4. Frequency dependence of $|S_{21}|$.

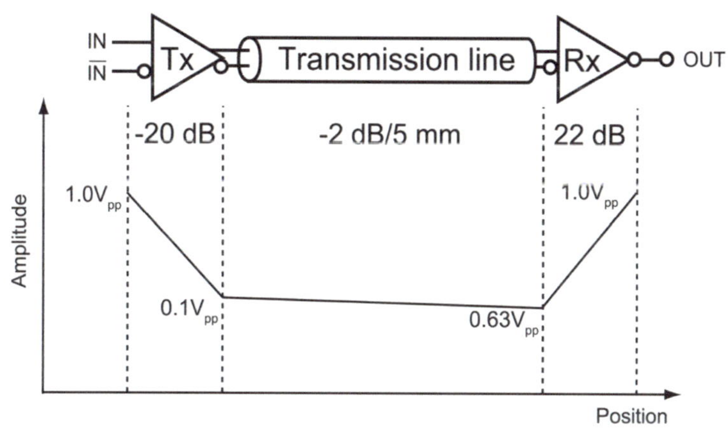

Fig. 5. Voltage level diagram.

In our design, signal amplitude attenuation had a low value of approximately 2 dB as shown in Figure 4. Small signal amplitude is suitable for high-speed signaling and low latency. However, when the signal amplitude is very small, it decays to zero due to attenuation. Taking into consideration the influence of noise, the amplitude of the voltage input to Rx is in the range of 20–60mV. Thus, we set the Tx output voltage at 100mV and the gain at -20 dB. Given the attenuation of the transmission line, Rx gain should be set approximately in the range of 20–22 dB.

3.2 The proposed PTLI

Schematics of the proposed on-chip pulsed-current mode transmission line interconnects (PTLI) are shown in Figure 6(a). PTLI consists of pre buffers that generate differential signals, stacked-switch Txs, an on-chip differential transmission line (DTL), and an Rx. Rail-to-rail signals are input into the PTLI, and Txs convert rail-to-rail signals into pulse-shaped differential RZ-signals. RZ signals propagate in the DTL at the speed of electromagnetic waves. V_{com} stabilizes common-mode voltages of the Tx output. Rx amplifies the pulse signals and converts the RZ signals into NRZ (non-return-to-zero) signals.

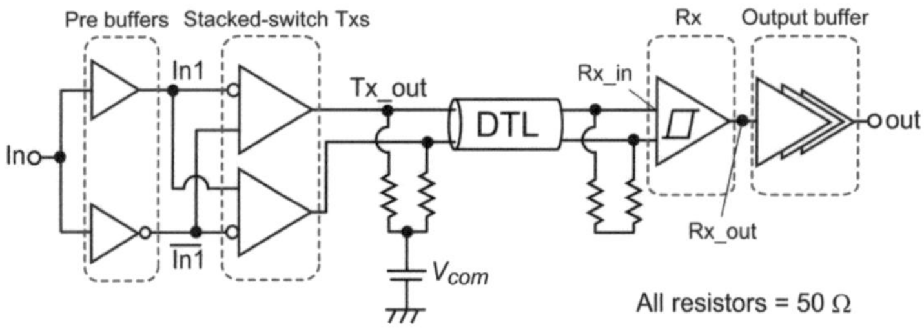

(a)Schematic of the proposed PTLI.

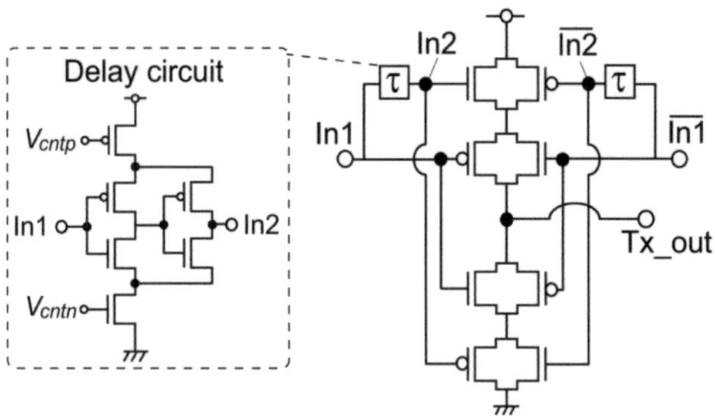

(b) Schematic of Tx.

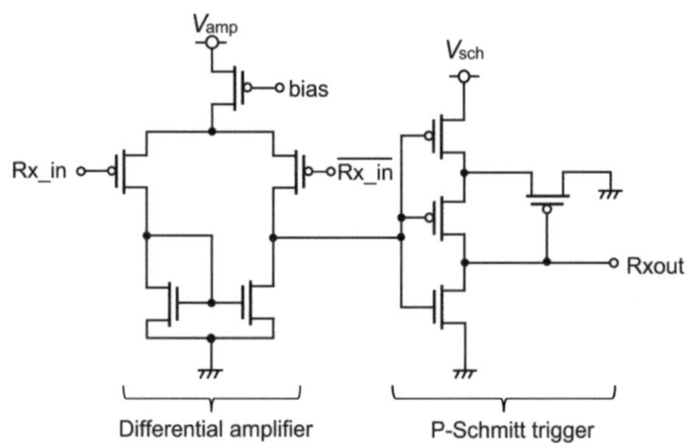

(c) Schematic of Rx.

Fig. 6. Schematics of the proposed PTLI.

3.2.1 Details of Tx

Tx consists of four CMOS switches and delay circuits, as shown in Figure 6(b). Let us first consider the stacked-switches. The output amplitude of Tx depends on the on-resistance of these CMOS switches. The on-resistance of NMOS and PMOS in the saturation region are described by equations (3) and (4), respectively.

$$R_{\text{on,N}} = \frac{1}{\mu_{\text{n}} C_{\text{ox}} \dfrac{W}{L} (V_{\text{DD}} - V_{\text{in}} - V_{\text{th,N}})}, \tag{3}$$

$$R_{on,P} = \frac{1}{\mu_p C_{ox} \frac{W}{L}(V_{in} - V_{th,P})}. \tag{4}$$

From these equations, it is seen that $R_{on,N(P)}$ depends on input voltage V_{in}, as shown in Figures 7(a) and (b). This makes it difficult to convert pulse-shaped RZ signals into NRZ signals. Although CMOS switches increase parasitic capacitance that limits maximum operating frequency, these switches can stabilize on-resistance $R_{on,C}$, as shown in Figure 7(c). In order to achieve better signal integrity, we choose CMOS stacked-switch topologies.

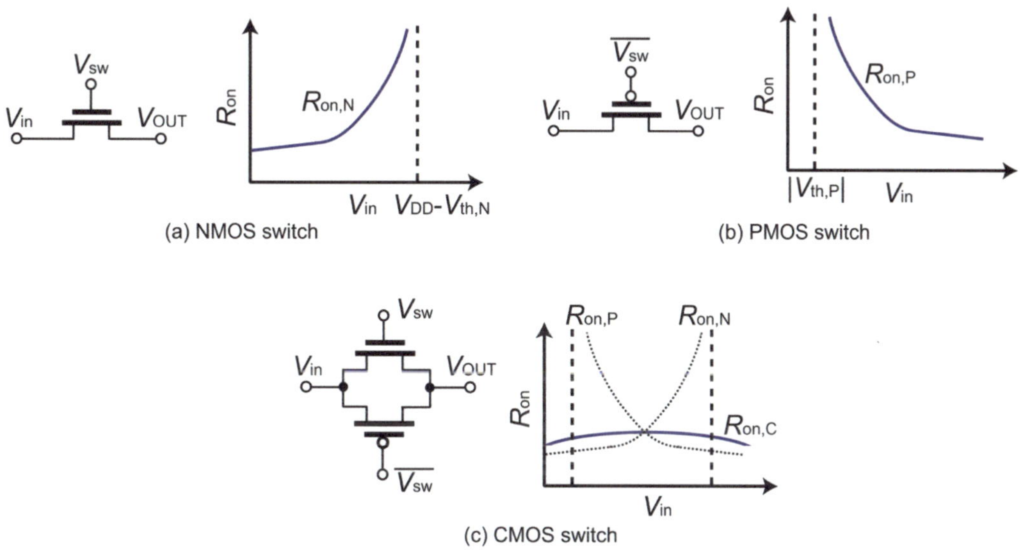

Fig. 7. On-resistances of NMOS, PMOS, and CMOS switches.

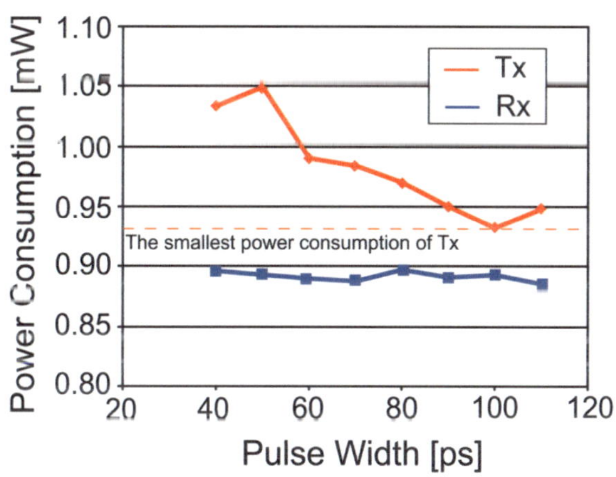

Fig. 8. Pulse-width dependence of simulated power consumption of Tx and Rx.

Next, let us discuss the delay circuits. Delay circuits output τ-lagged signals, and the delay time τ can be changed by controlling bias voltages V_{cntp} and V_{cntn}. τ determines the pulse width of signals. Pulse width, i.e. the delay time τ, should be set by considering a trade-off between power consumption and the spectra of signals. Narrow pulse signals are preferred to save power consumption. However, high-frequency spectral components become stronger as the pulse width decreases, and the attenuation in the case of narrow pulses is greater than that in the case of broad pulses. Figure 8 shows the result of a simulation of power consumption of Tx and Rx. The proposed PTLI is designed to achieve the lowest power with 100-ps-pulse-width signaling. 100 ps is the minimum pulse-width of RZ signals at 10 Gb/s.

The circuit operation of the proposed stacked-switch Tx is as follows. Figure 9 is used to explain the operation.

1. $0 \le t < t_1$ (Figure 9(b))
   One of the input voltages of Tx (In1) is assumed to be 1 (high level), as shown in Figure 9(b), and Tx is in a steady state. Tr1 and Tr3 are on, while Tr2 and Tr4 are off. There are no current paths connecting the power supply, Tx out, and the ground. Hence, Tx does not consume power. Output impedance is very high, and Tx out is represented by V_{com}.
2. $t_1 \le t < t_2$ (Figure 9(c))
   After In1 becomes 0 (low level), Tr2 and Tr3 are turned on and off, respectively. One of the outputs of the delay circuit (In2) is still 1, and the states of Tr1 and Tr4 do not change. Current flows into the DTL from the power supply, and Tx out becomes high.
3. $t_2 \le t < t_3$ (Figure 9(d))
   After time τ, In2 changes to 1. Then, Tr1 and Tr4 are turned off and on, respectively. The current from the power supply is blocked, and Tx out decreases to V_{com}.
4. $t_3 \le t < t_4$ (Figure 9(e))
   The input In1 becomes 1, and Tr2 and Tr3 change to off and on, respectively. The states of Tr1 and Tr4 remain unchanged because of the delay circuit. Current flows from the DTL into the ground, and the voltage level of Tx out decreases.
5. $t_4 \le t$ (Figure 9(b))
   The output voltage of the delay circuit (In2) becomes 1 at t_4, i.e., after τ of t_3. The current to the ground is blocked after Tr4 is turned off. Then, the voltage of Tx out increases to V_{com}, and Tx repeats above operations.

Current does not flow from the power supply to Tx out and the ground in the steady states. Thus, the proposed Tx can save power during low bit-rate transmissions and signaling with a low activity factor. The output impedance of Tx is high in the steady states; this enables multi-drop and bidirectional signal transmissions without the degradation of signal integrity (Ito et al., 2008).

Figure 10 shows a result of transient simulation at 10 Gb/s. Pseudo-random bit sequence (PRBS) of length $2^9 - 1$ is input to Tx. Simulation results show that Tx outputs pulse-shaped RZ-signals as expected by theory.

3.3 Details of Rx

Rx consists of a differential amplifier and a Schmitt trigger circuit as shown in Figure 6(c). The differential amplifier amplifies the pulse-shaped RZ signals and the Schmitt trigger converts these RZ signals into NRZ signals. The gain of the differential amplifier is almost equal to the overall gain of Rx because the gain of the Schmitt trigger is almost 0 dB.

Rx has to be suitably designed to achieve a gain of approximately 22 dB as specified in Section 3.1. The proposed PTLI can control the common-mode voltage V_{com} (Figure 6(a)) from the measuring equipment, and the gain of the differential amplifier depends on V_{com} to a certain extent. Figure 11 shows the dependence of Rx gain on V_{com}. This simulation result indicates that a gain of 22 dB gain can be achieved when V_{com} ranges between 0.4V and 0.5V. Since the main purpose of designing PTLI is to realize low-power operations, V_{com} is set to 0.4V at the time of measurement.

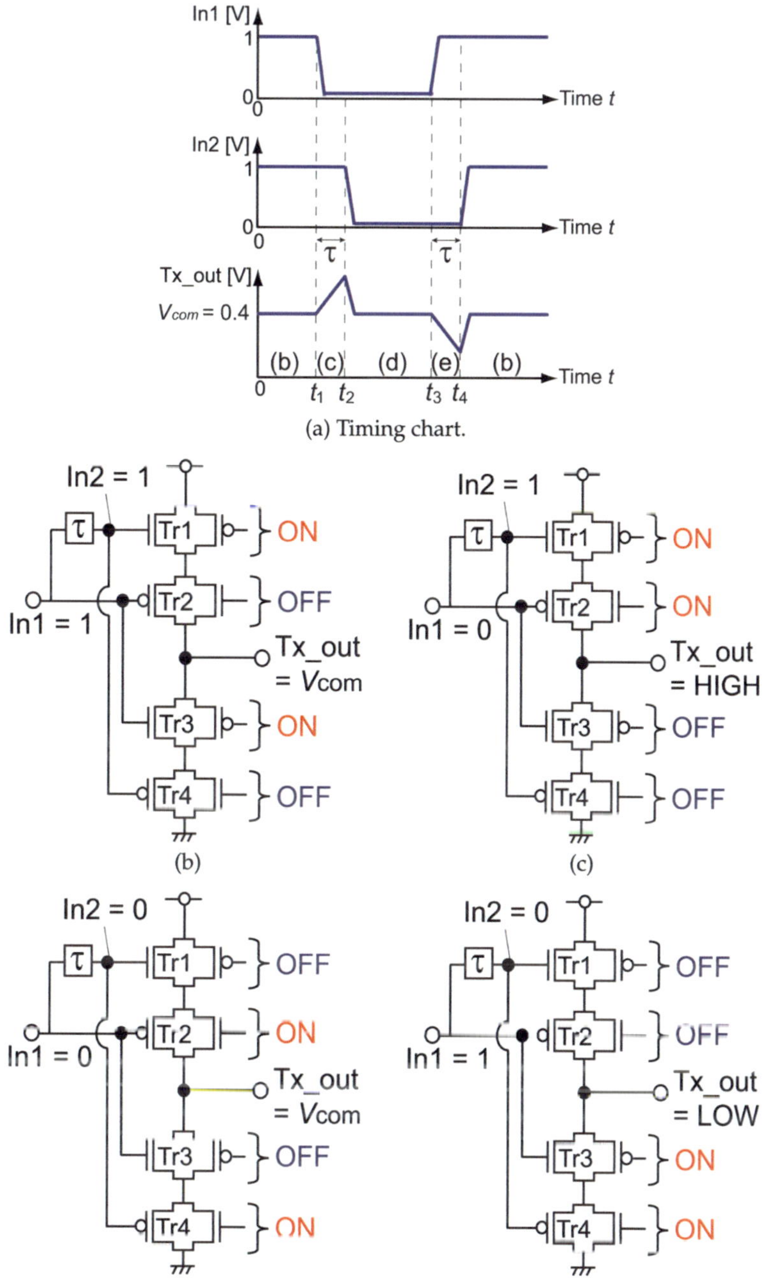

(a) Timing chart.

(b)

(c)

(d)

(e)

Fig. 9. Operation of the proposed Tx.

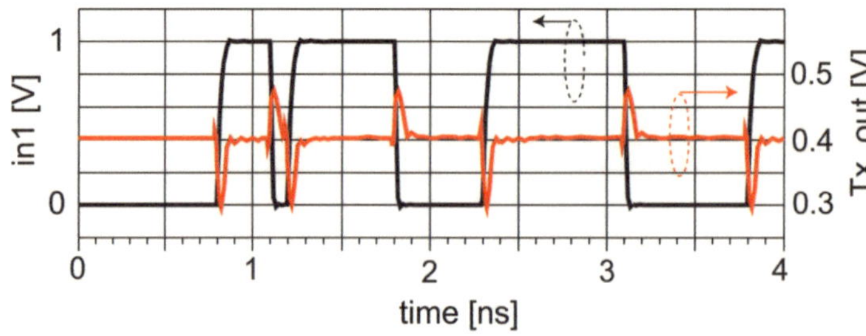

Fig. 10. Simulated output waveforms of Tx.

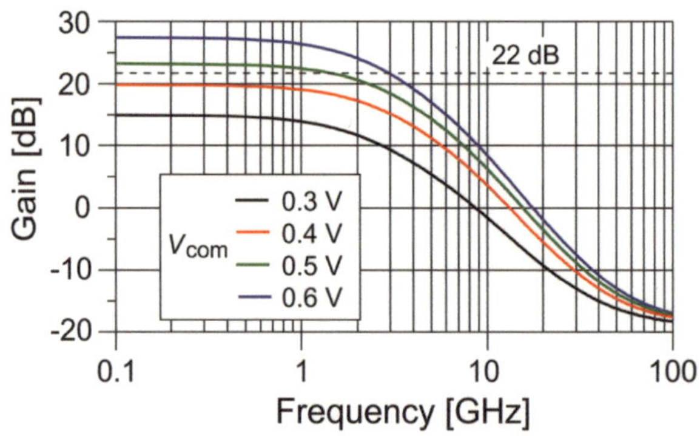

Fig. 11. Simulation of the dependence of Rx gain on V_{com}.

CMOS Schmitt trigger circuits are comparator circuits that incorporate positive feedback by using NMOS and PMOS transistors. The hysteresis characteristics of Schmitt triggers show two threshold voltages V_+ and V_-. When the input is higher than a certain chosen threshold V_+, the output is 1 (high level). When the input is below V_-, the output voltage is 0 (low level). When the input is between V_+ and V_-, the output retains its value.

Generally, the center voltage of two threshold voltages is close to half the voltage of power supply. The common-mode voltage V_{com} is set to 0.4V. However, we have to design a certain hysteresis loop whose central voltage is 0.4V. Thus, the P-Schmitt trigger circuit, which has only PMOS feedback system, as shown in Figure 12, is considered the best topology for this design. P-Schmitt trigger can achieve lower power consumption and smaller area as compared to CMOS Schmitt trigger circuits due to a reduction in the number of elements; this reduction is achieved because the P-Schmitt trigger does not have an NMOS feedback system.

The threshold voltages V_- and V_+ can be determined by choosing suitable transistor sizes. The low threshold voltage V_- is equal to that of inverter consisting of M1, M2, and M3 transistors, shown in Figure 12. The high threshold voltage V_+ can be calculated by using following equation.

$$V_{-}\frac{\sqrt{\frac{\beta_1}{\beta_4}}(V_{dd}-|V_{Tp}|}{1+\sqrt{\frac{\beta_1}{\beta_4}}}. \qquad (5)$$

Here, β_1 and β_4 are the aspect ratios of transistors M1 and M4, and V_{Tp} is the threshold voltage of the PMOS transistors. Figure 13 shows the hysteresis loop of the P-Schmitt trigger used in the proposed PTLI. From the figure, it is seen that the central voltage of the hysteresis loop is nearly 0.4V. Figure 14 shows the transient simulation result. Signals input into "Rx in", as in Figure 6 (a), are assumed to have an amplitude of 20mV and a pulse width of 100 ps. This result indicates that rail-to-rail NRZ signals can be achieved by using the proposed Rx at 10 Gb/s.

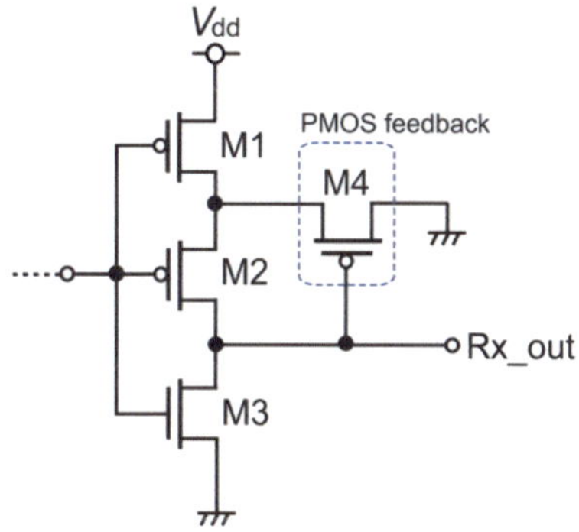

Fig. 12. P-Schmitt trigger circuit.

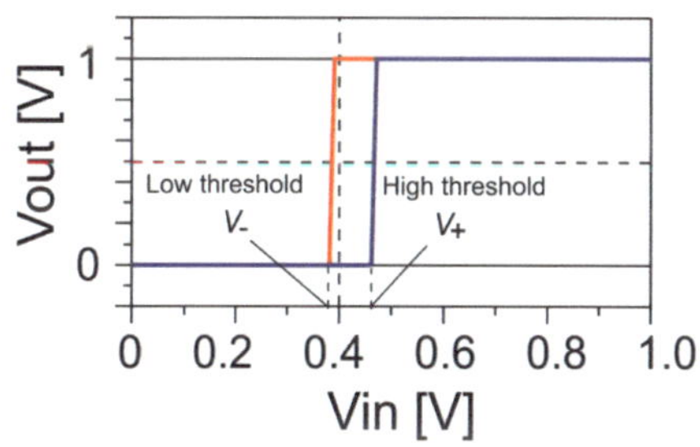

Fig. 13. Hysteresis loop.

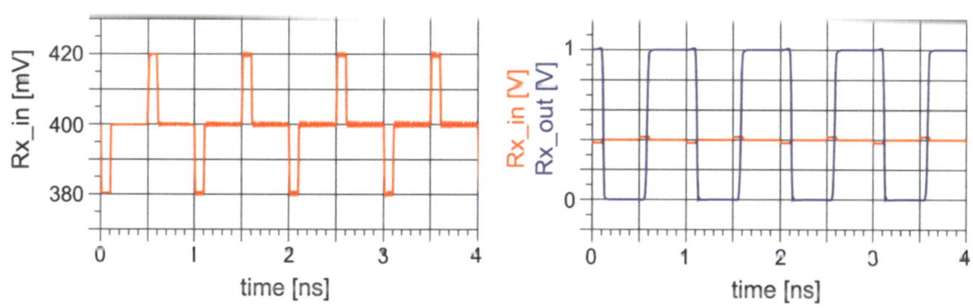

Fig. 14. Transient simulation result.

4. Measurements and discussions

4.1 Point-to-point and multi-drop PTLI

Point-to-point and multi-drop PTLI are fabricated with a 90nm Si CMOS process. Figure 15 shows chip micrographs of the test circuits. The line length is 5mm, and a buffer is used for measurement. Multi-drop PTLI has six I/Os. Txs and Rxs are connected every 1mm, and they share one differential transmission line.

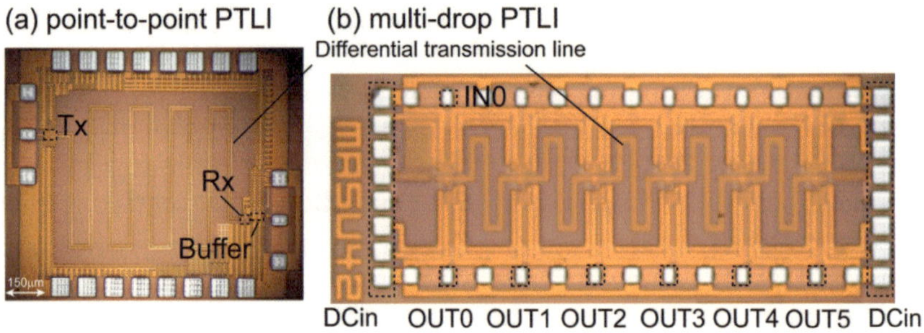

Fig. 15. Chip micrographs of the test circuits.

4.1.1 Point-to-point PTLI

PRBS of length $2^9 - 1$ is input to Tx through an RF probe, and the output signals from the buffer are measured. The eye diagram and the bathtub curve at 8 Gb/s are shown in Figures 16 and 17, respectively. The performance of point-to-point PTLI is summarized in Table 1.

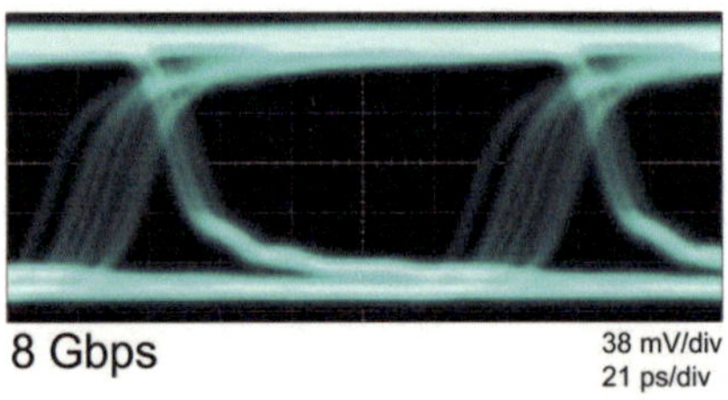

Fig. 16. Eye diagram at 8 Gb/s.

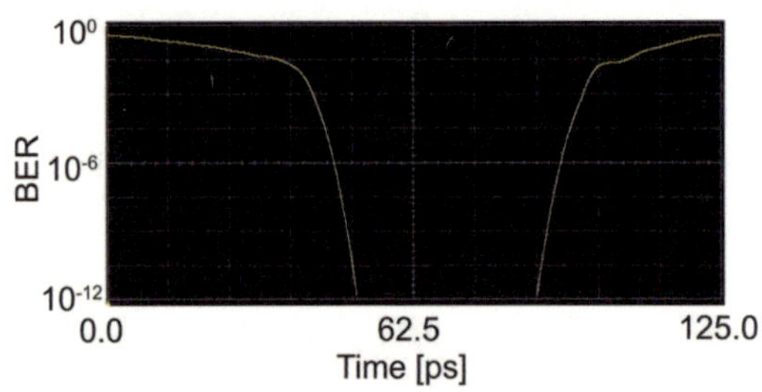

Fig. 17. Bathtub curve at 8 Gb/s.

Process	90 nm standard Si CMOS process
Maximum bit-rate	8.0 Gb/s
Power consumption power supply = 1.0 V, @8 Gb/s	Tx: 1.2 mW Rx: 1.3 mW Total: 2.5 mW
Energy per bit	0.31 pJ/bit
Delay (w/o buffer)	164 ps
Area	Tx: $48 \times 78\,\mu m^2$ Rx: $22 \times 32\,\mu m^2$

Table 1. Performance summary of point-to-point PTLI.

The maximum bit-rate is determined by the eye width, and it is 8 Gb/s. Eye-width margin is assumed to be greater than 20% of the input signal period at a bit error rate (BER) of 10^{-12}. Power consumption without the output buffer is 2.5mW, and the energy per bit is 0.31 pJ/bit. The delay time between "In" and "Rx out" shown in Figure 6(a) is 164 ps. This value is calculated by subtracting the simulated buffer delay from the measured delay between "In" and "out" shown in Figure 6(a).

4.1.2 Multi-drop PTLI

PRBS of length $2^9 - 1$ is also input into IN0, and the output signals from the buffer are measured. Measured eye diagrams at each Rx node and a bathtub curve at OUT5 at 8 Gb/s are shown in Figure 18. The proposed multi-drop PTLI can achieve 8 Gb/s signaling as in the case of the point-to-point PTLI. Since an attenuator is inserted only when performing the measurements for multi-drop PTLI, the jitter characteristics of the falling edge shown in Figures 16 and 18(a) are different. The performance of the multi-drop PTLI is summarized in Table 2. Power consumption of Tx increases slightly because we readjust the bias voltages V_{contp} and V_{contn} shown in Figure 6(b) at the time of measurement. The delay time of the multi-drop PTLI is longer than that of the point-to-point PTLI . This is attributed to the fact that the Txs and Rxs connected to the transmission lines increase the effective capacitance.

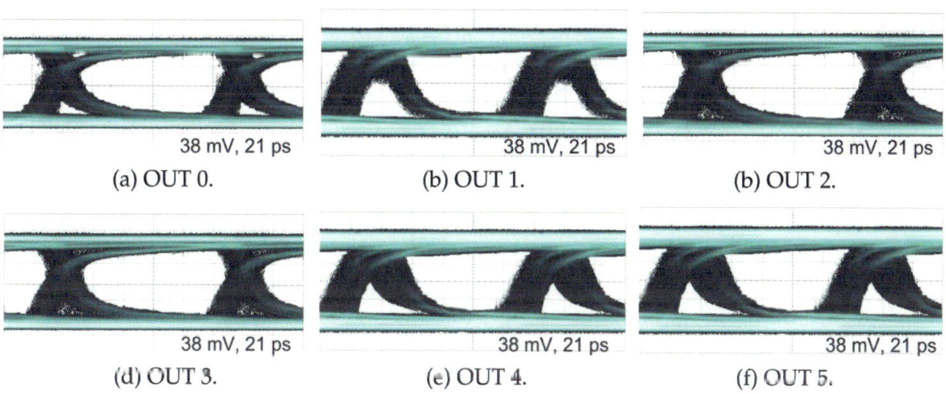

(a) OUT 0. (b) OUT 1. (b) OUT 2.

(d) OUT 3. (e) OUT 4. (f) OUT 5.

(a) Measured eye-patterns at Rx outputs.

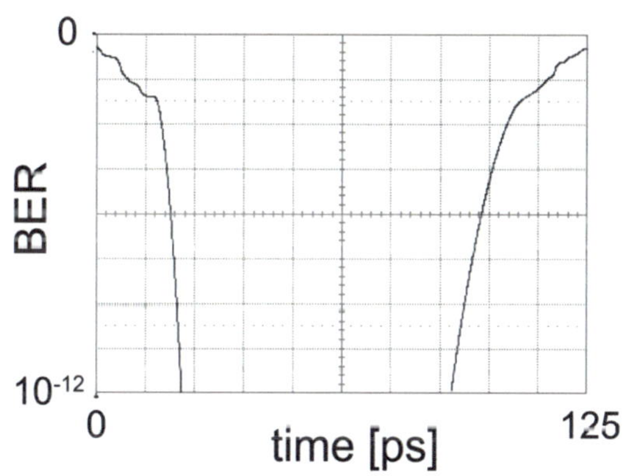

(b) Bathtub curve measured at OUT 5.

Fig. 18. Measured results at 8 Gb/s.

Process	90 nm standard Si CMOS process
Maximum bit-rate	8 Gb/s
Average power consumption (power supply = 1.0 V), @8 Gb/s	Tx: 1.4 mW Rx: 1.3 mW Total (Tx and six Rxs): 9.1 mW
Energy per bit	1.1 pJ/bit
Delay (w/o buffer)	177 ps
Area	Tx: $48 \times 78\,\mu m^2$ Rx: $22 \times 32\,\mu m^2$

Table 2. Performance summary of multi-drop PTLI.

The maximum bit-rates of both PTLIs are lower than those observed in the simulations because of deterministic jitter. The main cause of this jitter is the bandwidth of the output buffer; although the buffer operates up to around 10 Gb/s in measurements, excessive gain of the buffer cause jitter and limit the bandwidth. Buffer characteristics have a negative impact on the measured results, as seen above. Thus, we have to modify the buffer appropriately to achieve high-speed signaling and better signal integrity.

4.2 Discussions

First, let us compare our interconnects with the conventional on-chip transmission line interconnects (TLIs). As has been discussed in Section 1, the CML (current-mode logic) or LVDS (low-voltage differential signaling) amplifiers shown in Figure 19 are commonly used in conventional TLIs as Tx or Rx. The power consumption of the conventional TLIs does not depend on bit-rate, while that of the proposed PTLI reduces as the bit-rate reduces. This enables low-power operation and improves power efficiency, especially at low frequencies, as shown in Figure 20. The number of signal transitions decreases as bit-rate reduces; this is similar to the decrease in the activity factor. Hence, it is expected that as the activity factor decreases, the proposed PTLI has lesser energy per bit than the conventional TLIs. The proposed PTLI would be useful in improving the bandwidth and power efficiency of on-chip networks whose activity factors and bit-rates are changed frequently.

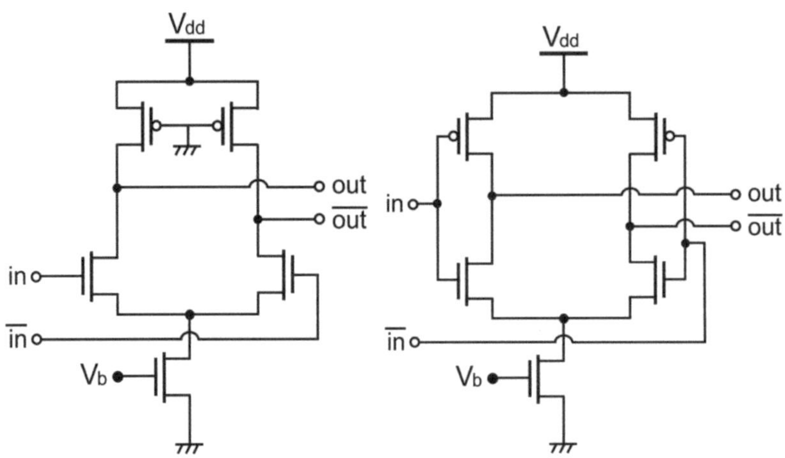

(a) CML-type Tx (b) LVDS-type Tx.

Fig. 19. CML and LVDS amplifiers that are commonly used as Tx or Rx.

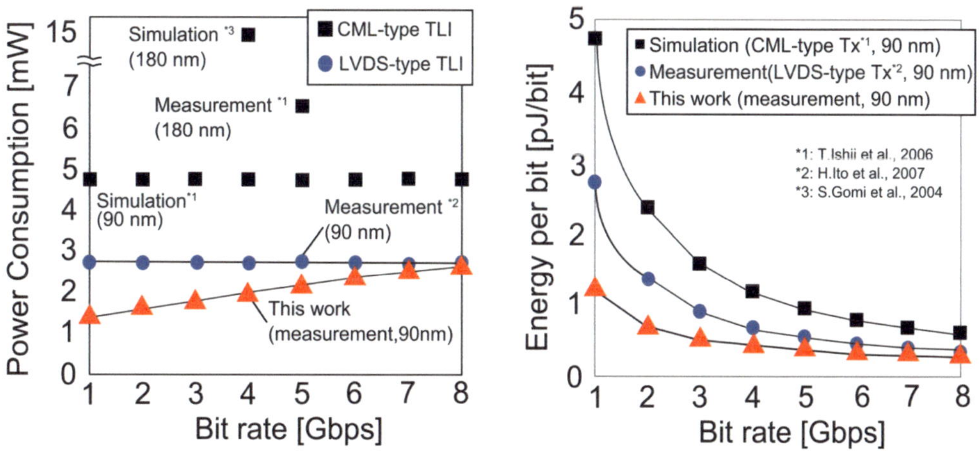

Fig. 20. Bit-rate dependence of power and energy per bit.

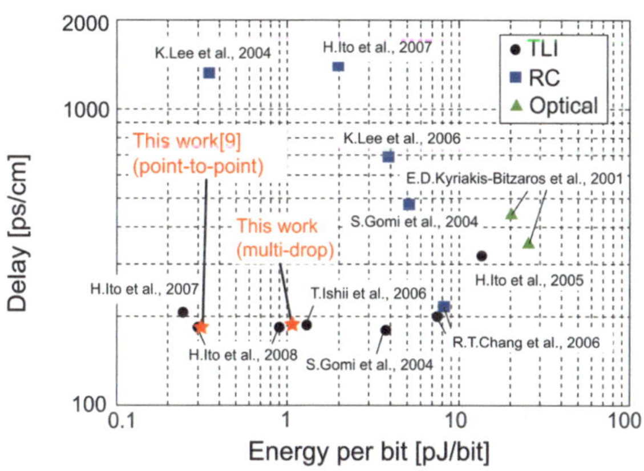

(a) Delay and energy per bit.

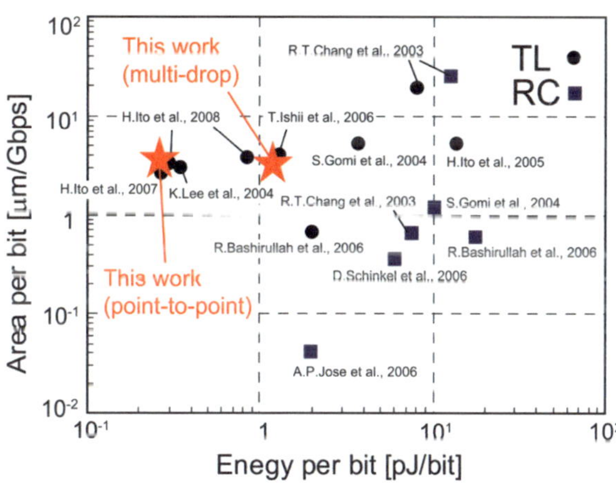

(b) Area and energy per bit.

Fig. 21. Performance comparisons.

Now, comparisons with other interconnects are discussed here. We compare the proposed PTLI with the conventional on-chip interconnects such as RC lines, optical interconnections, and TLIs. Comparison of the delay time and energy per bit are shown in Figure 21(a). Smaller delay and smaller energy per bit mean better performance. In order to compare the delay and power consumption of these interconnects for a uniform line length of 1 cm, we recalculate the values under the following conditions:

- Repeater-less interconnects: The delay time of the interconnect is the sum of that of Tx, on-chip wire, and Rx. A relative permittivity ε_r of 4 is assumed, and the delay of the transmission line is 6.7 ps/mm.
  Power consumption is assumed to be determined by Tx and Rx and does not depend on line lengths.
- Repeater-inserted interconnects:
  Delay and power consumption are proportional to line length.
- Delay of interconnect (Lee et al., 2004):
  Wire latency of 320 ps/5mm is reported in this paper, and delay of 640 ps is used for wiring portion.

The delay time of the proposed PTLIs and the other TLIs are almost equal. This indicates that the delay characteristics of the proposed PTLIs are better than those of other on-chip interconnects. Energy per bit of our interconnects is almost the same as that reported in a previous study (Ito et al., 2008) and is greater than that reported in another study (Ito et al., 2007). The operating frequency in the previous study (Ito et al., 2007) is higher than that of the proposed PTLI. However, the comparison of the power consumption at 8 Gb/s shows that the power consumption of PTLI is smaller than that in (Ito et al., 2007), as shown in Figure 20.

Wiring area is one of the most significant characteristics of on-chip interconnects. Since transmission lines usually occupy large area, on-chip TLIs are required to achieve better area per bit compared to the other on-chip interconnects. Energy per bit and area per bit of the on-chip high-speed interconnects are shown in Figure 21(b). Small energy per bit and small area per bit indicate good performance. The area occupied by the differential transmission line is assumed to be (line width) × 2 + (space between lines) × 3 and that occupied by single-ended RC line is (line width)×3. On-chip interconnects using RC lines have better characteristics in terms of area per bit than TLIs; this is because of the fine line width of the RC lines. However, the maximum bit-rate of RC interconnects is lesser than that of TLIs. Maximum bit-rate comparable to that of TLIs can be achieved by using RC lines in a bundle; however, this would imply that the RC lines will occupy virtually the same area as the TLIs. The area per bit of the proposed point-to-point PTLI is almost equal to that of the other point-to-point TLIs (Gomi et al., 2004; Ito et al., 2005; Ishii et al., 2006; Ito et al., 2007, 2008; Lee et al., 2004). The area per bit of the proposed multi-drop PTLI is the equal to the bidirectional and multi-drop transmission line interconnects (Ito et al., 2008). Thus, the proposed PTLIs are comparable to other on-chip TLIs in view of area and bit-rate.

Finally, we shall discuss the capability of the proposed PTLI. Although the measured maximum bit-rate is 8 Gb/s, this limitation may be overcome by the proposed PTLI if better buffers are used for measurement. In order to simplify the discussion, we do not consider buffers in the following.

Figure 22 shows the simulation result of the dependence of eye-width margin on signal bit-rate at a bit error rate of 10^{-12}. The eye diagram is observed at "Rx out" in Figure 6(a). The pulse width of the proposed PTLI is optimized to have the minimum pulse-width for each

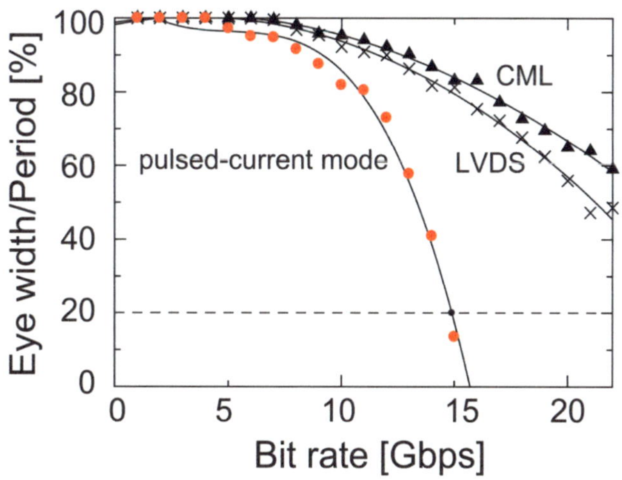

Fig. 22. Bit-rate dependences of eye-width margin.

signal frequency by controlling the delay circuits in this simulation. The differential amplifiers used in CML and LVDS interconnects have same topologies, as shown in Figure 6(c).

The eye-width margin of the PTLI decreases sharply beyond 12 Gb/s. One possible reason is the parasitic capacitance of Tx. The use of CMOS stacked-switches leads to an increase in the parasitic capacitance, and this increases the signal rise time. Another possible reason is the limitation of variable pulse width regulated by the delay circuits; in fact, this may be a dominant factor. The delay circuits shown in Figure 6(b) consist of CMOS inverters and NMOS/PMOS switches. Thus, the power consumption of the delay circuits is relatively-small at the cost of variable pulse-width. In order to achieve signaling at speeds greater than 15 Gb/s, the delay circuits used in the proposed PTLI may need further improvement.

5. Conclusion

This paper proposed highly energy-efficient on-chip PTLIs with stacked-switch Txs. The 5 mm-long point-to-point and multi-drop PTLIs were fabricated by 90nm Si CMOS process. The point-to-point PTLI achieved 8 Gb/s signaling with a power consumption of 2.5mW and a delay of 164 ps. The multi-drop PLTI with six I/Os also achieved 8 Gb/s signaling with a power consumption of 9.1mW. Our interconnects had superior power efficiency compared to the conventional on-chip high-speed interconnects at low bit-rate signaling and low activity factors as well as at high bit rate. The proposed PTLIs also had good area characteristics compared to the on-chip RC line interconnects. These facts indicate that our PTLIs enable the designing of multipoint-to-multipoint on-chip networks and would improve the extensibility of on-chip networks.

6. Acknowledgments

This work was partially supported by KAKENHI, MIC.SCOPE, STARC, Special Coordination Funds for Promoting Science and Technology, and VDEC in collaboration with Agilent Technologies Japan, Ltd., Cadence Design Systems, Inc., and Mentor Graphics, Inc.

7. References

Bakoglu, H. B., *Circuits, Interconnections, and Packaging for VLSI*, AddisonWesley, 1990.

Bashirullah, R.,Wentai, L., Cavin, R., and Edwards, D. (2004). A 16 Gb/s adaptive bandwidth on-chip bus based on hybrid current/voltage mode signaling, *IEEE Symposium on VLSI Circuits, Digest of Technical Papers*, pp. 392-393.

Chang, R. T., Talwalkar, N., Yue, C. P., and Wong, S. S. (2003). Near speed-of-light signaling over on-chip electrical interconnects, *IEEE Journal of Solid-State Circuits*, vol. 38, no. 5, pp. 834–838.

Gomi, S., Nakamura, K., Ito, H., Okada, K., and Masu, K. (2004). Differential transmission line interconnect for high speed and low power global wiring, *IEEE Custom Integrated Circuits Conference*, pp. 325–328.

Ishii, T., Ito, H., Kimura, M., Okada, K., and Masu, K. (2006). A 6.5-mW 5-Gbps on-chip differential transmission line interconnect with a low-latency asymmetric Tx in a 180nm CMOS technology, *IEEE Asian Solid-State Circuits Conference*, pp. 131–134.

Ismail, Y. I., Friedman, E. G., and Neves, J. L. (1998). Figures of merit to characterize the importance of on-chip inductance, *ACM/IEEE Design Automation Conference*, pp. 560- 565.

Ito, H., Inoue, J., Gomi, S., Sugita,H., Okada, K., and Masu, K. (2004). On-chip transmission line for long global interconnects, *IEEE International Electron Devices Meeting*, pp. 677–680.

Ito, H., Sugita, H., Okada, K. and Masu, K. (2005). 4 Gbps on-chip interconnection using differential transmission line, *IEEE Asian Solid-State Circuits Conference*, pp. 417–420.

Ito, H., Seita, J., Ishii, T., Sugita, H., Okada, K., and Masu, K. (2007). A low-latency and high-power-efficient on-chip LVDS transmission line interconnect for a RC interconnect alternative, *IEEE International Interconnect Technology Conference*, pp. 193-195.

Ito, H., Kimura, M., Miyashita, K., Ishii, T., Okada, K., and Masu, K. (2008). A bidirectional- and multi-drop-transmission-line interconnect for multipoint-to-multipoint on-chip communications, *IEEE Journal of Solid-State Circuits*, vol. 43, no. 4. pp. 1020–1029.

Jose, A. P., Patounakis G., and Shepard, K. L. (2006). Pulsed current-mode signaling for nearly speed-of-light intrachip communication, *IEEE Journal of Solid-State Circuits*, vol. 41, no. 4, pp. 772–780.

Kyriakis-Bitzaros, E. D., Haralabidis, N., Lagada,M., Georgakilas, A., Moisiadis, Y., and Halkias, G. (2001). Realistic end-to-end simulation of the optoelectronic links and comparison with the electrical interconnections for system-on-chip applications, *Journal of Lightwave Technology*, vol. 19, no. 10, pp. 1532-1542.

Lee, K., Lee, S., Kim, S., Choi, H., Kim, S., Lee, M. W., and Yoo, H.-J. (2004). A 51mW 1.6GHz on-chip network for low-power heterogeneous SoC platform, *IEEE International Solid-State Circuits Conference Digest of Technical Papers*, 2004, pp. 152-153.

Lee, K., Lee, S., and Yoo, H.-J. (2006). Low-power network-on-chip for high performance SoC design, *IEEE Transaction on Very Large Scale Integrated Systems*, vol. 14, no. 2, pp. 148-160

Schinkel, D., Mensink, E., Klumperink, A. M., Tuijl van, Ed (A.J.M.), and Nauta, B. (2006). A 3-Gb/s/ch transceiver for 10-mm uninterrupted RC-limited global on-chip Interconnects, *IEEE Journal of Solid-State Circuits*, vol. 41, no. 1, pp. 297-306

15

An Inductive-Coupling Inter-Chip Link for High-Performance and Low-Power 3D System Integration

Kiichi Niitsu and Tadahiro Kuroda
Keio University
Japan

1. Introduction

Three-dimensional (3D) system integration is one of the promising candidates for the next-generation high-performance and low-power LSI systems. In 3D system integration, we can implement analog and digital circuits in LSI chips in their optimal process and they are stacked and connected through vertical inter-chip link. Development of wide-band and low-power inter-chip link is the key factor to realize high-performance 3D system integration.

One of the most attractive applications of 3D system integration is processor-memory interface since memory capacity and bandwidth is a bottleneck of a processor system. Integrating large size memory on a processor increases die size (SRAM) or process steps (eDRAM), either way, raising cost and leakage. It is desired in low-power consumer electronics that a memory chip and a processor chip are each fabricated in their optimal process and integrated by heterogeneous chip stacking in a package. One of the technical challenges is a wide bandwidth between the processor and the memory. The gap between computing power and communication bandwidth can be filled if chip area is used for a data link rather than chip periphery only. A Micro-bump and a capacitive-coupling link (Fazzi et al., 2008) are area interfaces, but they can be used only for two chips that are placed face-to-face. A Through Silicon Via (TSV) (Koyanagi et al., 2009) has fewer limitations, but it requires additional process steps and production equipment. An inductive-coupling link (Miura et al., 2007) is used as a wireless TSV, but with small impact on cost. It is a circuit solution on a standard CMOS process, and hence is less expensive than TSV. It bears comparison with TSV in performance. The data rate is 11Gb/s/channel (Miura et al., 2009) and power efficiency is 65fJ/b (Niitsu et al., 2008). 1Tb/s aggregated bandwidth is achieved by arranging 1000 channels in 1mm^2 in 0.18μm CMOS, and BER is lower than 10^{-14} (Miura et al., 2007). Furthermore, it provides an AC-coupled interface, and therefore a level shifter is not needed. Power supply voltages can be different, and they can be changed for dynamic voltage scaling (DVS) and power gating with little impact on interface delay. An ESD protection device is not needed, either.

Figure 1 shows the concept of 3D system integration using an inductive-coupling link. Processor chips, memory chips and analog and RF front-end chips are implemented in each optimized process and stacked. In addition to data and clock, wireless inductive-coupling power delivery with high-frequency was demonstrated (Onizuka et al (2006)). By utilizing wireless inter-chip power delivery, we can omit conventional wire-bondings for power supply and achieve further cost reduction.

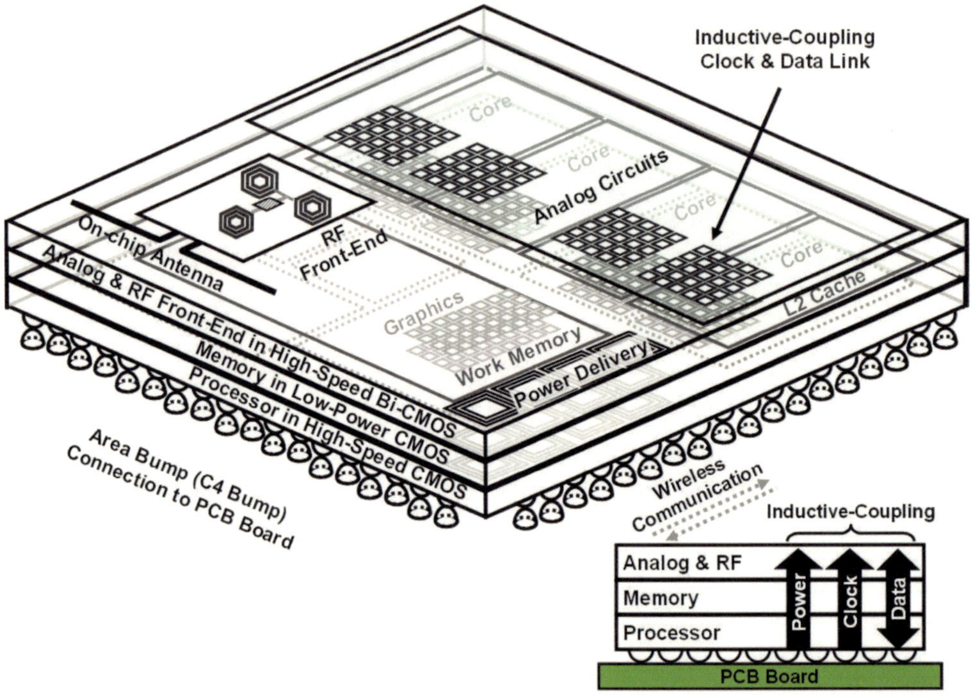

Fig. 1. 3D system integration using an inductive-coupling link.

This chapter is organized as follows. Section 2 introduces transceiver design of an inductive-coupling link. Section 3 reports the interference between an inductive-coupling link and other circuits. In Section 4, the modelling and experimental verification of tolerance to misalignment between stacked chips are introduced. In Section 5, application of an inductive-coupling link to processor-memory interface is shown. Section 6 concludes the chapter.

2. Inductive-coupling inter-chip link

Figure 2 shows the transceiver circuits for inductive-coupling inter-chip link. Bi-phase modulation is employed. Transmitter circuit consists of an H-bridge circuit which generates positive or negative pulse current, IT according to transmit data, Txdata. In every clock cycle, positive pulse is generated when Txdata is high, and negative pulse is generated when Txdata is low. In the receiver circuit, positive or negative pulse voltage, V_R that corresponds to the polarity of I_T is induced in the receiver inductor. Receiver circuit samples V_R, and then it recovers a binary receiver data, Rxdata. Since the complementary type latch is used as a sense amplifier, the receiver consumes power only at clock rising edge. Transmitter consumes power by I_T generation. Because of the weak coupling between transmitter inductors and receiver inductors, large pulse current I_T is necessary for generating enough V_R. Therefore, transmitter consumes higher power than receiver.

In our previous work (Miura et al., 2007), the transceiver consumes 3W at 1Tb/s. 80% of total power is consumed in transmitter. The power reduction of the transmitter is more critical in reducing the total power.
As explained above, an inductive-coupling link generates magnetic flux, which causes interference from/to other circuits. In Section 3, the investigation of this interference will be provided.
Besides, in order to achieve low-power operation, synchronous scheme is utilized in an inductive-coupling link. In Section 4, timing adjustment scheme is proposed and applied to processor-memory interface.

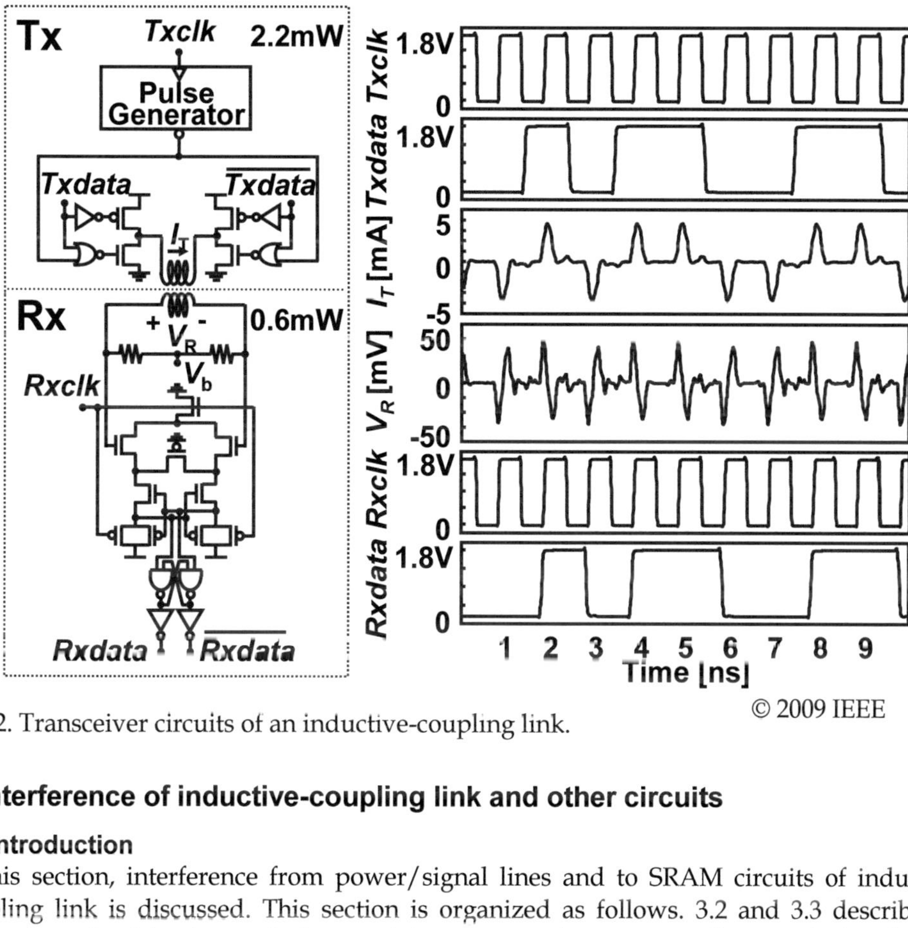

Fig. 2. Transceiver circuits of an inductive-coupling link.

3. Interference of inductive-coupling link and other circuits

3.1 Introduction

In this section, interference from power/signal lines and to SRAM circuits of inductive-coupling link is discussed. This section is organized as follows. 3.2 and 3.3 describe the analyses and mitigation techniques of interference from power lines and signal lines, respectively. 3.4 describes the analysis and mitigation technique of interference to SRAM circuits.

3.2 Interference from power lines to an inductive-coupling link

In state of the art LSI chips, the occupied area of power lines is increasing. However, power line degrades the performance of inductive-coupling link since eddy current in power line reduces magnetic flux as shown in Figure 3.

The shape of power line decides how much coupling degradation occurs. In this study, the dependence on the influence from the shape of power lines is compared. As the common power line, three types of power lines are investigated. Mesh type is employed in high-performance LSI chips such as microprocessor. The line and space type is employed for mobile application. The line and space type is classified into two types, without side line (type I) and with side line (type II). Type I has the loop of metal wire, while type II does not. Figure 4 shows the simulated trans impedance (the ratio of received voltage to the transmit current) at 1 GHz dependence on the metal density filled by the power lines.

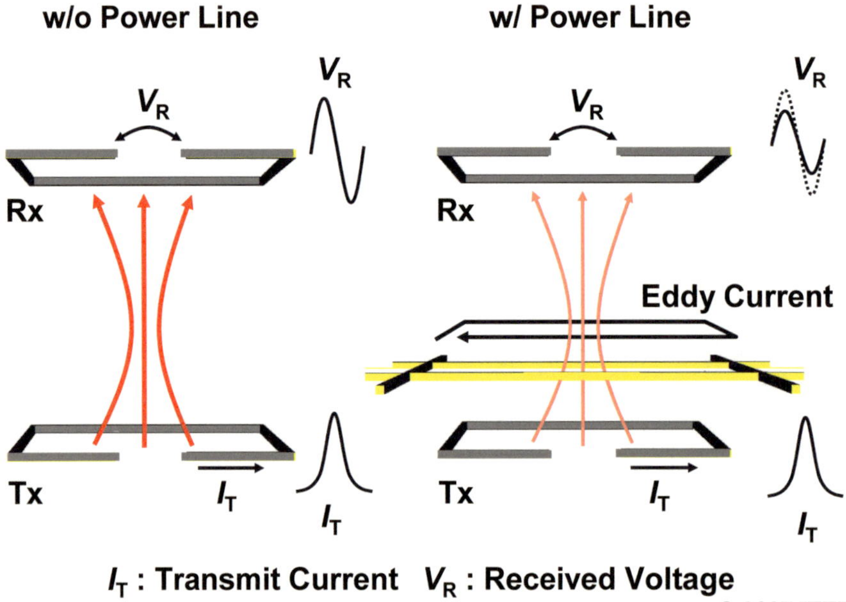

Fig. 3. Coupling degradation by eddy current in power lines.

For this simulation, three-dimensional electromagnetic solver is employed. The thickness of metal layer and the gap of metal layer are set to 0.5 μm.

From simulation results in Fig. 4, it can be seen that the mesh type of power line (for high-performance LSI) affects the performance of inductive-coupling link more significantly than others. The line and space (I) does not reduce the transimpedance since there is no loop of metal wire and hence eddy current does not flow. The line and space (II) affects the inductive-coupling link more seriously than line and space(I). The influence by Line and space (II) does not change between 20% and 50%. The reason is that long side of metal wire is dominant when power line diminishes magnetic flux.

In order to measure the interference from power lines, test chips were designed and fabricated in 65nm CMOS. Figure 5 depicts microphotograph of stacked chips fabricated in this research. The area of this test chip is 3.5 mm * 2.5 mm and 2.5 mm * 1.9 mm. It consists of 20 types of transceivers with different configurations. The transmitters have an on-chip metal inductor whose outer diameter is 160 μm and 80 μm. As shown in Fig. 5, test chips are stacked face to back (both face-up) and communication distance is 70 μm. The upper chip was polished and its thickness is 50 μm. The thickness of adhesive layer is 20 μm. Probe card was utilized for this measurement.

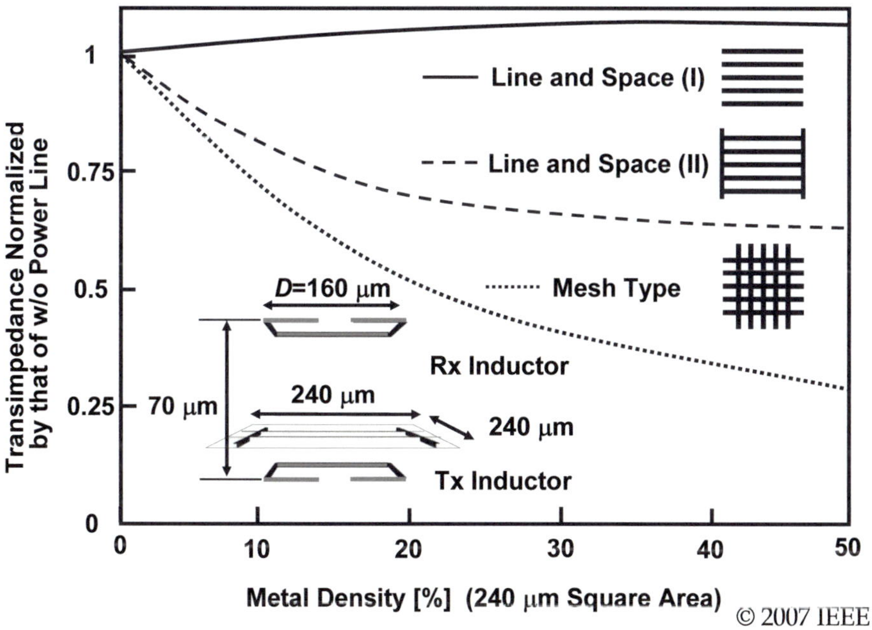

Fig. 4. Simulated transimpedance dependence on type of power line.

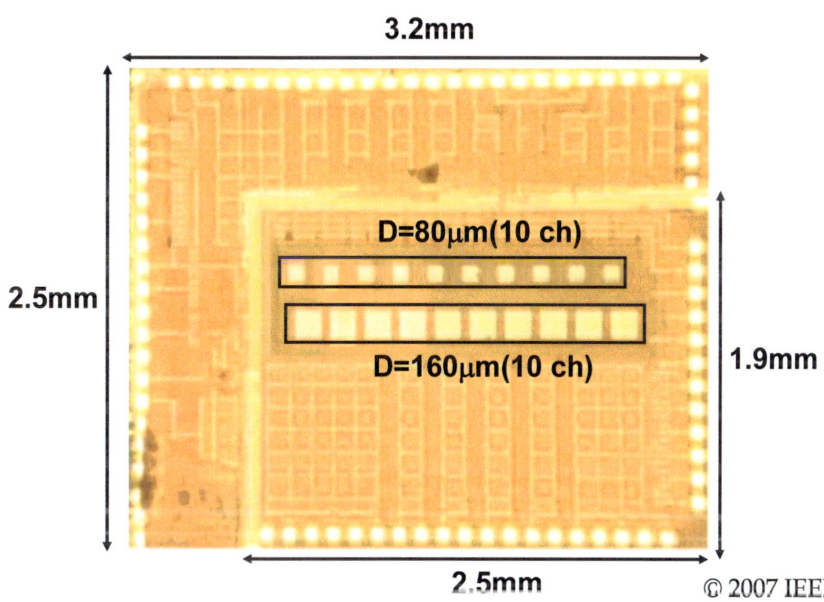

Fig. 5. Stacked chip microphotograph.

Figure 6 shows measured transmit power dependence on metal density. Transmit power is measured when achieved bit error rate (BER) is same. In Fig. 6, vertical axis is normalized by transmit power when there are not power lines but dummy metals above the inductors. Measured result matches well with simulation result in the case of line and space (I) and (II).

However, measured result of mesh type of power line is less than simulation result. The accuracy may improve by taking eddy current in substrate into consideration.

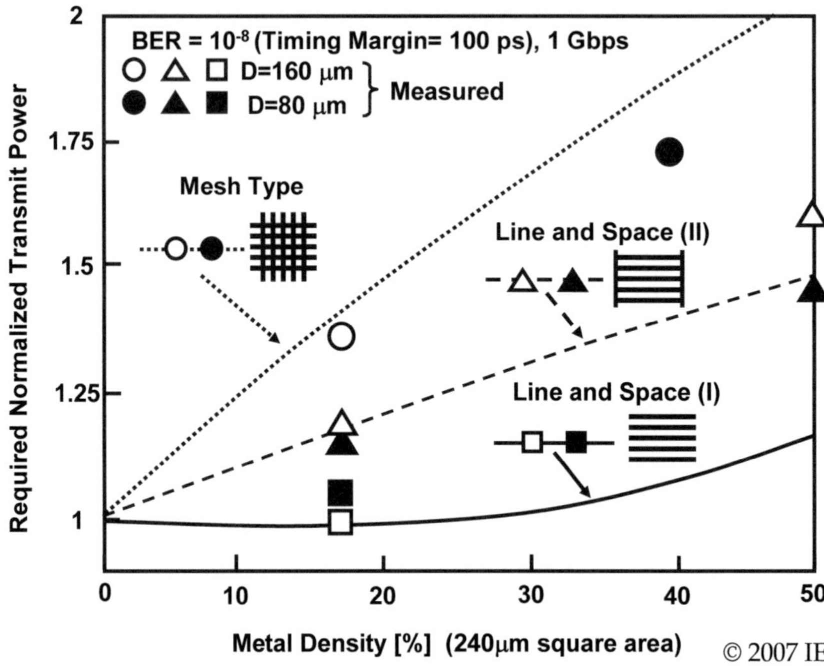

Fig. 6. Measured transmit power dependence on types of power line.

3.2 Interference from signal lines to an inductive-coupling link

Immunity to interference from high-speed signal lines is a very important issue in the implementation of the inductive-coupling link with recent LSI chips. Especially, receiver circuits may not have high immunity to interference from high-speed signal line since receiver circuits sample very small signal. In this work, we implemented high-speed signal lines near transceiver inductors of the inductive-coupling link. With this implemented module, influence to the operation of inductive-coupling link when high-speed signal lines drive large capacitance is measured.

For the purpose of measuring the influence from signal lines, we implemented signal line under the inductive-coupling link. In previous work, mutual inductance between signal line and inductor dependence on the position is simulated. From simulation result, mutual inductance is maximized when the signal line is allocated under the center between the lines of inductor. In this study, 3 mm length signal line is implemented under the transceiver inductors. Near the inductors, buffer is implemented to drive signal line with 3 mA peak-to-peak.

Fig. 7 depicts measured BER dependence on the timing between driving signal line and sensing transmit current in the inductive-coupling link. Influence on the transmitter inductor is smaller compared with that on the receiver inductor. The disadvantage of placing high-speed signal line near the inductive-coupling link is as small as 9% additional transmitter power consumption.

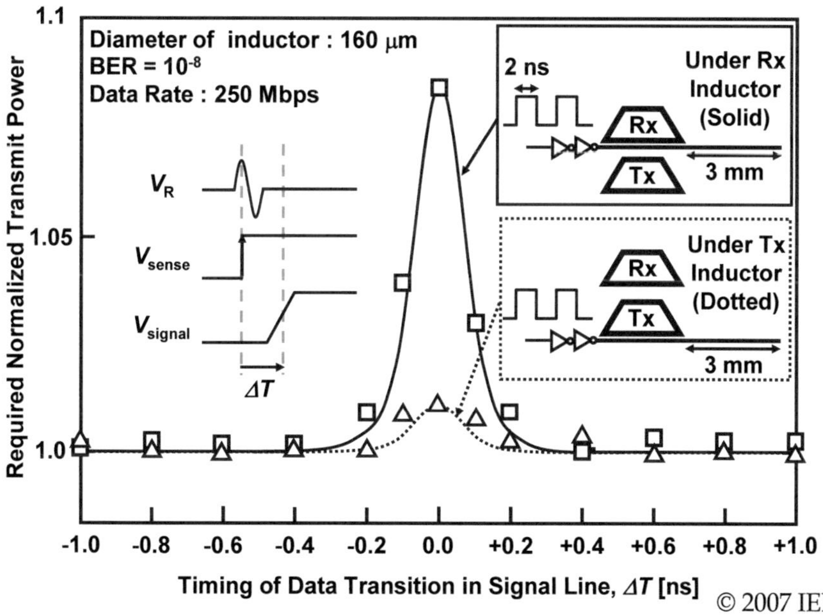

Fig. 7. Measured required transmit power to achieve communication performance dependence on the timing of data signal.

Since power of an inductive-coupling link has become lower than other interfaces by developing low-power techniques such as in the previous work (Niitsu et al., 2008), 9% additional transmitter power can be neglected. In implementing inductive-coupling link near the logic circuits, precise care for timing between them is not necessary.

3.2 Interference from an inductive-coupling link to SRAM array operation

In order to develop high-performance LSI system, large-size of SRAM is necessary. Recently, the proportion of SRAM area to whole chip size is increasing rapidly (Hattori et al., 2006 & Ito et al., 2007). However, large on-chip SRAM causes yield degradation and increase of leakage power. As a solution of this problem, SRAM will be implemented in another chip, and three-dimensionally stacked. Inductive-coupling link will be utilized as an interface between SRAM and processor core. In this situation, magnetic flux from inductive-coupling link will be very important issue from the view point of reliable SRAM operation. In this study, electromagnetic interference on SRAM was measured and investigated.

At first, we estimated interference to SRAM circuits from inductive-coupling link in the case of 32 Kbit modules. Figure 8 illustrates the simple model of the inductive-coupling link and SRAM module.

The scattering parameter between transmitter inductor and bit line is extracted with three-dimensional electromagnetic solver. Figure 9 shows the bit-line noise induced by transmitter of inductive-coupling link. As shown in this waveform, the voltage of bit-line noise from inductive-coupling link is less than 1mV. In SRAM circuits such as (Yamaoka et al., 2005), even small voltage affects the performance such as operation speed and power dissipation. However, the sensing voltage is almost 50 mV, and the bit-line noise from inductive-coupling link is less than 1 mV. The bit-line noise from inductive-coupling link is very small compared with sensing voltage.

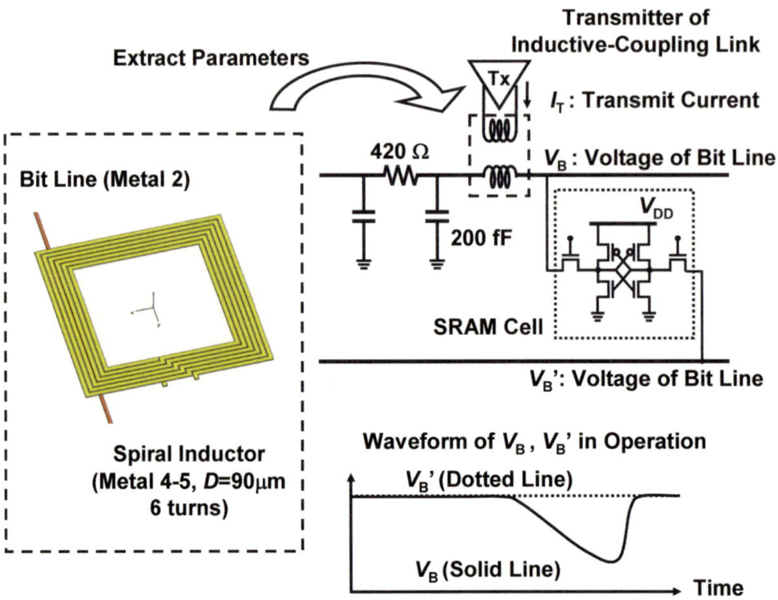

Fig. 8. Simple model of inductive-coupling link and SRAM circuits.

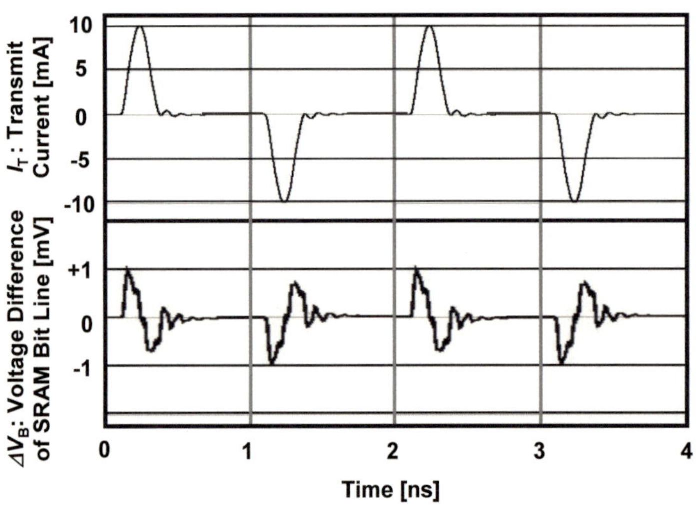

Fig. 9. Simulated waveform of transmit current and voltage induced in SRAM bit line.

For the measurement to investigate influence to SRAM, another test chip was fabricated. This test chip was also fabricated in 65 nm CMOS. In this test chip, inductor with transmitter circuit was implemented above the SRAM arrays. SRAM circuits were allocated as Fig. 10 for influence on the bit line to be maximized.

Figure 11 depicts the measured waveform of output voltage. In this measurement, SRAM module repeated read and write toggle data pattern. As shown in Fig. 11, error occurs only when the inductive-coupling link generates transmit pulse current.

Figure 12 shows measured error rate in read operation of SRAM dependence on supply voltage. The difference of minimum supply voltage to maintain operational performance is only 10 mV when supply voltage is much lower than typical range. In typical region of SRAM operation, there is no difference between with the inductive-coupling link and without it. It is clear that influence on SRAM from inductive-coupling link is negligible. Influence from the inductive-coupling link is less serious than that from soft errors. That is why the inductive-coupling link does not affect SRAM operation in typical region of supply voltage while soft errors may affect. Compared with influence from device variations, it is much smaller since the difference in supply voltage of 10 mV corresponds to the difference in threshold voltage variation of 1 mV (Yamaoka et al., 2004), which is much smaller than process variation. From this measurement result, we have reached to a conclusion that inductive-coupling link can be placed near the SRAM circuits.

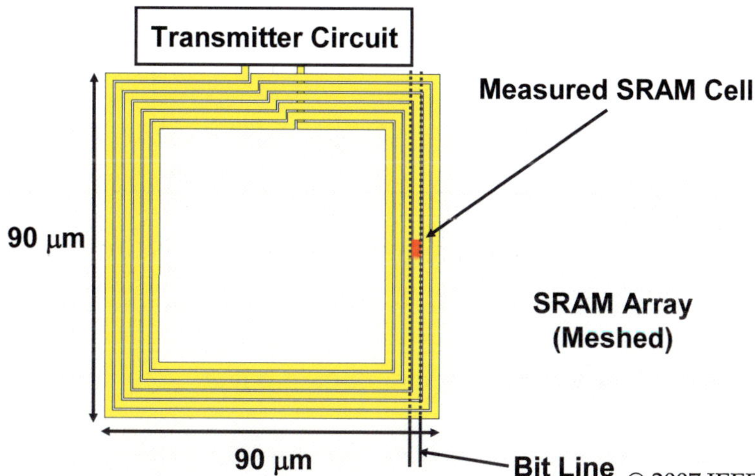

Fig. 10. Test element group to measure the influence to SRAM circuits.

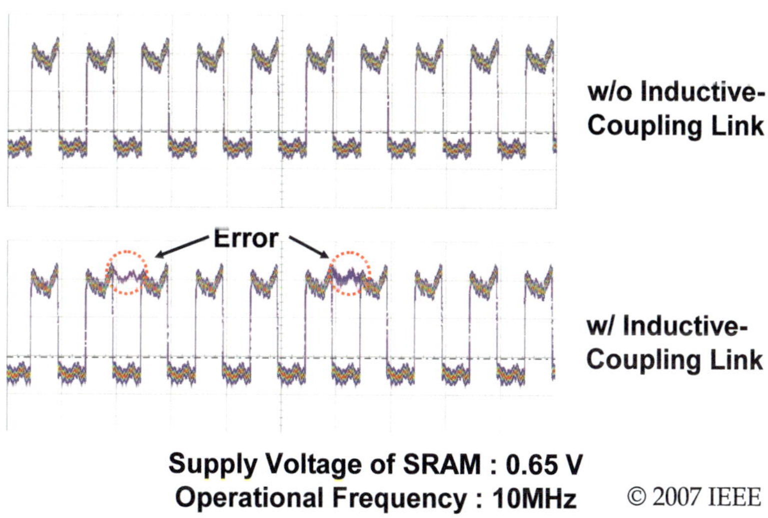

Fig. 11. Measured waveform.

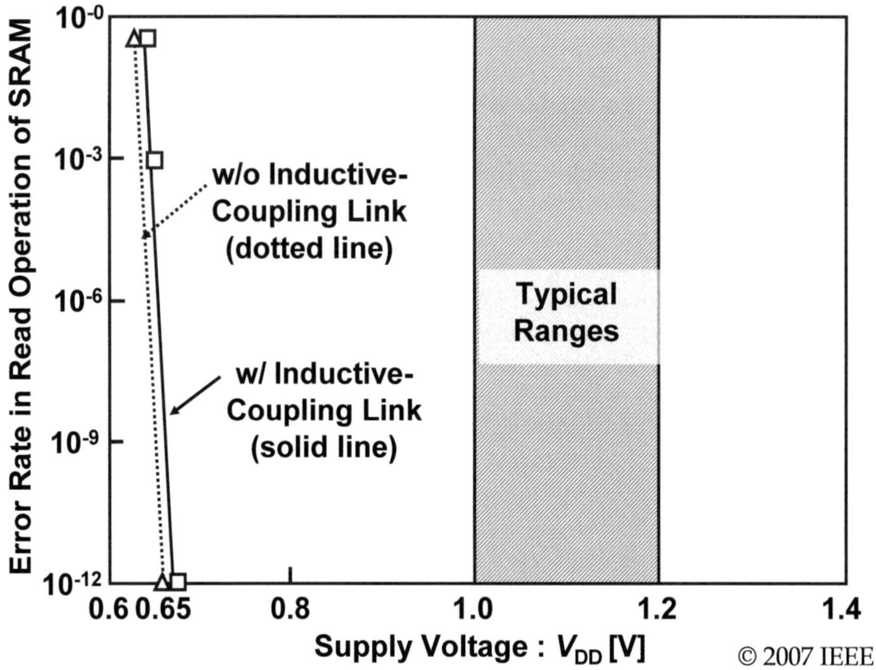

Fig. 12. Error rate in read operation dependence on supply voltage of SRAM circuits.

4. Misalignment tolerance of inductive-coupling link

4.1 Introduction

This section introduces modelling and investigation of misalignment tolerance of an inductive-coupling link. Figure 13 shows the conceptual image of increase in transmitter power due to misalignment between stacked chips. Because of misalignment, magnetic flux generated by the transmitter inductor can not be transferred to the receiver inductor. As a result, received voltage is attenuated. To keep received voltage constant under the misalignment, transmitter current must be increased and it causes increase in transmitter power. We proposed a model for estimating increased transmitter power due to misalignment.

4.2 Modeling of attenuation of received voltage

For the purpose of simplifying the analysis, self inductances of transmitter and receiver inductors are kept constant, same as that in magnetic field scaling (Mizoguchi et al., 2007?). This is achieved by adjusting the number of turns depending on the inductor's diameter. Pulse width is also kept constant for the timing margin to be constant. Under those conditions, received voltage is proportional to the coupling coefficient only (Finkenzeller, 2003).

Coupling coefficient, which is determined by communication distance and diameter of transceiver inductors, is reduced by misalignment since it increases communication distance equivalently as shown in Fig. 14. In order to compensate this reduction, transmitter power (here after, energy, which is normalized by data rate, for example, 1mW/1Gbps=1pJ/b) should be increased. Normalized required transmitter energy (E'/E) can be approximated as below.

$$E'/E = \frac{1}{\left\{1+(2Z/D)^2\right\}^{\frac{3}{2}}} \Bigg/ \frac{1}{\left\{1+(2Z'/D)^2\right\}^{\frac{3}{2}}}$$
$$= \left\{\frac{D^2+4(Z^2+\Delta X^2+\Delta Y^2)}{D^2+4Z^2}\right\}^{\frac{3}{2}} \quad \because Z' = \sqrt{Z^2+\Delta X^2+\Delta Y^2} \tag{1}$$

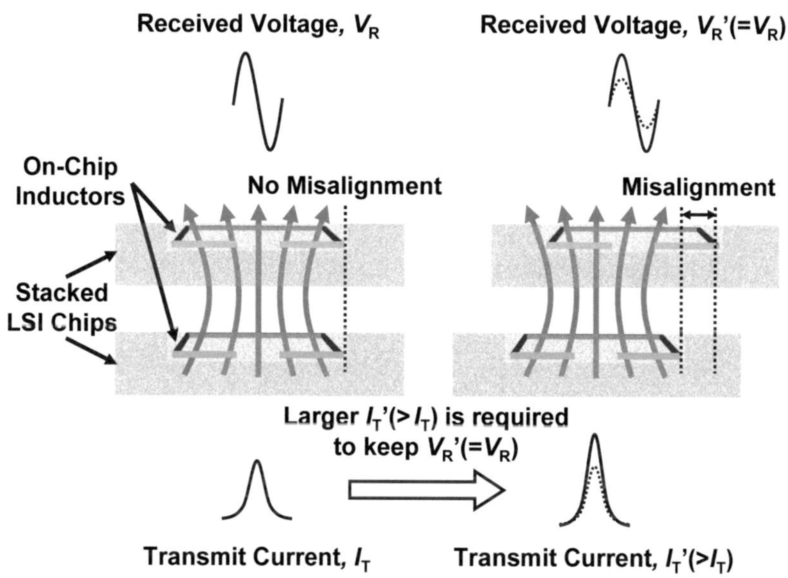

Fig. 13. Concept of increase of transmitter power due to chip-to-chip misalignment.

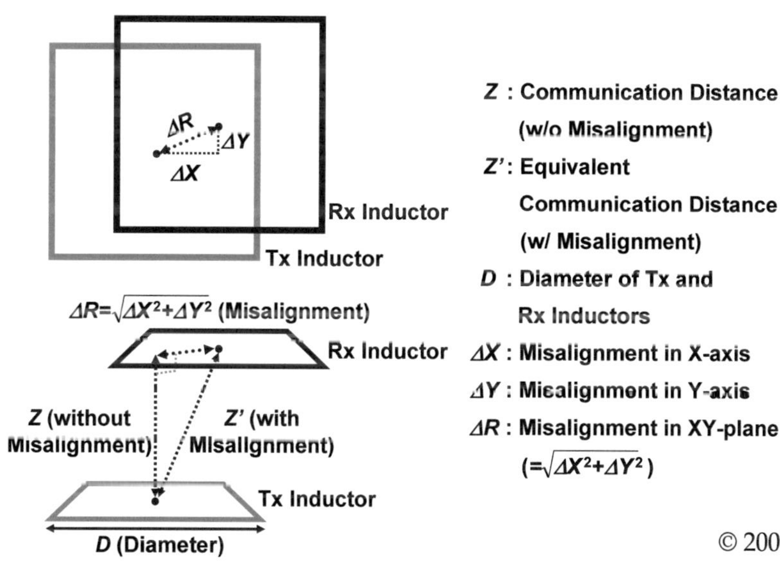

Fig. 14. Increase of communication distance due to misalignment.

Where, E′ and E are the transmitter energies in case of with and without misalignment, respectively. Z′ and Z are the equivalent communication distances with and without misalignment, respectively. D is the average between outer and inner diameter of inductors. ΔX and ΔY are the values of misalignment in X-axis and Y-axis, respectively.
Figure 15 shows the total transmitter energy dependence on the angle of the inductor where the misalignment value, ΔR is constant. The diameter and communication distance are 80μm and 70μm. respectively. As shown in this figure, the difference of transmitter energy for all angles is less than 5%. This result shows that proposed modeling can be applied to not only 1D analysis but also 2D analysis.

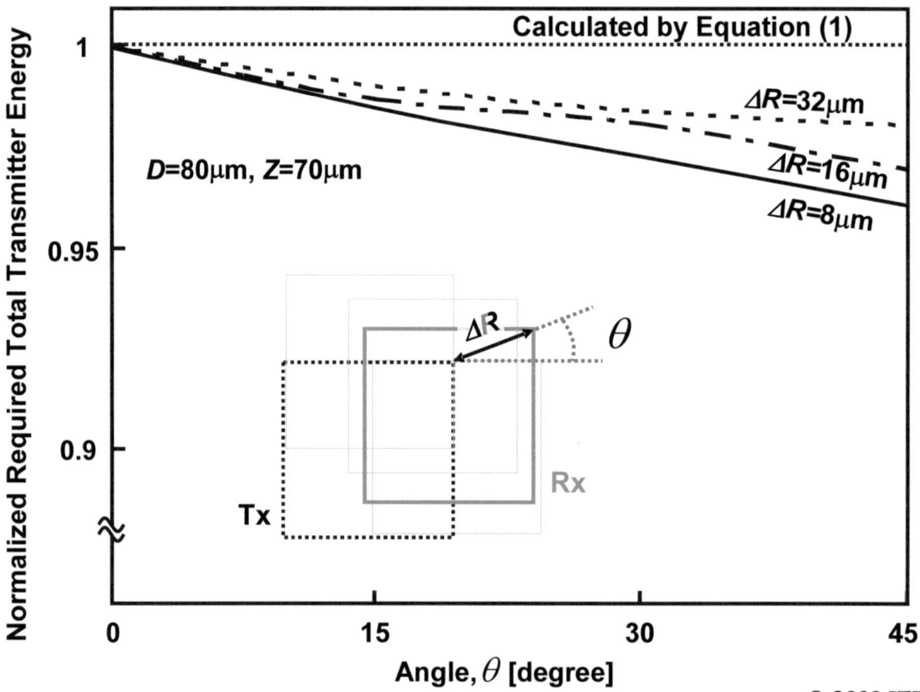

Fig. 15. Normalized total transmitter energy dependence the position of the inductor.

4.2 Estimation of transmitter energy under misalignment

From the above theoretical analysis, we can calculate the relationship between design parameters and misalignment, which is shown in Fig. 16. By referring to this figure, parameter design with taking misalignment into consideration becomes possible. In order to determine the specific value of transmitter energy, we targeted the BER and timing margin. However, the proposed model can be applied to any BER and timing margin by scaling the transmitter energy calculated by (1). The reason is that misalignment affects only coupling coefficiency and the relationship between BER, timing margin, transmitter energy and coupling coefficient is introduced in (Miura et al., 2007). In Fig. 16, the region where (1) is valid will be explained in the following discussion. As shown in Fig. 16, there are points where magnetic filed lines change the vertical direction. If the directions of all magnetic field lines in the receiver inductor are same, (1) is valid. Such points were calculated from the simulation by 3D electro-magnetic (EM) solver and plotted in Fig. 16. When $Z/\Delta X$ is more than approximately 0.8, (1) gives accurate value and its accuracy is confirmed by comparing with simulation results by EM solver and measurement results in the following sections.

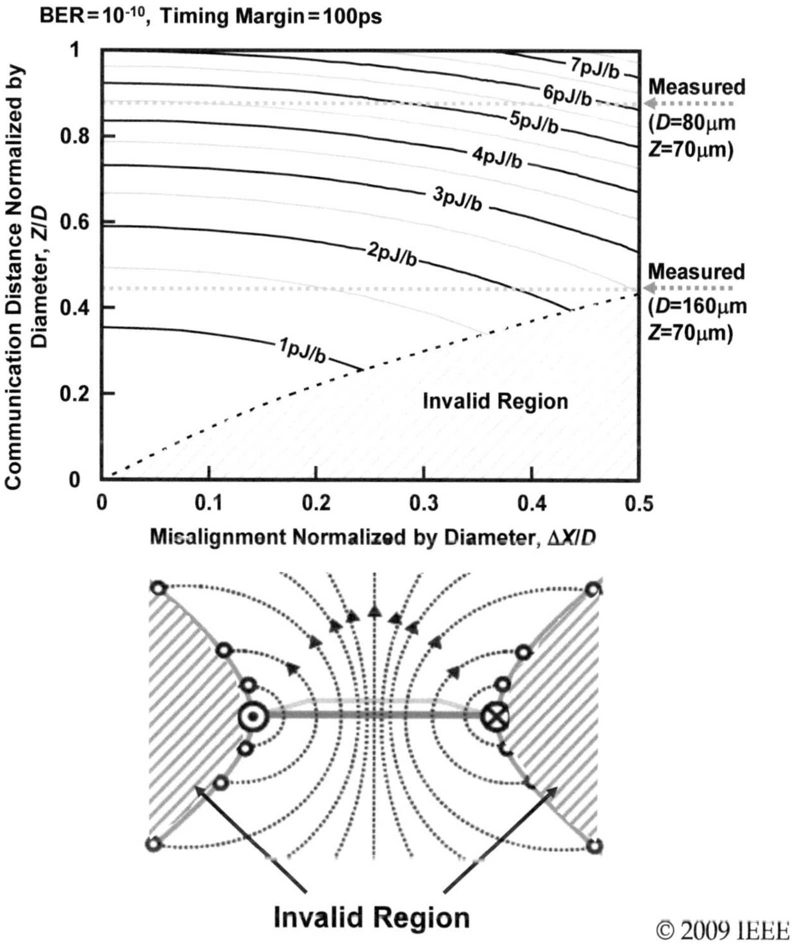

Fig. 16. Relationship among energy dissipation, normalized misalignment and communication distance.

4.3 Estimation of transmitter energy with consideration of crosstalk

Misalignment also affects the performance in array operation. In arrayed inductive-coupling link, bit error rate is given by the following equation (Miura et al., 2007).

$$\mathrm{BER} = \frac{1}{2}\mathrm{erfc}\left(\frac{\tau}{4\sqrt{2}\tau_{jrms}}\sqrt{\ln\frac{S-C-N}{N}}\right) \tag{2}$$

Note that erfc() is the error faction complement, τ is the pulse width of transmitter current, $\tau_{j,rms}$ is rms jitter of sampling clock in receiver, S is signal, N is ambient noise and C is crosstalk.

As in (2), in order to keep the same BER, the difference of signal(S) and crosstalk(C), has to be maintained. The value of ambient noise, N, is constant in both cases with and without misalignment. Since signal is attenuated and crosstalk is increased due to misalignment (Fig. 17), transmitter energy needs to be increased to maintain that difference.

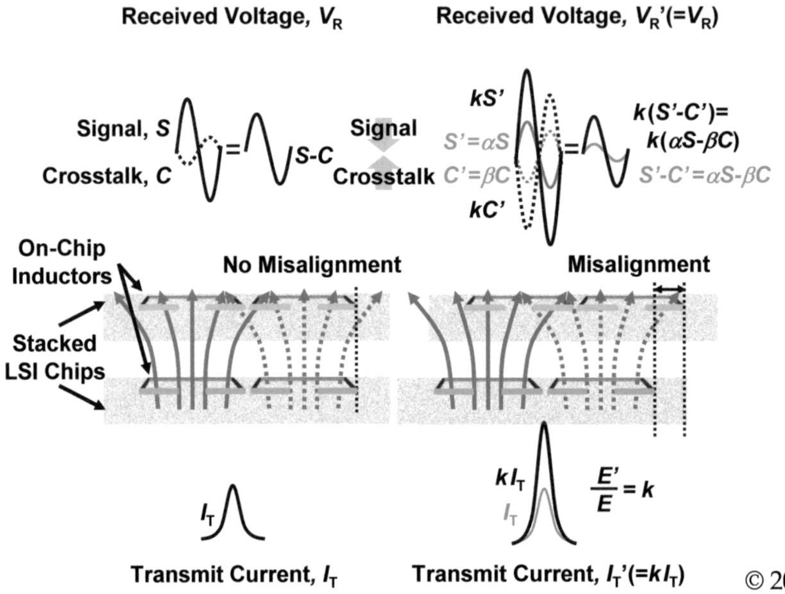

Fig. 17. Increase of crosstalk due to misalignment.

In order to estimate the transmitter energy with consideration of misalignment in array operation, we propose the simplified model. At first, crosstalk is assumed to be proportional to $1/R^3$ as reported in (Miura et al., 2004), where R is horizontal distance from the channel which causes crosstalk. The values of crosstalk from Tx1 and Tx2 have already been known to be C_1 and C_2 since they are essential for estimating transmitter energy even without consideration of misalignment (Fig. 18). With these values, we can get the relationship between crosstalk, C and horizontal distance, R, and then, between required transmitter energy and misalignment as in the following equations.

$$\begin{cases} C_1 = A\dfrac{1}{R_1^{\,3}} + B \\ C_2 = A\dfrac{1}{R_2^{\,3}} + B \end{cases} \Leftrightarrow \begin{cases} A = \dfrac{C_1 - C_2}{\dfrac{1}{R_1^{\,3}} - \dfrac{1}{R_2^{\,3}}} \\ B = C_1 - A\dfrac{1}{R_1^{\,3}} \end{cases} \quad (3)$$

$$C_i = A\frac{1}{R_i^{\,3}} + B, \quad C_i' = A\frac{1}{R_i'^{\,3}} + B \quad (4)$$

Where, C'_i and C_i are crosstalk from i-th transmitter channel with and without misalignment, respectively. R'_i and R_i are horizontal distances from i-th transmitter channel with and without misalignment, respectively. A, B is the constant.

Signal attenuation due to misalignment is modeled by (1) as explained previously. With the above conditions, required transmitter energy can be approximated as bellow.

$$E'/E = k = \frac{S-C}{S'-C'} = \frac{S-C}{\alpha S - \beta C}$$
$$where\ \alpha = \left\{\frac{D^2 + 4(Z^2 + \Delta X^2 + \Delta Y^2)}{D^2 + 4Z^2}\right\}^{-\frac{3}{2}}, \tag{5}$$
$$\beta = \frac{\sum_{i=1\sim 8} C_i'}{\sum_{i=1\sim 8} C_i}$$

Where, E′ and E are required transmitter energy with and without misalignment, respectively. α is the ratio of signal in the misaligned case to the signal in case with no misalignment, and β is the ratio of total crosstalk in 3×3 array between with and without misalignment as shown in Fig. 17.

Figures 18, 19 and 20 show the simulation condition, the absolute and normalized transmitter energy dependence on misalignment. The dependency on the angle is negligibly small and we investigated required transmitter energy with 1-D misalignment (X-Axis). Due to the increase in crosstalk, required transmitter energy for the same BER is increased. The gap between simulation results and calculation results by (5) is also increased.

In array operation, misalignment has to be taken into account more carefully especially when the channel pitch, P is small. Nevertheless, in usual conditions (D=80 μm, Z=70 μm, ΔX=16 μm, P=160 μm), increase in crosstalk due to misalignment is small enough to be ignored. A misalignment of 16 μm is found in commercial mass production.

From the above theoretical analysis, we can calculate the relationship between design parameters and misalignment, which is shown in Fig. 6.

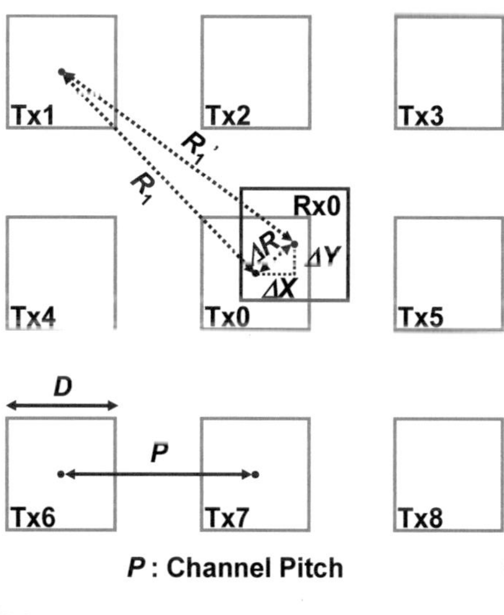

Fig. 18. Simulation condition.

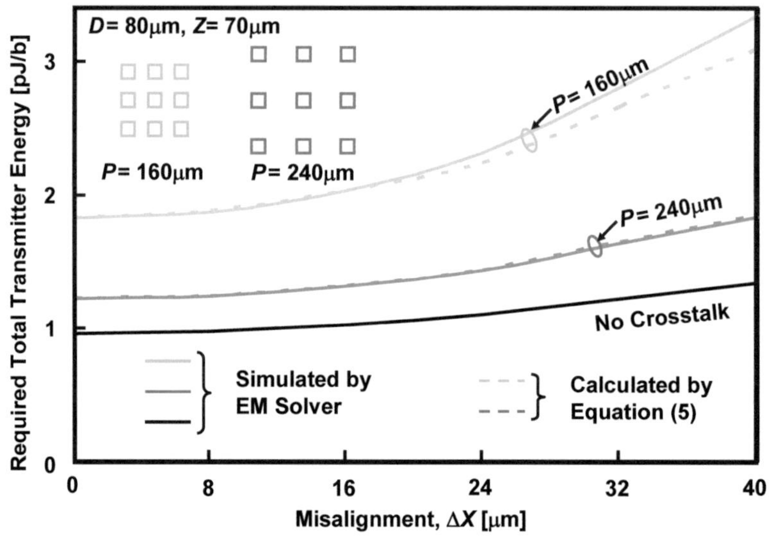

Fig. 19. Required total transmitter energy dependence on misalignment in array operation.

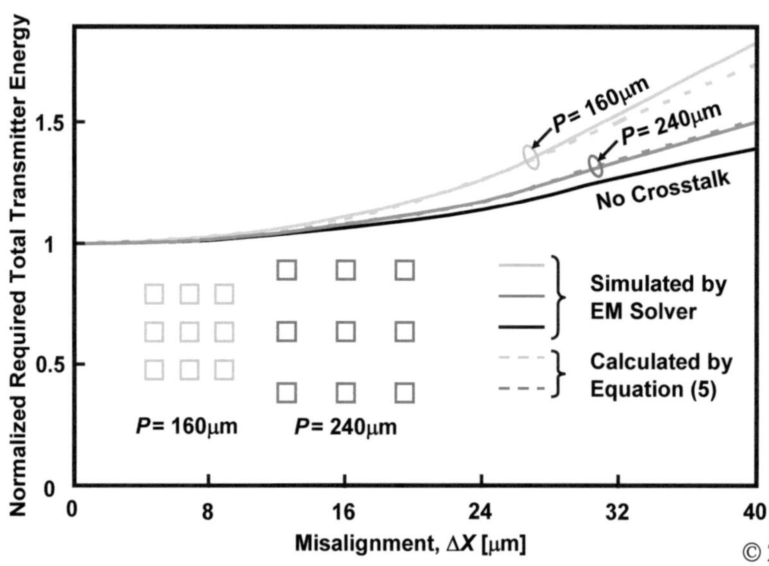

Fig. 20. Normalized required total transmitter energy dependence on misalignment in array operation.

4.3 Experimental verification

Test chips shown in Fig. 5 were utilized for measurement. Figure 21 illustrates the test chip configuration. The transmitter and receiver chips have twelve channels. Transmitter inductors and receiver inductors are arranged with different pitches to make a misalignment. The difference of pitches in larger inductors (D=160 μm) and smaller inductors (D=80 μm) are 16 μm and 8 μm, respectively. With this configuration, misalignments corresponding to 10%, 20%, 30%, 40%, 50% of the outer diameters of inductors are made.

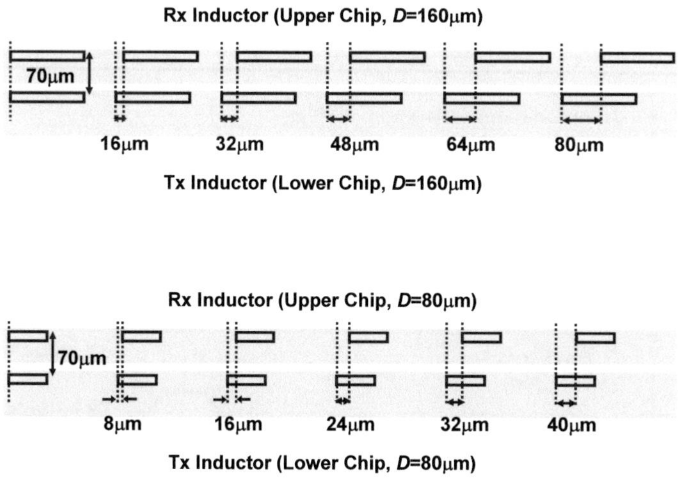

Fig. 21. Test chip configuration.

Figures 22 and 23 show the absolute and normalized measured and simulated transmitter power dependence on the misalignment. In simulation, 3D electro-magnetic solver was used. The power dissipation in this figure is normalized by that without misalignment.

In usual condition (D=80 μm, Z=70 μm), 16 μm of misalignment, while ±10 μm is available in commercial mass production, can be compensated with increasing transmitter power by only 6%. It means that misalignment tolerance of inductive-coupling inter-chip link is high enough. Besides, influence of misalignment is less serious than that of process variations. On the other hand, through-Si via (TSV) technology requires alignment accuracy of ±1 μm (Matsumoto et al., 1998).

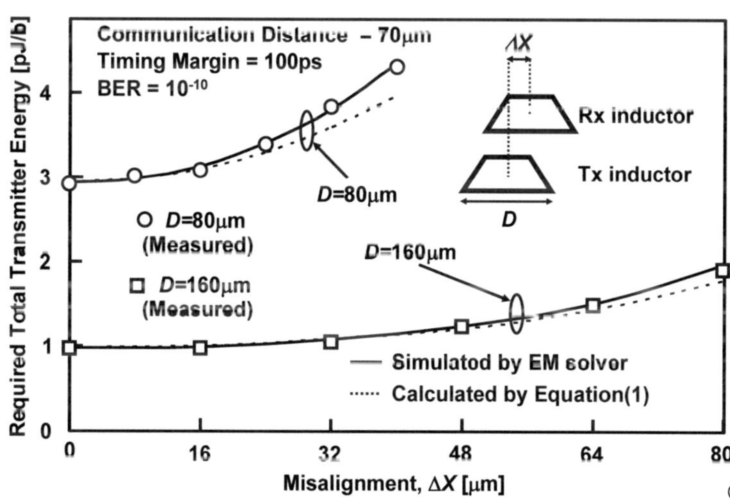

Fig. 22. Measured, simulated and calculated total transmitter energy dependence on the value of misalignment.

Measured results match well with both simulation results from electro-magnetic solver and calculated results from (1). As mentioned in Sect. II, (1) does not cover all of region and has an invalid region. The gap between measured and calculated results becomes larger as the result curves approach the invalid region.

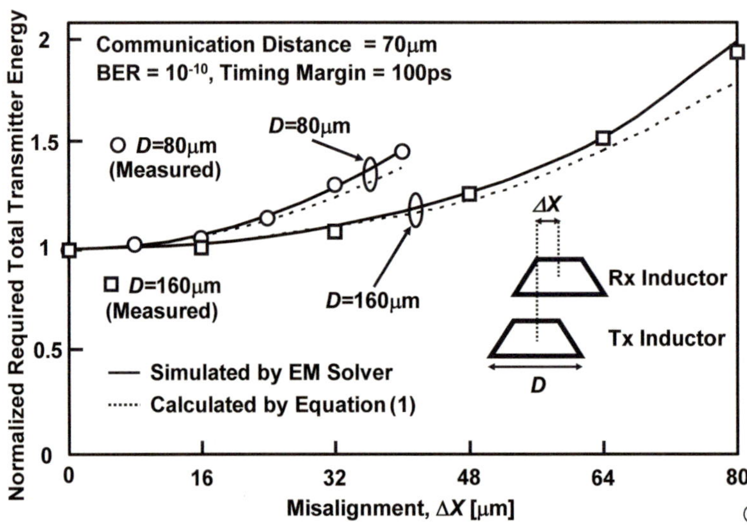

Fig. 23. Measured, simulated and calculated normalized total transmitter energy dependence on the value of misalignment.

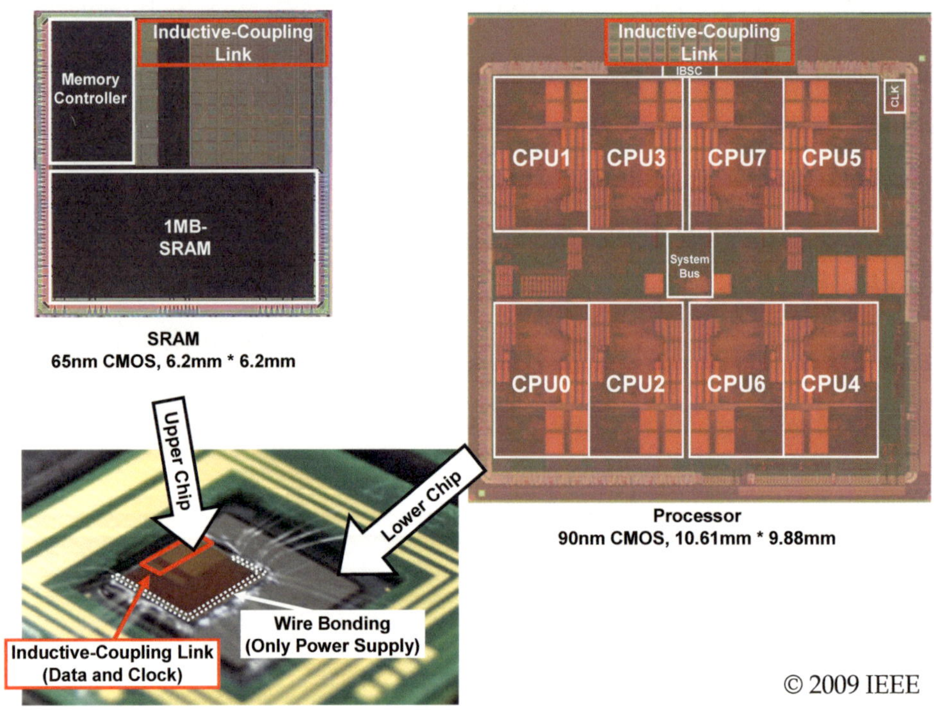

Fig. 24. Chip microphotograph and overhead view of stacked chips.

5. Inductive-coupling link for processor-memory interface

5.1 Introduction

This section presents a three-dimensional (3D) system integration of a commercial processor and a memory by using inductive coupling. A 90nm CMOS 8-core processor, back-grinded to a thickness of 50μm, is mounted face down on a package by C4 bump. A 65nm CMOS 1MB SRAM of the same thickness is glued on it face up, and the power is provided by conventional wire-bonding. The two chips under different supply voltages are AC-coupled by inductive coupling that provides a 19.2Gb/s data link. Measured power and area efficiency of the link is 1pJ/b and $0.15mm^2$/Gbps, which is 1/30 and 1/3 in comparison with the conventional DDR2 interface respectively (Ito et al., 2008). The power efficiency is improved by narrowing a transmission data pulse to 180ps. Reduced timing margin for sampling the narrow pulse, on the other hand, is compensated against timing skews due to layout and PVT variation by a proposed 2-step timing adjustment using an SRAM through mode. All the bits of the SRAM is successfully accessed with no bit error under changes of supply voltages (±5%) and temperature (25°C, 55°C).

5.2 Performance summary of developed 3D LSI system

Micrographs of the chips and their stacking are presented in Fig. 24. A 90nm CMOS processor is mounted face down on a package by C4 bump. A 65nm CMOS SRAM is glued on it face up, and the power is provided by conventional wire-bonding.

Figure 25 summarizes performance. The two chips are each fabricated in their optimal process and supplied with optimal voltages. Thickness of the chips is both 50μm. The radius of the inductors is the same as the communication distance, 120μm. There are 18 data channels for uplink and downlink each. In total 36 inductors are arranged in a 243μm by 320μm pitch. Both the rising and falling edges of a clock are used for 2 phase interleaving to reduce crosstalk between the adjacent channels (Miura et al., 2007). There are clock channels for source synchronous transmission (Miura et al., 2009). One size larger inductors are employed to strengthen the coupling coefficient for asynchronous channel. Total layout area for the inductive coupling link is $2.02mm^2$. Aggregated bandwidth is 19.2Gb/s. Area normalized by bandwidth is $0.15mm^2$/Gbps, which is 1/3 of a conventional DDR2 interface in the same technology (Ito et al., 2008). Since the previous designs of the processor and the memory were reused in large part, the inductive coupling channels are placed in the peripheral region. They can be distributed to each core if a chip layout is carried out from scratch. The circuitry alone occupies an area of $0.072mm^2$, which is only 2.6% of the total area for the inductive coupling link. The area efficiency of circuit alone is therefore $0.0038mm^2$/Gbps, which is 1/120 of the conventional DDR2 interface. Even if the inductor is placed above a bit line of an SRAM and transmits data, no interference is observed (Niitsu et al., 2007). The inductive coupling can be applied to DRAM as well. The inductor can be constructed using 2 metal layers.

5.2 System architecture design with adaptive timing adjustment

Figure 26 depicts a block diagram of the developed 3D LSI system. An inductive-coupling bus state controller (IBSC) supports packet-based communications by adding two signals (vld and eop). A control register in IBSC is used for timing adjustment. The timing

Chip	Processor	SRAM
Process (Property)	90nm CMOS (High Speed)	65nm CMOS (Low Power)
Supply Voltage	1.0 V	1.2 V
Stacking	Face-Down	Face-Up
Connection with PCB	Area Bump	Wire Bonding
Thickness	50 μm	50 μm
Data and Clock Link	Inductive-Coupling	
Communication Distance	120μm (Glue:20μm)	
Inductor Size	Data : 240μm, Clock : 350μm	
Channel Pitch	X: 243μm, Y: 320μm	
Total Bandwidth	19.2 Gbps	
Energy Efficiency	1pJ/b (1/30 of DDR2)	
Area Efficiency	0.15mm²/Gbps (1/3 of DDR2)	

Fig. 25. Performance Summary.

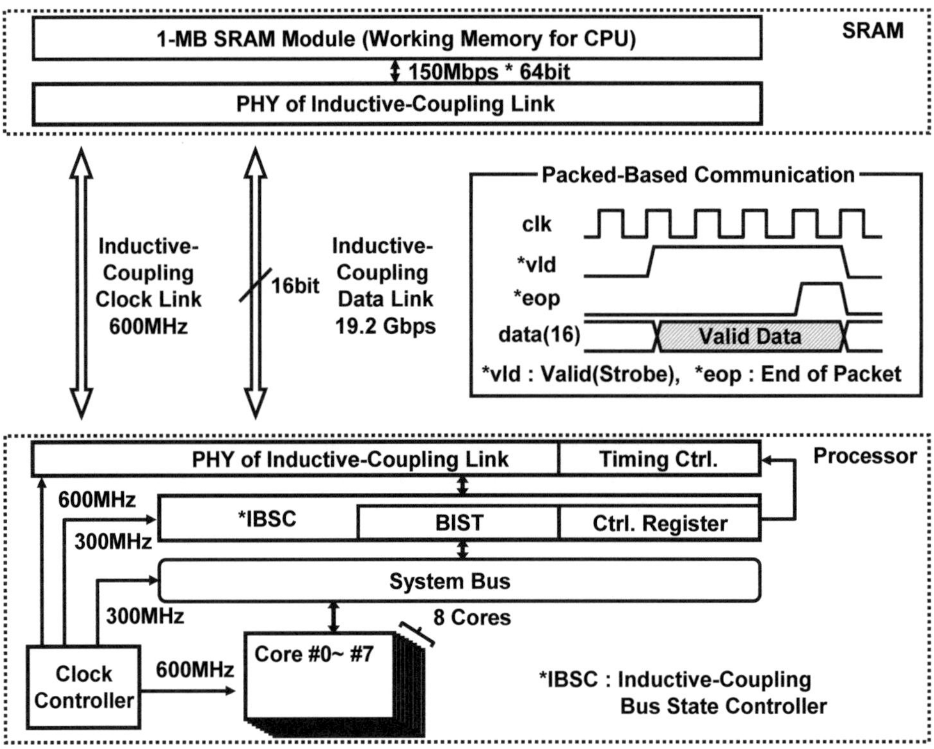

Fig. 26. Block diagram.

adjustment is essential for a practical application. There is a trade off between power dissipation and timing margin. Since power dissipation in a transmitter is in proportion to the square of the pulse width (Miura et al., 2008), the narrower the pulse, the smaller the power dissipation. The timing margin for sampling the narrow pulse, however, will be reduced. Low-power design requires accurate timing control.

Adaptive circuits and systems are required to adjust the timing for the following reasons: 1) timing jitter caused by PVT variations, especially in a clock path with long latency through another chip, 2) VDD changes by DVS, and 3) inter-channel skews, especially when the channels are distributed in a wide area. The timing jitter under PVT variations can be monitored and calibrated by a coarse timing control unit with the control register in IBSC (Fig. 27). Once the calibration result under each condition of DVS is stored in the control register, the timing control unit can adjust the timing for DVS instantly by digital control.

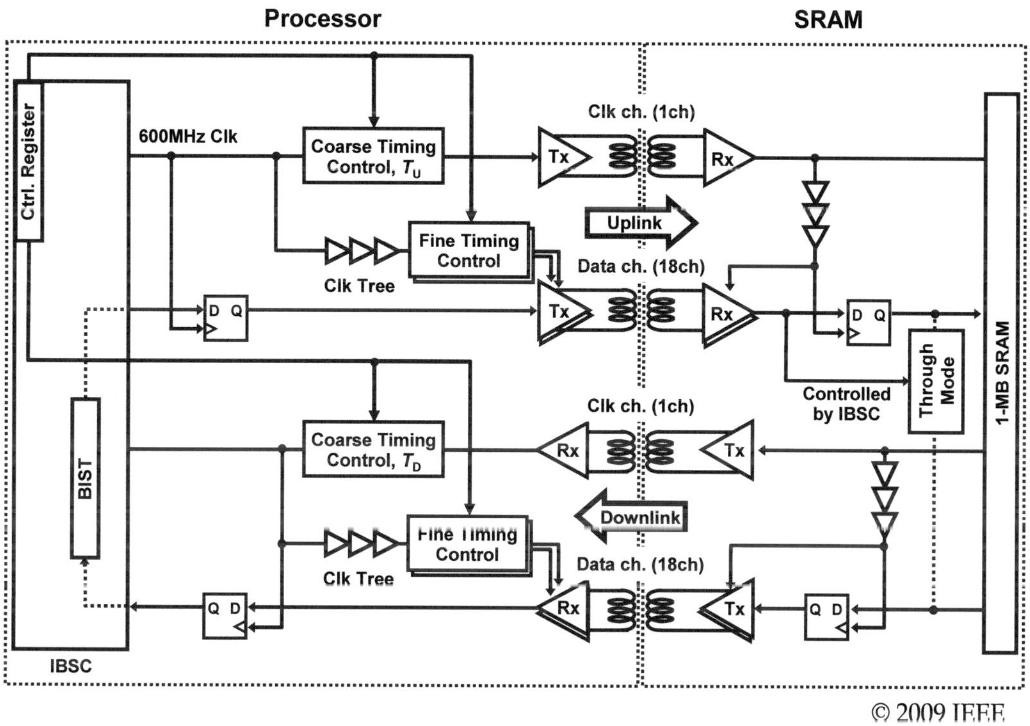

Fig. 27. Adaptive timing adjustment.

The inter-channel de-skew can be performed by a fine timing control unit that is implemented in each channel. Figure 28 shows the timing adjustment flow that is controlled by the processor. First, the control register sets a loopback path in the SRAM for a test mode (an SRAM through mode). Secondly, pass/fail information, much like a shmoo plot, is stored in a register for both the uplink and downlink by changing the coarse timing. Thirdly, the coarse timing is set such that the timing margin becomes the largest when all the channels pass. For each channel, fine timing is tuned next such that the timing margin becomes the largest.

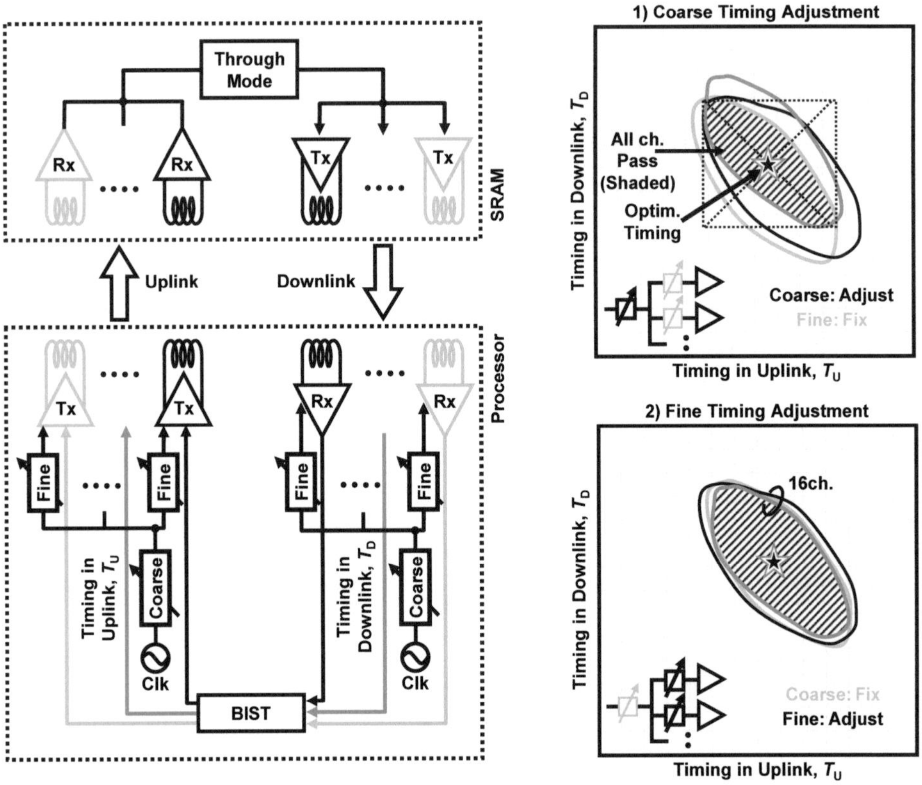

Fig. 28. Fine and coarse (2-step) timing adjustment.

5.4 Measurement results and discussions

The SRAM was accessed (read and write) from the processor and BER was measured by changing the control register. A timing shmoo plot is depicted in Fig. 29, a bathtub curve marked by a broken line is also depicted. A BER of lower than 10^{-14} is achieved with a 2^{31}-1 PRBS. After optimizing the timing by setting the control register at the center of the shmoo plot, tolerance against VDD and temperature changes was measured. The measured result is presented in Fig. 30. No single bit failed under ±5% VDD variations and temperature ranges from 25°C to 55°C. The VDD tolerance can be improved from ±5% to ±10% by widening the pulse width from 180ps to 320ps at a cost of an increase in power efficiency from 1pJ/b to 2.5pJ/b (still 1/12 of DDR2).

6. Conclusion

This chapter presents the fundamental investigation and application of an inductive-coupling link.

First, the interference from power/signal lines and to SRAM of an inductive-coupling link was investigated. Measurement result shows that influence from line and space (I) is none and required normalized transmit power is 1.10 (line and space, type II) and 1.27 (mesh type) when metal density is 16%. The line and space type of power line is better for the

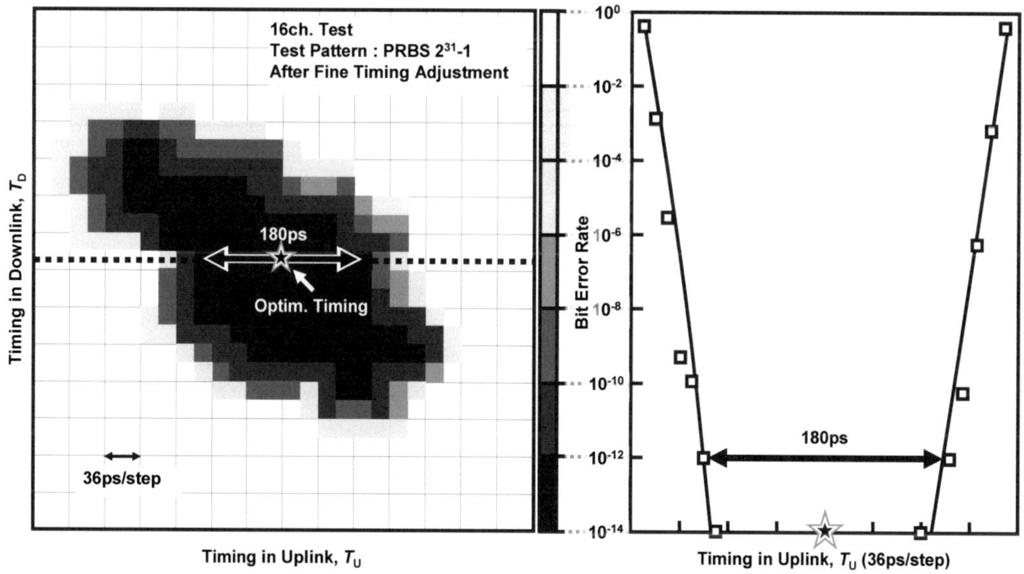

Fig. 29. Measured bit error rate.

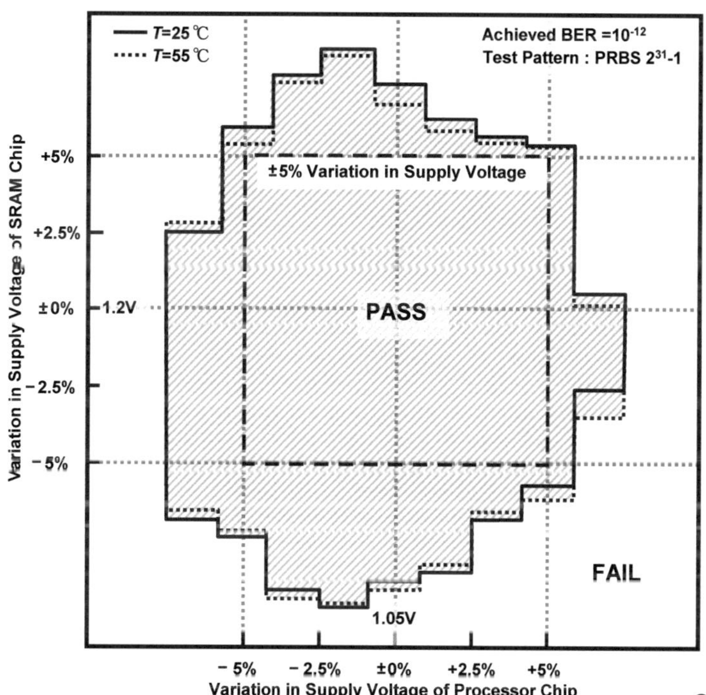

Fig. 30. Measured tolerance (BER<10^{-12}) to variations in supply voltages and temperature.

inductive-coupling link than mesh type. Additional power dissipation to achieve BER of 10^{-8} is only 9% when signal line drives interconnect of 3mm length. In typical ranges, SRAM array operation does not depend on existence of the inductive-coupling link.

Second, modeling of misalignment tolerance in inductive-coupling inter-chip link is introduced. By comparing the calculated result based on the proposed modeling with the measured result, the modeling was found to be accurate in common cases. The estimated and measured results show that misalignment tolerance of inductive-coupling inter-chip link is high enough to keep the performance under the existence of misalignment in usual condition.

Third, application of an inductive-coupling link to interconnection of commercial MPU and SRAM was performed. By exploiting proposed 2-step adaptive timing adjustment, reliable operation under PVT variation has become possible. Achieved performances are power efficiency of 1pJ/bit and area efficiency of $0.15mm^2$/Gbps, which are 1/30 and 1/3 of conventional DDR2 interface, respectively.

7. Acknowledgements

This work has been in part supported by the Grant-in-Aid for JSPS fellows and the Central Research Laboratory of Hitachi Limited.

8. References

Finkenzeller, K. (2003). RFID Handbook, Wiley, 2nd ed., 2003, pp 68-71

Fazzi, A., Canegallo, R., Ciccarelli, L., Magagni, L., Natali, F., Jung, E., Rolandi, P. & Guerrieri, R. (2008). 3-D Capacitive Interconnections With Mono- and Bi-Directional Capabilities, *IEEE Journal of Solid-State Circuits*, Vol. 43, No. 1, pp. 275-284

Hattori, T., lrita, T., Ito, M., Yamamoto, E., Kato, H., Sado, G., Yamada, Y., Nishiyama, K., Yagi, H., Koike, T., Tsuchihashi, Y., Higashida, M., Asano, H., Hayashibara, I., Tatezawa, K., Shimazaki, Y., Morino, N., Hirose, K., Tamaki, S., Yoshioka, S., Tsuchihashi, R., Arai, N., Akiyama, T. & Ohno, K. (2006). A Power Management Scheme Controlling 20 Power Domains for Single-Chip Mobile Processor, *Proceedings of IEEE International Solid-State Circuits Conference*, pp. 2210-2219, Feb., 2006

Ito, M., Hattori, T., Irita, T., Tatezawa, K., Tanaka, F., Hirose, K., Yoshioka, S., Ohno, K., Tsuchihashi, R., Sakata, M., Yamamoto, M. & Aral, Y. (2007). A 390MHz Single-Chip Application and Dual-Mode Baseband Processor in 90nm Triple-Vt CMOS, *Proceedings of IEEE International Solid-State Circuits Conference*, pp. 274-275, Feb., 2007

Ito, M., Hattori, T., Yoshida, Y., Hayase, K., Hayashi, T., Nishii, O., Yasu, Y., Hasegawa, A., Takada, M., Mizuno, H., Uchiyama, K., Odaka, T., Shirako, J., Mase, M., Kimura, K. & Kasahara, H. (2008). An 8640 MIPS SoC with Independent Power-Off Control of 8 CPUs and 8 RAMs by An Automatic Parallelizing Compiler, *Proceedings of IEEE International Solid-State Circuits Conference*, pp. 90-91, Feb., 2008

Koyanagi, M., Fukushima, T. & Tanaka, T. (2009). High-Density Through Silicon Vias for 3-D LSIs, *Proceedings of the IEEE*, Vol. 97, No. 1, pp. 49-59

Matsumoto, T., Satoh, M., Sakuma, K., Kurino, H., Miyakawa, N., Itani, H. & Koyanagi, M. (1998). New Three-Dimensional Wafer Bonding Technology Using the Adhesive Injection Method, *Japanese J. of Applied Physics*, Vol. 37, No. 3B, pp. 1217-1221, Mar. 1998.

Miura, N., Mizoguchi, Sakurai, T. & Kuroda, T. (2004). Cross Talk in Inductive Inter-Chip Wireless Superconnect, *Proceedings of IEEE Custom Integrated Circuits Conference*, pp. 99-102, Sept., 2004

Miura, N., Mizoguchi, D., Inoue, M., Niitsu, K., Nakagawa, Y., Tago, M., Fukaishi, M., Sakurai, T. & Kuroda, T. (2007). A 1 Tb/s 3 W Inductive-Coupling Transceiver for 3D-Stacked Inter-Chip Clock and Data Link, *IEEE Journal of Solid-State Circuits*, Vol. 42, No. 1, pp. 111-122

Miura, N., Ishikuro, H., Niitsu, K., Sakurai, T. & Kuroda, T. (2008). A 0.14pJ/bit Inductive-Coupling Transceiver with Digitally-Controlled Precise Pulse Shaping, *IEEE Journal of Solid-State Circuits*, Vol. 43, No. 1, pp. 285-291

Miura, N., Kohama, Y., Sugimori, Y., Ishikuro, H., Sakurai, T. & Kuroda, T. (2009). A High-Speed Inductive-Coupling Link With Burst Transmission, *IEEE Journal of Solid-State Circuits*, Vol. 44, No. 3, pp. 947-955

Mizoguchi, D., Miura, N., Ishikuro, H. & Kuroda, T. (2008). Constant Magnetic Field Scaling in Inductive-Coupling Data Link, *IEICE Transactions on Electronics*, vol. E91-C, no. 2, pp. 200-205, Feb., 2008.

Niitsu, K., Sugimori, Y., Kohama, Y., Osada, K., Irie, N., Ishikuro, H. & Kuroda, T. (2007)., Interference from Power/Signal Lines and to SRAM Circuits in 65nm CMOS Inductive-Coupling Link, *Proceedings of IEEE Asian Solid-State Circuits Conference*, pp. 131-134, Nov., 2007

Niitsu, K., Kawai, S., Miura, N., Ishikuro, H. & Kuroda, T. (2008). A 65 fJ/b inductive-coupling inter-chip transceiver using charge recycling technique for power-aware 3D system integration, *Proceedings of IEEE Asian Solid-State Circuits Conference*, pp. 97-100, Nov., 2008

Niitsu, K., Shimazaki, Y., Sugimori, Y., Kohama, Y., Kasuga, K., Nonomura, I., Saen, M., Komatsu, S., Osada, K., Irie, N., Hattori, T., Hasegawa, A. & Kuroda, T. (2009). An inductive-coupling link for 3D integration of a 90nm CMOS processor and a 65nm CMOS SRAM, *Proceedings of IEEE International Solid-State Circuits Conference*, pp. 480-481, Feb., 2009

Niitsu, K., Kohama, Y., Sugimori, Y., Kasuga, K., Osada, K., Irie, N., Ishikuro, H. & Kuroda, T. (2010)., Modeling and Experimental Verification of Misalignment Tolerance in Inductive-Coupling Inter-Chip Link for Low-Power 3D System Integration, *IEEE Transactions on VLSI Systems*, (in print)

Onizuka, K., Kawaguchi, H., Takamiya, M., Kuroda, T. & Sakurai, T. (2006). Chip-to-Chip inductive wireless power transmission system for SiP applications, *Proceedings of IEEE Custom Integrated Circuits Conference*, pp. 575-578, Sept., 2006

Yamaoka, M., Osada, K., Tsuchiya, R., Horiuchi, M., Kimura, S. & Kawahara, T. (2004). Low power SRAM menu for SOC application using Yin-Yang-feedback memory cell technology, *Proceedings of IEEE Symposium on VLSI Circuits*, pp. 288-291, Jun., 2004

Yamaoka, M., Maeda, N., Shinozaki, Y., Shimazaki, Y., Nii, K., Shimada, S., Yanagisawa & Kawahara, T. (2005). Low-power embedded SRAM modules with expanded margins for writing, *Proceedings of IEEE International Solid-State Circuits Conference*, pp. 480-481, Feb., 2005

16

Polycrystalline Silicon Piezoresistive Nano Thin Film Technology

Xiaowei Liu[1], Changzhi Shi[1] and Rongyan Chuai[2]
[1]*Harbin Institute of Technology*
[2]*Shenyang University of Technology*
China

1. Introduction

The piezoresistive effect of semiconductor materials was discovered firstly in silicon and germanium (Smith, 1954). Dissimilar to the piezoresistive effect of metal materials induced from the change in geometric dimension, the piezoresistive phenomenon in silicon is due to that mechanical stress influences the energy band structure, thereby varying the carrier effective mass, the mobility and the conductivity (Herring, 1955). The gauge factor (GF) is used to characterize the piezoresistive sensitivity and defined as the ratio of the relative resistance change and the generated strain (nondimensional factor). Usually, the GF in silicon is around 100 and changes with stress direction, crystal orientation, doping concentration, etc. Recently, the giant piezoresistances were observed in silicon nanowires (He & Yang, 2006; Rowe, 2008) and metal-silicon hybrid structures (Rowe, et al., 2008), respectively. Although these homogeneous silicon based materials or structures possess high piezoresistive sensitivity, there are still several issues influencing their sensor applications, such as, p-n junction isolation, high temperature instability, high production cost and complex fabrication technologies.

As another monatomic silicon material with unique microstructure, polycrystalline silicon has been investigated since the 1960s. The discovery of its piezoresistive effect (Onuma & Sekiya, 1974) built up a milestone that this material could be applied widely in field of sensors and MEMS devices. Moreover, polycrystalline silicon could be grown on various substrate materials by physical or chemical methods, which avoids p-n junction isolation and promotes further its applications for piezoresistive devices (Jaffe, 1983; Luder, 1986; Malhaire & Barbier, 2003). Among numerous preparation methods, the most popular technology is chemical vapour deposition (CVD), which includes APCVD, LPCVD, PECVD, etc. The PECVD method can deposit films on substrates at lower temperatures, but the stability and uniformity of as-deposited films are not good, and the samples could contain a large number of amorphous contents. Subsequently, the metal-induced lateral crystallization (MILC) technique was presented (Wang, et al., 2001). By enlarging grain size and improving crystallinity, the gauge factor of MILC polycrystalline silicon was increased to be about 60. But the MILC polycrystalline silicon-based devices could suffer the contamination from the metal catalyst layer (e.g. Ni, Al, etc.). Compared with the aforementioned technologies, the LPCVD process is a mature and stable CVD method with advantages of good product uniformity, low cost, IC process compatibility, etc. Therefore, the preparation method in this work is mainly based on LPCVD, while the magnetron sputtering technology will be utilized as a reference result.

The experimental results reported by other researchers indicate that the gauge factor of polycrystalline silicon thicker films (around 400nm in thickness generally) has a maximum as the doping concentration is at the level of 10^{19} cm^{-3} and then degrades rapidly with the further increase of doping concentration (Schubert, et al., 1987; French & Evens, 1989; Gridchin, et al., 1995; Le Berre, et al., 1996). Moreover, the gauge factor of highly doped polycrystalline silicon thicker films is only 20-25. It results in that the research works were emphasized on the medium doped polycrystalline silicon thicker films. However, the lower doping concentration brings the higher temperature coefficients of resistance and gauge factor. This limits the working temperature range of polycrystalline silicon thicker film-based sensors.

In our research work, when the film thickness is reduced to nanoscale and the doping concentration is elevated to the level of 10^{20} cm^{-3}, the enhanced piezoresistance effect is observed, and the temperature coefficients of resistance and gauge factor are reduced further. These phenomena are different from the polycrystalline silicon thicker films and can not be explained reasonably based on the existing piezoresistive theory. The unique properties of polycrystalline silicon nano thin films (PSNFs) could be useful for the design and fabrication of piezoresistive sensors with miniature volume, high sensitivity, good temperature stability and low cost. In the following sections, the details of sample fabrication, microstructure characterization, experimental method and measurement results will be provided. In order to analyze the experimental results, the tunnelling piezoresistive theory is established and predicts the experimental results with a good agreement.

2. Film preparation technologies

2.1 Low pressure chemical vapor deposition

Due to the aforementioned advantages, the low pressure chemical vapour deposition (LPCVD) technology is utilized to prepare the polycrystalline silicon films. According to the difference of technological parameters, three groups of film samples were prepared (Group A – different thicknesses; Group B – different doping concentrations; Group C – different deposition temperatures).

a. Group A – Firstly, by controlling deposition time, the polycrystalline silicon thin films with different thicknesses were deposited on 500 μm-thick (100) and (111) silicon substrates (4 inch diameter) coated with 1μm-thick thermally grown SiO_2 layers by LPCVD at 620 °C at 45~55 Pa, respectively. For the (100) substrates, the thicknesses of as-deposited films are in the range of 30~90 nm; for the (111) substrates, the film thicknesses are ranged from 123 nm to 251 nm. Then, the solid-state boron diffusion was performed at 1080 °C in N_2 atmosphere with a flow rate of 2L/min to obtain the doping concentration of 2.3×10^{20} cm^{-3}.
b. Group B – Subsequently, according to the piezoresistive sensitivities of polysilicon thin films with different thicknesses, the optimal film thickness was extracted. The experimental results show that the ~80 nm-thick films possess the highest gauge factor (discussed later). Therefore, the thickness of polysilicon thin films with different doping concentrations was selected to be 80 nm. After the same LPCVD process, the obtained polysilicon thin films were ion-implanted by boron dopants with doses of

9.4×10^{13}~8.2×10^{15} cm^{-2}. Then, the post-implantation annealings were carried out in N_2 at 1080 °C for 30 min to activate dopants and eliminate ion-implantation damages. Finally, the doping concentrations were in the range of 8.1×10^{18} ~ 7.1×10^{20} cm^{-3}.

c. Group C – Before preparing films, a 1 μm-thick SiO_2 layer was grown on the 500 μm-thick (111) Si wafers (4 inch diameter) by thermal oxidization at 1100 °C. Then, the 80 nm-thick PSNFs were deposited on the thermally oxidized Si substrates by LPCVD at a pressure of 45~55 Pa over a temperature range of 560~670 °C. The reactant gas was SiH_4 and the flow rate was 50 mL/min. Since the films deposited at 560~600 °C exhibited amorphous appearance mixed with polycrystals, the pre-annealing was performed on them in dry N_2 at 950° C for 30 min to induce the recrystallization of amorphous regions. For the dopant implantation, boron ions were implanted into the samples at a dose of 2×10^{15} cm^{-2} at 20 keV. For the sake of dopant activation and ion implantation damage elimination, the post-implantation annealing was carried out in N_2 atmosphere at 1080 °C for 30 min. Then, the doping concentration was estimated to be 2×10^{20} cm^{-3}.

2.2 Magnetron sputtering

As a reference, a group of samples were prepared by magnetron sputtering. Before preparing films, a 1 μm-thick SiO_2 layer was grown on the 500 μm-thick (100) Si wafers (4 inch diameter) by thermal oxidization at 1100 °C. Then, the polycrystalline silicon films were prepared by magnetron sputtering system from an undoped silicon target and the substrate temperature was 300 °C. The base pressure of system was maintained at 0.12 Pa. The discharge current on the magnetron was held constant at 0.3 A, while the substrate bias voltage was 500 V. The sputtering time was 10 min, and the thickness of films was 200 nm. Through the SEM observation, it can be seen that the obtained films are amorphous. Thus, the annealing of 1080 °C was carried out in N_2 atmosphere for 60 min to obtain the lowest film resistivity. After annealing, the solid-state boron diffusion was performed at 1080 °C in N_2 with a flow rate of 2 L/min to obtain the doping concentration of 2.3×10^{20} cm^{-3}.

3. Microstructure characterization

3.1 Samples with different thicknesses

In order to analyze the surface morphology, the film samples with different thicknesses were characterized by SEM. The SEM images of samples with different thicknesses are given in Fig. 1. For the characterization of grain orientation, the XRD experiment was performed. The XRD patterns of samples with different thicknesses are shown in Fig. 2. From the SEM images in Fig. 1, it can be seen that the grain size of the samples increases with increasing film thickness. For 30, 40, 60, 90, 123, 150, 198, 251 nm-thick samples, their grain sizes are 11, 30, 37, 48, 48, 58, 69, 80 nm, respectively. By XRD analysis, the (111) peaks of the films thicker than 120 nm and the (400) peaks of the films thinner than 100 nm are attributed to the crystal orientation of substrates. It can be also observed that the (220), (400) and (331) peaks appear as the films are thicker than 120 nm and the intensities of these diffract peaks increase with the increase of film thickness. Moreover, the (311) peak is observed in 251 nm-thick films. It indicates that the increase in film thickness improves the crystallinity and enhances the preferred growth. However, no obvious diffract peaks are observed in 60 and 90 nm-thick films, so they could be considered to be randomly oriented. Noticeably, the (201) peaks appear in 30 and 40 nm-thick samples. According to the report (Zhao et al., 2004), this preferred orientation occurs in nanocrystalline silicon and corresponds to tetragon microstructure. It indicates that these two samples exhibit the structural characteristic of nanocrystalline silicon. For the sake of brevity, the 60-100 nm-thick films are called polysilicon nano thin films (PSNFs), while the films thicker than 120 nm are called polysilicon common films (PSCFs). The films thinner than 50 nm are called nanocrystalline-like polysilicon thin films (NL-PSTFs).

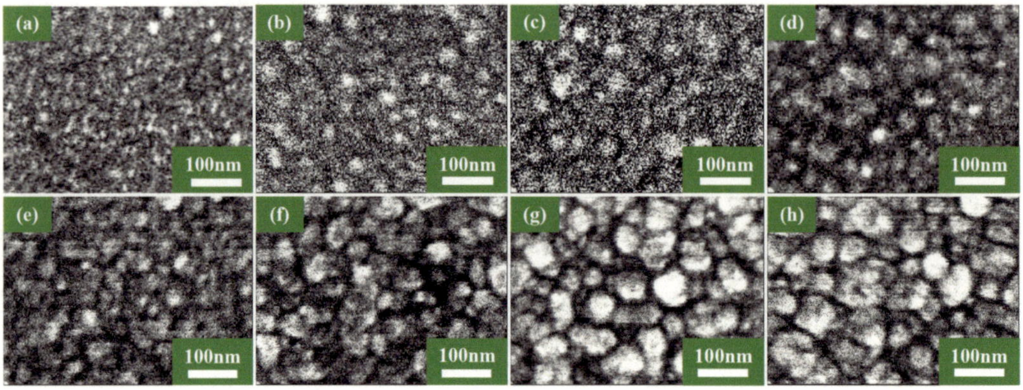

Fig. 1. SEM images of polycrystalline silicon thin film samples with different thicknesses

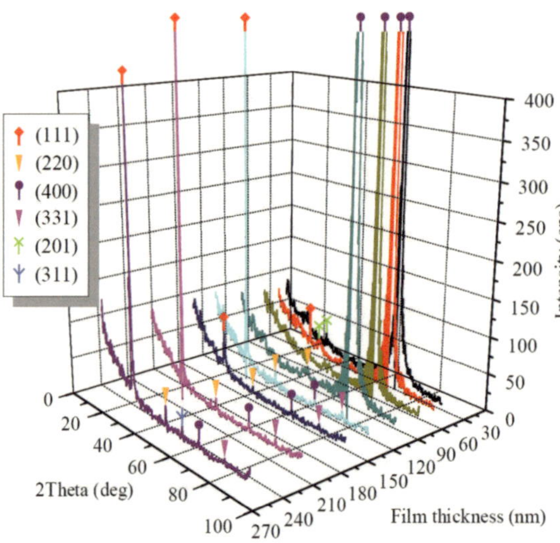

Fig. 2. XRD patterns of polycrystalline silicon thin films with different thicknesses

3.2 Samples with different doping concentrations

Fig. 3 provides the SEM and TEM images of the 80 nm-thick PSNFs with doping concentrations of 2×10^{19} cm^{-3}, 4.1×10^{19} cm^{-3} and 4.1×10^{20} cm^{-3}. It can be observed that the variation of doping concentration does not influence the grain size obviously. Thus, the grain size of the samples with different doping concentrations is considered to be constant. In the XRD pattern of Fig. 4, only the weak (220) peak is observed and the strong (111) peak is attributed to the crystal orientation of substrates. It indicates that these samples are randomly oriented.

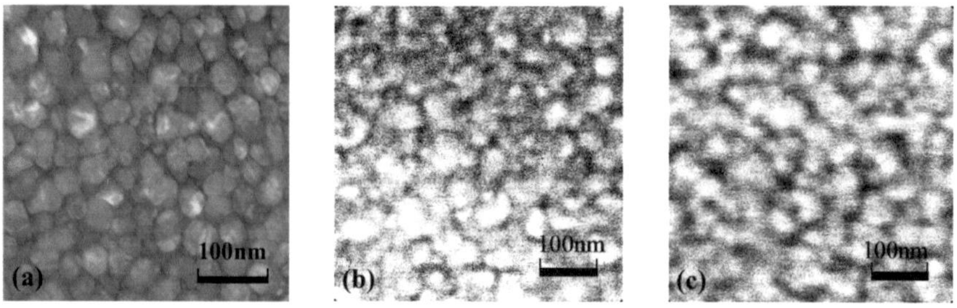

Fig. 3. TEM and SEM images of 80 nm-thick PSNF samples with different doping concentrations. (a) 2×10^{19} cm^{-3} TEM; (b) 4.1×10^{19} cm^{-3} SEM; (c) 4.1×10^{20} cm^{-3} SEM

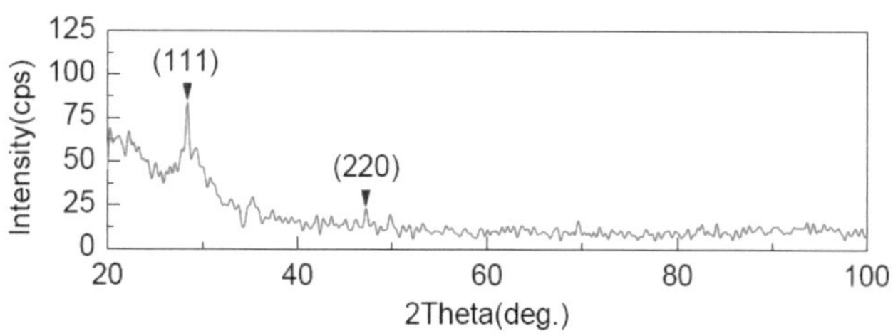

Fig. 4. XRD spectrum of 80 nm-thick polycrystalline silicon nano thin films

3.3 Samples with different deposition temperatures

The surface morphology of PSNFs was characterized by SEM, as shown in Figs. 5(a)-(e). It can be seen that the grain size increases with elevating deposition temperature. This indicates that the crystallinity of PSNFs can be improved by raising deposition temperature. The grain size can be determined by TEM, as shown in Fig. 5(f). The mean grain size of 620 °C samples is estimated to be 40 nm approximately. With the deposition temperature varying from 560 °C to 670 °C, the mean grain size increases from 30 nm to 70 nm. For the sake of clarity, the 560~600 °C films undergoing the preannealing of 950 °C are called

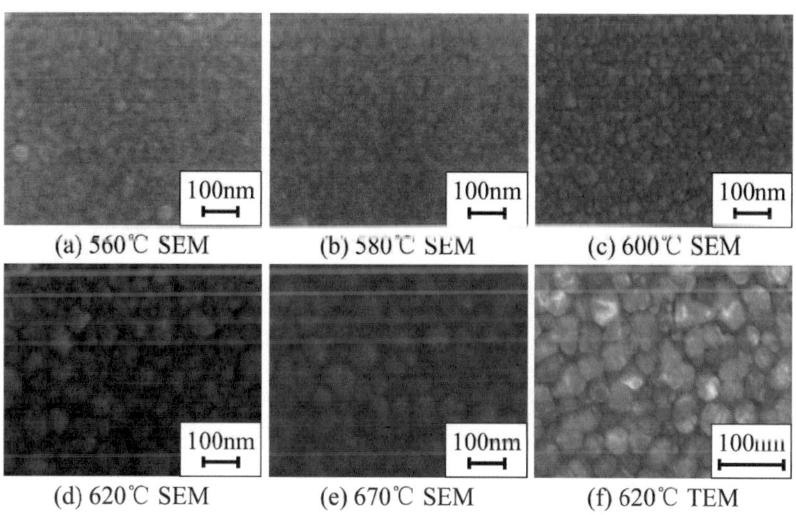

Fig. 5. SEM and TEM images of PSNFs deposited at different temperatures

recrystallized (RC) PSNFs, while the 620~670 °C films are called directly crystallized (DC) PSNFs. From Fig. 5, it can be seen that the borders between grain boundaries and grains of RC PSNFs are obscure as well as the 670 °C samples. It shows that the grain boundaries of the abovementioned samples contain a large number of amorphous phases.

In order to analyze the film microstructure, the XRD experiment was performed on the samples. In the XRD spectra shown in Fig. 6, all the (111) peaks are attributed to Si substrates. The clear (220) peak of 670 °C PSNFs is due to the preferred grain growth along (220) orientation, while the other PSNFs are oriented randomly. Furthermore, it should be noted that the broad peaks (2θ=85~100 °) related to amorphous phases appear on the spectra of RC and 670 °C PSNFs, thereby testifying the existence of amorphous phases at grain boundaries. Because amorphous phases in the 620 °C PSNFs are much fewer, no remarkable broad peak is observed. The peak intensity and FWHM of RC PSNFs are larger than those of the 670 °C ones. It demonstrates that the crystallinity of RC PSNFs is lower than DC ones. The broad peak of 670 °C samples is likely due to the preferred growth aggravating disordered states of grain boundaries.

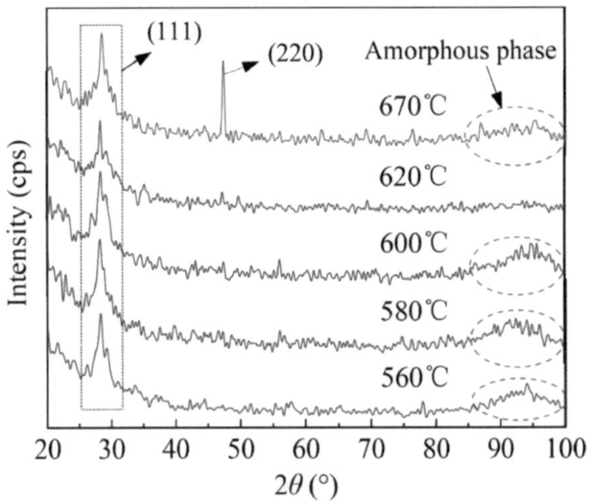

Fig. 6. XRD spectra of PSNF samples deposited at different temperatures

3.4 Magnetron sputtering samples

Fig. 7 provides the SEM images of polycrystalline silicon films prepared by magnetron sputtering before and after the annealing of 1080 °C. From Fig. 7(a), we can see that the film is amorphous and has no micrograined texture. After high temperature annealing, the

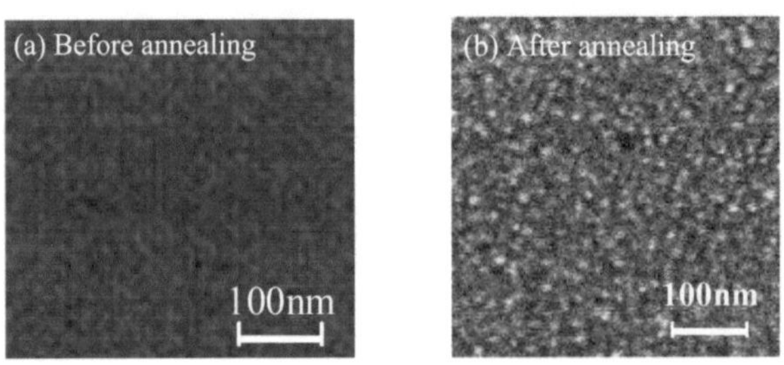

Fig. 7. SEM images of polycrystalline silicon films prepared by magnetron sputtering

recrystallization occurs in the film, which make the film transfer from amorphous state to polycrystalline state, as shown in Fig. 7(b). By calculation, the grain size of magnetron sputtering films is around 10 nm. It indicates that the crystallinity of magnetron sputtering films is very low and the recrystallization induced by high temperature annealing is limited for the improvement of film crystallinity.

4. Fabrication of cantilever beam samples

4.1 Piezoresistors

For measuring gauge factor, the cantilever beams were fabricated based on photolithography and etching technologies. Firstly, the sample wafers were ultrasonically degreased with methylbenzene, acetone and ethanol for 5 min in each and then rinsed repeatedly in de-ionized water. The cleaned samples were pre-baked at 120 °C for 15 min. Next, after spin-coating with positive photoresist and a soft-bake at 90°C for 10 min, the samples were exposed for 90 s using the mask plate as shown in Fig. 8(a) and developed in the 0.5% NaOH solution. Then, a hard-bake for 25 min was performed at 120 °C for the successive etching process. After photolithography, the samples were etched in $HNO_3/HAc/HF$ (4:1:1) solution to form PSNF resistors and then rinsed in de-ionized water. The photoresist was removed by acetone to obtain the sample wafers with PSNF resistors as shown in Fig. 8(b).

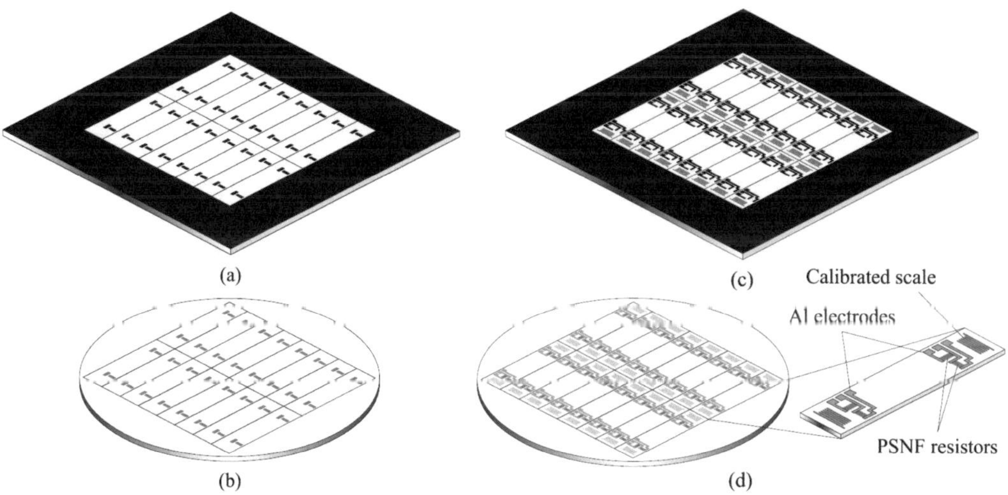

Fig. 8. Schematic diagram of mask plates and sample wafers in the fabrication of cantilever beams. (a) The mask plate for patterning resistors. (b) The sample wafer after patterning resistors. (c) The mask plate for patterning electrodes and calibrated scales. (d) The sample wafer and the cantilever beam after fabricating electrodes and calibrated scales.

4.2 Metal contact electrodes

Here, the aluminium is used as the metal electrode material. In order to measure the contact resistance between PSNFs and metal electrodes, the ohmic contact test patterns based on linear transmission line model (LTLM) were also fabricated on the samples. Before depositing metal, the samples were dipped in HF/H_2O (1:10) for 8 s to remove the native oxide. The Al layer was evaporated onto the samples by vacuum evaporation. Then, the positive photoresist was coated and patterned in the same process as the resistor fabrication. The schematic diagram of mask plate is shown in Fig. 8(c). The Al layer was etched in concentrated phosphorous acid at 80~100 °C to form electrodes. The electrode fabrication was completed by removing the photoresist left.

4.3 Alloying and scribing

After scribing, the sample wafers were divided into individual cantilever beams of 26 mm×4 mm, as shown in Fig. 8(d). Then, the samples were alloyed at 410 °C, 450 °C and 490 °C for 20 min in N_2 to form ohmic contact. By measuring the LTLM test patterns, the *I-V* characteristic curves after alloying at different temperatures are provided in Fig. 9. From Fig. 9, it can be seen that the samples annealed at 450 °C have a linear *I-V* curve, which indicates that the good ohmic contact is formed. The specific contact resistivity is about $2.4\times10^{-3}\ \Omega\cdot cm^2$.

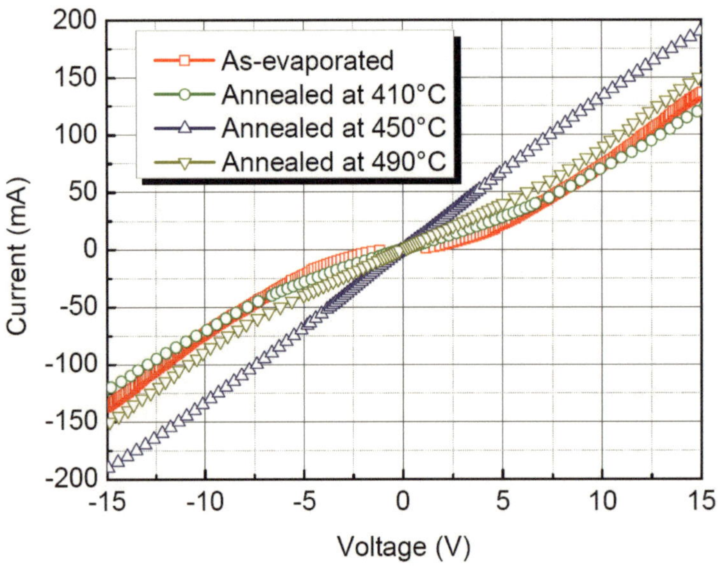

Fig. 9. *I-V* characteristic curves of metal contact electrodes after annealed at different alloying temperatures

Finally, on the actual cantilever beam sample given in Fig. 10, two groups of PSNF piezoresistors were fabricated. Each group consists of three sets of longitudinal and transversal piezoresistors with length-width ratios of 1:4, 2:1 and 8:1, respectively. And the current directions through longitudinal resistors were aligned with the (110) orientation. Fig. 10(b) and (c) are the micrographs of a PSNF resistor taken by laser scanning microscope. Also, the Al calibrated scales were fabricated near both ends of cantilever beams for measuring the arm of applied force.

5. Gauge factor measurement

The gauge factor test setup is shown in Fig. 11. Either end of the cantilever beam is fixed by the clamp. The piezoresistors are connected to the electric instruments through Al electrodes.

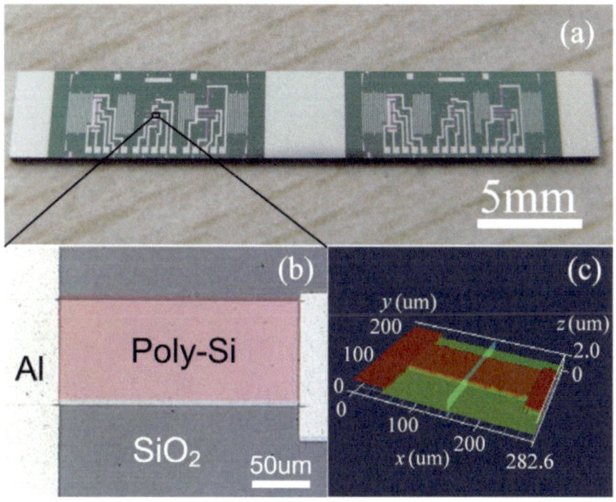

Fig. 10. (a) Photo of a cantilever beam sample; (b) Laser scanning microscope 2D image of a polysilicon piezoresistor; (c) Laser scanning microscope 3D image of a polysilicon piezoresistor

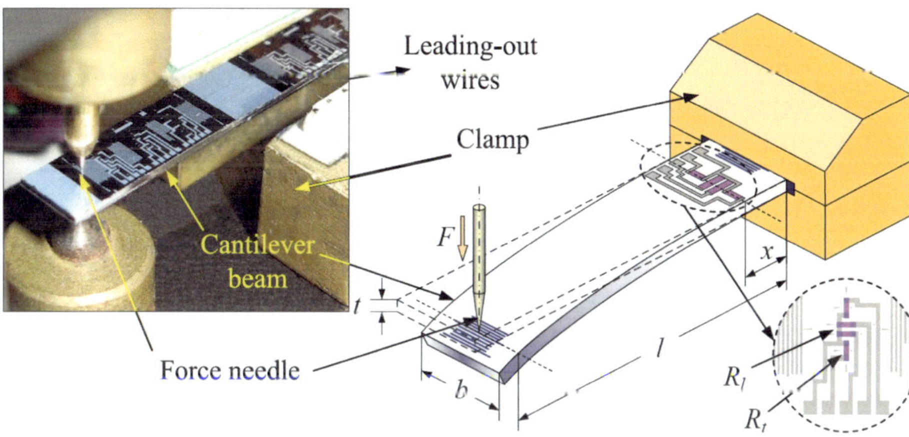

Fig. 11. Strain loading setup for measuring gauge factor

When an axial force F is applied to the free end of the cantilever beam, the strain $\varepsilon(x)$ produced at x can be expressed as

$$\varepsilon(x) = \frac{6(l-x)\cdot F}{bt^2Y} \tag{1}$$

where l is the force arm of the axial force F, b and t are the width and the thickness of the cantilever beam (b, t << l here), respectively. Y is Young's modulus of silicon. The initial resistance R_0 (without strain) and the varied resistance R (with strain) were measured by a Keithley 2000 digital multimeter. The gauge factor can be calculated by:

$$GF = \frac{R-R_0}{R_0 \cdot \varepsilon} = \frac{\Delta R}{R_0 \cdot \varepsilon} \tag{2}$$

6. Tunneling piezoresistive theory

6.1 Analysis of existing theories

The existing piezoresistive theories of polysilicon were established during 1980s~1990s and used to ameliorate the process steps for the optimization of device performance. In the early models proposed (Mikoshiba, 1981; Erskine, 1983; Germer & Tödt, 1983), the contribution of grain boundaries to piezoresistive effect was neglected, thereby resulting in the discrepancy between experimental data and theoretical results at low doping levels. To tackle this issue, Schubert et al. took the piezoresistive effect of depletion region barriers (DRBs) arising from carrier trapping at grain boundaries into account and established a theoretical model for calculating gauge factors (Schubert, et al., 1987). Thereafter, French et al. suggested that the piezoresistive effect of p-type polysilicon is not only due to the shift in heavy and light hole band minima relative to each other, but also due to the warpage of two sub-bands (French & Evens, 1989). Moreover, the barrier effect of grain boundaries was introduced into the model, achieving the good agreement with the experimental data. Noticeably, it was considered in these models that the PRCs of grain boundaries and DRBs are much lower than that of grain neutral regions. Based on this viewpoint, since the PRC of grain neutral regions (bulk Si) falls off rapidly at high doping concentrations (Toriyama & Sugiyama, 2002), it has been considered that the gauge factor of polysilicon could be degraded sharply with increasing doping concentrations. Accordingly, the optimization of fabrication technologies was emphasized on improving crystallinity and controlling doping concentration to prepare the films with larger grain sizes and lower trap densities. It results in that the research works have been mainly focused on polycrystalline silicon thicker films and scarcely involving PSNFs.

6.2 Carrier transport mechanisms through grain boundaries

Polysilicon can be considered as composed of small crystals joined together by grain boundaries. Each crystal is viewed as a Si single crystal, while the grain boundaries are full of defects and dangling bonds and form extremely thin amorphous layers. The forbidden band width of grain boundaries is larger than that of monocrystalline silicon (1.12 eV) (Mandurah, et al., 1981; Kamins, 1971) and approaches that of amorphous silicon (1.5-1.6 eV) (Taniguchi, et al., 1978). The Fermi level is pinned near the midgap at grain boundaries. In this case, the grain boundary barriers are formed to hinder carriers from traversing grain boundaries. Moreover, dangling bonds at grain boundaries can be occupied by carriers and dopant atoms, so the DRBs are created on the sides of grain boundaries. As a result, the grain boundary barriers and the DRBs form the composite grain boundary barriers.

Theoretically, carriers pass through grain boundaries by two transport mechanisms of thermionic emission and tunneling. For simplification, the carrier transport is considered to be one-dimensional. So, according to the kinetic energy E_x of carriers, there are three current components in the conduction current of carriers traversing grain boundaries (Fig. 12), where w, δ, $q\phi$ and qV_b are the DRB width, the grain boundary width, the grain boundary barrier height and the DRB height, respectively. At very low temperatures, $E_x<qV_b$, carriers traverse the composite grain boundaries only by tunneling, forming the field emission current J_1; At intermediate temperatures, $qV_b<E_x<q\phi$, carriers cross the DRBs by thermionic emission and penetrate the grain boundary barrier by tunneling, forming the composite current J_2; At very high temperatures, $E_x>q\phi$, carriers traverse the composite grain boundary completely by thermionic emission, forming the thermionic emission current J_3. In the temperature range of polysilicon devices working, J_2 is dominant, and J_1 and J_3 could be neglected (Mandurah, et al., 1981).

In our tunneling piezoresistive model, the piezoresistive effect of grain boundaries is due to that the stress induced deformation gives rise to the split-off of the degenerate heavy and light hole sub-bands, thereby causing the carrier transfer between two bands and the conduction mass shift. Inside each grain, due to the single crystal nature of grain neutral regions, the gauge factor of this regions, GF_g, is dependent on the PRC of Si single crystals, π_g. The gauge factor of composite grain boundaries, GF_b, is dependent on the PRC of DRBs (π_d) and the PRC of grain boundary barriers (π_δ). Hence, in order to explain the piezoresistive behavior of PSNFs theoretically, it is necessary to deduce the relationship between π_g, π_d and π_δ.

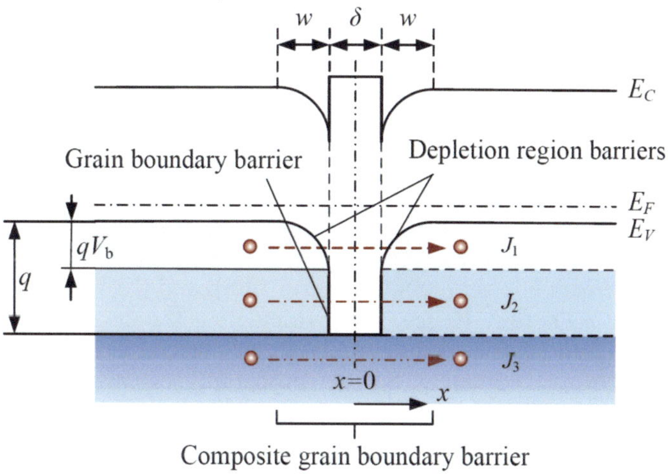

Fig. 12. Energy band structure and carrier transport mechanisms near grain boundaries

6.3 Tunneling current through grain boundary barriers

For DRBs, based on the dependence of thermionic emission current on strain, the relational expressions of longitudinal PRC π_{dl} and transversal PRC π_{dt} in the <111> orientation have been derived in our previous work (Liu, et al., 2004) and expressed as:

$$\pi_{dl} = 0.525\,\pi_{gl} \tag{4}$$

$$\pi_{dt} = 0.616\,\pi_{gt} \tag{5}$$

where π_{gl} and π_{gt} are the longitudinal and transversal PRCs of p-type monocrystalline silicon in the <111> orientation, respectively.

Before deducing the PRC π_δ, the conduction current of carriers penetrating grain boundary barriers must be determined. Fig. 13 provides the energy band diagram and tunneling mechanism of grain boundary barrier omitting DRBs. It is assumed that the voltage drop over the grain boundary barrier is V_δ. Using Fermi-Dirac statistics, the number of holes having energy within the range dE_x incident from left to right on the grain boundary barrier per unit time per unit area is (Murphy & Good, 1956):

$$N(T,\xi,E_x)dE_x = \frac{4\pi \cdot m_d kT}{h^3}\ln\left\{1+\exp\left[\frac{-(E_x+\xi)}{kT}\right]\right\}dE_x \tag{6}$$

where m_d is the effective mass of holes for state density, $\xi = E_F - E_V$, is the difference of Fermi level and valence band edge, h is Planck's constant, k is Boltzmann's constant, T is the absolute temperature.

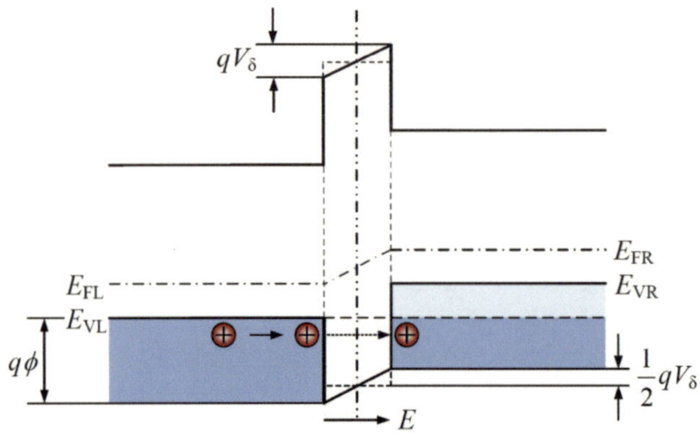

Fig. 13. Energy band diagram and tunneling mechanism of grain boundary barrier omitting depletion region barriers.

The grain boundary width δ is very small (around 1nm), and the number of the holes with high energies around $q\phi$ is few. Hence, when calculating the current density, the oblique distribution of energy band at the top of grain boundary barrier in Fig. 13 can be substituted by the horizontal line approximately. So, the probability of carriers with the energy E_x ($0 \le E_x \le q\phi - qV_\delta / 2$) tunneling the GB barrier is given by:

$$D(E_x) = \exp\left\{\frac{-4\pi\delta}{h}[2m_i(a - E_x)]^{1/2}\right\} \tag{7}$$

$$a = q\phi - \frac{1}{2}qV_\delta \tag{8}$$

where m_i is the hole effective mass in the tunneling direction. In Fig. 13, the left valence band edge E_{VL} is taken to be the zero point of energy. By deducing from Eqs. (6)-(8), the current density of holes tunneling grain boundary barrier from left to right is:

$$S_{LR} = \int_0^a N(T,\xi,E_x)\cdot D(E_x)dE_x \tag{9}$$

The current density of holes tunneling grain boundary barrier from right to left is:

$$S_{RL} = \int_0^a N(T,\xi',E_x)\cdot D(E_x)dE_x \tag{10}$$

$$\xi' = \xi + qV_\delta \tag{11}$$

By simplifying the logarithmic function term in Eq. (6) into the exponential form, Eq. (9) can be expressed as:

$$S_{LR} = \frac{4\pi \cdot m_d kT}{h^3}\int_0^a \exp\left[\frac{-(E_x+\xi)}{kT}\right]\cdot D(E_x)dE_x\,. \tag{12}$$

Considering the fact that the holes gather mostly near the valence band edge, when solving the integration in Eq. (12), the square root term is expended by the Taylor's series as follows:

$$(a-E_x)^{1/2}=\sqrt{a}-\frac{E_x}{2\sqrt{a}}+\cdots. \tag{13}$$

Substituting Eq. (13) into Eq. (7), Eq. (12) can be solved out by integrating:

$$S_{LR}=\frac{4\pi m_d k^2T^2}{h^3c_1}\left[\exp\left(-\frac{2\pi\delta}{h}\sqrt{2m_ia}-\frac{a+\xi}{kT}\right)-\exp\left(-\frac{4\pi\delta}{h}\sqrt{2m_ia}-\frac{\xi}{kT}\right)\right], \tag{14}$$

where

$$c_1=\frac{2\pi\delta}{h}kT\sqrt{\frac{2m_i}{a}}-1. \tag{15}$$

Similarly,

$$S_{RL}=\frac{4\pi m_d k^2T^2}{h^3c_1}\left[\exp\left(-\frac{2\pi\delta}{h}\sqrt{2m_ia}-\frac{a+\xi'}{kT}\right)-\exp\left(-\frac{4\pi\delta}{h}\sqrt{2m_ia}-\frac{\xi'}{kT}\right)\right]. \tag{16}$$

Then, the current density of tunneling boundary barrier region can be given by:

$$\begin{aligned}J_\delta&=q(S_{LR}-S_{RL})\\&=q\frac{4\pi m_d k^2T^2}{h^3c_1}\exp\left(-\frac{\xi}{kT}\right)\cdot\left[\exp\left(c_2-\frac{a}{kT}\right)-\exp\left(c_2-\frac{a+qV_\delta}{kT}\right)-\exp(2c_2)+\exp\left(2c_2-\frac{qV_\delta}{kT}\right)\right],\end{aligned} \tag{17}$$

where

$$c_2=-\frac{2\pi\delta}{h}\sqrt{2m_ia}. \tag{18}$$

In the case of low voltage bias ($qV_\delta << kT$), the exponential terms in Eq. (17) can be expanded by using the Taylor's series. After taking the first order approximation, it yields:

$$J_\delta=\frac{4\pi q^2m_dkT}{h^3c_1}\exp\left(-\frac{\xi_p}{kT}\right)\cdot\left[\exp\left(c_2-\frac{a}{kT}\right)-\exp(2c_2)\right]\cdot V_\delta. \tag{19}$$

Considering the hole concentration formula:

$$p=N_V\exp\left(\frac{-\xi}{kT}\right)=2\left(\frac{2\pi m_dkT}{h^2}\right)^{\frac{3}{2}}\exp\left(\frac{E_V-E_F}{kT}\right), \tag{20}$$

and then Eq. (19) can be rewritten as:

$$J_\delta=\frac{pq}{c_1}\left(\frac{kT}{2\pi m_d}\right)^{1/2}\left[\exp\left(c_2-\frac{a}{kT}\right)-\exp(2c_2)\right]\cdot\frac{qV_\delta}{kT}=pJ_{\delta0}. \tag{21}$$

When two sub-bands split off under an axial stress, the total tunneling current (J_δ) consists of tunneling currents of heavy holes ($J_{\delta 1}$) and light holes ($J_{\delta 2}$) and can be expressed as:

$$J_\delta = \sum_{j=1}^{2} J_{\delta j} = J_{\delta 1} + J_{\delta 2}, \tag{22}$$

$$J_{\delta j} = p_j \left(J_{\delta 0}\right)_j, \tag{23}$$

where $J_{\delta j}$ is the tunneling current component of degenerate sub-band, p_j is the corresponding hole concentration, the subscript j=1, 2, represents the heavy and light hole sub-bands, respectively.

6.4 Piezoresistance coefficient of grain boundary barriers

When the heavy and light hole sub-bands split off under stress, the band shift ε' is defined as the shift of two split-off sub-bands (E_{V1} and E_{V2}) relative to the initial degenerate band (E_V). For the sake of simplification, the applied axial stress is assumed to be along the <111> orientation. According to the result of the cyclotron resonance experiment (Hensel & Feher, 1963), the effective mass of holes under an axial stress is obtained in Table 1, where m_{lj} and m_{tj} are the longitudinal and transversal effective mass of holes at the sub-band E_{Vj}, respectively.

m_{l1}	m_{t1}	m_{d1}	m_{l2}	m_{t2}	m_{d2}
0.870	0.170	0.293	0.135	0.369	0.264

Table 1. Hole effective mass in highly stressed silicon (unit: free-electron mass m_0)

The split-off heavy and light hole sub-bands are E_V+ε' and E_V-ε', respectively. By differentiating Eq. (20) and substituting dE_V by the band shift ε', the concentration changes of two sorts of holes are, respectively:

$$\Delta p_1 = N_{v1} \exp\left(\frac{E_V - E_F}{kT}\right) \cdot \frac{\varepsilon'}{kT}, \tag{24}$$

$$\Delta p_2 = -N_{v2} \exp\left(\frac{E_V - E_F}{kT}\right) \cdot \frac{\varepsilon'}{kT}. \tag{25}$$

When the uniaxial stress is $\bar{\sigma}$, the band shift ε' is (Hensel & Feher, 1963):

$$\varepsilon' = \frac{1}{3} D_u C_{44}^{-1} \bar{\sigma}, \tag{26}$$

where D_u is deformation potential constant, C_{44} is the corresponding elastic stiffness constant. Due to the different effective mass of heavy and light holes, the change in the corresponding hole concentrations can lead the tunneling current J_δ to vary, which is the principle of tunneling piezoresistive effect. The relative change of the equivalent tunneling resistivity is:

$$\frac{\Delta\rho_\delta}{\rho_\delta} = -\frac{\Delta J_\delta}{J_\delta} = -\frac{\Delta p_1 (J_{\delta 0})_1 + \Delta p_2 (J_{\delta 0})_2}{p_1 (J_{\delta 0})_1 + p_2 (J_{\delta 0})_2}. \tag{27}$$

Substituting Eqs. (20), (24) and (25) into Eq. (27), it yields:

$$\frac{\Delta\rho_\delta}{\rho_\delta}=\frac{1-\left(\frac{m_{d1}}{m_{d2}}\right)^{3/2}\cdot\frac{(J_{\delta 0})_1}{(J_{\delta 0})_2}}{1+\left(\frac{m_{d1}}{m_{d2}}\right)^{3/2}\cdot\frac{(J_{\delta 0})_1}{(J_{\delta 0})_2}}\cdot\frac{\varepsilon'}{kT}. \tag{28}$$

From Eqs. (21)-(23), it results in:

$$\frac{(J_{\delta 0})_1}{(J_{\delta 0})_2}=\frac{(c_1)_2\left\{\exp\left[(c_2)_1-\frac{a}{kT}\right]-\exp\left[2(c_2)_1\right]\right\}}{(c_1)_1\left\{\exp\left[(c_2)_2-\frac{a}{kT}\right]-\exp\left[2(c_2)_2\right]\right\}}, \tag{29}$$

where $(c_1)_j$ and $(c_2)_j$ can be determined by Eqs. (15) and (18), respectively. According to the experimental data (Mandurah, et al., 1981), the grain boundary width δ is set to be 1nm and the grain boundary barrier height $q\phi$ is about 0.6eV. Thus, for heavy holes, j=1, $(c_1)_1$ and $(c_2)_1$ are calculated to be -0.84 and -3.72, respectively; for light holes, j=2, $(c_1)_2$ and $(c_2)_2$ are calculated to be -0.94 and -1.46, respectively. In general, $qV_\delta<<q\phi$, it can be obtained from Eq. (8) that $a\approx q\phi$. From Eq. (29) and Table 1, it yields:

$$\left(\frac{m_{d1}}{m_{d2}}\right)^{3/2}=\sqrt{\frac{m_{l1}m_{t1}^2}{m_{l2}m_{t2}^2}}=1.180, \tag{30}$$

$$\frac{(J_{\delta 0})_1}{(J_{\delta 0})_2}=1.22\times10^{-2}. \tag{31}$$

Finally, using Eqs. (26), (28)-(31), the longitudinal PRC of grain boundary barriers along the <111> orientation is expressed as:

$$\pi_{\delta l}=\frac{\Delta\rho_\delta}{\rho_\delta\bar{\sigma}}=\frac{0.972}{3}\cdot D_u C_{44}^{-1}\cdot\frac{1}{k_0T}. \tag{32}$$

Similarly, the transversal PRC along <111> orientation is:

$$\pi_{\delta t}=\frac{\Delta\rho_\delta}{\rho_\delta\bar{\sigma}}=\frac{-0.684}{3}\cdot D_u C_{44}^{-1}\cdot\frac{1}{k_0T}. \tag{33}$$

From our previous research results, the longitudinal and transversal PRCs of p-type monocrystalline silicon (grain neutral regions) under a uniaxial stress $\bar{\sigma}$ applied along the <111> orientation can be expressed as follows, respectively (Liu, et al., 2004):

$$\pi_{gl}=\frac{0.695}{3}\cdot D_u C_{44}^{-1}\cdot\frac{1}{k_0T}, \tag{34}$$

$$\pi_{gt} = -\frac{0.435}{3} \cdot D_u C_{44}^{-1} \cdot \frac{1}{k_0 T} \,. \tag{35}$$

Comparing Eqs. (32) and (33) with (34) and (35) correspondingly, it yields:

$$\pi_{\delta l} = 1.4\pi_{gl} \,, \tag{36}$$

$$\pi_{\delta t} = 1.6\pi_{gt} \,. \tag{37}$$

From Eqs. (36) and (37), it can be seen that the PRCs π_δ and π_g present a proportional relationship and the PRC π_δ is larger than π_g.

6.5 Piezoresistance coefficient of composite grain boundaries

From the above theoretical analysis, it can be seen that both the PRC of depletion region barriers and the PRC of grain boundary barriers are proportional to the PRC of grain neutral regions. Noticeably, according to Eqs. (4), (5), (36) and (37), the PRC of depletion region barriers π_d is lower than π_g, while the PRC of grain boundary barriers π_δ is higher than π_g. Therefore, the relationship between the PRC of composite grain boundaries π_b and the PRC of grain neutral regions π_g is dependent on the weights of the equivalent resistivities ρ_d (for depletion region barriers) and ρ_δ (for grain boundary barriers) in the equivalent resistivity ρ_b of composite grain boundaries. In this case, π_b can be expressed as:

$$\pi_b = \frac{\Delta\rho_b}{\rho_b \bar{\sigma}} = \frac{\rho_d}{\rho_b} \cdot \frac{\Delta\rho_d}{\rho_d \bar{\sigma}} + \frac{\rho_\delta}{\rho_b} \cdot \frac{\Delta\rho_\delta}{\rho_\delta \bar{\sigma}} = \frac{\rho_d}{\rho_b}\pi_d + \frac{\rho_\delta}{\rho_b}\pi_\delta \,, \tag{38}$$

where

$$\rho_b = \rho_d + \rho_\delta \,. \tag{39}$$

If the potential drops across depletion regions on the left and right hand sides of the grain boundary are denoted by V_L and V_R, respectively; then the potential drops on depletion region barriers and grain boundary barrier are V_L+V_R and V_δ, respectively. So, Eq. (38) can be expressed as follows:

$$\pi_b = \frac{V_L + V_R}{V_0}\pi_d + \frac{V_\delta}{V_0}\pi_\delta \,, \tag{40}$$

$$V_0 = V_\delta + V_L + V_R \,, \tag{41}$$

where V_0 is the potential drop over the composite grain boundary. Thus, it can be seen that determining the proportional relationship between V_L+V_R and V_δ is the key to solve out π_b. Because the polysilicon usually work under low current and low voltage bias, the condition of $V_L+V_R<4V_b$ can be always satisfied. Then, the relationship of V_L, V_R, V_b and V_δ can be obtained (Mandurah, et al., 1981):

$$2V_b^{1/2} = (V_b + V_R)^{1/2} + (V_b - V_L)^{1/2} \,, \tag{42}$$

$$V_\delta = \delta\left(\frac{qN_A}{2\varepsilon_s\varepsilon_0}\right)^{1/2}\left[(V_b+V_R)^{1/2}-(V_b-V_L)^{1/2}\right]. \tag{43}$$

According to the approximation of depletion region, it yields

$$V_b = \frac{qN_A W^2}{2\varepsilon_s\varepsilon_0}, \tag{44}$$

$$W = \frac{N_t}{2N_A}, \tag{45}$$

where N_A is the boron doping concentration, N_t is the trap density at grain boundary, ε_s and ε_0 are the relative and vacuum dielectric constants of silicon, respectively. In this paper, the trap density N_t is taken to be 1.0×10^{13} cm^{-2}. By calculating, the distribution of the voltage V_δ normalized to the voltage V_0 as a function of N_A is provided in inset of Fig. 14.

The experimental results indicate that the longitudinal piezoresistive sensitivity is twice larger than the transversal sensitivity. Hence, the following derivations are based on the longitudinal piezoresistive effect. Considering that the fundamental cubic piezoresistance coefficients π_{11}, π_{12} and π_{44} satisfy $\pi_{11}+2\pi_{12}<<2\pi_{44}$ for p-type single crystal silicon, the longitudinal piezoresistance coefficient π_{gl} can be taken to be $2\pi_{44}/3$. Therefore, combining Eqs. (4), (36), (40) and (41), the piezoresistance coefficient of composite grain boundary can be expressed as:

$$\pi_{bl} = (0.525+0.875\frac{V_\delta}{V_0})\pi_{gl} = (0.35+0.583\frac{V_\delta}{V_0})\pi_{44} \tag{46}$$

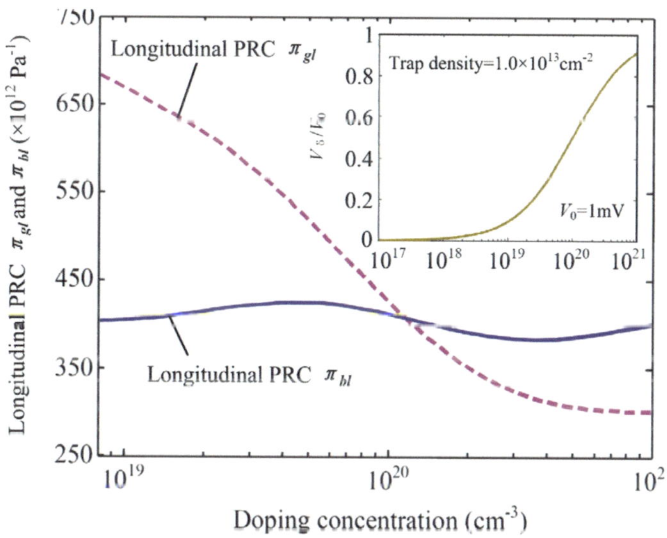

Fig. 14. Dependences of the longitudinal piezoresistance coefficients (π_{gl} and π_{bl}) and the potential drop ratio V_δ/V_0 on the doping concentration. The potential drop V_0 on composite grain boundary and the trap density at grain boundary is taken to be 1 mV and 1.0×10^{13} cm^{-2}.

In virtue of the dependences of the potential drop ratio V_δ/V_0 and π_{44} on the doping concentration (Tufte & Stelzer, 1963; Toriyama & Sugiyama, 2002; Shi, et al., 2009), the relational curves of longitudinal piezoresistance coefficients π_{gl} and π_{bl} on the doping concentration are obtained as shown in Fig. 14. Seen from the inset of Fig. 14, the potential drop ratio V_δ/V_0 increases sharply at high doping concentrations, therefore resulting in that the longitudinal PRC of composite grain boundary becomes much larger than that of grain neutral regions when $V_\delta/V_0 > 0.543$ (the doping concentration> $1.1\times10^{20}cm^{-3}$).

7. Experimental results and analyses

7.1 Magnetron sputtering samples

After high temperature annealing, micrograins are formed and the films transfer from the amorphous state to polycrystalline state. The dependence of the film resistivity on the annealing time is provided in Fig. 15. From Fig. 15, it can be seen that when the annealing time is 60 min, the film resistivity is the lowest. Noticeably, if the annealing is performed after the boron diffusion, the dopants are almost not activated and the corresponding resistivity is very high. It is likely due to that the most of boron dopants are captured by the dangling bonds, and the post-annealing is difficult to break the covalent bonds.

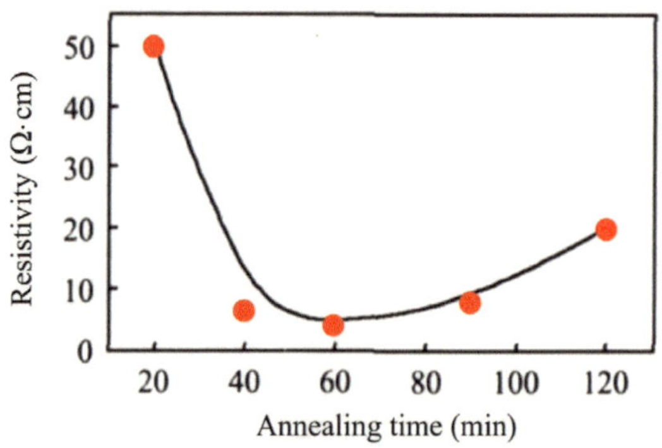

Fig. 15. Dependence of the resistivity of magnetron sputtering films on the annealing time

The measurement results show that the gauge factor of magnetron sputtering films is ranged from 10 to 80. The uniformity of experimental data is poor, so that this film preparation could not be suited for the fabrication of sensors. Once the crystallization of films is improved, the magnetron sputtering may be a favourable method of preparing polycrystalline silicon films.

7.2 LPCVD films with different thicknesses

The dependences of longitudinal gauge factor and grain size on film thickness are provided in Fig. 16. For PSCFs, the average value of gauge factor is between 20 and 25. It is due to that the larger grain size and better crystallinity reduce greatly the proportion of composite grain boundary resistivity to film resistivity, resulting in that the gauge factor depends on the piezoresistance coefficient of grain neutral regions. For PSNFs, the average gauge factor is enhanced to be 32~34. Due to the reduction of film thickness, the film crystallinity and the grain size are both diminished clearly; at the same time, the grain boundary width and the trap density at grain boundaries increase correspondingly. Therefore, the proportion of composite grain boundary resistivity to film resistivity is improved further. In this case, the gauge factor depends on the piezoresistance coefficient of composite grain boundary. According to the above tunneling piezoresistive theory, when the trap density at grain boundaries is not too high, the piezoresistance coefficient of composite grain boundary is larger than that of grain neutral regions at high doping levels. Hence, the PSNFs exhibit the enhanced piezoresistive sensitivity (about 50%), compared with the PSCFs. The intervenient films are ascribed to the transition type.

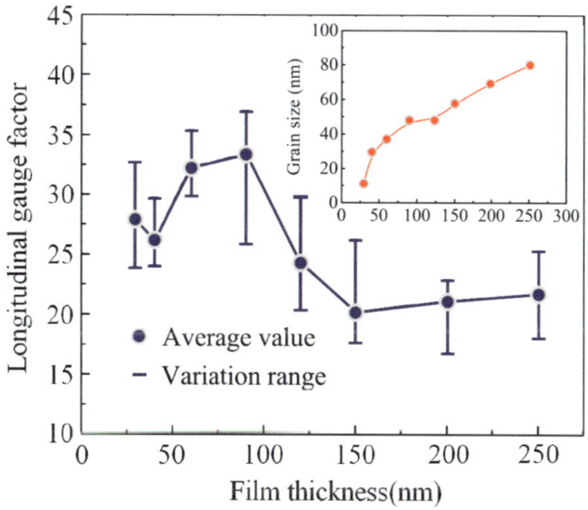

Fig. 16. Longitudinal gauge factor and grain size of polysilicon thin films with different thicknesses.

When the film thickness is further reduced (thinner than 50 nm), the film crystallinity and the grain size are diminished greatly, thereby resulting in the increase of trap density at grain boundaries. At the same doping level, this makes more carriers captured by traps at grain boundaries and broadens the width of DRBs beside the grain boundary. When the current flows through the composite grain boundary, the potential drop on broadened DRBs increases. According to Eqs. (40) and (41), it increases the proportion of DRB piezoresistance coefficient to composite grain boundary piezoresistance coefficient. Based on the tunneling piezoresistive theory, the piezoresistance coefficient of DRBs is much smaller than that of grain neutral regions and grain boundary barriers. Consequently, the composite grain boundary piezoresistance coefficient of NL-PSTFs is reduced compared with the PSNFs. Although the increase in trap density improves the proportion of the composite grain boundaries of NL-PSTFs, the decrease of the composite grain boundary piezoresistance coefficient makes the longitudinal gauge factor of NL-PSTFs smaller than that of PSNFs. Additionally, the NL-PSTFs is the transition type towards the nanocrystalline silicon. Due to the high gauge factor of nanocrystalline silicon (He, et al., 1996), the gauge factor of NL-PSTFs increases slightly with the reduction of film thickness.

7.3 LPCVD films with different doping concentrations

The relationship between the longitudinal gauge factor of 80 nm-thick PSNFs and doping concentration is shown in Fig. 17. It can be seen from Fig. 17 that the gauge factor reaches the maximum as the doping concentration is about 4×10^{19} cm^{-3}; however, when the doping concentration is higher than 2×10^{20} cm^{-3}, the gauge factor increases again with the increase of doping concentration. This exceptional increase of gauge factor has not been observed in polysilicon thicker films and can not be explained reasonably by the existing piezoresistive theory. Here, the phenomenon is analyzed based on the tunneling piezoresistive theory.

The longitudinal gauge factor of PSNFs could be expressed as the weighted superposition of gauge factors of grain neutral regions and composite grain boundaries, and the weight factors are the products of the resistivity ratios and width ratios of grain neutral regions and composite grain boundaries to films, respectively. Thus, the longitudinal gauge factor of PSNFs can be given by:

$$GF_l = \frac{L-(2w+\delta)}{L}\cdot\frac{\rho_g}{\rho}GF_{gl} + \left[1-\frac{L-(2w+\delta)}{L}\cdot\frac{\rho_g}{\rho}\right]GF_{bl} \tag{47}$$

where ρ_g and ρ are the resistivity of grain neutral regions and the film resistivity, respectively; L is the grain size; GF_{gl} and GF_{bl} are the longitudinal gauge factors of grain neutral regions and composite grain boundaries, respectively.

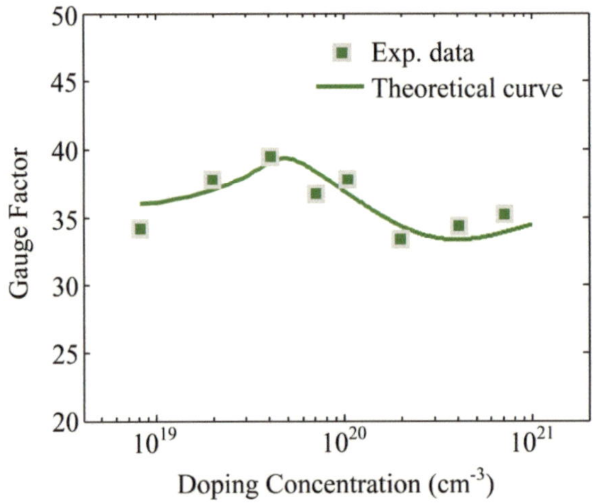

Fig. 17. Experiment data and theoretical curve of the longitudinal gauge factor of the PSNFs with different doping concentrations

Due to the random orientation of PSNFs, the gauge factors GF_l, GF_{gl} and GF_{bl} should be substituted by the average gauge factors $<GF_l>$, $<GF_{gl}>$ and $<GF_{bl}>$ derived along the all orientations. Based on the method presented by Schubert (Schubert, et al., 1987), the proportion factor of the average gauge factors ($<GF_{gl}>$ and $<GF_{bl}>$) with random grain orientations and the gauge factors ($GF_{gl,<111>}$ and $GF_{bl,<111>}$) along the <111> orientation was calculated to be 0.537. Therefore, the average gauge factors can be expressed as:

$$<GF_{gl}> = 0.537GF_{gl,<111>} = 0.537\cdot(1+2\upsilon+f_Y\cdot Y_{Si<111>}\pi_{gl,<111>}) \tag{48}$$

$$<GF_{bl}> = 0.537GF_{bl,<111>} = 0.537\cdot(1+2\upsilon+f_Y\cdot Y_{Si<111>}\pi_{bl,<111>}) \tag{49}$$

where υ is Poisson's ratio and taken to be 0.25, $Y_{Si<111>}$ is Young's modulus of single-crystal silicon along <111> orientation and taken to be 1.87×10^{20} Pa, f_Y is a correction factor and

taken to be 0.85, $\pi_{gl,<111>}$ and $\pi_{bl,<111>}$ are the piezoresistance coefficients of grain neutral regions and composite grain boundaries along the <111> orientation, respectively. The introduction of the correction factor f_Y is due to that the Young's modulus of polysilicon is lower than that of single-crystal silicon (Greek, et al., 1999; Yi & Kim, 1999).
According to Eq. (47), in order to figure out the relationship between the longitudinal gauge factor and doping concentration, it is necessary to determine the resistivity ratio ρ_g/ρ. Generally, when calculating the gauge factor, the resistivity of grain neutral regions ρ_g is taken to be the value of monocrystalline silicon (Schubert, et al., 1987; Mosser, et al., 1991), and the resistivity of polysilicon ρ is taken to be the actual measured value. For the samples with different doping concentrations, the fitting dependence of ρ on N_A at room temperature is:

$$\rho = 15.651e^{-1.73\times10^{-19}N_A} + 0.014\ (\Omega\ \text{cm}) \tag{50}$$

In semiconductor physics, the resistivity ρ_g at room temperature can be expressed as

$$\rho_g = a_s N_A^{-1} + a_i \tag{51}$$

where a_s and a_i are the constants determined by the scattering probability resulted from the acoustic phonons and impurity scattering, respectively. According to the relational curve of the resistivity of p-type monocrystalline silicon versus doping concentration, in the doping concentration range of 5×10^{18}~10^{21} cm^{-3}, by fitting in the function form of Eq. (51), the relationship between ρ_g and N_A at room temperature is

$$\rho_g = 6.8741\times10^{16} N_A^{-1} + 2\times10^{-3}\ (\Omega\ \text{cm}) \tag{52}$$

Although the grain neutral regions possess monocrystal structure, the defect density is very high and increases with the grain size reducing. Therefore, it is necessary to take the scattering process of lattice defects into consideration when calculating the resistivity. If the defect density is independent of doping concentration, a modifying factor a_d determined by defect density can be introduced, thus Eq. (52) can be modified as

$$\rho_g = (6.8741\times10^{16} + a_d) N_A^{-1} + 2\times10^{-3}\ \ (\Omega\ \text{cm}) \tag{53}$$

For the PSNFs mentioned here, the value of a_d is 2×10^{16} Ω cm^{-2}. Here, the scattering process of lattice defects is equivalent to the ionized impurity scattering at the doping concentration of 10^{19} cm^{-3}. This is comprehensible for the PSNFs (the thickness is 80 nm and the average grain size is 27 nm). When it is assumed that 1% of the bonds on the grain surface are dangling bonds and these dangling bonds are regarded as defects, the defect density is at the level of 10^{19} cm^{-3}. Using Eqs. (47)-(50) and (53) obtained from the above analysis and combining the dependences of piezoresistance coefficients π_{gl} and π_{bl} on N_A, the theoretical curve of gauge factor versus doping concentration for PSNFs was gained in Fig. 17. Thus, it was seen that the calculating results of tunnelling piezoresistive model were greatly in agreement with the experiment data.

7.4 LPCVD films with different deposition temperatures

Fig. 18 provides the resistivity of highly boron doped PSNFs versus deposition temperature. It can be seen that the resistivity changes from 1.54×10^{-1} to 4.9×10^{-3} Ω cm with elevating deposition temperature. Considering the experiment results that the grain size increases with raising deposition temperature, it indicates that the weight of the resistivity of composite grain boundaries ρ_b in the resistivity of PSNFs ρ is reduced by increasing deposition temperature. Because ρ_b is dependent on the resistivity of grain boundary barriers ρ_δ and the resistivity of depletion region barriers ρ_d (i.e., $\rho_b = \rho_\delta + \rho_d$), the elevation of deposition temperature might reduce either of ρ_δ and ρ_d. According to the SEM and XRD results, there are more amorphous contents in RC PSNFs (560~600°C samples) than in DC PSNFs (620~670°C samples). The existence of amorphous phases at grain boundaries could increase the resistivity ρ_δ. On the other hand, the high doping concentration narrows the width of depletion region barriers to a few angstroms, so that the contribution of ρ_d to ρ_b could be neglected. Therefore, at high doping concentration, the resistivity of composite grain boundaries ρ_b is dependent on the resistivity of grain boundary barriers ρ_δ, and the reduction of amorphous contents at grain boundaries caused by elevating deposition temperature is responsible for the falloff of the film resistivity ρ. However, the resistivity of 620°C samples is slightly higher than that of 600°C ones. It is likely due to the recrystallization of 600°C samples after the pre-annealing at 950°C.

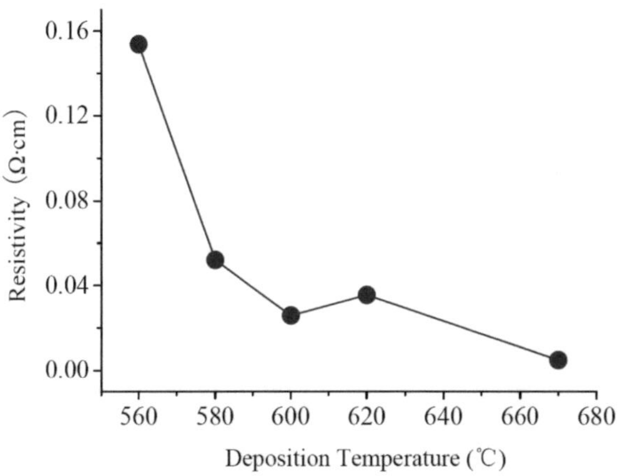

Fig. 18. Resistivity of boron doped PSNFs versus deposition temperature

The dependences of the resistance change $\Delta R/R_0$ in longitudinal and transversal piezoresistors on the strain ε with different deposition temperatures are shown in Figs. 19(a) and (b), respectively. Obviously, the longitudinal and transversal piezoresistances vary linearly with the strain. From the insets of Figs. 19(a) and (b), it can be seen that RC PSNFs and DC PSNFs exhibit different piezoresistive properties. And the critical deposition temperature differentiating RC PSNFs and DC PSNFs is around 605°C. The samples deposited below this critical temperature present amorphous appearance mixed with polycrystals, while the samples above this value present better polysilicon appearance. This critical value is consistent with the reported result (French & Evens, 1989).

For the longitudinal piezoresistive sensitivity, it can be seen in Fig. 19(a) that the gauge factors of RC or DC PSNFs decrease with elevating deposition temperature. As discussed above, the amorphous contents at grain boundaries are reduced by raising deposition temperature. For RC PSNFs, when the deposition temperature is lowered, the crystallinity of samples is aggravated and there are more amorphous phases existing at grain boundaries. The increase of amorphous phases raises the resistivity ρ_δ. Moreover, the deficient crystallinity increases the width of grain boundary barriers δ and further increases the weight of ρ_b in the film resistivity

ρ. According to the existing piezoresistive model, the PRCs of DRBs (π_d) and grain boundary barriers (π_δ) are lower than that of grain neutral regions π_g. Thus, it implies that the piezoresistive sensitivity of PSNFs with more amorphous contents and smaller grain size should be much lower. However, it is obvious that the deduction is inconsistent with the experiment results. Based on the tunneling piezoresistive theory presented here, the longitudinal PRC of composite grain boundaries π_{bl} is much larger than that of grain neutral regions π_{gl} at high doping concentration. As a result of lowering deposition temperature, both the resistivity ρ_b and the weight of ρ_b in the film resistivity ρ increase. It enhances the contribution of the PRC π_{bl} on the piezoresistive sensitivity, thereby increasing longitudinal gauge factors. For DC PSNFs, the XRD analysis indicates that there are more amorphous phases in the 670°C samples than in the 620°C ones, which is likely due to the <110> preferred growth aggravating disordered states of grain boundaries. It makes the resistivity ρ_δ of 670°C samples higher than 620°C samples. However, the SEM results show that the grain size of 670°C samples is ~70nm and much larger than that of 620°C ones. This reduces severely the weight of the resistivity ρ_b in the film resistivity ρ and weakens the contribution of the PRC π_{bl} on the piezoresistive sensitivity. Therefore, the longitudinal gauge factor of 670°C samples with larger grains is much lower than that of 620°C ones.

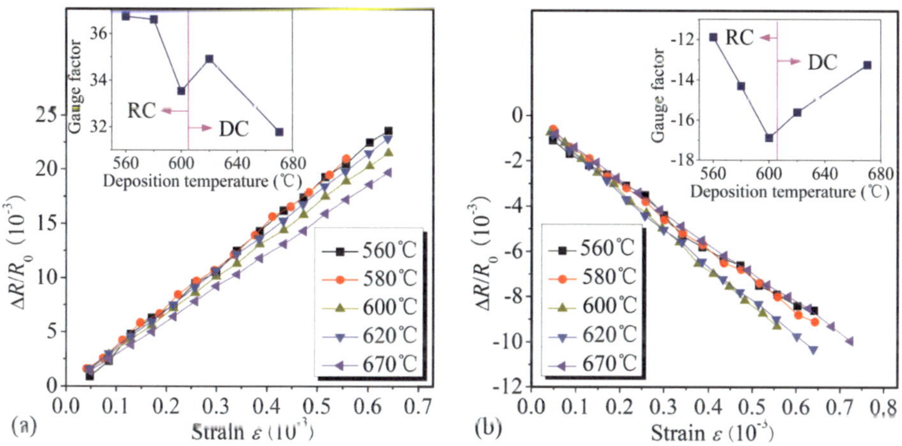

Fig. 19. (a) Dependences of the resistance change $\Delta R/R_0$ in longitudinal piezoresistors on strain ε with different deposition temperatures, and the longitudinal gauge factor vs. deposition temperature. (b) Dependences of $\Delta R/R_0$ in transversal piezoresistors on strain ε with different deposition temperatures, and the transversal gauge factor vs. deposition temperature.

For the transversal piezoresistive sensitivity, the inset of Fig. 19(b) shows that the magnitude of the transversal gauge factor in DC PSNFs increases with lowering deposition temperature, similar to the longitudinal gauge factor dependence; while the magnitude of the transversal gauge factor in RC PSNFs falls off drastically with lowering deposition temperature. Comparing the insets of Figs. 19(a) and (b), it can be seen that the longitudinal gauge factor of DC PSNFs is about twice larger than the transversal one. However, the longitudinal and transversal gauge factors of RC PSNFs do not satisfy the above proportional relation. On the contrary, the transversal gauge factor of RC PSNFs decreases from 1/2 to 1/3 of the longitudinal one with lowering deposition temperature, which might be due to the degradation of transversal piezoresistive effect in amorphous silicon.

It is noteworthy that the stress-induced modulation of surface depletion region width in silicon nanowires (He &Yang, 2006) is not fit for the explanation of enhanced piezoresistive effect in PSNFs. For silicon nanowires, the surface depletion regions are parallel to the direction of carrier transport, and the change in surface potential barrier caused by stress only influences the conducting channel width of carriers along silicon nanowires. However, for PSNFs, the depletion regions are perpendicular to the direction of carrier transport and the carriers have to traverse them by thermionic emission or tunneling. Moreover, the depletion region width is reduced greatly at high doping concentration and can be neglected. So the tunneling effect of carriers becomes dominant.

8. PSNF-based pressure sensor

Fig. 20 provides the photo of a PSNF-based pressure sensor. In Fig. 20, there are 4 sets of half Wheatstone bridge; precise matching of piezoresistors can be obtained by selecting proper half bridge. Some main performance characteristics of the PSNF-based pressure sensor are listed in Table 2. It can be seen that the PSNF-based pressure sensor possesses favourable sensitivity and temperature stability.

Fig. 20. Photos of PSNF-based pressure sensor chip and packaging

Parameter	Value	
Working temperature (°C)	25	200
Sensitivity (mV/V/MPa)	22.23	18.27
Full scale output (mV)	66.38	54.82
Offset (mV)	9.63	9.49
Temperature coefficient of sensitivity (%/°C)	-0.098	-0.098
Temperature coefficient of offset (%/°C)	-0.017	-0.017
Linearity (%FS)	0.06	0.38
Hysteresis (%FS)	0.49	0.93
Repeatability (%FS)	1.08	2.07

Table 2. Performance characteristics of the PSNF-based pressure sensor

9. Summary

Our research group has been spending a great effort on the investigation of polycrystalline silicon-based sensors. Through the alteration of technological conditions, the enhanced piezoresistance effect of heavily doped polycrystalline silicon nano thin films was discovered, and this characteristic could be used in the design and fabrication of piezoresistive sensors with miniature volume, high sensitivity and wide working temperature range. In the experiments, the influences of film thickness, doping concentration and deposition temperature on the piezoresistive properties of polycrystalline silicon films were studied. The results indicate that the optimal technological parameters are: the thickness of polycrystalline silicon film is in the range of 80-90 nm; the doping concentration is $2\text{-}3\times10^{20}$ cm^{-3}; the deposition temperature is set to be 620 °C. Additionally, in order to explain reasonably the unique piezoresistive phenomenon, the tunnelling piezoresistive model was established. In this model, the contribution of grain boundaries to the piezoresistive effect is taken into consideration. By calculation and derivation, it is proved that the piezoresistance coefficient of composite grain boundaries is much higher than that of grain neutral regions at high doping levels. The experimental data and the theoretical results gain a good agreement. Finally, the PSNF-based pressure sensor was fabricated successfully. The test results show that the sensor provides high sensitivity and very low temperature coefficients. Therefore, it can be concluded that the polycrystalline silicon nano thin films could be potential for the application of MEMS-based piezoresistive sensors.

10. References

Erskine, J.C. (1983). Polycrystalline silicon-on-metal strain gauge transducers. *IEEE Trans. Electron Dev.*, Vol. ED-30, 796-801, 0018-9383

French, P.J.; Evens, A.G.R. (1989). Piezoresistance in polysilicon and its applications to strain gauges. *Solid-State Electron*, Vol. 32, 1-10, 0038-1101

Germer, W.; Tödt, W. (1983). Low-cost pressure/force transducer with silicon thin film strain gauges. *Sens. Actuators*, Vol. 4, 183-189, 0250-6874

Greek, S.; Ericson, F.; Johansson, S.; Furtsch, M.; Rump, A. (1999). Mechanical characterization of thick polysilicon films. Young's modulus and fracture strength evaluated with microstructures. *J. Micromech. Microeng.*, Vol. 9, 245-251, 0960-1317

Gridchin, V.A.; Lubimsky, V.M.; Sarina, M.P. (1995). Piezoresistive properties of polysilicon films. *Sens. Actuators A*, Vol. 49, 67-72, 0924-4247

He, R.; Yang, P. (2006). Giant piezoresistance effect in silicon nanowires. *Nature Nanotech.*, Vol. 1, 42-46, 1748-3387

He, Y.L.; Liu, H.; Yu, M.B.; Yu, X.M. (1996). The structure characteristics and piezo-resistance effect in hydrogenated nanocrystalline silicon films. *Nanostructured Materials*, Vol. 7, 769-777, 0965-9773

Hensel, J.C.; Feher, G. (1963). Cyclotron resonance experiments in uniaxially stressed silicon: valence band inverse mass parameters and deformation potentials. *Phys. Rev.*, Vol. 129, 1041-1062, 0031-899X

Herring, C. (1955). Transport properties of a many-valley semiconductor. *Bell Syst. Tech. J.*, Vol. 34, 237-290, 0030-4018

Jaffe, J.M. (1983). Monolithic polycrystalline silicon pressure transducer. *IEEE Trans. Electron. Dev.*, Vol. ED-30, 420-421, 0018-9383

Kamins, T.I. (1971). Hall mobility in chemically deposited polycrystalline silicon. *J. Appl. Phys.*, Vol. 42, 4357-4365, 0021-8979

Le Berre, M.; Kleimann, P.; Semmache, B. ; Barbier, D. ; Pinard, P. (1996). Electrical and piezoresistive characterization of boron-doped LPCVD polycrystalline silicon under rapid thermal annealing. *Sens. Actuators A*, Vol. 54, 700-703, 0924-4247

Liu, X.W.; Huo, M.X.; Chen, W.P.; Wang, D.H.; Zhang, Y. (2004). Theoretical research on piezoresistive coefficients of polysilicon films. *Chin. J. Semiconduct.*, Vol. 25, 292-296, 0253-4177

Luder, E. (1986). Polycrystalline silicon-based sensors. *Sens. Actucators*, Vol. 10, 9-23, 0250-6874

Malhaire, C.; Barbier, D. (2003). Design of a polysilicon-on-insulator pressure sensor with original polysilicon layout for harsh environment. *Thin Solid Films*, Vol. 427, 362-366, 0040-6090

Mandurah, M.M.; Saraswat, K.C.; Kamins, T.I. (1981). A model for conduction in polycrystalline silicon-Part I: theory. *IEEE Trans. Electron Dev.*, Vol. ED-28, 1163-1171, 0018-9383

Mikoshiba, H. (1981). Stree-sensitive properties of silicon-gate MOS devices. *Solid-State Electron*, Vol. 24, 221-232, 0038-1101

Mosser, V.; Suski, J.; Goss, J.; Obermeier, E. (1991). Piezoresistive pressure sensors based on polycrystalline silicon. *Sens. Actuators A*, Vol. 28, 113-131, 0924-4247

Murphy, E.L.; Good, R.H. (1956). Thermionic emission, field emission and the transition region. *Phys. Rev.*, Vol. 102, 1464-1469, 0031-899X

Onuma, Y.; Sekiya, K. (1974). Piezoresistive properties of polycrystalline silicon thin film. *Jpn. J. Appl. Phys.*, Vol. 11, 420-421, 0021-4922

Rowe, A.C.H. (2008). Silicon nanowires feel the pinch. *Nature Nanotech.*, Vol. 3, 311-312, 1748-3387

Rowe, A.C.H.; Donoso-Barrera, A.; Renner, Ch.; Arsott, S. (2008). Giant room-temperature piezoresistance in a metal-silicon hybrid structure. *Phys. Rev. Lett.*, Vol. 100, 145501-1-4, 0031-9007

Schubert, D.; Jenschke, W.; Uhlig, T.; Schmidt, F.M. (1987). Piezoresistive properties of polycrystalline and crystalline silicon films. *Sens. Actuators*, Vol. 11, 145-155, 0250-6874

Shi, C.-Z.; Liu, X.-W.; Chuai, R.-Y. (2009). Piezoresistive sensitivity, linearity and resistance time drift of polysilicon nanofilms with different deposition temperatures. *Sensors*, Vol. 9, No. 2, 1141-1166, 1424-8220

Smith, C.S. (1954). Piezoresistance effect in germanium and silicon. *Phys. Rev.*, Vol. 94, 42-49, 0031-899X

Taniguchi, M.; Hirose, M.; Osaka, Y. (1978). Substitutional doping of chemically vapor-deposited amorphous silicon. *J. Cryst. Growth*, Vol. 45, 126-129, 0022-0248

Toriyama, T.; Sugiyama, S. (2002). Analysis of piezoresistance in p-type silicon for mechanical sensors. *J. Microelectronmech. Syst.*, Vol. 11, 598-604, 1057-7157

Tufte, O.N.; Stelzer, E.L. (1963). Piezoresistive properties of silicon diffused layers. *J. Appl. Phys.*, Vol. 34, 313-318, 0021-8979

Wang, M.X.; Meng, Z.G.; Zohar, Y.; Wong, M. (2001). Metal-induced laterally crystallized polycrystalline silicon for integrated sensor applications. *IEEE Trans. Electron. Dev.*, Vol. 48, 794-800, 0018-9383

Yi, T.-C.; Kim, C. -J. (1999). Measurement of mechanical properties for MEMS materials. *Meas. Sci. Technol.*, Vol. 10, 706-716, 0957-0233

Zhao, Z.X.; Cui, R.Q.; Meng, F.Y.; Zhao, B.C.; Yu, H.C.; Zhou, Z.B. (2004). Nanocrystalline silicon thin films prepared by RF sputtering at low temperature and heterojunction solar cell. *Materials Letters*, Vol. 58, 3963-3966, 0167-577X

17

Sputtered AlN Thin Films for Piezoelectric MEMS Devices - FBAR Resonators and Accelerometers

*Friedel Gerfers[1], Peter M. Kohlstadt[1], Eyal Ginsburg[1], Ming Yuan He[1], Dean Samara-Rubio[1], Yiannos Manoli[2] and Li-PengWang[1]

[1]*Intel Corporation, Microsystems Technology, Santa Clara*
[2]*Albert-Ludwigs University Freiburg*
[1]*USA*
[2]*Germany*

1. Introduction

Over the past two decades, significant advances have been made in the field of micromachined sensors and actuators. As microelectromechanical systems (MEMS) have become mainstream, a clear need for the integration of materials other than silicon and its compounds into micromachined transducers has emerged. MEMS devices based on piezoelectric materials take advantage of the high energy transduction that scales very favorably upon miniaturization leading to an ever-growing interest in piezoelectric films for MEMS applications.

Piezoelectric materials provide a direct transduction mechanism to convert signals from mechanical to electrical domains and vice versa. The reversible and linear piezoelectric effect manifests as the production of a charge (voltage) upon application of stress (direct effect) and/or as the production of strain (stress) upon application of an electric field (converse effect). Transducers using piezoelectric materials can be configured either as actuators, when the design of the device is optimized for generating strain or stress using the converse piezoelectric effect, or as sensors when the design of the device is optimized for the generation of an electric signal, using direct piezoelectric effect, in response to mechanical input. Furthermore, piezoelectric devices also allow the integration of sensing and actuating elements in one device (Xu et al., 2002). The elementary piezoelectric effects are given by

$$D_i = d_{ij}\sigma_j + \varepsilon^T_{ii} E_i \quad \text{or} \quad D_i = e_{ij} S_j + \varepsilon^S_{ii} E_i \tag{1}$$

$$S_j = s^E_{ij}\sigma_j + d_{ij} E_i \quad \text{or} \quad T_j = c^E_{ij} S_j - e_{ij} E_i \tag{2}$$

* Friedel Gerfers was with Intel Corporation, Microsystems Technology, Santa Clara, USA. He is now with Aquantia Inc., USA. Peter M. Kohlstadt is now with Solyndra Inc., USA. Li-Peng Wang is now with Technologies Inc., USA.

where S_j is the mechanical strain, σ_j is the mechanical stress, E_i is the electric field, D_i is the electrical displacement, c_{ij} is the elastic stiffness constant, s_{ij} is the elastic compliance coefficient, and ε_{ii} is the permittivity. The piezoelectric coefficients, d_{ij} and e_{ij}, are third rank tensors which in reduced tensor notation correspond to a 3×6 matrix (Nye, 1995; Giacovazzo, 2002).
Furthermore, the indices (i = 1. . . 3) define normal electric field or displacement orientation, (j = 1. . .3) define normal mechanical stresses or strains and (j = 4. . .6) represent shear strains or stresses.
In the direct effect using equation (1), a mechanical stress σ_j or strain S_j causes a net electrical displacement, D_i, on i faces of the material, the magnitude of which depends on d_{ij} and e_{ij} respectively. Similarly, the converse effect expressed by equation (2) relates the induced normal and shear stress or strain to the applied electric field via the piezoelectric coefficient tensor. As a result, large piezoelectric coefficients d_{ij} are desired in actuator applications whereas sensor applications take advantage of large e_{ij} coefficients.
The piezoelectric coefficients are not the only material parameters of interest. In resonant structures the electromechanical coupling coefficient (3) and the dielectric loss angle (tanδ) (4) are essential measures in piezoelectric materials. The coupling coefficient represents the effectiveness of the energy transformation from the mechanical (electrical) to the input electrical (mechanical) energy. The definition of coupling coefficient depends on the orientation. The planar coupling coefficient, k_p, describes the radial coupling in a thin disc, when the electrical field is applied through the thickness, whereas the thickness coupling coefficient, k_t, is identical to k_{33} when the element is clamped laterally.

$$k_t^2 = \frac{d_{33}^2}{\varepsilon_{33}^T S_{33}^E} \tag{3}$$

$$\tan\delta = \omega C \times R \tag{4}$$

Part of the electrical energy is dissipated and transformed to heat. The dielectric loss angle tanδ (4) (the inverse denotes the dissipation factor η) quantifies this phenomenon. The term refers to the angle in a complex plane between the resistive (lossy) component (R) of an electromagnetic field and its reactive (lossless) component (C). The resistive component in (4) generates also a noise current or voltage, which limits directly the signal-to-noise ratio of the sensors.

$$SNR = \frac{e_{31}^2}{\sqrt{\varepsilon_0 \varepsilon_{33} \tan\delta}} \tag{5}$$

Among the piezoelectric films used, Aluminum Nitride (AlN) films have been less explored than Lead zirconate titanate (PZT) and Zinc oxide (ZnO) films due to its smaller piezoelectric constant. However, its temperature/humidity stability (Lakin et al., 2000), higher signal-to-noise ratio (Trolier-McKinstry & Muralt, 2004; Setter, 2005) and the compatibility with CMOS processing are attractive (Gerfers et al., 2006; Oestman et al., 2006). Furthermore, AlN is a large band gap material (6eV) with a large resistivity, whereas ZnO is really a semiconductor (3eV) with the inherent risk of increased conductivity due to off-stoichiometry. This low DC resistivity translates into a high dielectric loss at low frequencies, which is specially harmful for sensor and actors operating at frequencies below 10kHz (Setter, 2005).

Two types of piezoelectric MEMS devices, surface-micromachined piezoelectric resonators and bulk-micromachined accelerometers, which utilize longitudinal (d_{33} mode) and transverse (d_{31} mode) piezoelectric effects, will be presented in this chapter. Film bulk acoustic resonator (FBAR) resonators have proven advantages of low loss, high power handling, small form factor, and easy silicon integration compared to conventional ceramic and surface acoustic wave (SAW) structures (Ruby et al., 2001; Weigel et al., 2002). Thus, FBAR resonators are getting popular as a transmitter and/or receiver filters e.g. in GSM/CDMA/UMTS applications (Ruby & Merchant, 1994; Ueda et al., 2005), replacing the bulky SAW filters.

For the reconfigurable RF front ends, the integration of adjacent-band filters in one chip is attractive from the cost and form factor perspective. But the resonance frequency of a FBAR is determined by the thickness of the film stack, which is equal to the corresponding half wavelength of the first fundamental mode (Wang et al., 2006). However, it is impractical from a manufacturing perspective to have multiple thicknesses of film stacks in order to obtain multiple-frequency resonators/filters. Therefore, we propose in this chapter a solution for this shortcoming (Wang et al., 2006).

Piezoelectric MEMS accelerometers have been successfully used in many applications, such as automotive, mobile phones, consumer electronics, and aerospace. An advanced application of condition-based maintenance (CBM) in equipment vibration monitoring and diagnostics is still solely relying on conventional bulk piezoelectric transducers due to the stringent dynamic range requirements. In recent years, CBM systems have naturally progressed from traditional data collector and wired systems to wireless sensor networks due to lower cost and ease of use (McLean & Wolfe, 2002). As a result, availability of high-performance, low-cost, and small form factor vibration sensors become a limiting factor to proliferate the number of monitoring points. Bulk-micromachined accelerometers (Gerfers et al., 2006) offer an appealing solution for such applications providing several benefits over conventional capacitive accelerometers (Monajemi & Ayazi, 2006; Kulah et al., 2006) in terms of form factor and noise. AlN based accelerometers provide an inherently small dissipation factor (around 0.1%) resulting in a excellent low frequency noise performance (Setter, 2005). Based on the initial sensor design (Gerfers et al., 2006), this chapter introduces the technique of stress concentration to improve the overall accelerometer SNR (Gerfers et al., 2007). In the first section, the deposition of highly c-axis oriented AlN films is described, since optimal piezoelectric and crystal properties of AlN films are essential for devices' performance (Ruby & Merchant, 1994; Naik et al., 2000). Furthermore, Section 2 discusses processing details and experimental results of the sputtered piezoelectric AlN films. The design of AlN surface-micromachined piezoelectric resonators, its fabrication and experimental results is presented in Section 4. Based on these FBAR resonators an on-chip tuning mechanism is proposed in Section 5, in order to obtain a multiple-frequency resonator (filter) in a single device.

Next, the design and characterization of low noise AlN based accelerometers is presented. Section 7 introduces the technique of stress concentration for ultra low noise piezoelectric AlN accelerometers. Experimental results demonstrate the SNR improvements. Finally, concluding remarks complete this chapter.

2. Piezoelectric AlN films

Both, surface-micromachined piezoelectric resonators and bulk-micromachined accelerometers require thin piezoelectric layers in the order of a few μm (Loebl et al., 1999;

Xu et al., 2002). Attractive piezoelectric films are beside aluminum nitride (AlN) (Wang et al., 2006), ZnO and $PbZr_xTi_{1-x}O_3$ (PZT)(Loebl et al., 1999). With AlN films, electromechanical coupling coefficients k_t^2 of > 6% and low losses can be achieved, if strongly c-axis oriented AlN films are grown (Loebl et al., 1999; Wang et al., 2006).

To achieve high piezoelectric coupling in AlN layers, they have to be strongly oriented in (002) direction (Loebl et al., 1999; Wang et al., 2006). Therefore, columnar AlN grains with the c-axis perpendicular to the substrate are needed. The texture of sputtered AlN films depends on the sputter deposition conditions, on the substrate, and on the deposition temperature. Also the AlN layer thickness has an influence on the quality and orientation of the grains. The formerly mentioned effects of substrate temperature, substrate and AlN layer thickness are investigated in detail in (Loebl et al., 1999).

The used AlN films were prepared by reactive sputtering (Wang et al., 2006), a technique with advantages of low deposition temperature, easy process control and low cost when compared to alternatives such as metal-organic chemical vapor deposition (MOCVD) and molecular beam epitaxy (MBE). The AlN films were sputtered at various conditions - substrate temperature from room temperature to 300°C, N_2/Ar gas ratio of 9 to 1, pressure of 1 to $6mTorr$, and power of 1 to $4kW$ - to obtain optimized films' properties. X-ray diffraction (XRD) rocking curve analysis was used to characterize the sputtered films since it has been shown that there is a correlation between the full width at half maximum (FWHM) of the XRD rocking curve and piezoelectric properties (Naik et al., 2000; Liaw & Hickernell, 1995). Fig. 1(a) shows an AlN film sample, prepared at optimized conditions, have desirable c-axis textured crystalline structure. The XRD rocking curve in Fig. 1(b) also shows the FWHM of (002) AlN as low as 1.57°.

The performance characteristics of the AlN devices, such as insertion loss, effective coupling coefficient, and the quality factor are highly related to the quality of both the piezoelectric and electrode materials (Ueda et al., 2005). In this work, Molybdenum (Mo) has been used as an electrode material due to its low resistivity and high acoustic impedance. Beyond, the quality of AlN films is affected by the surface roughness of the underlying film (Ueda et al., 2005).

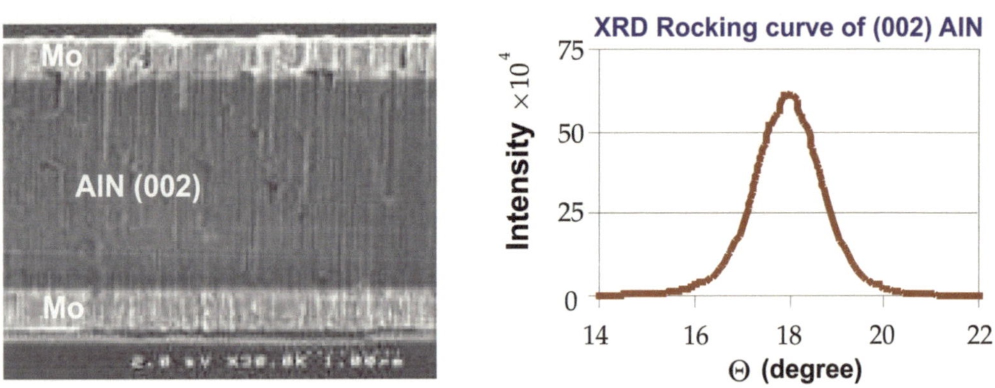

(a) SEM micrograph (b) AlN XRD rocking curve

Fig. 1. (a) SEM micrograph of highly c-axis textured AlN films; (b) (002) AlN XRD rocking curve with FWHM of 1.57°.

3. Material properties of sputtered polycrystalline AlN thin films

Since material properties of sputtered polycrystalline AlN films can vary over a wide range depending on the deposition conditions, it's important to characterize elastic and piezoelectric properties of the AlN films for future design and simulation. Phase velocity in the longitudinal direction (therefore, c_{33}) and e_{33}, were extracted from measured resonance frequencies and k_t^2 of the resonators with different thickness. To obtain transverse properties, various cantilever-beam test patterns were fabricated in the same chip of the accelerometers. Impedance resonance measurements were performed to obtain resonance frequencies of the beams and these were fit to FEA values to extract the stiffness matrix of the AlN films (see Fig. 2(a)). The transverse piezoelectric constant, e_{31}, was determined by measuring mechanical response (velocity) to an electrical drive signal using a laser vibrometer. Good matching between experimental and FAE results for all testing patterns were shown in Fig. 2(a)-(b). The extracted material parameters are given in Table. 1.

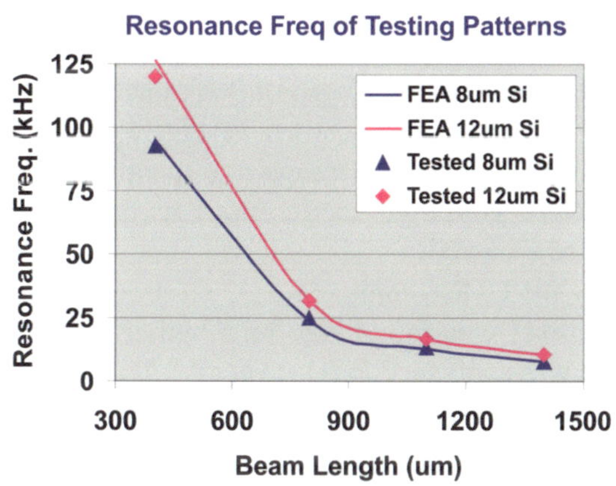

(a) Measured resonance frequencies

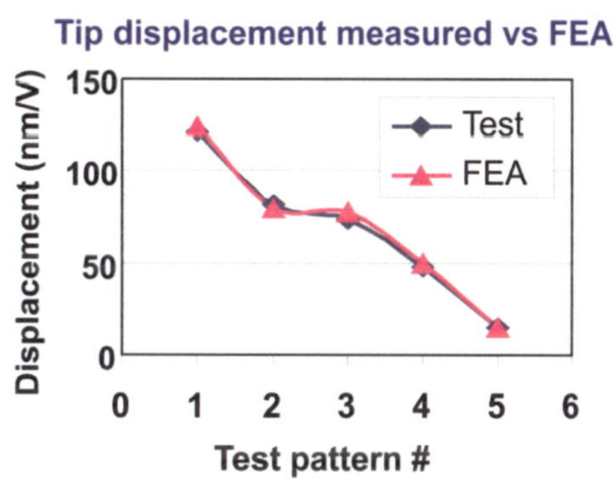

(b) Transverse piezoelectric constant

Fig. 2. (a) Measured resonance frequencies of all cantilever beams fit well with FEA-simulated values. (b) Transverse piezoelectric constant (e_{31}) was extracted by fitting measured and FEA-simulated values.

Materials	Materials Properties
Mo (sputtered)	$E = 503GPa, \nu = 0.33,$ $\rho = 7000kg/m^3$
AlN (elastic properties)	$c_{11} = 375GPa, c_{12} = 125GPa,$ $c_{13} = 120GPa, c_{33} = 435GPa,$ $c_{44} = 118GPa, \rho = 2700kg/m^3$
AlN (piezoelectric properties)	$e_{31} = -0.58C/m^2,$ $e_{33} = 1.55C/m^2,$ $e_{15} = -0.48C/m^2$

Table 1. Summary of the extracted properties of piezoelectric film stack.

4. Piezoelectic resonators

The basic FBAR resonator structure is a piezoelectric film sandwiched between two electrode films, as shown in Fig. 1(a). The fabrication process flow of such an AlN piezoelectric resonator is shown in Fig. 3. The substrates were 6-in diameter (100) silicon wafers. The process started with a silicon trench etch, followed by trench fill using silicon dioxide as a sacrificial layer. The surface was then planarized by a chemical-mechanical polishing (CMP) step. It is important to have a very smooth surface for the following piezoelectric film stack deposition. A bottom Mo layer ($0.32\mu m$) was sputtered and patterned to define the electrode area. An AlN layer ($1.1\mu m$) was deposited by reactive sputtering. A second Mo layer was deposited and patterned to define top electrode area. The AlN layer was dry etched to open contact windows on bottom Mo electrodes and form release holes. Finally, the sacrificial layer was etched with BOE to release the membrane. Fig. 4(a) shows optical and (b) cross-section SEM micrographs of a resonator after completion of the fabrication.

4.1 Experimental results

The resonators were tested using an Agilent network analyzer. One-port S-parameters were measured to obtain the input impedance of the resonators, shown in Fig. 4(c). Then Butterworth-Van Dyke equivalent circuit was used to extract resonator's quality factor, series/parallel resonant frequency and effective coupling coefficient. For the $2GHz$ resonator, a Q as high as 1000 and k_t^2 of 6.5% were achieved.

In addition to the Mo electrodes, resonators with CMOS-compatible Al electrodes were investigated. However, a lower Q (3 times worse) was found due to a higher acoustic propagation loss of Al (see Fig. 4(d)). This result was confirmed in (Ueda et al., 2005).

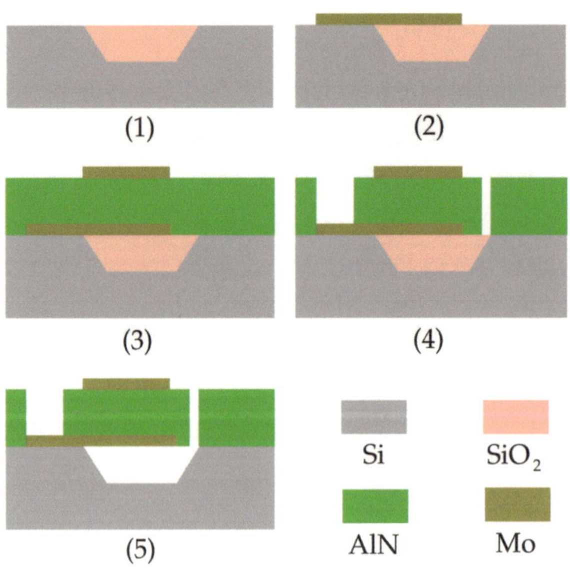

Fig. 3. Fabrication process flow of AlN piezoelectric resonators

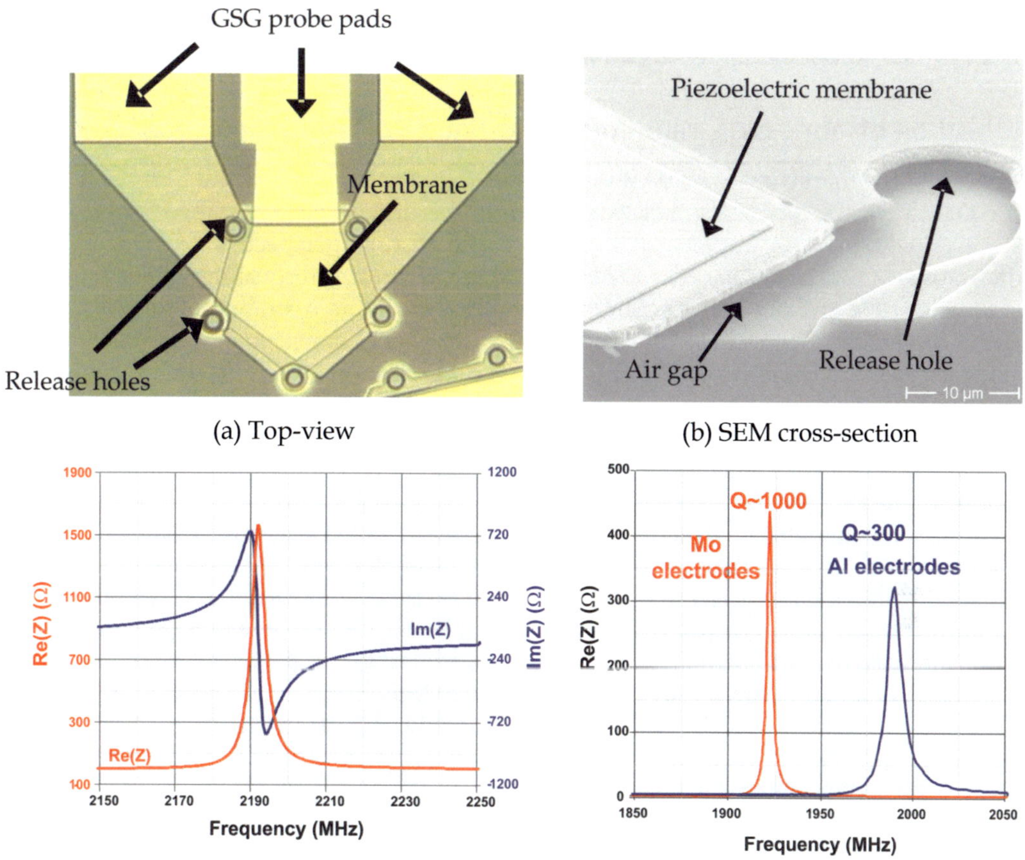

(a) Top-view

(b) SEM cross-section

(c) Measured input impedance

(d) Impact of Mo/Al electrodes

Fig. 4. Surface-micromachined AlN piezoelectric resonator, (a) top-view optical graph, (b) cross-section SEM graph. (c) Measured input impedance of the 2*GHz* resonator achieving a Q of 1000 and k_t^2 of 6.5%; (d) Resonators with Mo electrodes have higher Q than the ones with Al electrodes.

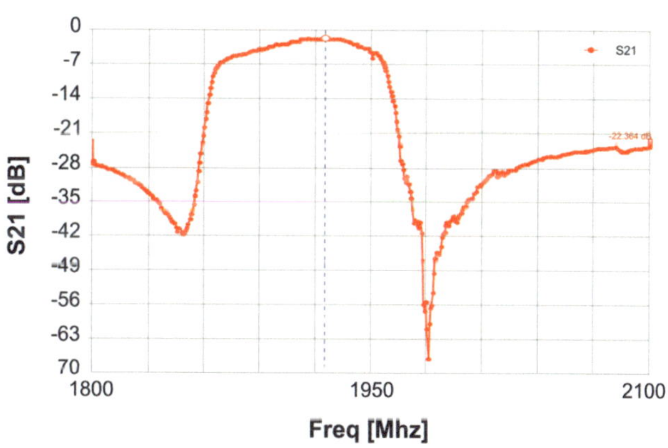

Fig. 5. Measured frequency response of the FBAR filter.

The measured frequency response of the FBAR filter is shown in Fig. 5. The insertion loss is 1.7*dB* and the lower and upper stopband rejection is 27*dB* and 23*dB* respectively.

5. Method of fabricating multi-frequency film FBAR in a single chip

The resonance frequency of a FBAR is determined by the thickness of the film stack, which is equal to the corresponding half-wavelength of the first fundamental mode (Lakin et al., 1995). However, it is impractical from a manufacturing perspective to have multiple thicknesses of film stacks in order to obtain multiple-frequency resonators/filters. In (Piazza et al., 2005) the authors explore AlN piezoelectric resonators operating in contour modes with resonance frequencies determined by in-plane dimensions. A much lower electromechanical coupling coefficient, which is a key parameter for the use of FBARs as a front-end RF filter, was obtained. In the following we describe an approach for integrating multiple-frequency FBARs in a single chip (Huang et al., 2005; Wang et al., 2006). The presented experimental results verify the novel concept and show the performance of the modified FBARs.

Two main configurations, air-gap FBAR and solidly mounted resonator (SMR), have been used to create low acoustic impedance terminations; therefore, the acoustic energy is confined within the piezoelectric film stack, which is critical for achieving a high mechanical quality factor Q. The air-gap FBAR has the piezoelectric film stack suspended with air on both sides. On the other hand, the SMR has an air interface on the free surface side and a quarter-wavelength acoustic mirror on the substrate side. In this work, only air-gap FBARs were used to demonstrate the concept, which should also be applicable to SMR FBARs (Wang et al., 2006).

The key concept of the frequency tuning is based on the mass loading effect. An additional tuning layer was added on top of the conventional FBAR membrane, Mo/AlN/Mo film stack. Then the tuning layer was patterned with a pitch of S and width of L. A schematic of the modified FBAR is shown in Fig. 6(a). Therefore, different mass loading (different L/S ratios) effects can be obtained by controlling the width and the pitch of the tuning patterns. The trimming Mo layer was patterned using different photomasks to define loading patterns. Figure 6(b) shows a SEM micrograph of a modified FBAR after completion of the fabrication.

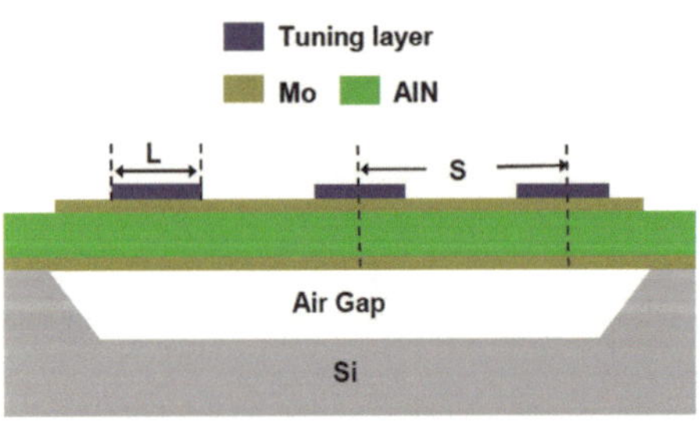

(a) FBAR using Mo tuning layer

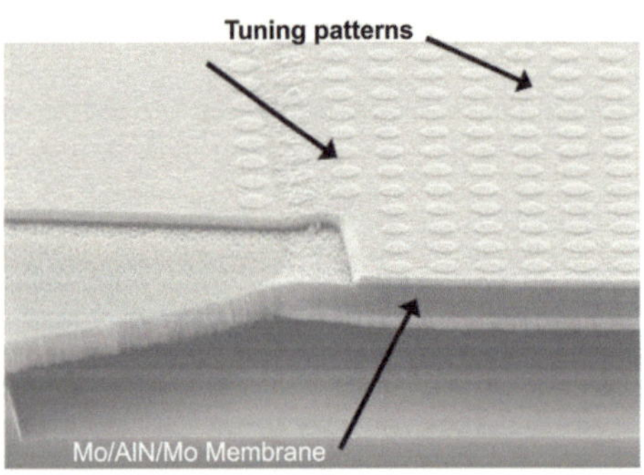

(b) SEM micrograph

Fig. 6. (a) Schematic of a modified FBAR - a tuning layer (Mo) is added and patterned on top of a conventional FBAR. (b) SEM micrograph of a released FBAR membrane, showing a Mo tuning layer that was deposited and patterned on top of the membrane.

5.1 Experimental results

Both, the series resonance frequency (f_S) and parallel resonant frequency (f_p) were analyzed from the one-port S-parameter measurements. The k_t^2 was calculated using

$$k_t = \frac{\pi}{2}\sqrt{1 - \frac{f_S}{f_p}}. \tag{6}$$

The comparison of different loading FBARs was made on the same die to minimize the frequency variation caused by the thickness variation of the deposited films (typically < 0.2% within die variation). When the pitch of tuning patterns ($40\mu m$) was much larger than the membrane thickness ($1.8\mu m$), two distinct resonant peaks were measured (Wang et al., 2006). The lower-frequency peak corresponded to the thickness mode with the tuning pattern and the higher-frequency one corresponded to the thickness mode without the tuning layer. The two resonant peaks agreed with the prediction from the finite element analysis, where stress-contours reveal two distinct resonant modes at the corresponding frequencies.

To achieve a desirable frequency response, the tuning patterns with $1.5\mu m$ pitch and five different loading percentages, 0, 20, 42, 64, and 100%, were designed and fabricated. The tested results, shown in Fig. 7, demonstrate that the resonant frequencies were modulated in relation to the corresponding loading percentages. More importantly, the modulated resonance peaks maintain the same shape as the non-modulated one; that is, a pure frequency shift was achieved. Three different thicknesses of tuning layers (75, 100, and $150nm$ Mo) were fabricated to study total tuning ranges and sensitivities (Wang et al., 2006). The results demonstrate that the thicker tuning layer provides, as expected, a larger loading effect and a wider total tuning range. For all modified FBARs, the effective coupling coefficient, k_t^2, was maintained within 90% of the non-modified one.

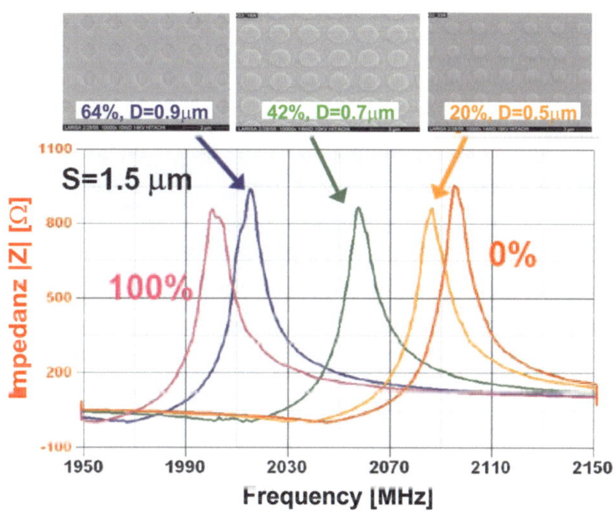

Fig. 7. Resonance frequencies were modulated according to the loading percentages. The resonant peaks of modified FBARs maintain a desirable response, a pure frequency shift, when the pitch ($S = 1.5\mu m$) is smaller than the membrane thickness ($1.8\mu m$).

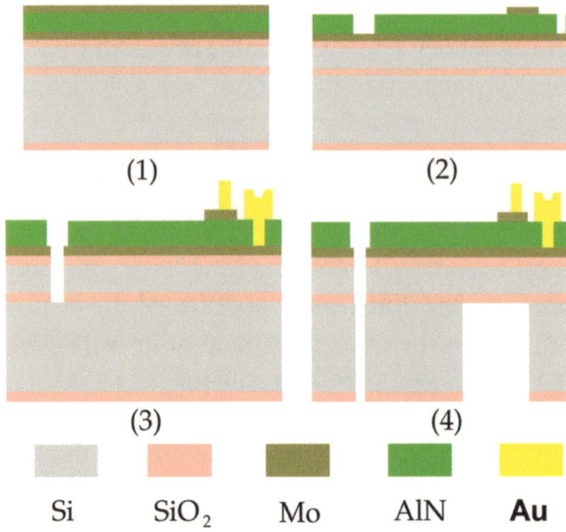

Fig. 8. Process flow of bulk micromachined piezoelectric AlN accelerometer using SOI wafer.

6. Bulk-micromaschined accelerometers

MEMS accelerometers using piezoelectric AlN thin film as read-out have been attracting a great deal of attention due to their simple structure, high SNR, small dielectric loss angle ($\tan\delta$), temperature/humidity stability and compatibility with CMOS processing (Lakin et al., 2000; Trolier-McKinstry & Muralt, 2004; Wang et al., 2006; Gerfers et al., 2007). The fabrication process of AlN piezoelectric accelerometers is based on bulk micromachining and silicon on insulator (SOI) techniques in order to have a large proof mass for low noise floor as well as precise device thickness control for minimal performance variation. The fabrication process flow is outlined in Fig. 8. First, piezoelectric films stack ($0.25\mu m$ Mo / $1.5\mu m$ AlN / $0.25\mu m$ Mo) was deposited on 6" silicon-on-insulator (SOI) wafers which have active Si thickness of $7.75\mu m$ and $12.09\mu m$, $2\mu m$ buried SiO_2, and $600\mu m$ bulk Si (shown in Fig. 8-(1)). Then the top Mo layer was patterned to define the electrode area and followed by AlN patterning to open contact windows on bottom Mo electrodes (see Fig. 8-(2)). The film stack of bottom Mo, oxide and active Si as well as buried oxide was etched to define flexible sensing structures of different accelerometer designs - cantilever beams, clamp-clamp beams, and annular membranes. The final front-side process was electroplating $3\mu m$ Au on the contact area (see Fig. 8 (3)). Finally, the wafers were completed with the backside process of DRIE $600\mu m$ bulk Si to release the accelerometers. Fig. 9(a) shows SEM graphs of two initial accelerometer designs, an annular membrane and a clamp-clamp beams after completion of the fabrication.

6.1 PE accelerometer testing

The differential accelerometer designs were epoxied and wire-bonded to a special printed circuit boards (PCB), such that the proof masses are free to move (Gerfers et al., 2006). This sensor PCB was directly mounted on the top of the reference sensor, in order to minimize out-of-axis acceleration effects. A Dytran accelerometer with sensitivity of $0.1V/g$ and wide bandwidth of $5kHz$ (with a max. ±2% pass-band variation and a resonance frequency of

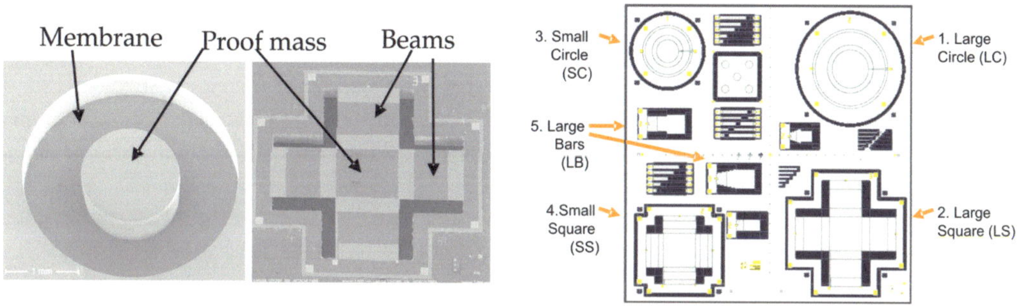

(a) SEM graphs of two fabricated accelerometers (b) AlN PE accelerometer testchip

Fig. 9. (a) SEM graphs of two fabricated accelerometer designs (left) backside of an annular diaphragm and (right) clamp-clamp beams. (b) Layout of the AlN PE accelerometer testchip. Besides the test structures that allow monitoring of the processing, a large and a small annular and a large and small quad-beam accelerometer structures are realized.

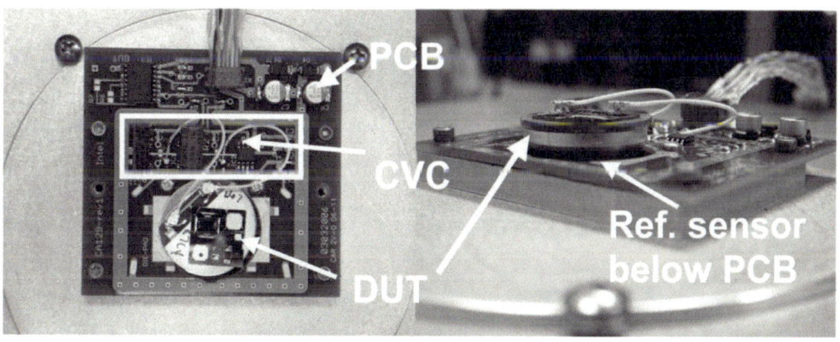

Fig. 10. Dynamic shaker setup.

f_0=25kHz) was used as a reference. Both the reference accelerometer and the shaker are set up such that the DUT on top of the reference is horizontally on the same level as the electronics PCB, see Fig. 10.

The overall optimization of the electronic sensor readout architecture and PCB design was focused on minimizing the electrical noise floor, which is imperative for testing these low noise accelerometers. Hence, the differential piezoelectric sensors are connected to the charge-to-voltage converters input (CVC) with two very short shielded low-capacitance coaxial cables in order to minimize noise coupling from the environment. Furthermore, the stiffness of the used coaxial cables and the way these cables are mounted between the sensor and the electronics is very important to obtain undistorted measurement results. Proper grounding of the electronics PCB and shaker setup is also mandatory in order to avoid ground loops and related 60Hz noise issues.

6.2 Low-noise PE accelerometer readout

PE MEMS accelerometers in the literature have noise floors as low as a few $\mu g/\sqrt{Hz}$ e.g. (Wang et al., 2003; Levinzon, 2004; 2005). In almost all of these surface or bulk micromachined accelerometers, the total noise floor is dominated by electronic noise (of the 1st stage). Thus, in the following all dominant noise and error sources are reviewed and the impact on the overall acceleration noise density is calculated, in order to specify some hands-on design guidelines for the integrated CMOS readout solution to achieve actually sub-$\mu g/\sqrt{Hz}$ levels (Gerfers, Ginsburg, Samara-Rubio, He, Manoli & Wang, 2007). The PE accelerometer mechanical–thermal noise and noise due to the losses in the PE material are given in (9) (Levinzon, 2004; Gerfers et al., 2007). A low-noise FET charge amplifier is used as a front-end stage to amplify the small transducer's output charge signals. Thus, in order to detect an estimated noise floor of a few hundred $ng/\sqrt{Hz}$ @ $10Hz$ all dominant noise sources have to be considered. In addition, the noise characteristics of the FET amplifier depends on the source impedance which is why the total noise floor of the PE accelerometer is determined by both noise characteristics of PE transducer and the FET amplifier (Levinzon, 2005). A simplified schematic of the front-end CVC shown in Fig. 11, which introduces three additional noise sources; a shot noise current density $\overline{i}_{n,ota}$ (caused by the input bias current I_{GS} of the amplifier), a thermal and a $1/f$ voltage noise source $\overline{v}_{n,ota}$. Referring these intrinsic amplifier noise sources to the sensor input, one obtains

$$\frac{\overline{a}^2_{n,ota}}{\Delta f} = \frac{\overline{i}^2_{n,ota}|Z|^2}{V_s^2} = \frac{2qI_{GS}}{(\omega C_s)^2}\frac{C_s^2}{Q^2_{T,\vec{z}}} \tag{7}$$

$$\frac{\overline{a}^2_{n,ota}}{\Delta f} \approx \left(\frac{\gamma_1 8kT}{3gm} + \frac{\gamma_2 K_F I_{DS}}{f^{AF} C_{OX} L^2_{eff}}\right)\frac{C_s^2}{Q^2_{T\vec{z}}} \tag{8}$$

where I_{GS} denotes the amplifier input gate-source leakage current, γ_1,γ_2 the noise excess factors, K_F the $1/f$ transistor noise coefficient, I_{DS} the drain-source bias current of the input devices, AF the $1/f$ noise slope coefficient, C_{OX} the gate oxide capacitance and L_{eff} the effective drawn transistor length. Please note, that the sensor capacitance is significantly larger than the integrator capacitance in order to obtain the required CVC gain. The lower bound for C_{int} is given by the full scale sensor signal and the maximum signal swing of the fist CVC stage whereas the upper bound of C_{int} is given by the dynamic range requirements. The penalty of the 2nd stage amplifier on the overall noise floor is rather small, since it is divided by the first stage gain. The differential CVC was built from a pair of ultra low-leakage current low-noise single-ended JFET amplifiers featuring a voltage noise floor ($\overline{v}_{n,ota}$) as low as $5nV/\sqrt{Hz}$ and simultaneously an extremely low current noise of $1f$ $A/\sqrt{Hz}$. For an expected sensor capacitance of C_S = $500pF$ and a sensitivity of $Q_{T,\vec{z}}$= $5.2pC/g$, the equivalent acceleration noise floor ($\overline{a}_{n,ota}$) is $670ng/\sqrt{Hz}$. The expected sensor Brownian noise floor is around $10ng/\sqrt{Hz}$, which is still much lower than the detectable noise floor, limited by interface readout electronics.

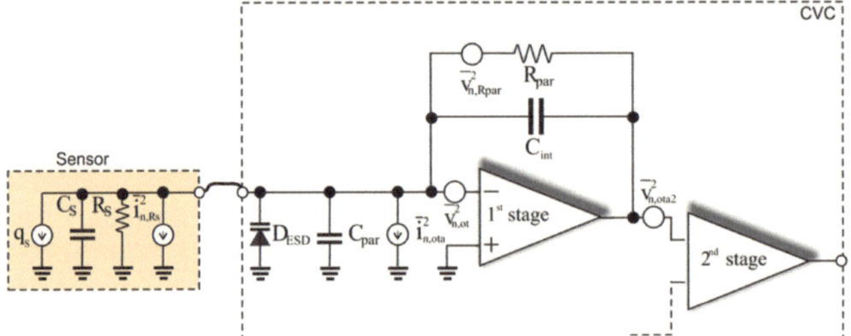

Fig. 11. PE MEMS sensor readout architecture. All dominant noise and error sources of the charge amplifier configuration are illustrated.

6.3 Measurements and characterizations

The formerly presented accelerometer structures have been analyzed in detail theoretically and numerically. Static and modal simulations with Finite Element Analysis (FEA) simulator are done to analyze the mechanical response given in at different applied accelerations. The symmetric quad-beam structure and the annular beam structure were fabricated using CMOS compatible piezoelectric AlN thin films and silicon-on-insulator (SOI) wafers using different proof mass weights and SOI thickness ranging from 7.75 – $12\mu m$ (Gerfers et al., 2006). The complete layout of the different accelerometers and process monitor structures is shown in Fig. 9(b).
The measured dynamic charge responses for two annular accelerometers with 8.6 and $12\mu m$ silicon beam thickness are shown in Fig. 12(a). The quality factor of these devices equals Q = 290. Figure 12(b) presents the total noise floor of the accelerometer plus interface electronics. The tested accelerometer sensitivities range from 0.06 to $1.73pC/g$ with acceleration noise floors from 0.8 to $17\mu g/\sqrt{Hz}$. The complete test results of two high performance accelerometers; the large annular design (LC) and the large quad-beam structure (LS) both with $7.75\mu m$ thick SOI beams is given in Tab. 2. Obviously, the simulated FEA values match well with the measured values validating our design methodology for future projects. Although, in order to meet the stringent specifications of condition-based maintenance and vibration monitoring the accelerometer performance has to be further improved especially in terms of signal-to-noise ratio.

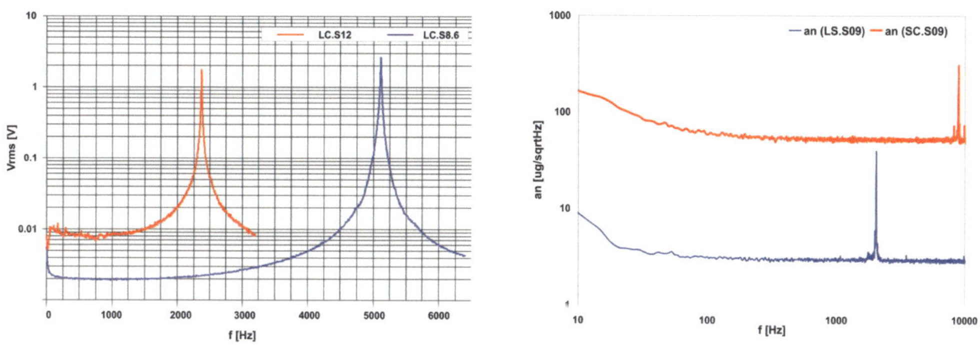

(a) Dynamic responses (b) Total noise floor

Fig. 12. (a) Measured dynamic charge responses for two annular accelerometers with 8.6 and $12\mu m$ silicon beam thickness. The quality factor of this device equals Q = 290. (b) Measured total noise floor spectrum of the accelerometer plus interface electronics.

		LC, $7.75\mu m$ SOI FEA	LC, $7.75\mu m$ SOI Meas.	LS, $7.75\mu m$ SOI FEA	LS, $7.75\mu m$ SOI Meas.
f_0	$[kHz]$	5.5	5.3	1.3	1.8
Sensitivity	$[pC/g]$	0.27	0.22	1.77	1.73
Sensor Capacitance	$[pF]$	110		195	
Size (Area)	$[mm^2]$	20		36	
Proof mass	$[mgram]$	4.39		4.37	

Table 2. Measured and FEA-simulated sensitivities of the two high performance accelerometers, the large annular design (LC) and the large quad-beam structure (LS) both with $7.75\mu m$ thick SOI beams.

7. Optimized AlN piezoelectric accelerometer

In the following, a new accelerometer design accomplishing an optimized performance in terms of charge sensitivity per unit area is introduced. Besides the dynamic range improvement we further introduce a sensing structure, which shifts higher order resonance modes to higher frequencies without affecting the charge sensitivity and fundamental resonance.

7.1 Review of the piezoelectric accelerometer noise performance

The total noise of a piezoelectric sensor in terms of equivalent acceleration noise is described in (9)-(10), which consists of two noise sources: mechanical-thermal or Brownian noise ($\bar{a}^2_{n,mech}$) and electrical-thermal noise ($\bar{a}^2_{n,ele}$),

$$\frac{\bar{a}^2_{n,tot}}{\Delta f} = \frac{\bar{a}^2_{n,mech}}{\Delta f} + \frac{\bar{a}^2_{n,ele}}{\Delta f} \tag{9}$$

$$\frac{\bar{a}^2_{n,tot}}{\Delta f} = 4k_B T \left(\frac{\omega_0}{mQ} + \frac{\eta C_S}{\omega Q^2_{T,\vec{z}}} \right) \tag{10}$$

where k_B, T, ω_0, m, Q are Boltzmann's constant, absolute temperature, resonant frequency, effective sensor mass, quality factor; and C_S, η, ω, $Q_{T,\vec{z}}$ are the sensor capacitance, dissipation factor of the piezoelectric material, operating frequency, and longitudinal sensitivity, respectively. Thus, the noise spectrum is dominated by Brownian noise at high frequencies and by electrical-thermal noise at low frequencies since $\bar{a}^2_{n,ele}$ has a $1/f$ relation.

As mentioned before, vibration condition monitoring requires a very low noise floor at low frequency because very little vibration amplitude in terms of acceleration is produced at low frequency. Therefore, reducing the electrical-thermal noise $\bar{a}^2_{n,ele}$ is our objective. According to (10) to minimize $\bar{a}^2_{n,ele}$, one strategy would be to increase the charge sensitivity $Q_{T,\vec{z}}$ and maintain or even reduce at the same time C_S and η. The charge sensitivity of the piezoelectric accelerometer based on the bending mode is described by

$$Q_{T,\vec{z}} = \int_{\substack{Electrode\\Area}} (d_{31} \times \sigma) \mathrm{d}A. \tag{11}$$

But from this relation it is obvious that for a given piezoelectric material (i.e. d_{31} is determined) and given electrode area, the charge sensitivity, the accelerometer capacitance as well as the dielectric loss are constrained. As a result, increasing the charge sensitivity with an improved sensing structure is the main design objective outlined in the following.

7.2 SNR optimized piezoelectric accelerometer structure

Maximizing the accelerometer charge sensitivity and therewith the overall sensor signal-to-noise ratio has been accomplished by revising the original beam structure (illustrated in Fig. 15) used in the initial studies (Wang et al., 2003; Gerfers et al., 2006). By remodeling the rectangular beam shape into a trapezoidal beam shape structure, an applied external force causes stress that is concentrated on a smaller active PE area. This way, a higher stress magnitude is obtained on a smaller electrode area accomplishing both design tasks in terms of increasing the charge sensitivity $Q_{T,\vec{z}}$ and reducing the total PE electrode area. Furthermore, as a result of the reduced active PE area, both the sensor capacitance C_S as well as the accelerometer dissipation factor η are reduced as well. Figure 13 illustrates four different sensor designs exploring the concept of stress concentration. All sensing structures are designed to have the same resonance frequency and weight of the proof mass for fair comparison. Their sensitivities have been analyzed by FEA simulation, which are plotted as functions of the electrode area in Fig. 14.

Compared to the original designs `I` and `II` introduced in Sec. 6, which are using the conventional trampoline and the annular diaphragm sensing structure, device `III` and `IV` clearly show an improved charge sensitivity because the tapered-beam design results in higher stress concentration on the electrode area. As a result, the new devices permit to use smallest die size due to its distinguished area utilization. Moreover, Fig. 14 features a relative linear relationship between the charge sensitivity $Q_{T,\vec{z}}$ and the electrode area for designs `I` and `II`, whereas sensor structures `III` and `IV` reveal a point of maximum sensitivity per area due to the non-linear slope of $Q_{T,\vec{z}}$. This optimal point can be found by calculating the derivative of $(\partial\, Q_{T,\vec{z}})/(\partial A)$ and is obtained for an electrode area of $3.5mm^2$.

Besides redesigning the beam structure in order to improve the overall signal-to-noise ratio, additional effort was spent reducing the transversal sensitivities $Q_{T,\vec{x}}$ and $Q_{T,\vec{y}}$. FEA simulations prove that structure `IV` is less susceptible to transversal accelerations than device `III`. In addition, by introducing four balanced bars designed to connect the four sensing beams at node positions, as illustrated in Fig. 14(a), the whole structure is even more reliable and is stiffer in the X-Y directions (minimizing the transversal sensitivity) without impairing the resonance frequency and sensitivity. This can be visually confirmed by analyzing the stress contours of the structure with balanced bars given in Fig. 14(a).

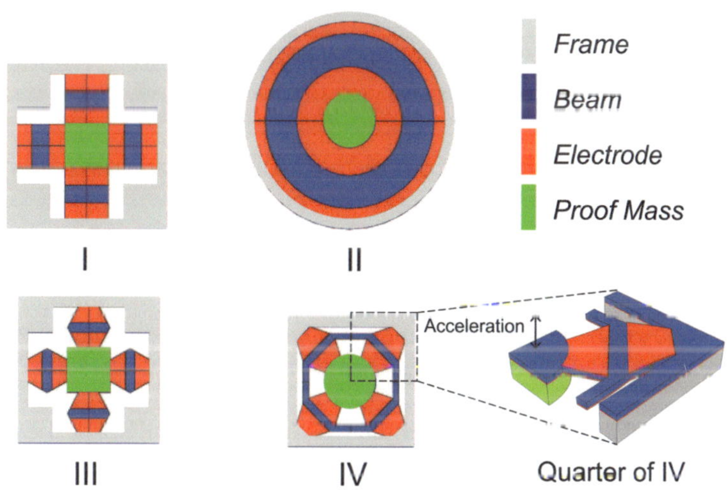

Fig. 13. Different sensing structures were investigated to have optimized performance. They are designed to have same resonance frequency and weight of the the proof mass for fair comparisons. Design `I` and `II` use conventional beam and membrane structures, while device `III` and `IV` employ a tapered-beams design.

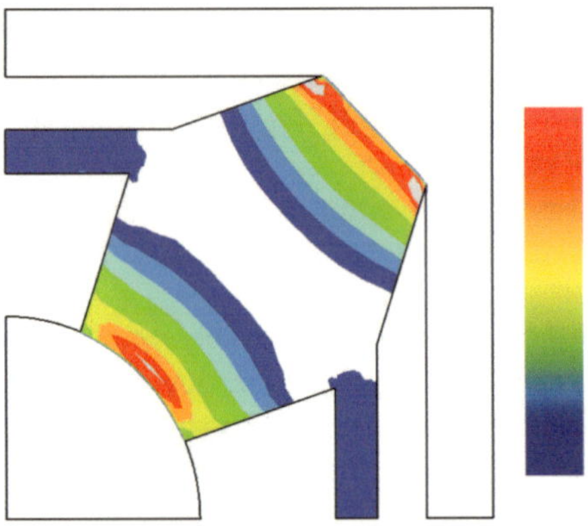

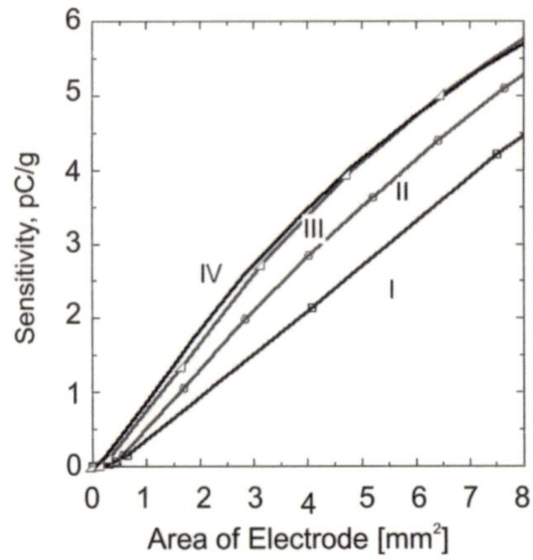

(a) Stress contours of design IV beams (b) Simulated charge sensitives per unit area

Fig. 14. (a) Stress contours of design IV with balanced bars, which connect the sensing beams at the node positions (zero stress point) to provide better stability and less off-axis sensitivity. (b) FEA simulated charge sensitives per unit area.

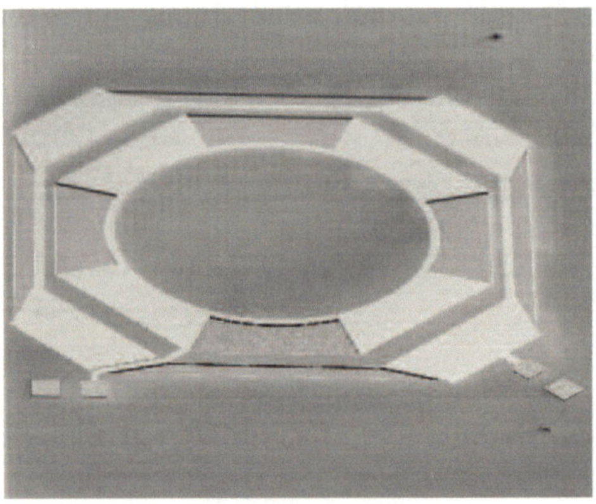

Fig. 15. SEM micrograph of design IV front-side before DRIE processing.

7.3 Experimental results

The dynamic frequency responses of the accelerometers were measured using mechanical shaker in an open-loop setup as shown in Fig. 10. The tested output spectra of PE accelerometer is shown in Fig. 16(a). The mean of the tested charge sensitivity for design IV is $5.2pC/g$ with a Q of 160 (Gerfers, Bar, Northemann, Manoli, Kohlstadt & Wang, 2007). The measured sensor linearity from $1mg$ to $10g$ of design IV is shown in Fig. 16(b) with maximum deviation of less than 0.3% over the entire g-range. The lower measurement limit was given by the power amplifier of the shaker, while the upper level is restricted by the chosen CVC gain and output voltage swing. Measurements with a reduced CVC gain show reliable operation beyond $20g$ acceleration amplitudes. The measured noise spectrum, shown in Fig. 16(c), demonstrats a total wideband noise floor of $670ng/\sqrt{Hz}$. The low frequency noise ($1/f$ noise) equals $4.2g/\sqrt{Hz}$ @ $10Hz$ such that the noise corner frequency is below $100Hz$ (Gerfers, Ginsburg, Samara-Rubio, He, Manoli & Wang, 2007). Please note, that the low frequency noise slope is less than f^{-1}. The spread of the measured accelerometer capacitance C_S is within 2.5% (3σ) and the dissipation factor $\eta = 1/(2\, fC_SR_p)$ is 0.11% @ 230Hz. The performance of the proposed accelerometer design IV is summarized in Table 3.

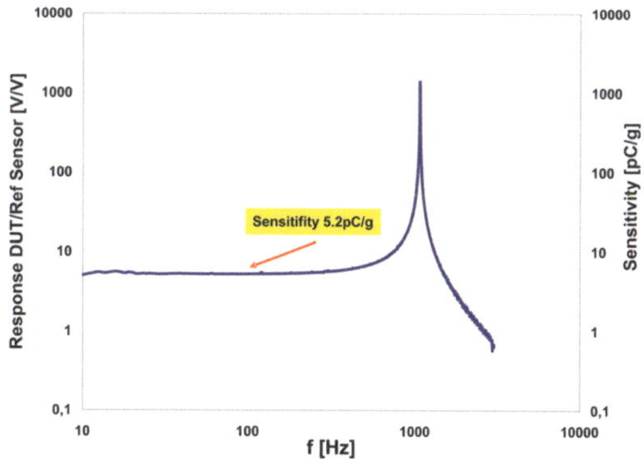

(a) Dynamic frequency response of design IV

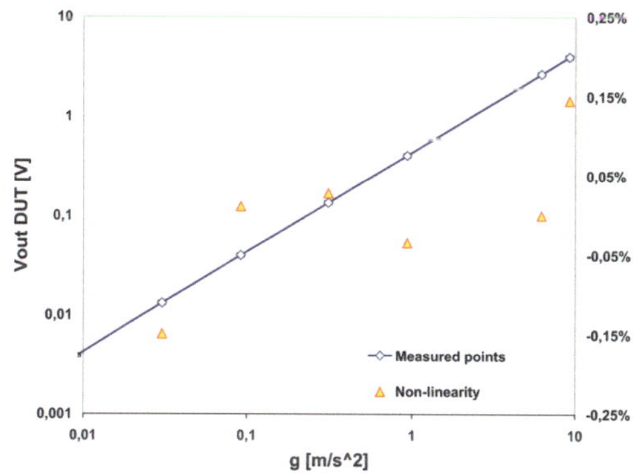

(b) Linearity of design IV

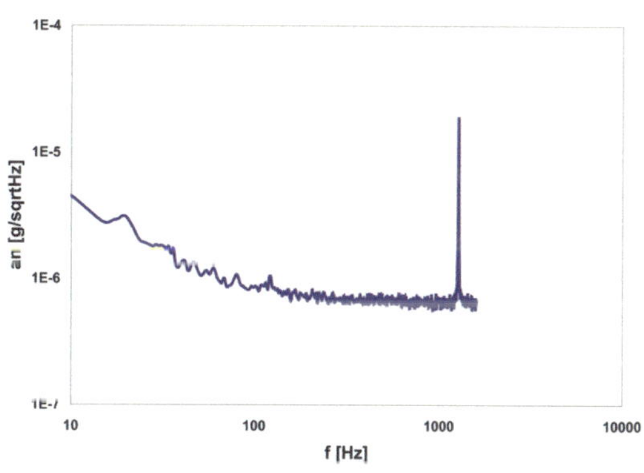

(c) Noise spectrum

Fig. 16. (a) Measured frequency response of the design IV PE AlN accelerometer.
(b) Measured linearity showing less than 0.3% deviation from 0.01 − 10g acceleration level.
(c) measured total noise floor spectrum of sensor and interface electronics.

		Design IV
f_0	$[kHz]$	1.1
Sensitivity	$[pC/g]$	5.2
Sensor Capacitance	$[pF]$	500
Dissipation factor	[1]	0.11%
Sensor Size	$[mm^2]$	45.8
Proof mass	$[mgram]$	10.1

Table 3. Device parameter summary.

8. Conclusion

Polycrystalline AlN films were prepared by reactive sputtering; optimized piezoelectric and crystalline properties were obtained at right sputtering conditions. Two types of piezoelectric MEMS devices, surface-micromachined resonators and bulk-micromachined accelerometers, utilizing longitudinal (d_{33} mode) and transverse (d_{31} mode) piezoelectric effects were fabricated and characterized.

We demonstrate a unique approach to integrating multiple-frequency FBARs in a single chip. By controlling in-plane dimensions of the periodic tuning patterns, resonance frequencies of modified FBARs are modulated corresponding to the mass loading percentages. As a result, multiple-frequency FBARs can be lithographically defined by a single deposition/patterning processing sequence. To obtain a desirable frequency response, a pure frequency shift, the pitch of the tuning patterns needs to be smaller than the membrane thickness. This approach provides a potential solution for integrating multiple-frequency FBAR filters of adjacent bands or frequency trimming. The fabricated piezoelectric resonators achieve a Q of 1000 and an electromechanical coupling of 6.5% at $2GHz$. The effective coupling coefficient k_t^2 for all frequency-tuned FBARs, was maintained within 90% of the non-modified one.

Vibration condition monitoring requires a very low acceleration noise floor at low frequency in order to be able to detect small acceleration amplitudes. In this chapter we have presented a new sensing structure for piezoelectric accelerometers improving the overall signal-to-noise ratio such that the fabricated accelerometers fulfill the vibration condition monitoring requirements. The devices take advantage of tapered beams resulting in stress concentration on the electrode area. As a result, the accelerometer charge sensitivity increases while at the same time both the sensor capacitance and the dissipation factor can be reduced therewith improving the electrical-thermal acceleration noise. Thus, the novel sensing structure features smallest die size due to its distinguished area utilization. The sensing structure has been designed to have low transverse sensitivity and to be reliable introducing four balanced bars designed to connect the four sensing beams at node positions therefore, the whole structure is stiffer in the X-Y direction but without impairing the resonance frequency and charge sensitivity.

Experimental results confirm the significantly improved sensitivity of the accelerometers obtained with the proposed sensing structures. The tested charge sensitivity is $5.2pC/g$ and the measured total noise floor of sensor plus interface electronics is as low as $670ng/\sqrt{Hz}$.

9. References

Gerfers, F., Bar, H., Northemann, T., Manoli, Y., Kohlstadt, M. & Wang, L.-P. (2007). An Ultra Low-Noise Vibration Monitoring System, *IEEE Sensors Conf.* pp. 880–883.

Gerfers, F., Ginsburg, E., Samara-Rubio, D., He, M. Y., Manoli, Y. & Wang, L.-P. (2006). Fabrication and Characterization of Bulk-micromachined Accelerometers Based on AlN Piezoelectric Sensing and SOI Wafers, *20th Eurosensors* pp. –.

Gerfers, F., Ginsburg, E., Samara-Rubio, D., He, M., Manoli, Y., & Wang, L.-P. (2007). Sub-μg Ultra Low Noise MEMS Accelerometers based on CMOS-Compatible Piezoelectric AlN Thin Films, *Int. Conf. on Solid-State Sensors, Actuators and Microsystems* pp. 1191- 1194.

Giacovazzo, C. (ed.) (2002). *Fundamentals of crystallography*, Oxford University Press.

Huang, Z., Suo, Z., Wang, L.-P., Shim, D. & Ma, Q. (2005). A Novel Approach to Integrate Multiple Film Bulk Acoustic Resonators (FBAR) with Different Frequencies in A Single Chip, *Proc. of Nano Science and Technology Institute (NSTI) Nanotechnology Conf.*, Vol. 3, pp. 435–438.

Kulah, H., Chae, J., Yazdi, N. & Najafi, K. (2006). Noise Analysis and Characterization of a SD Capacitive Microaccelerometer, *IEEE J. Solid-State Circuits* 41(2): 352–361.

Lakin, K. M., Kline, G. R., & McCarron, K. T. (1995). Development of Miniature Filters for Wireless Applications, *IEEE Trans. Microw. Theory Tech.* 43(12): 2933–2939.

Lakin, K.M.,McCarron, K. T. & McDonald, J. (2000). Temperature Compensated Bulk Acoustic Thin Film Resonators, *Proc. IEEE Ultrasonics Symposium*, pp. 855–858.

Levinzon, F. A. (2004). Fundamental noise limit of piezoelectric accelerometer, *IEEE Sensors J.* 4(1): 108 - 111.

Levinzon, F. A. (2005). Noise of piezoelectric accelerometer with integral FET amplifier, *IEEE Sensors Conf.* pp. 1235 - 1242.

Liaw, H. M. & Hickernell, F. S. (1995). Characterization of Sputtered Polycrystalline Aluminum Nitride on Silicon by Surface Acoustic Wave Measurements, *IEEE Trans. Ultrason., Ferroelectr., Freq. Control* 42(2): 404.

Loebl, H. P., Klee, M., Kiewitt, O.W. R., Dekker, R. & Pelt, E. V. (1999). Piezo-electric AlN and PZT Films for micro-electronic Applications, *IEEE Ultrasonics Symposium*, pp. 1031- 1036.

McLean, C. & Wolfe, D. (2002). Intelligent Wireless Condition-Based Maintenance. (Machine Monitoring/Networking)., *Sensors Magazine* 19(6): 14–17.

Monajemi, P. & Ayazi, F. (2006). Design Optimization and Implementation of a Microgravity Ccapacitive HARPSS Accelerometer, *IEEE J. Solid-State Circuits* 41(6): 39–46.

Naik, R. S., Lutsky, J. J., Reif, R. & Sodini, C. (2000). Measurements of the Bulk, C-axis Electromechanical Couplingconstant as a Function of AlN Film Quality, *IEEE Trans. Ultrason., Ferroelectr., Freq. Control* 47: 292.

Nye, J. F. (1995). Physical Properties of Crystals: Their Representation by Tensors and Matrices, Oxford University Press.

Oestman, K. B., Sipil, S. T., Uzunov, I. S. & Tchamov, N. T. (2006). Novel VCO Architecture Using Series Above IC FBAR and Parallel LC Resonance, *IEEE J. Solid-State Circuits* 41(1): 2248–2246.

Piazza, G., Stephanou, P. J., Black, J. P., White, R. M. & Pisano, A. P. (2005). Single-chip multiple-frequency RF microresonators based on aluminum nitride contour-mode and FBAR technologies, *IEEE Ultrasonics Symposium*, Vol. 2, pp. 1187–1190.

Ruby, R., Bradley, P., Oshmyansky, Y. & Chien, A. (2001). Thin Film Bulk Acoustic Resonators (FBAR) forWireless Applications, *IEEE Ultrasonics Symposium*, pp. 813–821.

Ruby, R. & Merchant, P. (1994). Micromachined Thin Film Bulk Acoustic Resonators, *IEEE Frequency Control Symposium* pp. 135–138.

Setter, N. (2005). *Electroceramic-based MEMS: fabrication-technology and applications*, Springer. Trolier-McKinstry, S. & Muralt, P. (2004). Thin Film Piezoelectrics for MEMS, *Kluwer, Journal of Electroceramics*, pp. 7–17.

Ueda, M., Nishihara, T., Tsutsumi, J., Taniguchi, S., Yokoyama, T., Inoue, S., Miyashita, T. & Satoh, Y. (2005). High-Q Resonators using FBAR/SAW Technology and their Applications, *IEEE Int. Microwave Symposium Digest*, pp. 209–212.

Wang, L.-P., Ginsburg, E., Diamant, D., Ma, Q., Huang, Z. & Suo, Z. (2006). Method to Fabricating Multiple-Frequency Film Bulk Acoustic Resonators in a Single Chips, *IEEE Int. Frequency Control Symposium and Exposition*, pp. 793–796.

Wang, L.-P., Wolf, R., Yu, W., Deng, K., Zou, L., Davis, R. & Trolier-McKinstry, S. (2003). Design, Fabrication, and Measurement of High-Sensitivity Piezoelectric Microelectromechanical Systems Accelerometers, *J. Microelectromech. Syst.* 4(12): 433 - 439.

Weigel, R., Morgan, D. P., Owens, J. M., Ballato, A., Lakin, K. M., Hashimoto, K. & Ruppel, C. W. (2002). Microwave Acoustic Materials, Devices, and Applications, *IEEE Microwave Acoustic Materials, Devices, and Applications* 50(3): 738–749.

Xu, F.,Wolf, R. A., Yoshimurs, T. & Trolier-McKinstry, S. (2002). Piezoelectric Films for MEMS Applications, *Proc. IEEE 11th International Symposium on Electrets*, pp. 386–396.

18

Micromachined Arrayed Capacitive Ultrasonic Sensor/Transmitter with Parylene Diaphragms

Seiji Aoyagi
Kansai University
Japan

1. Introduction

For the external environment recognition of a robotic field, an ultrasonic sensor has advantages in cost performance compared with other sensors such as vision devices. In particular, in the spaces where vision devices cannot be used (e.g., in the dark, smoky situation such as in the disaster site), ultrasonic sensors are effective. For the purpose of using ultrasonic devices in microrobot applications (Aoyagi, 1996), and/or for the purpose of imitating the dexterous sensing functions of animals such as bats and dolphins (Mitsuhashi, 1997; Aoyagi, 2001), it is necessary to miniaturize the current ultrasonic sensors/transmitters (Haga et al., 2003).

The effectiveness of miniaturization is discussed herein from the viewpoint of directivity. Let us assume a piston-type ultrasonic device, the radius of which is R. The angle $\theta_{1/2}$ at which the sound pressure level becomes half of the maximal level achieved on the centerline of the piston (θ =0) is expressed as follows (Mitsuida, 1987):

$$\theta_{1/2} = \sin^{-1}(0.353\lambda / R), \tag{1}$$

where λ is the wavelength. The schematic explanation of this angle is shown in Fig. 1. This equation indicates that directivity becomes wider as the radius becomes smaller. Using many miniaturized transmitters/sensors in an array, the electrical scanning of directivity based on the delay-and-summation principle (Fig. 2) (Ono et al., 2005; Yamashita et al., 2002a; Yamashita et al., 2002b) and acoustic imaging based on the synthesis aperture principle (Guldiken & Degertekin, 2005) are possible, which could be effectively used for robotic and medical applications. Miniaturizing one sensing/transmitting element is useful both for realizing an arrayed device in a limited space and for realizing a device with omnidirectional characteristics, since the directivity of each element becomes wider as its diaphragm area becomes smaller based on equation (1).

There are two types of available ultrasonic sensor, one is piezoelectric, and another is capacitive. The working principle and the typical received waveform of piezoelectric type are schematically shown in Fig. 3. This type is further classified to thin film type and bimorph type. The former uses a micromachined thin film as a diaphragm, on which piezoelectric material such as lead zirconate titanate (PZT) is deposited using sol-gel method or sputtering. The latter uses a rather thick bulk plate as an elastic body of receiving and/or transmitting ultrasound. In case of the thin film type, piezoelectric constant d_{31} is rather

small, so it can act only as a receiver and cannot transmit ultrasound. Although the bimorph type can transmit ultrasound, its size is comparatively large.
The merit of these piezoelectric types is that they do not require bias voltage for their operation. The drawback of piezoelectric types is that the received waveform is burst one, i.e., the waveform continues during several tens cycles, since they are usually operated at their resonant frequencies with small damping. In the ranging system for airborne use (see Section 4.5), the precise arrival time of the ultrasound is difficult to detect for the burst waveform with dull rising, since the first peak is difficult to detect by setting a threshold level.

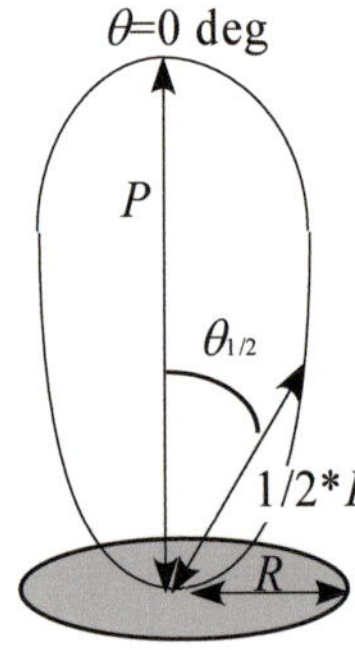

Fig. 1. Definition of $\theta_{1/2}$.

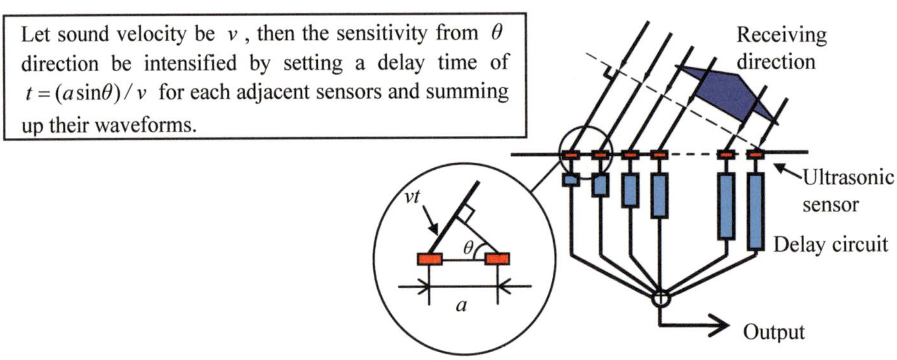

Fig. 2. Electrical scanning of directivity.

By contrast, although the capacitive type needs bias voltage for its operation, it can detect the arrival time of ultrasound accurately by setting an appropriate threshold level, since the received waveform is impulsive and well-damped, as schematically shown in Fig.4. A capacitive sensor can also act as a transmitter by applying an impulsive high voltage between two electrodes (Sasaki & Takano, 1988; Diamond et al., 2002), i.e., a diaphragm and a backing plate, both of which are conductive or coated by thin metal films.
As an example of conventional commercially available capacitive microphones, B&K-type 4138 (Brüel & Kjær, 1982) can receive sound pressure in the ultrasonic frequency range, and can be approximated to be nondirectional by virtue of the small area of its diaphragm. The structure of this microphone is shown in Fig. 5. The diameter, sensitivity, and frequency bandwidth of this microphone are 1/8 in. (3.175 mm), 0.9 mV/Pa, and 100 kHz, respectively. However, this microphone has the drawback of being expensive due to its

complicated and precise structure, i.e., it is composed of a thin nickel diaphragm of 1.6 µm thickness, a support rim, and a nickel backing plate facing the diaphragm surface with a small gap of 20 µm.

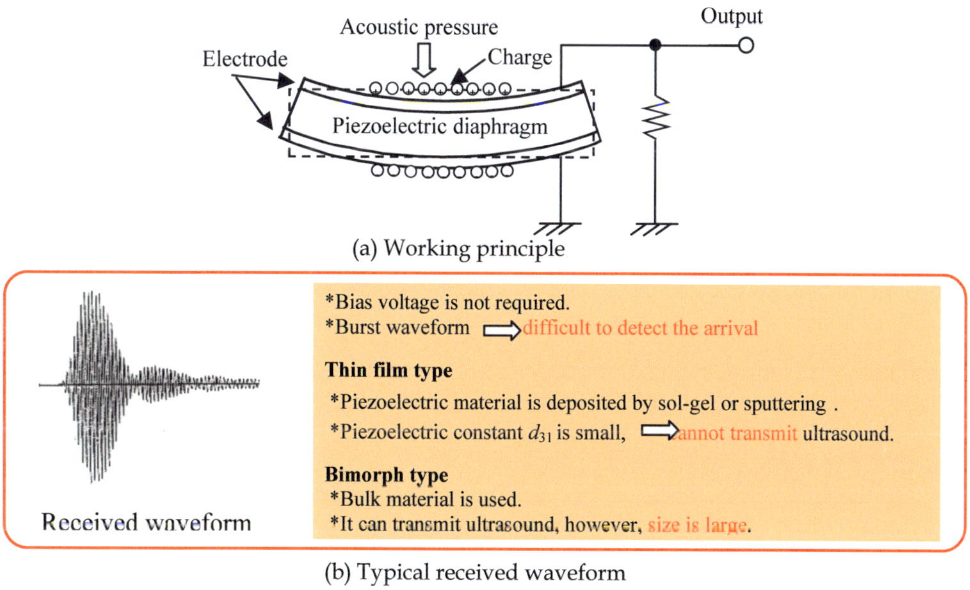

Fig. 3. Piezoelectric type ultrasonic sensor.

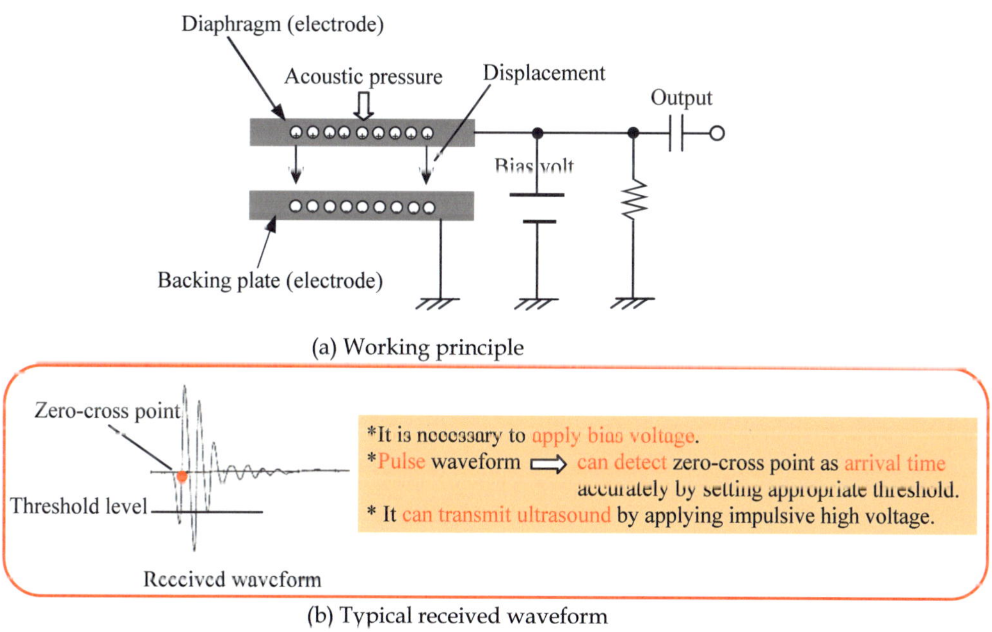

Fig. 4. Capacitive type ultrasonic sensor.

A capacitive sensor can also transmit ultrasound by applying impulsive high voltage as mentioned above: however, this B&K microphone is not applicable for the use of a transmitter because of the possibility of diaphragm fracture, taking into account its high cost.

In contrast, several studies on a capacitive microphone with a silicon diaphragm (Scheeper et al., 1992; Bergqvist & Gobet, 1994; Ikeda et al., 1999; Chen et al., 2002; Martin et al., 2005; Khuri-Yakub et al., 2000; Zhuang et al., 2000) have been conducted using micromachining technology (Kovacs, 1998), and some of them have been commercialized (Knowles Acoustics, 2002). Using this technology, numerous arrayed miniaturized ultrasonic sensors with uniform performance can be fabricated on a silicon wafer with a fine resolution of several microns and a comparatively low cost, which may make it possible to fabricate an arrayed-type sensor (Yamashita et al., 2002a; Yamashita et al., 2002b; Guldiken & Degertekin, 2005; Khuri-Yakub et al., 2000; Zhuang et al., 2006) and to activate it as a transmitter or speaker (Diamond et al., 2002; Khuri-Yakub et al., 2000).

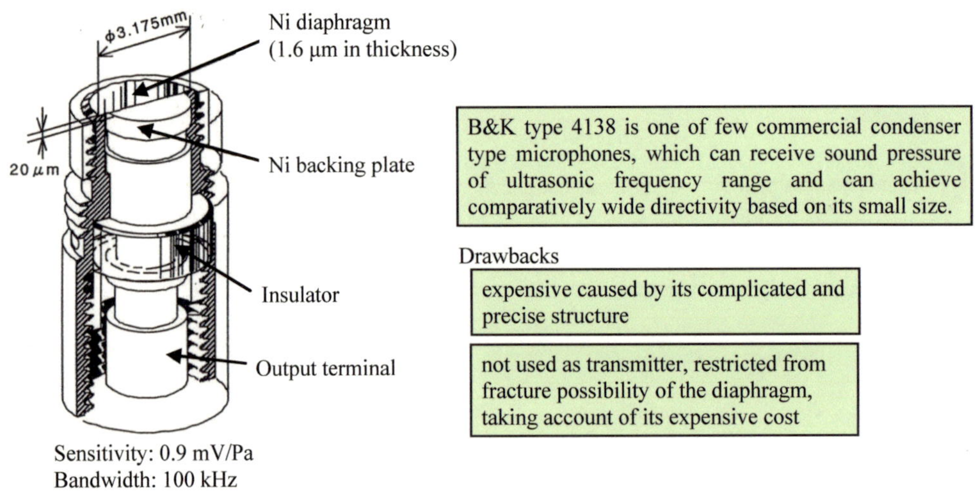

Fig. 5. Stracture of Brüel & Kjær 4138 microphone.

In micromachined capacitive microphones, the diaphragms are generally made of a silicon-based material, such as polysilicon and silicon nitride. In a few studies a polymer material was used for the diaphragms, such as polyimide (Pederson et al., 1998; Schindel et al., 1995), poly(tetrafluoroethylene) (trade name: Teflon) (Hsieh et al., 1999), and poly(ethylene terephthalate) (PET; trade name: Mylar) (Schindel et al., 1995). Since polymer materials have high durability due to their flexibility and nonbrittleness compared with silicon-based materials, their use in transmitters or speakers is thought to be possible. That is, the possibility of survival of a polymer diaphragm would be higher compared with that of a silicon diaphragm even when the applied high impulsive voltage for transmission passes instantaneously over the collapse voltage (Yaralioglu et al., 2005), at which the diaphragm is strongly pulled by an electrostatic attractive force to adhere to the substrate, causing the collapse of the device structure. Since a large displacement of the diaphragm per sound pressure is obtained due to the flexibility of the polymer diaphragm, the high sensitivity of the microphone can be realized. This is because the mechanical impedance of the diaphragm theoretically becomes low as the Young's modulus of the diaphragm's material decreases, provided that the radius, thickness, and input frequency are constant (Khuri-Yakub et al., 2000).

An ultrasonic transducer with a Mylar diaphragm has been commercialized (MicroAcoustic Instruments, trade name: BAT), and is often used in the ultrasonic research field (Hayashi et al., 2001); however, although the pits on the backing plate of this transducer are fabricated by micromachining technology, the polymer diaphragm film is assembled by pressing it to the backing plate with adequate pre-tension using a holder, the assembly of which appears as complicated as that of the above-mentioned B&K-type 4138 microphone.

Polyparaxylene (trade name: Parylene) is one of the polymer materials expected to be applied in the polymer micro-electro-mechanical-systems (MEMS) field (Tai, 2003). The deposition of Parylene is based on chemical vapor deposition (CVD), which is suitable for MEMS diaphragm fabrication. The mechanical properties of silicon, silicon nitride, Parylene, and Mylar are compared, as shown in Table 1. In addition to its flexible and nonbrittle characteristics compared with common polymer materials, Parylene has several excellent characteristics as follows. 1) It is a biocompatible material, which allows medical applications of the device. 2) It is chemically stable, i.e., it has high resistivity to acid, base, and organic solvents, which protects the device from external chemical environments. 3) It has high complementary metal oxide semiconductor (CMOS) compatibility compared with other polymer materials, since it can be deposited at room temperature. This characteristic makes the integration of a device with electrical circuits possible; such a device is called a smart device. 4) Its CVD deposition is conformal, thus the deposition of a domeshaped diaphragm is possible, which is effective for realizing a real spherical sound source/receiver. Due to these characteristics, an ultrasonic device utilizing a Parylene diaphragm has great potential in future applications. The principal aim of this study is to develop a capacitive microphone with a Parylene diaphragm (Aoyagi et al., 2007a).

	Young's modulus (GPa)	Shear modulus (GPa)	Density (kg/m^3)	Poisson ratio
Silicon*1	131	80	2,330	0.27
Silicon nitride*2	290	—	3,290	0.27
Parylene	3.2	—	1,287	0.4
PET (Mylar)	2.8	—	1,370	0.4

*1 Crystal silicon in (100) plane.
*2 LP CVD Si_3N_4 (Tabata et al., 1989).
— Not cleared.

Table 1. Comparison of mechanical properties of silicon and polymer materials.

The reported capacitive microphones focus on audio applications, in which bandwidth is below 15-20 kHz, where the important issues include sensitivity, linearity, and noise floor. In contrast, the present Parylene transducer focuses on ultrasonic applications in air, in which bandwidth is as high as 100 kHz, where the important issue is the accuracy of the distance measurement between the transmitter and the receiver. The directivity of the sensor is also the important issue in these applications. The second aim of this research is to characterize the fabricated Parylene ultrasonic receiver from the viewpoints of the accuracy of distance measurement and the directivity (Aoyagi et al., 2007a).

As the third aim of this research, an arrayed sensor device comprising 5×5 developed sensors is fabricated, and its receiving performance is characterized to prove the possibility of the electrical scanning of directivity based on delay-and-summation principle (Aoyagi et al., 2008a). As the fourth aim of this research, we confirm that each developed sensor can act as a transmitter by applying a high impulsive voltage, which means that the scanning of transmitting directivity is also possible. In this research, the scanning performance as the arrayed transmitter is also characterized (Aoyagi et al., 2008b).

2. Structure design of a sensor with Parylene diaphragm

2.1 Resonant frequency considering intrinsic stress

The resonant frequency of a Parylene diaphragm is investigated to define the size of the sensor and the bandwidth herein. The shape of the diaphragm is assumed to be a circle. Since Parylene has intrinsic tensile stress influenced by the temperature history of the fabrication (Harder et al., 2002), the relationship between the tensile stress and the resonant frequency is investigated herein.

Assume that the diaphragm has membrane characteristics, in which internal tensile stress plays an important role. Then, the following theoretical expression exists according to the theory of elastic vibration (Sato et al., 1993):

$$\omega_n = \lambda_{ns} \frac{1}{R} \sqrt{\frac{\sigma}{\rho}}, \tag{2}$$

where ω_n is the resonant frequency (rad/s), λ_{ns} is the eigenvalue (2.405), σ is the intrinsic tensile stress in the diaphragm (N/m^2), ρ is the density of the diaphragm material (kg/m^3), and R is the radius of the diaphragm (m).

In FEM (Finite Element Method) simulation, σ is applied in the cross section area of the boundary, i.e., the rim, which stretches the diaphragm. The modal FEM simulation is carried out for this stretched diaphragm. ANSYS is employed as the FEM software. In case the diaphragm radius R is 500 μm, theoretical and FEM simulated values of resonant frequency are obtained by changing the value of tensile stress in the range of 0-30 MPa. The result is shown in Fig. 6. This result shows that the influence of tensile stress on the resonant frequency is large. In the following part of this paper, it is assumed that the tensile stress σ is 25 MPa, based on the experimental data using rotation tip measurement (see Section 3.2). Under this condition, the relationship between the radius and the resonant frequency is shown in Fig. 7. Considering that the aimed bandwidth is in the ultrasonic range of 40-100 kHz, a radius R in the range of 500-1,200 μm is employed in this research according to this figure.

2.2 Influence of acoustic holes on damping ratio

In microphones, acoustic holes are generally set in the backing plate to control air damping. In the case of a simple square diaphragm, the viscous damping coefficient is calculated analytically (Scheeper et al., 1992; Bergqvist & Gobet, 1994; Škvor, 1967) in relation to the number of acoustic holes and to the surface fraction occupied by the acoustic holes. However, there has been no research on air damping for an arbitrary diaphragm shape. Thus, the damping ratio of a circular diaphragm is simulated using the FEM software.

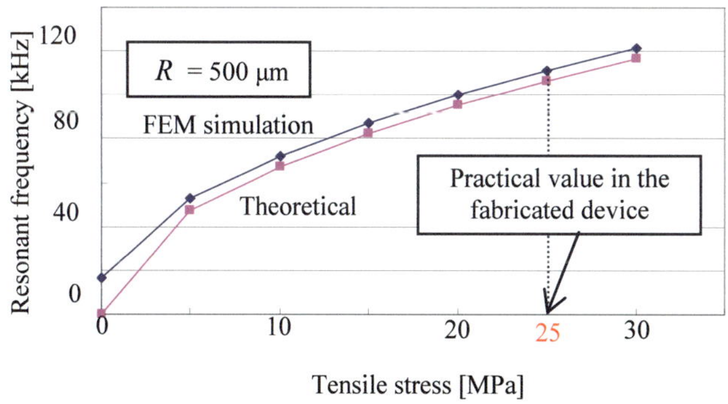

Fig. 6. Relationship between tensile stress and resonant frequency.

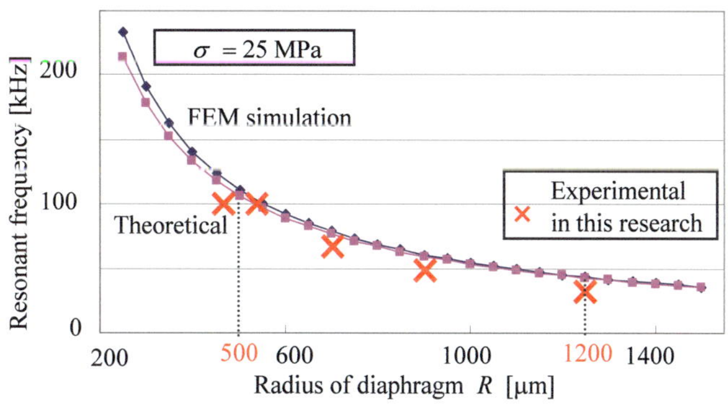

Fig. 7. Relationship between diaphragm radius and resonant frequency.

The flow distribution inside the air gap between the diaphragm and the backing plate, and the flow distribution inside the acoustic holes are simulated by FEM. Taking symmetry into account, a quarter model is employed. An example of the simulation model and its result are shown in Fig. 8. The transition of the displacement distribution, which is based on the first-order resonant vibration mode of a circular diaphragm, was given to the diaphragm. Then, the distribution of vertical flow velocity under the diaphragm was simulated. Total force F was obtained by summing up the pressures of all the elements just below the diaphragm. Flow velocity u^* was obtained by averaging the velocities of all the elements inside the air gap. Then, the damping ratio ζ was obtained as follows:

$$\zeta = \frac{\lambda}{2m\omega_n} = \frac{\Gamma / u^*}{2m\omega_n} \quad ,(3)$$

where m is the mass of the diaphragm, ω_n is the resonant frequency of the diaphragm, λ is the viscous damping coefficient.

The effects of the radius of the acoustic hole r and the number of holes n on the damping ratio ζ were investigated. The simulation result is shown in Fig. 9. Three cases in which the radii of the diaphragm (R) were 500, 700, or 1,200 μm are focused on. Considering the practical fabrication condition, the air gap and thickness of the backing plate are assumed to be 1.5 and 150 μm, respectively.

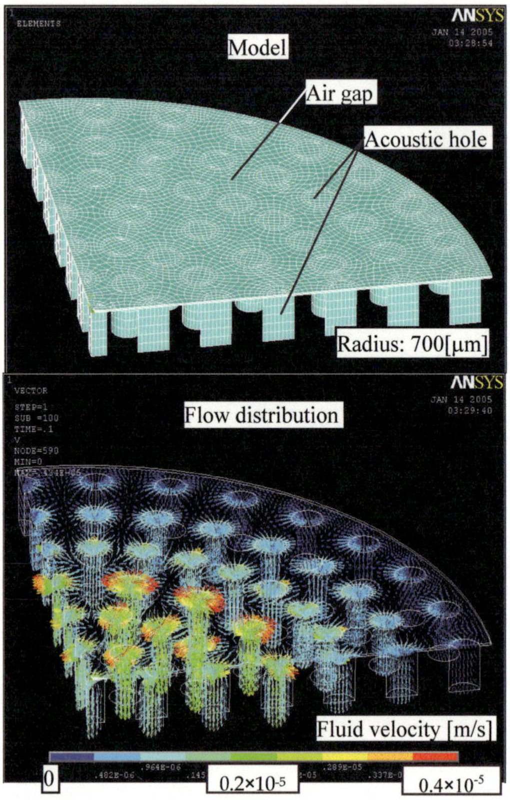

Fig. 8. FEM simulation for influence of acoustic holes on damping ratio.

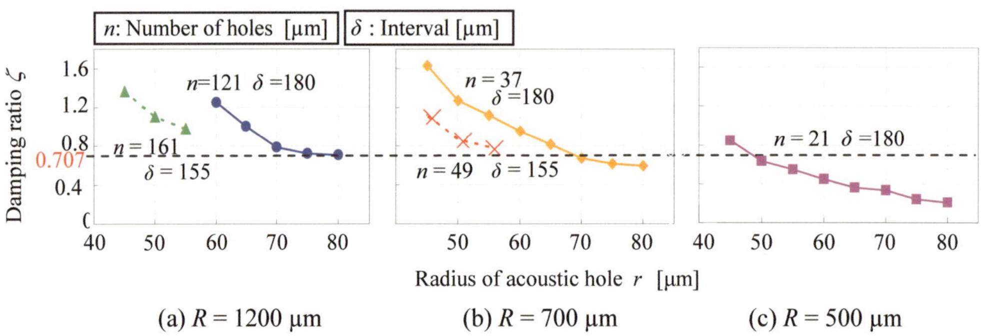

Fig. 9. Damping ratio by FEM simulation.

Also, considering the practical fabrication condition, several combinations of r and δ (the interval of adjacent acoustic holes) are tested to realize the optimal damping ratio of $\zeta = 1/\sqrt{2}$ =0.707 through trial and error.

In this figure, the damping ratio ζ is inversely proportional to r and n. Also, ζ decreases as R decreases, indicating that air damping is less effective for smaller diaphragms. For example, in the case of R =1,200 µm, the condition in which n =121 and r =80 µm with δ = 180 µm is suitable for realizing the optimal damping ratio. Photomasks for a micromachining fabrication of the sensor structure including acoustic holes are designed on the basis of the simulation results explained herein.

3. Fabrication process of a sensor

3.1 Fabrication process

The ultrasonic sensor was fabricated by depositing Parylene (2 µm in thickness) on a Si wafer (150 µm in thickness) with a thermally grown oxide (1 µm in thickness). Parylene deposition was based on chemical vapor deposition (CVD), and a coating apparatus (PDS-2010, Specialty Coating Systems) was used. The schematic overview of the developed sensor is shown in Fig. 10. The process flow is shown in Fig. 11 and proceeded as follows:

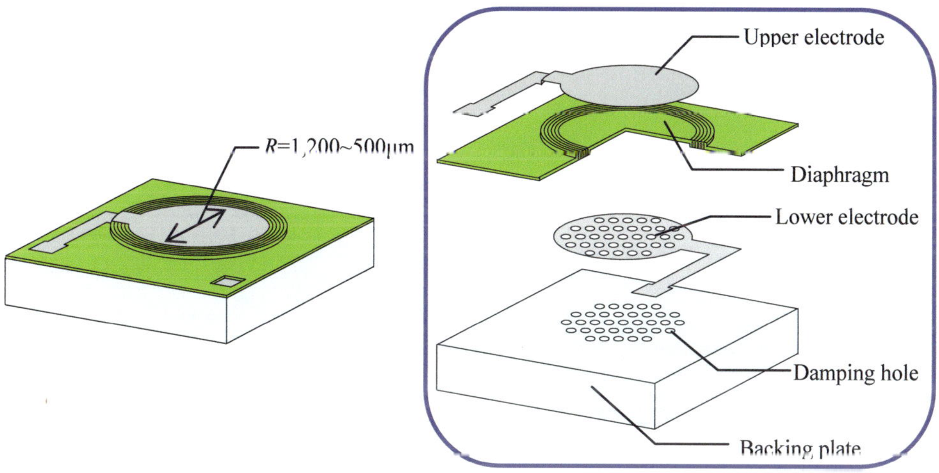

Fig. 10. Schematic overview of parylene ultrasonic sensor.

Aluminum (0.2 µm in thickness) was sputtered onto the oxidized silicon wafer, and patterned for the lower electrode and the bonding pad (see Fig. 11(1)).
As a sacrificial layer, amorphous silicon (1.5 µm in thickness) was deposited by plasma-enhanced CVD, followed by etching using SF_6 plasma to make slots, the function of which is explained later (see Fig. 11(2)).
The Parylene (2 µm in thickness) layer was deposited and patterned using O_2 plasma to reveal a bonding pad area (see Fig. 11(3)). In this patterning, a photoresist of 5 µm (AZP-4903) was used as the etching mask. Since the etching ratios of Parylene and the photoresist are almost the same, the mask made of the photoresist is gradually consumed during O_2 plasma etching. Therefore, a rather thick photoresist was employed.
The slots on the amorphous Si layer were filled with Parylene, providing anchor contact between Parylene and the substrate. Considering the mechanical strength at the edge of the diaphragm, it is desirable that the height of Parylene is the same at the anchor and the diaphragm. If the anchor contact area is large, the height of Parylene at the anchor will be smaller than that at the diaphragm by the thickness of the sacrificial layer, as schematically

shown in Fig. 12(a). To cope with this problem, slots were created and the anchor contact area was minimized. The height of the anchor was maintained at the same level as that of the diaphragm, since Parylene deposition is so conformal as to fill up these slots, as schematically shown in Fig. 12(b). The shapes and sizes of the slots for the anchor are shown in Fig. 12(c).

Aluminum (0.5 μm in thickness) was sputtered and patterned for the upper electrode using the liftoff process. This electrode must surpass the step height of Parylene and amorphous silicon layer (totally 3.5 μm in thickness) to reach the bonding pad, so a comparatively thick aluminum layer is necessary (see Fig. 11(4)).

The backside of the silicon wafer was dry etched by Inductively-Coupled Plasma Deep Reactive Ion Etching (ICP-DRIE) to produce acoustic holes (see Fig. 11(5)). These holes also play a role as the etching holes for the sacrificial amorphous silicon layer, inside which XeF_2 etching gas was later introduced.

The oxide layer at the bottom of the acoustic holes was etched using CHF_3 plasma (see Fig. 11(6)). The sidewalls of the acoustic holes were covered by Parylene (1 μm in thickness) to protect them from the XeF_2 etching gas used later. The conformal deposition of Parylene assists this process (see Fig. 11(7)). The Parylene at the bottom of the holes was etched using O_2 plasma. The vertical etching characteristic of the reactive ion etching (RIE) assists the selective etching of the bottom area.

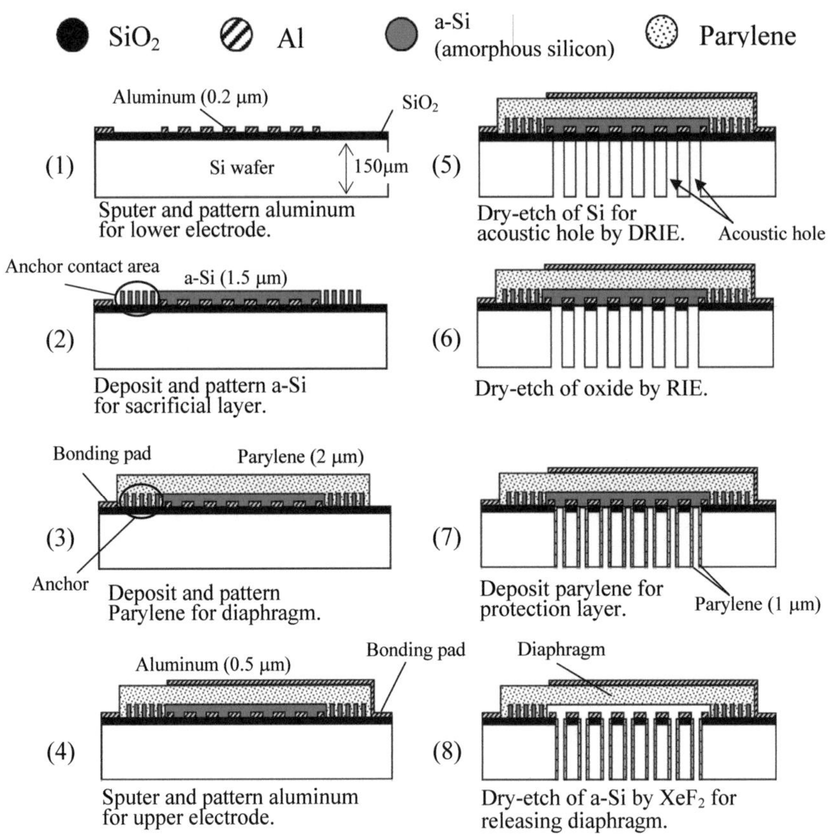

Fig. 11. Process flow of ultrasonic sensor.

Finally, the sacrificial amorphous silicon layer was dry etched away using XeF_2 gas in order to release the diaphragm (see Fig. 11(8)). This dry etching process is effective for preventing stiction (Yao et al., 2001).

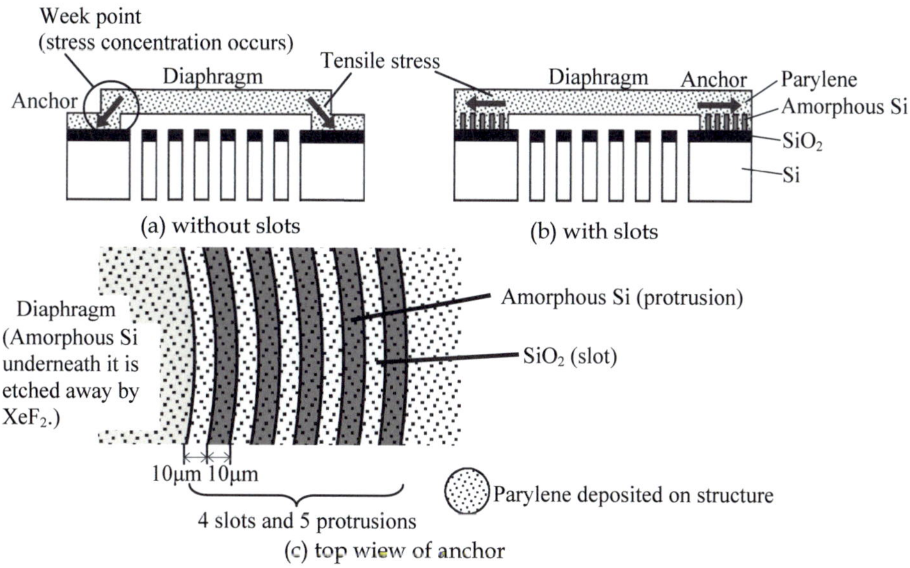

Fig. 12. Reducton of stress concentration using slots.

3.2 Fabrication results and intrinsic stress

An overview and schematic cross section of the fabricated sensor are shown in Fig. 13. Scanning Electron Microscope (SEM) images of fabricated sensors are shown in this figure. In this example, the radius of the diaphragm is 1,200 μm, and that of the acoustic hole is 50 μm. Looking at the back-side and cross section views of SEM images, it is proven that the acoustic holes were successfully fabricated. In the front-side view of SEM image, the Parylene circular diaphragm over the acoustic holes is seen. The aluminum upper electrode crossing the anchor is seen.

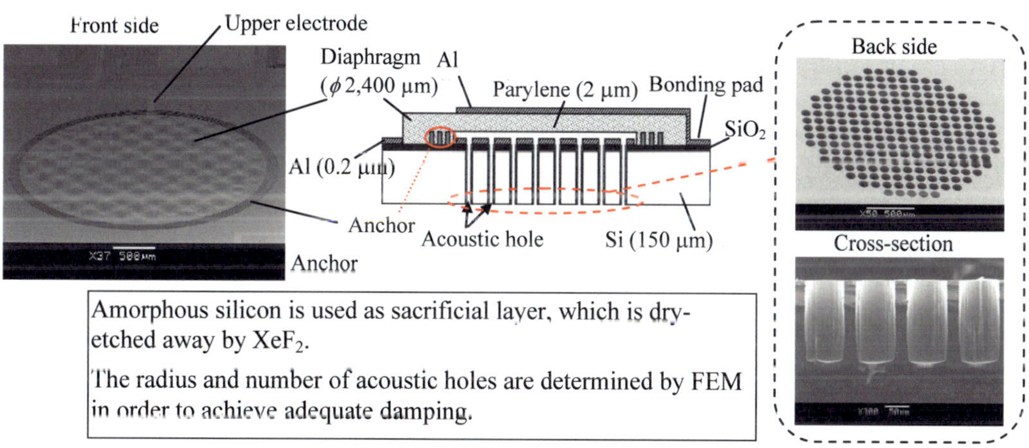

Fig. 13. Overview and schematic cross section of fabricated sensor.

A rotation tip was fabricated in the same substrate in order to estimate the actual tensile stress of Parylene, as shown in Fig. 14. The shrinkage of the beams supporting the tip is $H \cdot \tan\alpha$, and the strain in the film is calculated as $H \cdot \tan\alpha / (L_A + W + L_B)$, using symbols in Fig. 14. Multiplying the strain by Young's modulus of Parylene (3.2 GPa), the stress is obtained, which is proven to be approximately 25 MPa.

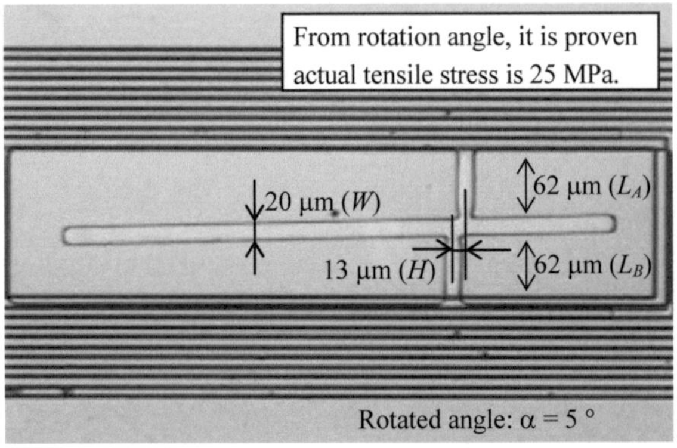

Fig. 14. Optical image of rotation tip.

4. Receiving performance of a sensor

4.1 Detecting circuitry for capacitance change

The circuitry used to detect the capacitance change due to the diaphragm displacement caused by ultrasonic sound pressure is documented herein. A bias voltage of 100 V was applied to the fabricated Parylene capacitive sensor. This value has an effect on the sensitivity, resonant frequency, and bandwidth (Schindel et al., 1995; Yaralioglu et al., 2005). In this study, this value is defined on the basis of values in references, in which 150 V (Sasaki et al., 1988), 100 V (Khuri-Yakub et al., 2000), 100-400 V (Schindel et al., 1995), and 50-135 V (Yaralioglu et al., 2005) were employed. In this study, the values of 150 and 200 V were experimentally tested; however, it was observed that the diaphragm was broken when a high impulsive voltage of 700 Vpp was applied during the transmitter use (the detail of which is explained in Section 6), although this failure rate is small. Thus, considering the safety factor, the value of 100 V was employed, under which condition neither diaphragm failure nor the disconnection of wiring was encountered.

Upon being supplied with a constant electrical charge due to the bias voltage, the diaphragm displacement was transformed to the voltage change at the sensor's electrode, and it was amplified by a factor of 30 (29.5 dB). The circuitry used for capacitance-to-voltage (CV) transformation and amplification is shown in Fig. 15, in which the high-frequency component of the voltage change is extracted by a bias-cut condenser, and it is input to an operational amplifier by a shunt resistor. Only the range within ±0.7 V is dealt with for amplification by virtue of a voltage limiter using two diodes, considering noise reduction.

4.2 Experimental setup for characterizing receiving performance

The experimental setup for characterizing the receiving performance of the developed sensor is schematically shown in Fig. 16. An electric spark discharge was used as an ultrasonic transmitter.

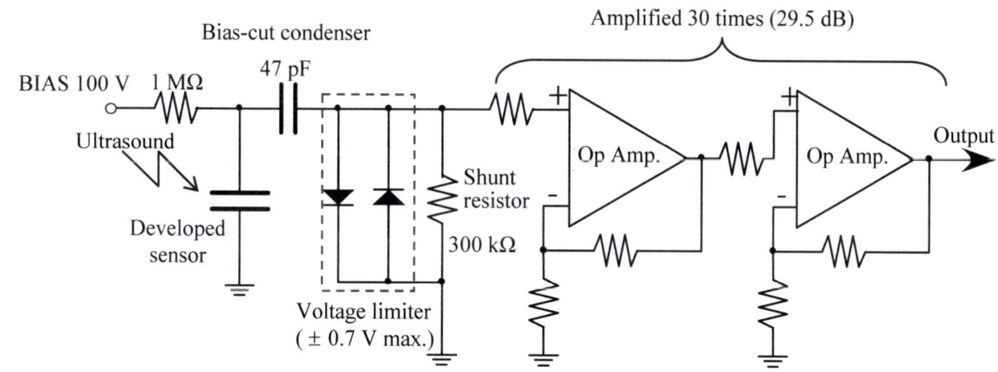

Fig. 15. CV transforming and amplifying circuit.

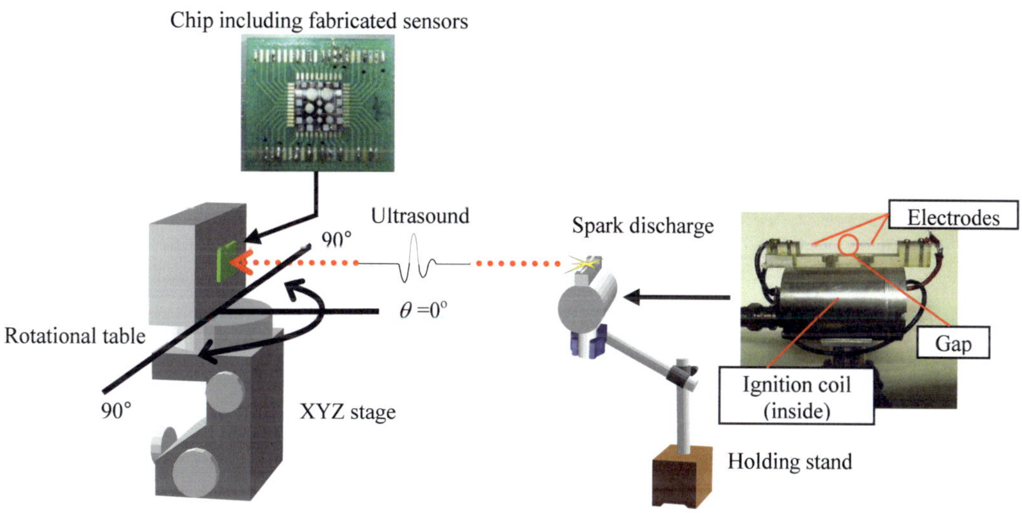

Fig. 16. Experimental condition for characterizing receiving performance.

Transmitted ultrasound is impulsive, the power spectrum of which is distributed over a broad frequency range (Aoyagi et al., 1992). The developed Parylene sensor was set on a rotational table. The distance between the transmitter and the sensor was set to 150 mm. As a reference, a microphone to estimate the sound pressure at the same position where the sensor was set, B&K type 4138 (already detailed in Section 1) was used.

4.3 Received pulse waveform, sensitivity, and resonant frequency of one sensor

An example of an ultrasonic pulse waveform received by the developed sensor, whose radius is 1,200 μm, is shown in Fig. 17. In this figure, the waveform received by the B&K microphone is also shown for reference. In the output signal of the developed sensor, there was electrical noise caused by the spark discharge, which could be suppressed by shielding the circuit completely in the future.

Considering that the sensitivity of the B&K microphone is 0.9 mV/Pa, and that the gain of amplification for the developed sensor is 30, the open-circuit sensitivity of the developed sensor was estimated to be 0.4 mV/Pa. The value of typical commercial microphone is in the range from 1 to 50 mV/Pa for the audio range (Brüel & Kjær, 1982; Knowles Acoustics,

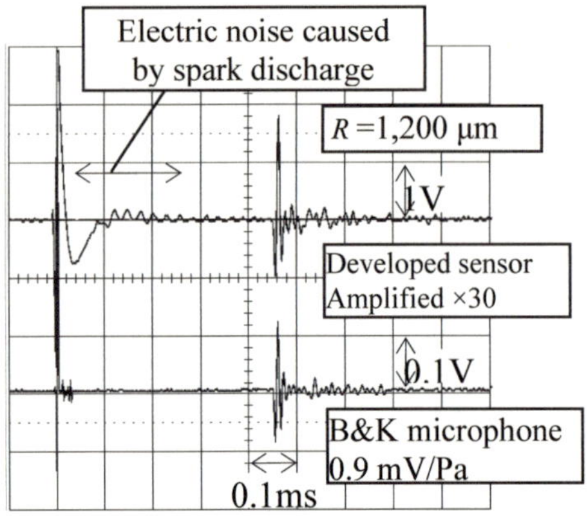

Fig. 17. Received ultrasonic waveforms by developed sensor and reference microphone.

2002). Considering that the diaphragm of the developed sensor is smaller than that of a commercial microphone, the realized sensitivity is reasonable. In the end, the high sensitivity, the order of which is comparable with the B&K microphone, was achieved.

In this study, the resonant frequency is defined as the reciprocal of the period between the first negative peak and the second one of the received waveform in a time domain, as shown in Fig. 18(a). An example of the power spectrum of the received waveform is shown in Fig. 18(b), which was obtained using a fast Fourier transform (FFT) analyzer. The resonant frequency measured based on the definition shown in Fig. 18(a) coincides well with the peak frequency in Fig. 18(b), which is 43 kHz in the case of the sensor used. This value agrees well with FEM simulated value, as shown in Fig. 7, in which experimental data of resonant frequency of the developed sensors having different diaphragm sizes are plotted.

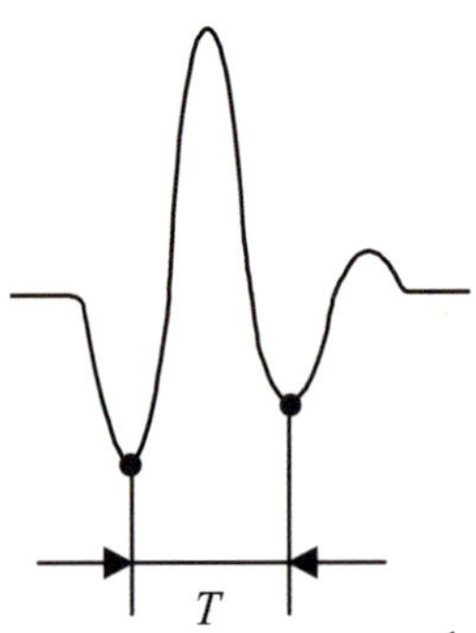

(a) Definition of resonant frequency in time domain

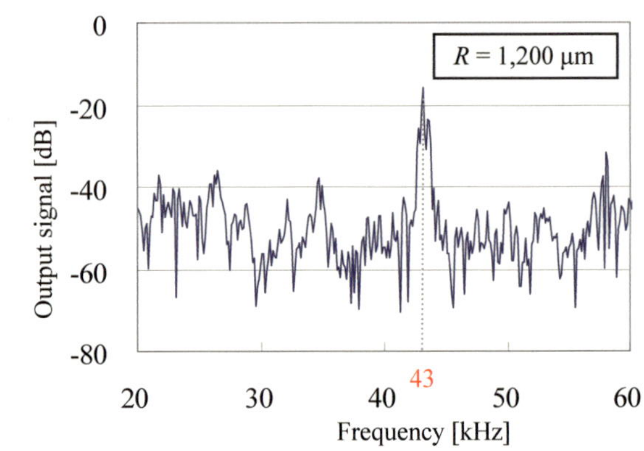

(b) Power spectrum of received waveform

Fig. 18. Measurement of resonant frequency.

4.4 Fidelity for sound pressure and damping ratio

The developed sensors with different sized acoustic holes, whose diaphragm radius is 1,200 μm, were employed. The radius of an acoustic hole (r) was 80, 65 or 50 μm. The ultrasonic pulse waveforms received by the sensors are shown in Figs. 19(a)-(c). To estimate the fidelity, three waveforms for each sensor are shown. The waveform received by the B&K microphone is also shown in Fig. 19(d) for reference.

The three waveforms in Fig. 19(a) resemble each other, as do those in Figs. 19(b) and (c). Thus, the reproducibility of the waveforms is good. In case that r is 80 μm, the residual vibration of the waveform is seen, whereas there are no residual vibrations, i.e., the waveform is well damped, in case that r is 65 and 50 μm. According to the FEM simulation results already shown in Fig. 9, the ζ values are 0.7, 1.0, and 1.1 for r values of 80, 65, and 50 μm, respectively. When ζ exceeds 1.0, there are no residual vibrations theoretically, which does not strongly contradict the experimental results, as shown in Figs. 19(b) and (c).

The waveforms received by the developed sensors shown in Figs. 19(b) and (c) coincide well with that received by the B&K microphone shown in Fig. 19(d), which confirms the high fidelity of the developed sensor for sound pressure in the ultrasonic frequency range, provided that an appropriate damping is given to it.

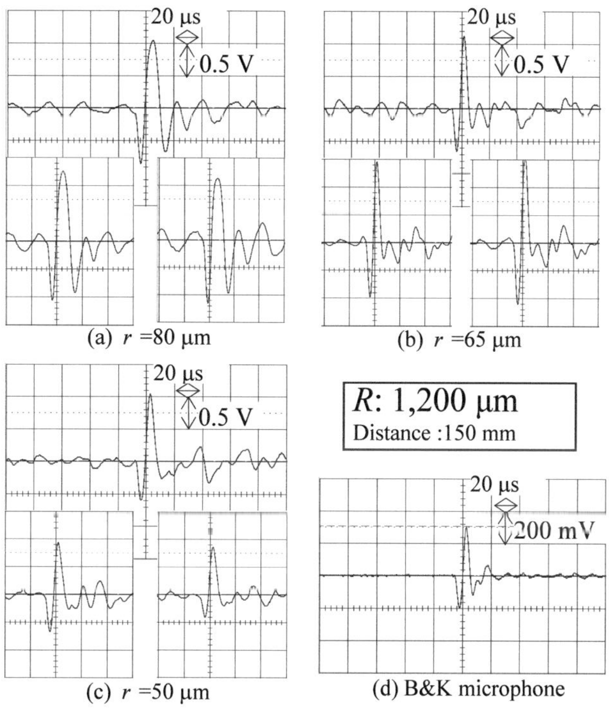

Fig. 19. Received ultrasonic pulse waveforms by changing the radius r of acoustic hole.

4.5 Distance measurement

The distance is measured by multiplying the arrival time of the first zero-cross point of the ultrasonic pulse by the sound velocity of 343.6 m/s (at 20°C), as shown in Fig. 20. This point is stable and gives high resolution to the ranging system even when the amplitude varies according to the change in the distance. The sensor, whose diaphragm radius is 1,200 μm, was used. By changing the distance between the transmitter and the developed sensor, the arrival time was measured. The results for distance from 0 to 1,000 mm are shown in Fig. 21.

The measured arrival time shows good linearity with the distance of the source, and error is within 0.1 % of the full range, i.e., this ranging system can detect the distances up to 1 m with an error of less than 1 mm. This ranging system could be effective for mobile robot devices for purposes such as detecting obstacles and recognizing the environment.

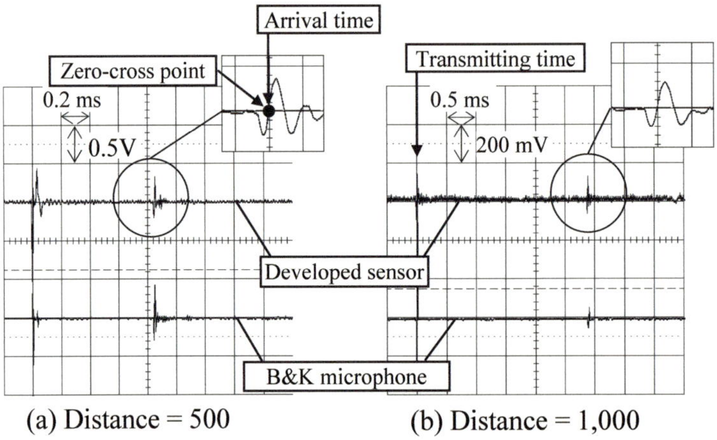

Fig. 20. Distance measurement by multiplying arrival time of zero-cross point by sound velocity.

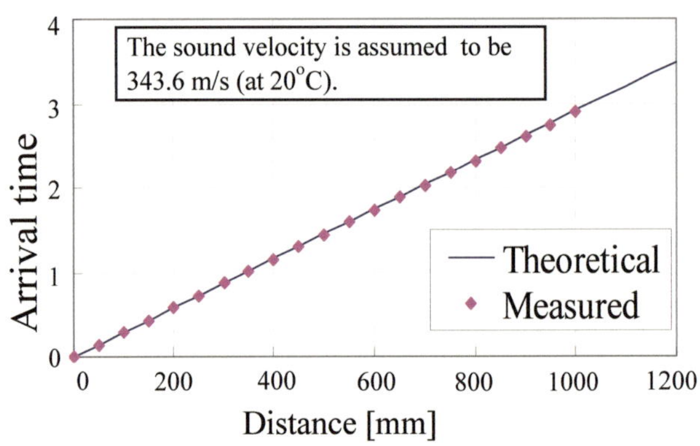

Fig. 21. Relationship between distance and measured arrival time.

4.6 Receiving directivity of one sensor

The directivity of the developed sensor was estimated using the experimental setup as already shown in Fig. 16. The peak voltage of received pulse waveform was estimated by changing the angle of the sensor using a rotational table. Results are shown in Fig. 22. From these results, the directivity becomes wide as the diaphragm radius decreases, which implies that miniaturizing the sensor size by micromachining is useful for achieving wide directivity.

It was confirmed that all the sensors used in this experiment can receive ultrasound from a wide area, which ranges from θ=-80 to 80°, with an attenuation level of less than -6 dB compared with the case θ =0°, i.e., $\theta_{1/2}$ (see equation (1) in Section 1) is approximately 80°.

This wide directivity is effective for realizing the omnidirectional characteristics of the arrayed device comprising many sensors, the detail of which is explained in the following section.

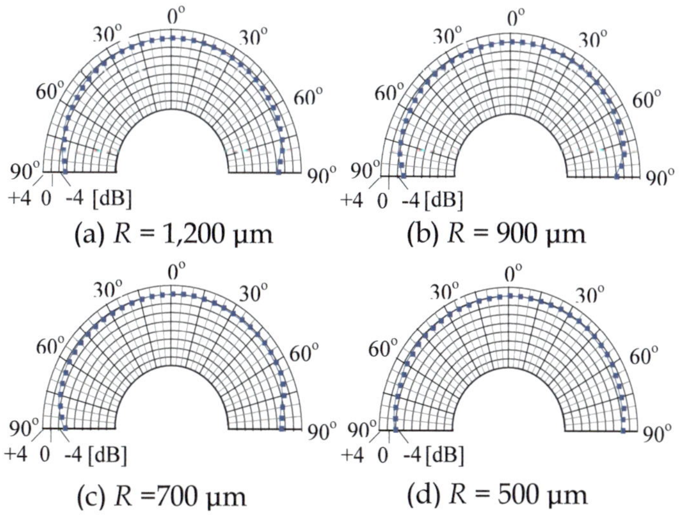

Fig. 22. Receiving directivity of developed sensor.

5. Arrayed sensor device and electrical scanning of receiving directivity

5.1 Detecting circuitry for capacitance change

An arrayed device comprising 5×5 developed sensors was fabricated. A photograph and its actual size are shown in Fig. 23. The specification of one sensor in the array is as follows: the radius (R) of the diaphragm is 1,200 μm, its thickness is 2 μm, the distance between adjacent diaphragms (a) is 3,000 μm, the radius of the acoustic hole (r) is 60 μm, and the number of holes (n) is 121.

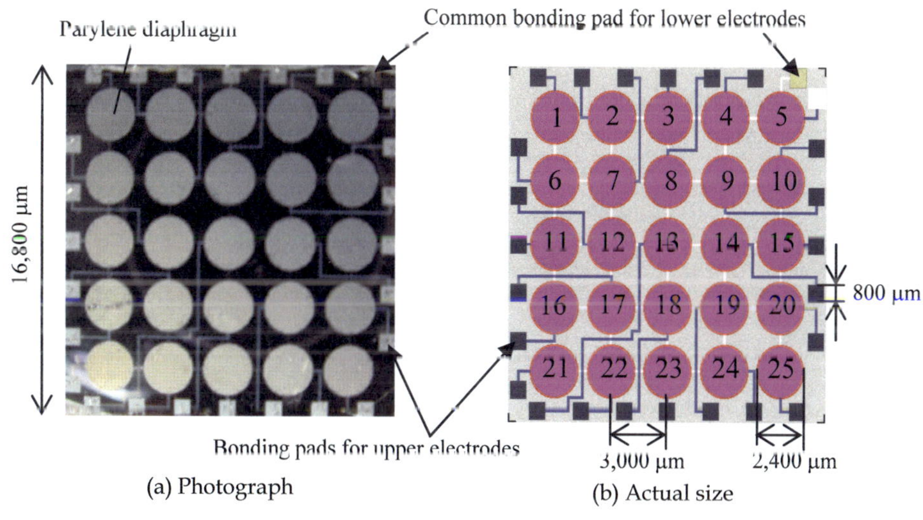

Fig. 23. Fabricated device of ultrasonic sensor array.

The capacitance (C), the dissipation factor ($\tan\delta$), and the impedance (Z) of individual sensors were measured using an LCZ meter (NF type 2341), examples of which are shown in Table 2. In this table, the wiring length for sensor no. 3 is the minimum and that for sensor no. 13 is the maximum among all the sensors, causing the difference of C between them.

Sensor no.	Capacitance C [pF]	Loss factor $\tan\delta$	Impedance at 100 kHz Z [kΩ]
3	36.0	0.02	41.9
7	43.6	0.019	29.5
13	69.5	0.024	19.5

Table 2. Examples of electrical properties of one sensor.

5.2 Dispersion of individual sensors' properties in arrayed device

The distribution of sensitivity of individual sensors in the developed arrayed device was estimated, where the peak voltage of the received ultrasonic waveform is taken as the index of the sensitivity. The experimental results are shown in Fig. 24(a), the values of which do not strongly contradict the anticipated value of 67 mV (see Section 4.3 and Fig. 17). There is dispersion of experimental sensitivity; however, it is not significant. Thus, the first zero-cross point of the received pulse waveform can be detected in all the sensors by setting an appropriate threshold level, i.e., the time-of-flight measurement of ultrasound for determining the distance can be generally performed for all the sensors.

The distribution of the resonant frequency of individual sensors was also estimated. The experimental results are shown in Fig. 24(b), the values of which do not strongly contradict the target value of 43 kHz, which is confirmed by both FEM simulation (see Section 2.1 and Fig. 7) and experiments (see Section 4.3 and Fig. 18). However, the uniformity of resonant frequency is unsatisfactory.

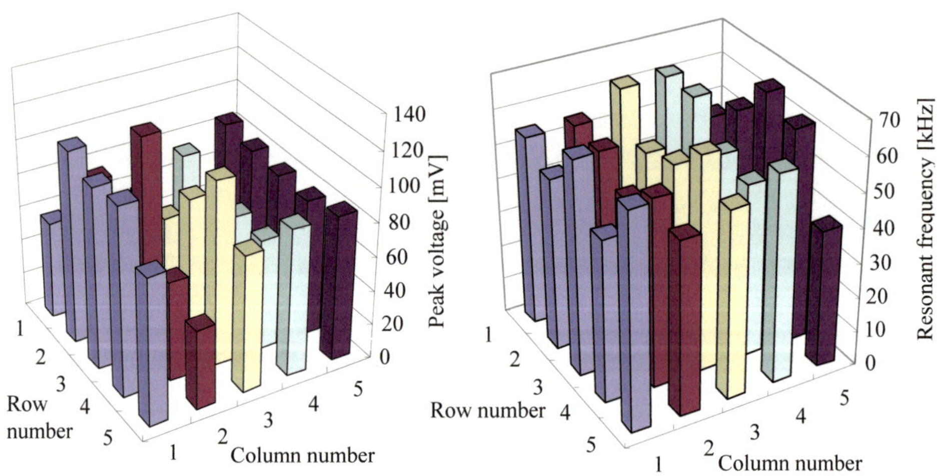

(a) Distribution of sensitivity (b) Distribution of resonant frequencies

Fig. 24. Result of sensitivity and resonant frequency.

One reason for the dispersion of resonant frequencies is due to the fabrication, i.e., the Young's modulus, thickness, and the intrinsic tensile stress of the Parylene diaphragm were not uniform all over the fabricated arrayed sensor area, since it is difficult to keep the process conditions strictly the same irrespective of the position inside the arrayed device. Because of this problem, the resonant frequency varied from one sensor to another, since the resonant frequency depends on these mechanical parameters (Khuri-Yakub et al., 2000; Aoyagi et al., 2007b). The process uniformity should be improved in future studies.

5.3 Electrical scanning of receiving directivity

The electric scanning of receiving directivity based on the delay-and-summation principle is possible by using many of sensors. Among totally twenty five sensors in the fabricated arrayed device, five sensors lying in one line were selected, and they were used for an experiment of performing the electrical scanning of receiving directivity, as shown in Fig. 25. The fabricated arrayed device was rotated using a rotational table, the center of which was set apart from an ultrasonic transmitter of electric spark discharge by 150 mm. Let the rotational angle be θ. Then the difference of sonic path length for two adjacent sensors is expressed as $a\sin\theta$, where a is interval between the sensors (a=3,000 μm in this case).
The procedure of the experiment is schematically shown in Fig. 26, which is as follows: Received pulse waveforms for the five sensors are schematically shown in Fig. 26(a). Their arrival times have differences based on the differences in sonic path length. After recording the waveforms in a computer, the positive peak of each waveform is detected. Taking this peak as the center, a rectangular pulse wave with 5 μs width is generated, as shown in Fig. 26(a). Then, each pulse is shifted by a delay time of $\{(n-1)\cdot a\sin\alpha\}/v$, where α is the scanning angle of directivity, v is the sound velocity (343.6 m/s is employed in this experiment), and n is the number of the sensor which takes $1, 2, \cdots, 5$. The shifted pulses are summed, and the area inside the width of pulse no. 1 is extracted from the summed result, which is the hatched area shown in Fig. 26(b). The average height of this area is estimated as the index of sensitivity.

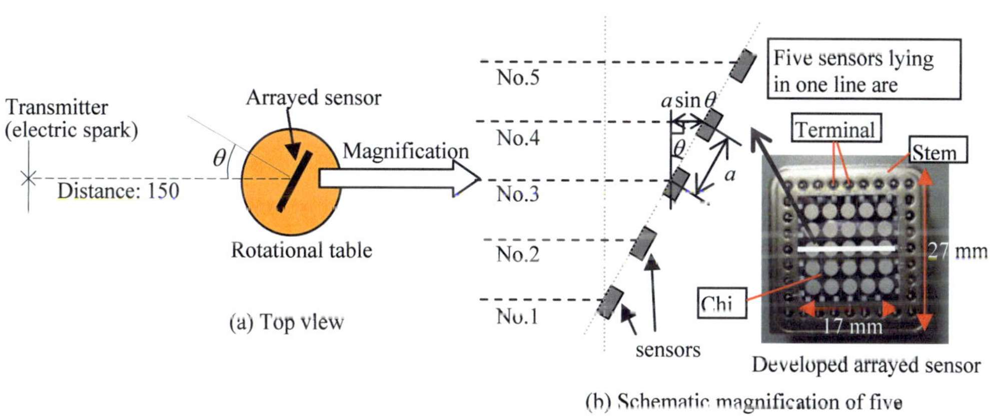

Fig. 25. Experimental conditions for electrical scanning of receiving directivity using arrayed sensor device.

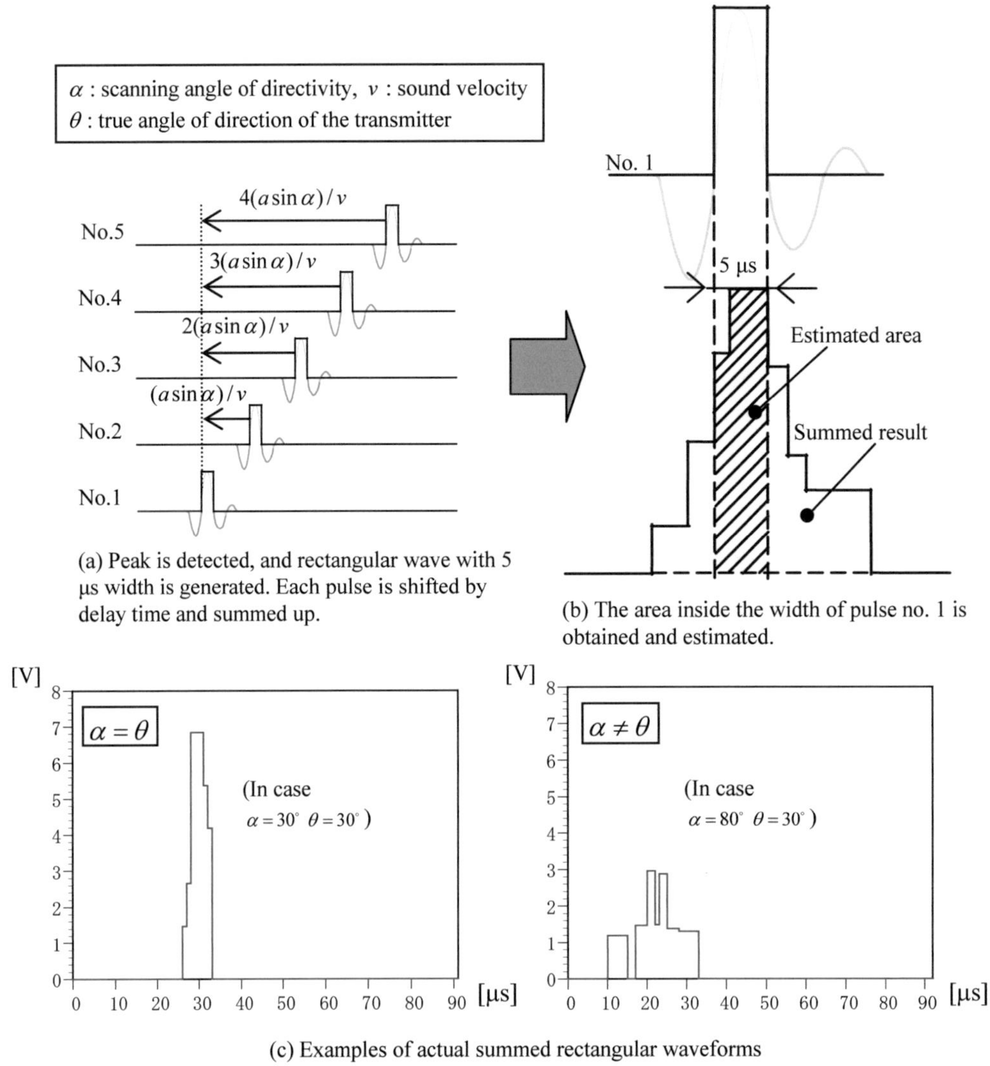

Fig. 26. Procedure of electrical scanning of receiving directivity using arrayed sensor.

Examples of actual summed rectangular waveforms are shown in Fig. 26(c). Looking at this figure, the width of the summed result almost coincides with that of pulse no. 1, i.e., it almost fits inside a 5 μs width in the case of $\alpha = \theta$, while it does not do so in the case of $\alpha \neq \theta$. Namely, the sensitivity is maximized in the former case.

These processes, i.e., detecting peaks, generating pulses, shifting them, summing them, and extracting the area for estimation, were performed by developed computer software. In the experiment, θ was set at $0, 10, \cdots, 90$ °. For each θ, a scanning angle α of $0, 10, \cdots 90$ ° was tested computationally, and the sensitivity of each combination of θ and α was estimated.

The results of electrical scanning performance of receiving directivity are shown in Fig. 27. In this figure, each data is normalized to a relative value in dB units, so that the sensitivity when $\theta = \alpha$ is 0 dB. The absolute value of the sound pressure level (SPL) for the case of 0 dB for each θ angle is shown in Table 3. Looking at this table, the SPL does not decrease as θ increases, i.e., it takes almost the same value irrespective of θ.

According to Fig. 27, the sensitivity is increased when $\alpha = \theta$, i.e., when the scanning angle (α) is coincident with the angle of direction of the transmitter (θ), except for only the two cases of θ=70 and 80°. Even in these two cases, the error is small, within 10°. Note that when θ is in the range from 0 to 50°, a sharp peak of directivity at the target scanning angle is obtained, which may be effective for detecting an angle at which a target object exists in microrobot applications. To conclude, it was proven that the directivity can be scanned electrically based on the delay-and-summation principle using the fabricated Parylene arrayed device. It was also proven that a wide scanning angle of at least 50° can be achieved. This omnidirectional characteristic is due to the wide directivity of the individual sensor, which was already characterized in Section 4.6.

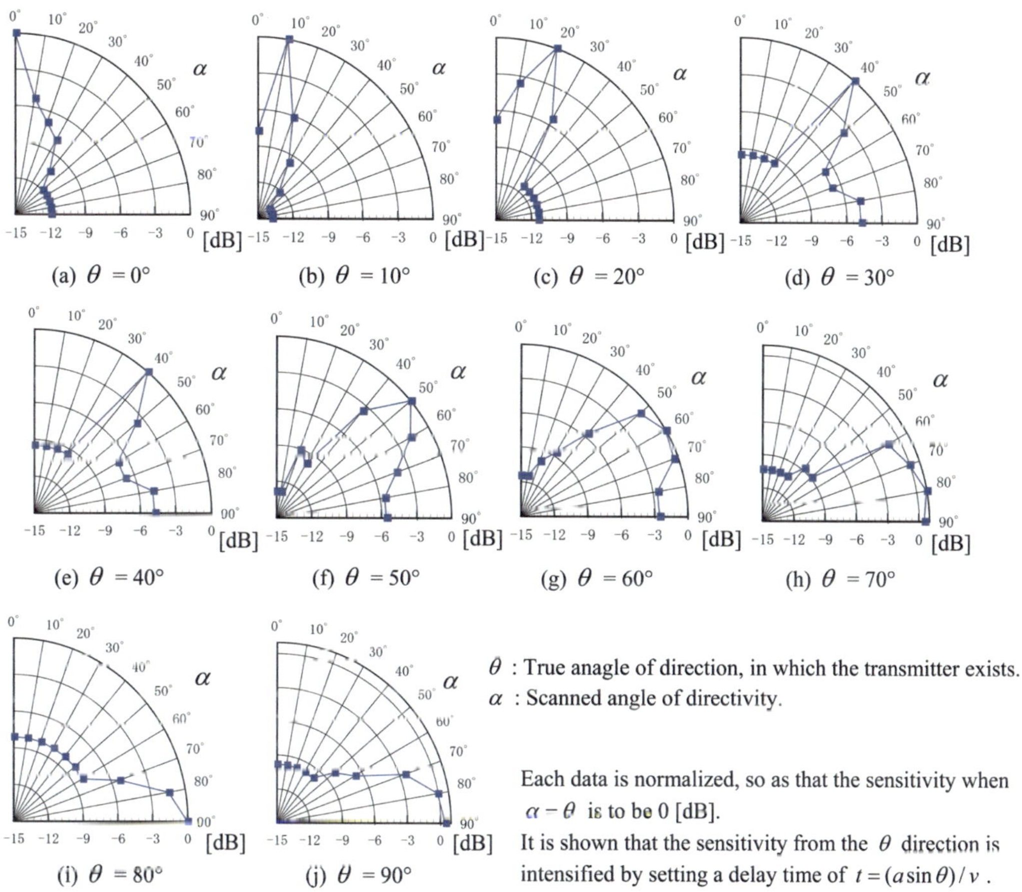

Fig. 27. Results of electric scanning of receiving directivity using arrayed sensor.

θ [°]	SPL [dB]
0	152
10	150
20	145
30	148
40	142
50	147
60	145
70	141
80	140

Table 3. Sound pressure level (SPL) for 0 dB case in Fig. 27 for each θ.

6. Transmitting performance of one sensor and electrical scanning of transmitting directivity

6.1 Transmitting circuitry

Because of the flexibility and durability of Parylene, one capacitive sensor with a Parylene diaphragm can also be used as a transmitter by applying a high impulsive voltage. A transmitting circuit was developed, as shown in Fig. 28(a), in which the same bias voltage of 100 V as that used in the receiving circuitry is employed. When the transistor is triggered, a condenser C_T of 0.1 μF is discharged and an electric current is instantaneously supplied to the primary side of the ignition coil. Then a high impulsive voltage is generated at the secondary side of this coil, as shown in Fig. 28(b), which exhibits a peak-to-peak voltage of approximately 700 Vpp (the positive voltage of 400 Vop and negative one of 300 Vop, both of which are values relative to the bias voltage of 100 V). The power spectrum of this voltage is shown in Fig. 28(c). In this figure, the peak frequency is 310 kHz, which is far larger than the resonant frequency of the developed device (43 kHz). This fact indicates that the response of the diaphragm's displacement at the transmission can be approximately regarded as an impulse response, on which the resonant frequency of the diaphragm has a large effect rather than the peak frequency of the input voltage.

6.2 Experimental setup for characterizing transmitting performance

The transmitting performance of the developed Parylene device was characterized. The experimental setup is schematically shown in Fig. 29. The device was set on a rotational table. Each sensor in the arrayed device was activated as a transmitter. In addition to the arrayed device, a device including several sensor/transmitters with different radii of the diaphragm and different radii of the acoustic hole was prepared. This device was used to investigate the effect of the area of the diaphragm on the transmitted sound pressure and the effect of the acoustic holes on damping of the transmitted waveform.

The B&K-type 4138 reference microphone (with sensitivity 0.9 mV/Pa) was used as a receiver. The distance between the center of the arrayed transmitter device and the receiver was set to several values ranging from 10 to 1,000 mm to characterize the performance of one transmitter, and 40 mm to perform the electrical scanning of the arrayed transmitter. In the case that the transmitted acoustic pressure is small, the received signal obtained by the reference microphone was amplified by a factor of 3,000 (69.5 dB) using an instrumentation amplifier (ACO type 6030).

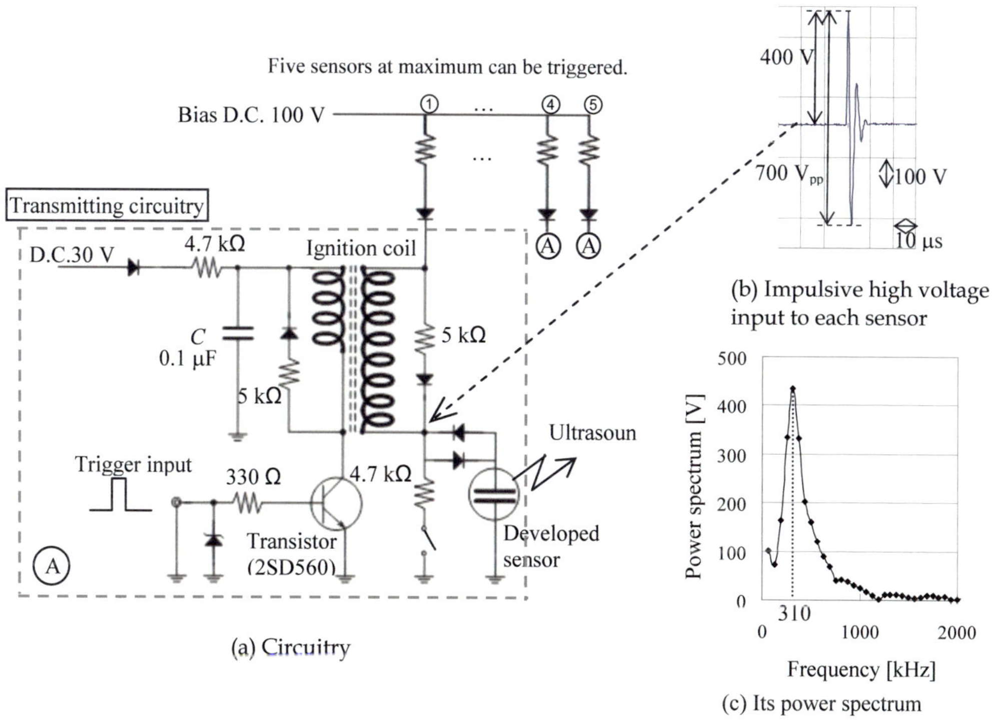

Fig. 28. Transmitting circuitry of generating a high impulsive voltage.

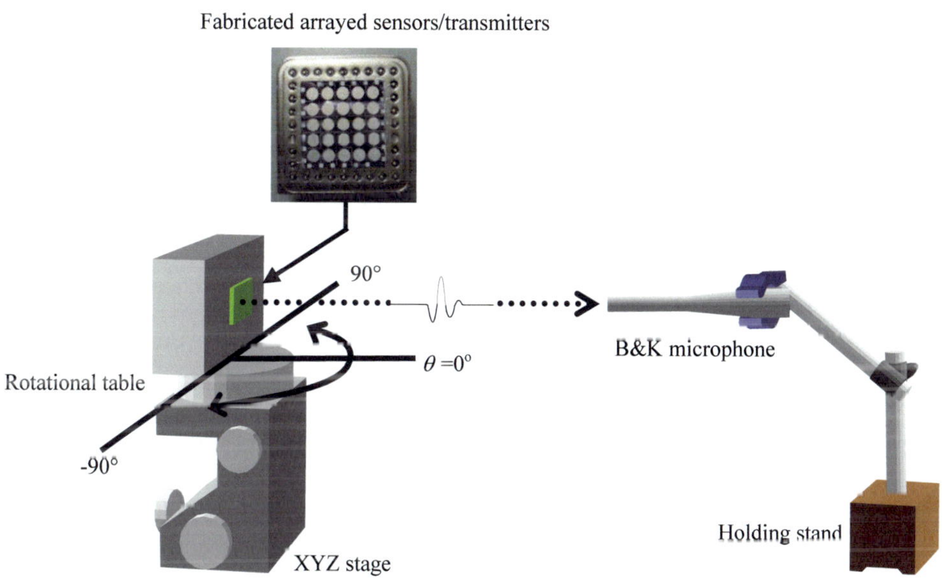

Fig. 29. Experimental conditions for characterizing transmitting performance.

6.3 Transmitted pulse waveform and detectable distance

The ultrasonic waveform, which is emitted by the developed transmitter and received by the B&K-type 4138 reference microphone, is shown in Fig. 30(a). The acoustic pressure obtained at a distance of 10 mm was 13 Pa, which is rather small. Therefore, the signal was amplified using an instrumentation amplifier. The amplified received waveform obtained at a distance of 150 mm is shown in Fig. 30(b). By this amplification, the maximum distance at which the transmitted waveform is detectable was extended. The experimental results of the relationship between the distance and the peak voltage of the transmitted waveform are shown in Table 4, which indicates that the transmitted waveform can be detected as far as 1,000 mm away by setting an appropriate threshold level. It was confirmed that the developed transmitter is useful for the application of ranging the distance based on the time-of-flight measurement in the air.

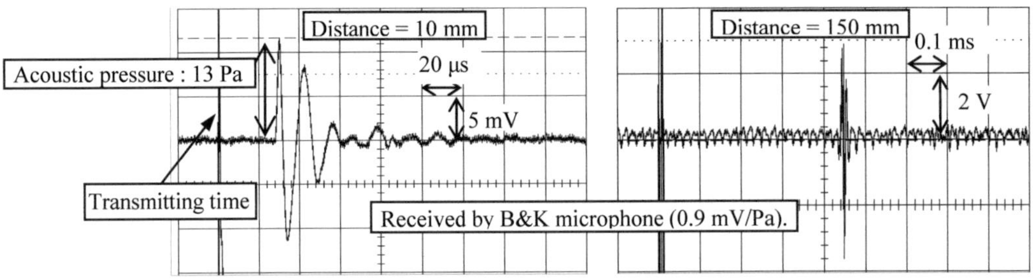

(a) Without amplification

(b) Amplified by 3,000 times (69.5 dB) by instrumentation amplifier.

Fig. 30. Emitted waveforms by developed transmitter (R = 1,200μm).

Distance [mm]	100	300	600	1,000
Peak voltage [V]	2.2	1	0.8	0.4

Note: B&K microphone output was amplified by 69.5 dB and estimated.

Table 4. Relationship between distance and peak voltage of transmitted waveform.

6.4 Effect of diaphragm area on transmitted sound pressure

The pulse waveforms emitted by the developed transmitters, of which the diaphragm radii are 500, 700, 900, and 1,200 μm, were obtained, and their peak voltages were transformed to the sound pressure. The relationship between the diaphragm area and the transmitted sound pressure at 150 mm distance is shown in Fig. 31. It was proven that the sound pressure increases proportionally with the diaphragm area.

6.5 Effect of acoustic holes on damping of transmitted waveform

We have theoretically investigated the effects of the radius of the acoustic hole r and the number of holes n on the diaphragm's damping ratio ζ in Section 2.2. It was proven that ζ is inversely proportional to r and n, which was also experimentally confirmed by the ultrasonic waveform received by the developed sensor as explained in Section 4.4. In this section, we aim to confirm this effect of acoustic holes by the ultrasonic waveform emitted by the developed transmitter.

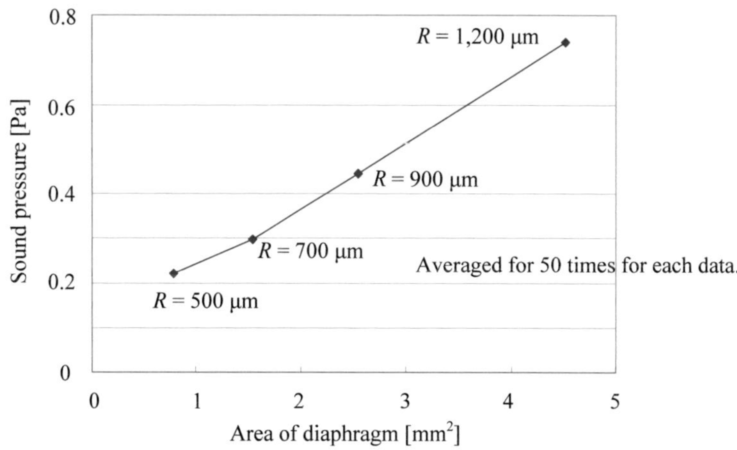

Fig. 31. Relationship between area of diaphragm and transmitted sound pressure.

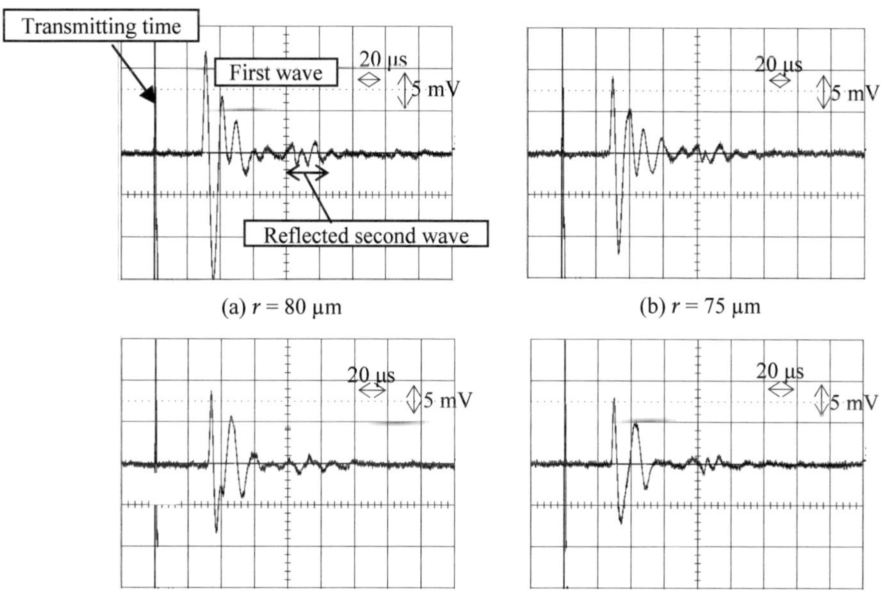

Fig. 32. Transmitted waveforms at distance of 10 mm by changing the radius *r* of acoustic hole.

The developed transmitters with different sizes of acoustic holes, of which diaphragm radius is 1,200 μm, were employed. The radii of the acoustic holes *r* are 80, 75, 55, and 50 μm. The ultrasonic pulse waveforms emitted are shown in Figs. 32(a)- (d). The distance was set to 10 mm, and the waveform was detected by the B&K microphone with no amplification. Note that a second small waveform is also observed in this figure, which is reflected by the B&K microphone, returns to the transmitter, reflected by the transmitter, and again returns to the microphone.

According to this figure, a well-damped transmitted waveform is obtained when r is 55 or 50 μm, whereas a residual vibration is seen when r is 80 or 75 μm. Namely, it was confirmed that ζ is inversely proportional to r. The effect of acoustic holes on the diaphragm damping confirmed here using the transmitted waveform does not contradict that confirmed using the received waveform.

6.6 Directivity of one transmitter

The directivity of the developed transmitter was estimated using the experimental setup shown in Fig. 29. The distance between the transmitter and the sensor was set to 150 mm, and the peak voltage of the received pulse waveform was estimated by changing the angle of the transmitter using a rotational table. Results are shown in Fig. 33. From these results, the directivity becomes wide as the diaphragm radius decreases. It was confirmed that both of the transmitters used in this experiment can emit ultrasound over a wide direction, which ranges from θ=-80 to 80°, with an attenuation level of less than -4 dB compared with the case where θ =0°. Namely, the developed transmitter can be approximated to be nondirectional.

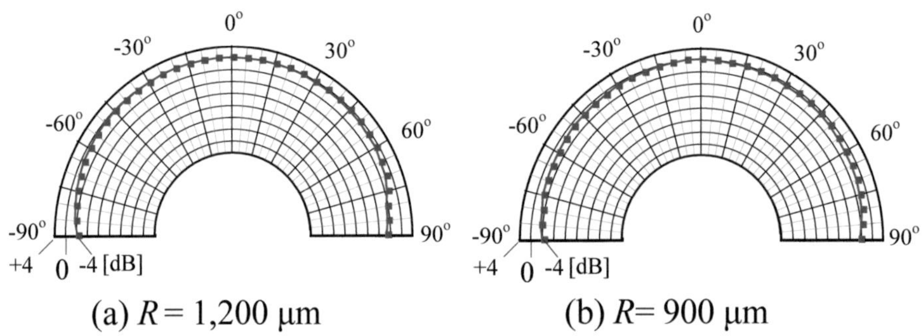

Fig. 33. Transmitting directivity of developed transmitter.

6.7 Electrical scanning of transmitting directivity

Five collinear transmitters were selected, and they were used for an experiment of performing the electrical scanning of transmitting directivity. The experimental conditions are schematically shown in Fig. 34(a). The fabricated arrayed device was rotated using a rotational table. Let the rotational angle be θ. Then the difference of the sonic path length for two adjacent transmitters is expressed as $a\sin\theta$, where a is interval between the transmitters. The procedure, based on the delay-and-summation principle, is as follows. Trigger input pulses for the five transmitters are schematically shown in Fig. 34(b). When the frequency of these pulses is set to $f = v/(a\sin\alpha)$, the transmitted waves are theoretically intensified in the α direction, where α is the scanning angle of directivity, and v is the sound velocity (343.6 m/s is employed in this experiment).

For each θ, the scanning angle (α) was set by changing the frequency (f) of the trigger pulses, which were input to the transmitting circuitry. The peak voltage of transmitted waveform, which is received by the B&K microphone, was estimated at each combination of θ and f.

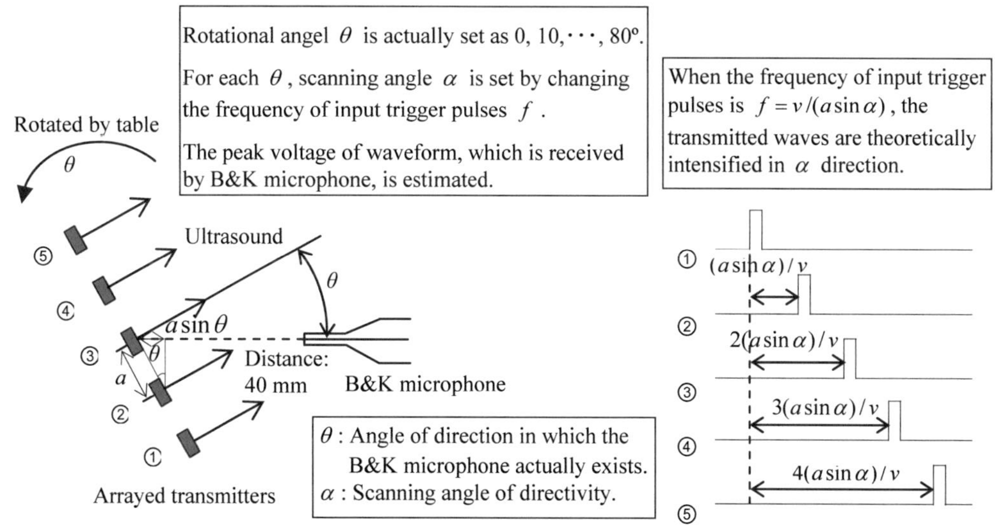

(a) Schematic of experimental setup using rotational table and reference microphone (B & K 4138)

(b) Shifting trigger inputs to each sensor for controlling the transmitting directivity

Fig. 34. Experimental conditions for electrical scanning of transmitting directivity using arrayed device.

The results of electrical scanning performance of transmitting directivity are shown in Fig. 35. In this figure, each data is normalized, so that the detected peak voltage when $f = v/(a \sin\theta)$, i.e., $\alpha = \theta$, is 0 dB. According to this figure, the transmitted waveform was intensified at $f = v/(a \sin\theta)$, i.e., it was intensified when the scanning angle (α) was coincident with the angle of the direction (θ) of the receiver. However, the directivity when θ =30° was less sharp than that in the other conditions in this figure. This may be caused by an experimental problem, the improvement of which is a possible future study. To conclude, although further study is necessary, the possibility of controlling the transmitting directivity was preliminarily shown in this experiment using the fabricated arrayed device.

7. Conclusions

An arrayed device comprising 5×5 ultrasonic sensors/transmitters featuring polymer Parylene diaphragms was fabricated, and its performance was characterized. In addition to the durability and high sensitivity due to polymer nonbrittleness and flexibility, merits attributable to Parylene, such as biocompatibility, chemical resistivity, CMOS compatibility, and conformal deposition, are expected to be achieved in the future.

The contents of this study are briefly summarized as follows. 1) An ultrasonic sensor with Parylene diaphragm was developed. The sensor was found to be able to receive an impulsive ultrasonic pulse transmitted by a spark discharge. The open-circuit sensitivity was 0.4 mV/Pa. 2) A well-damped waveform was obtained by setting appropriate acoustic

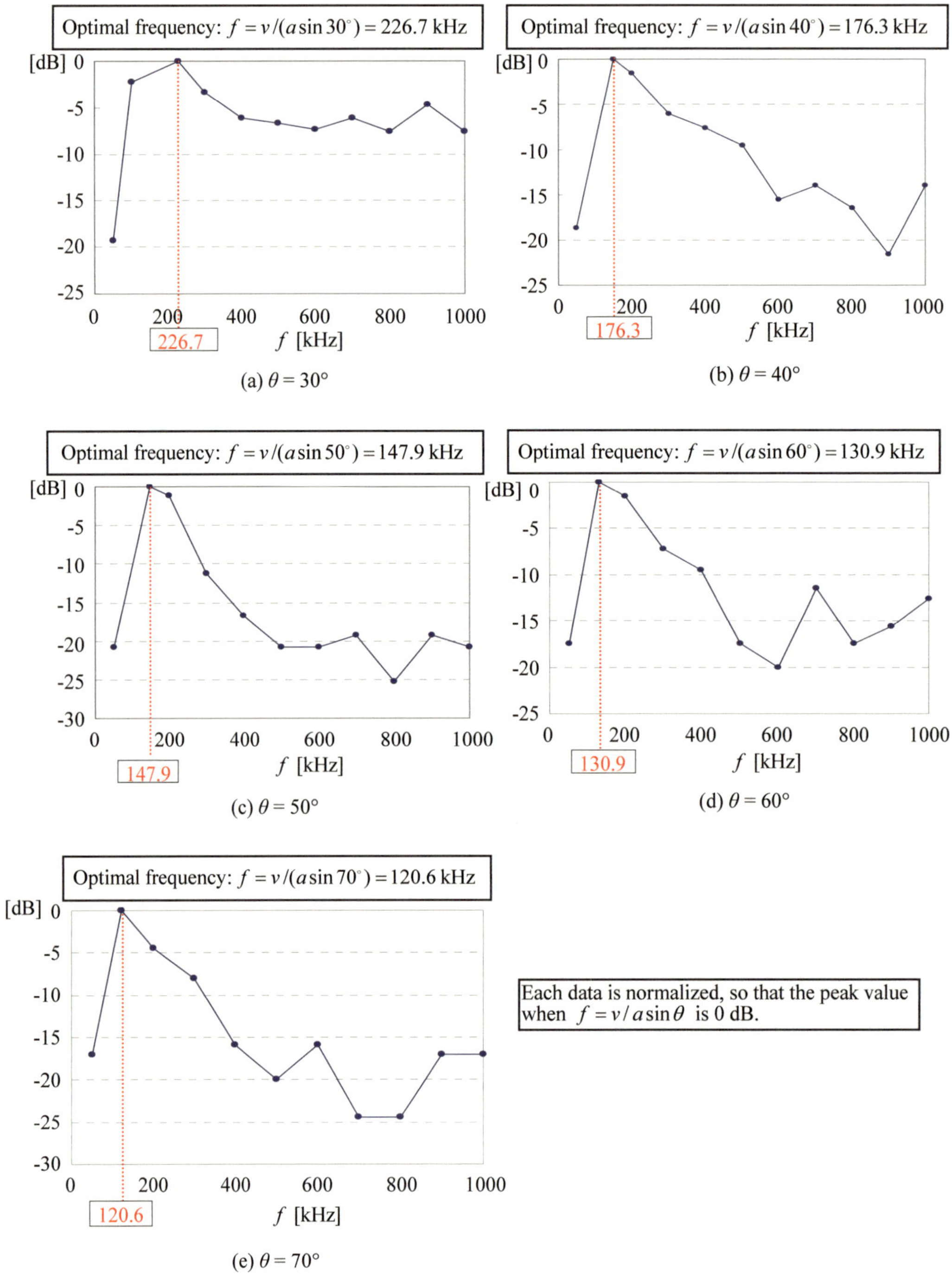

Fig. 35. Results of electric scanning of transmitting directivity using arrayed device.

holes. 3) The ranging system using this sensor can detect distances up to 1 m with an error of less than 1 mm. 4) The developed sensor can receive ultrasound from a wide area, which ranges from θ=-80 to 80°. 5) An arrayed ultrasonic device was developed by the micromachining technique. The dispersion of individual sensors' properties, i.e., the sensitivity and the resonant frequency, was proven to be tolerable. 6) The electrical scanning of receiving directivity was performed on the basis of the delay-and-summation principle. A wide scanning angle of at least 50° was achieved. 7) Each developed sensor was activated as a transmitter by applying a high impulsive voltage. The transmitted waveform was detectable as far as 1,000 mm away. The ultrasound was transmitted over a wide direction ranging from θ=-80 to 80°. 8) The possibility of electrical scanning of transmitting directivity was preliminarily confirmed using the developed arrayed device.

By scanning both the transmitting directivity and the receiving directivity of the developed arrayed device, detecting the direction in which objects or obstacles exist is a future study. In this study, by detecting the time-of-flight of an ultrasonic pulse reflected by objects or obstacles, the distance from them is also detectable. By using the information on both the direction and the distance, the positions of objects or obstacles may be obtained in the future. Further quantitative investigation of the merits of the developed Parylene ultrasonic arrayed sensors/transmitters compared with other reported silicon or polymer devices is also a planned future study.

8. Acknowledgements

This work was supported by a Grant-in-Aid for Scientific Research (17656090) from the Ministry of Education, Culture, Sports, Science and Technology (MEXT) and the "High-Tech Research Center" Project for Private Universities: Matching Fund Subsidy from MEXT, 2005-2009, and the Kansai University Special Research Fund, 2007 and 2008. The author is grateful to Professors Hiromitsu Kozuka and Yasuhiko Arai at Kansai University for their invaluable advice and for supplying measuring apparatus. The author is grateful to Mr. Go Kawai for his drawing figures.

9. References

Aoyagi, S.; Kamiya, Y. & Okabe, S. (1992). Development of Powerful Airborne Ultrasonic Transmitter for Robot Metrology. *Proc. the 12th Symposium on Ultrasonic Electronics, Japanese J. Applied Physics*, Vol. 31, Suppl. 31-1, pp. 263-265.

Aoyagi, S. (1996). Application of Ultrasonic Sensors to Robot Measurement. *J. the Japan Society for Precision Engineering*, Vol. 62, No. 3, pp. 373-376 (in Japanese).

Aoyagi, S. & Takehata, K. (2001). Study on Object Shape Recognition Using an Ultrasonic Sensor. *Integrated Computer-Aided Engineering*, Vol. 8, pp. 105-117.

Aoyagi, S.; Furukawa, K.; Yamashita, K.; Tanaka, T.; Inoue, K. & Okuyama, M. (2007a). Development of Capacitive Ultrasonic Sensor with Parylene Diaphragm Using Micromachining Technique. *Japanese J. Applied Physics*, Vol. 46, pp. 4595-4601.

Aoyagi, S.; Yoshikawa, D; Isono, Y & Tai,Y,C. (2007b). Development of a Capacitive Accelerometer Using Parylene (Part 1) -Study on Resonant Frequency of Parylene Suspended Structure-, *IEEJ Trans. SM*, Vol.127, No.6, pp. 314-320.

Aoyagi, S; Furukawa, K; Ono, D; Yamashita, K; Tanaka, T; Inoue, K & Okuyama, M. (2008a). Development of a capacitive ultrasonic sensor having parylene diaphragm and characterization of receiving performance of arrayed device. *Sensors and Actuators A*, Vol. 145-146, pp. 94-102.

Aoyagi, S; Ono, D; Kawai, G; Yamashita, K; Okuyama, M. (2008b). Micromachined Arrayed Capacitive Ultrasonic Sensor/Transmitter with Parylene Diaphragms, *Japanese J. Applied Physics*, Vol. 47, No. 8, pp. 6513-6525.

Bergqvist, J. & Gobet, J. (1994). Capacitive Microphone with a Surface Micromachined Backplate Using Electroplating Technology. *J. Microelectromechanical Systems*, Vol. 3, No. 2, pp. 69-75.

Brüel & Kjær (1982). *Condenser Microphones Data Handbook*, Brüel & Kjær, Nærum, Denmark.

Chen, J.; Liu, L.; Li, Z.; Tan, Z.; Xu, Y. & Ma, J. (2002). Single-Chip Condenser Miniature Microphone with a High Sensitive Circular Corrugated Diaphragm. *Proc. MEMS'02*, pp. 284-287, Las Vegas, USA, January, 2002.

Diamond, B, M.; Neumann, J, J. & Gabriel, K, J. (2002). Digital Sound Reconstruction Using Arrays of CMOS-MEMS Microspeakers. *Proc. MEMS'02*, pp.292- 295, Las Vegas, USA, January, 2002.

Guldiken, R, O. & Degertekin, F, L. (2005). Micromachined Capacitive Transducer Arrays for Intravascular Ultrasound Imaging. *Proc. MEMS'05*, pp. 315-318, Miami, USA, January, 2005.

Haga, Y.; Fujita, M.; Nakamura, K.; Kim, C, J. & Esashi, M. (2003). Batch Fabrication of Intravascular Forward-Looking Ultrasonic Probe. *Sensors and Actuators A*, Vol. 104, pp. 40-43.

Harder, T, A.; Yao, T, J.; He, Q.; Shih, C, Y. & Tai, Y, C. (2002). Residual Stress in Thin-Film Parylene-C. *Proceeding of MEMS'02*, pp. 435-438, Las Vegas, USA, January, 2002.

Hayashi, T.; Kawashima, K. & Endoh, S. (2001). The Generation and Detection of Fundamental Lamb Modes in Plastic Plates by Air-coupled Transducers. *Proc. American Institute of Physics Conference*, Vol. 557, pp. 105-110, New York, USA, 2001.

Hsieh, W, H.; Yao, T, J. & Tai, Y, C. (1999). A High Performance MEMS Thin-film Teflon Electret Microphone. *Tech. Digest Transducers'99*, pp. 1064-1067, Sendai, Japan, June 1999.

Ikeda, M.; Shimizu, N. & Esashi, M. (1999). Surface Micromachined Driven Shielded Condenser Microphone with a Sacrificial Layer Etched from the Backside. *Tech. Digest Transducers'99*, pp. 1070-1073, Sendai, Japan, June 1999.

Khuri-Yakub, B, T.; Cheng, C, H.; Degertekin, F, L. & Ergun, S. (2000). Silicon Micromachined Ultrasonic Transducers. *Japanese J. Applied Physics*, Vol. 39, pp. 2883-2887.

Knowles Acoustics (2002). *Surface Mount Microphones*, Knowles Acoustics, Itasca, IL, USA.

Kovacs, G, T, A. (1998). *Micromachined Transducers Sourcebook*, McGraw-Hill, ISBN 0-07-290722-3, New York, USA.

Martin, D, T.; Kadirval, K.; Liu, J.; Fox, R, M.; Sheplak, M. & Nishida, T. (2005). Surface and Bulk Micromachined Dual Back-Plate Condenser Microphone. *Proc. MEMS'05*, pp. 319-323, Miami, USA, January 2005.

Mitsuhashi, W. (1997). Target Parameter Estimation on the basis of Phase Histograms of the Outputs of Constant-Q Filter Bank. *IEEJ Trans. Sensors and Micromachines*, Vol. 117-E, pp. 201-8 (in Japanese).

Mitsuida, Y. (1987). *Onkyo Kogaku (Acoustic Engineering)*, Shokodo, p. 64, ISBN978-4-7856-0114-0 ,Tokyo, Japan (in Japanese).

Ono, N.; Arita, T.; Senjo,Y. & Ando, S. (2005). Directivity Steering Principle for Biomimicry Silicon Microphone. *Tech. Digest Transducers'05*, Vol. 1, pp. 792-795, Seoul, Korea, June 2005.

Pederson, M.; Olthuis, W. & Bergveld, P. (1998). High-performance Condenser Microphone with Fully Integrated CMOS Amplifier and DC-DC Voltage Converter. *J. Microelectromechanical Systems*, Vol. 7, No. 4, pp. 387-394.

Sato, H.; Okabe, S. & Iwata, Y. (1993). *Kikai Shindogaku (Mechanical Vibration Theory)*, Kogyo Chosakai , ISBN 4-7693-2105-8, Tokyo, Japan (in Japanese).

Sasaki, K.; Takano, M. & Akeno, K. (1988). A New Method of Object Recognition and Sensory Feedback Control by High Accuracy Ultrasonic Sensor. *J. The Faculty of Engineering, The University of Tokyo, Ser.* B Vol. 49, pp. 209-240.

Scheeper, P, R.; Donk, A, G, H.; Olthuis, W. & Bergveld, P. (1992). Fabrication of Silicon Condenser Microphones Using Single Wafer Technology. *J. Microelectromechanical Systems*, Vol. 1, No. 3, pp. 147-154.

Schindel, D, W.; Hutchins, D, A.; Zou, L. & Sayer, M. (1995). The Design and Characterization of Micromachined Air-Coupled Capacitance Transducers. *IEEE Trans. Ultrasonics, Ferroelectrics, and Frequency control*, Vol. 42, pp. 42-50.

Škvor, Z. (1967). On the Acoustic Resistance Due to Viscous Losses in Air Gap of Electrostatic Transducers. *Acoustica*, Vol. 19, pp.295-299.

Tabata, O.; Kawahata, K.; Sugiyama, S. & Igarashi, I. (1989). Mechanical Property Measurements of Thin Films Using Load-Deflection of Composite Rectangular Membranes. *Sensors and Actuators A*, Vol. 20, pp. 135–141.

Tai, Y, C. (2003). Parylene MEMS: Material, Technology and Applications. *Proc. of 20th Sensor Symposium*, pp. 1-8, Tokyo, Japan, October 2003.

Yamashita, K.; Katata, H.; Okuyama, M.; Miyoshi, H.; Kato, G.; Aoyagi, S. & Suzuki, Y. (2002a). Arrayed Ultrasonic Microsensors with High Directivity for in-Air Use Using PZT Thin Film on Silicon Diaphragms. *Sensors and Actuators A*, Vol. 97-98, pp. 302-307.

Yamashita, K.; Murakami, H.; Fukunaga, T.; Okuyama, M.; Aoyagi, S. & Suzuki, Y. (2002b). Ultrasonic Phased Array Micro Sensor Using Piezoelectric PZT Thin Film and Resonant Frequency Tuning by Poling. *Proc. 13th IEEE Int. Symp. Applications of Ferroelectrics*, pp. 487-490, Nara, Japan, May 2002

Yao, T, J.; He, Q.; Yang, X. & Tai, Y, C. (2001). BrF_3 Dry Release Technologies for Large Freestanding Parylene, *Tech. Digest. Transducers'01*, pp. 652-655, Munich, Germany, June 2001.

Yaralioglu, G, G.; Ergun, A, S. & Khuri-Yakub, B, T. (2005). Finite-Element Analysis of Capacitive Micromachined Ultrasonic Transducers. *IEEE Trans. Ultrasonics, Ferroelectrics, and Frequency control*, Vol. 52, pp. 2185-2198.

Zhuang, X.; Ergun, A, S.; Oralkan, O.; Wygant, I, O. & Khuri- Yakub, B, T. (2006). Interconnection and Packaging for 2D Capacitive Micromachined Ultrasonic Transducer Arrays Based on Through-Wafer Trench Isolation, *Proc. MEMS'06*, pp. 270-273, Istanbul, Turk, January, 2006.

19

Application of Microsystems Technology in the Fabrication of Thermoelectric Micro-Converters

L.M. Goncalves and J.G. Rocha
University of Minho, Guimarães, Portugal

1. Introduction

The use of thin-film deposition techniques with microsystems technologies renewed the interest in the thermoelectricity in the last years. Integration of efficient solid-state thermoelectric (TE) microdevices with microelectronics is desirable for local cooling and, since they can be used to stabilise the temperature of devices, decrease noise levels and increase operation speed. Their use in thermoelectric microgeneration (energy harvesting) can also supply energy to low power consumption electronic devices. In this chapter, the fabrication of thermoelectric microconverters is compared, both on materials from thin-film composites to supperlattice structures, and on its fabrication techniques.

Various materials can be used for this type of converters. However, for room temperature application, Bi/Sb/Te compounds are still the most efficient thermoelectric materials. Recently, efforts were made to apply quantum confinement to thermoelectric materials, and the results are thin-film superlattice structures and nanowires and even more recently, bulk nanocomposites. Some of these materials proved the ability to double efficiency of current thermoelectric devices. Several deposition techniques can be used for the fabrication of Bi/Sb/Te thin-films: co-sputtering, electrochemical deposition, metal-organic chemical vapor deposition or flash evaporation are some examples compared here.

The patterning process must use photolithography techniques to create the small dimensions of these devices. Despite these techniques are commonly used in microelectronic devices, mainly with silicon based substrates, its application in other thermoelectric alloys is still under development.

The patterning of thermoelectric structures for the fabrication of thermoelectric microconverters can be done using common microsystems technologies. Techniques used in MEMS fabrication, namely wet-etching, lift-off (with SU-8 photoresist), Reactive Ion Etching (RIE) and Lithography-Electroplating-Molding (LIGA) are here compared for the fabrication of thermoelectric microsystems.

2. Theory behind thermoelectric devices

There are two groups of applications for thermoelectric materials based on Seebeck and Peltier effects respectively. In the Seebeck effect, a temperature difference between the junctions of two different materials produces an electric voltage (figure 1), and an electric current flows when the electric circuit is closed (Seebeck, 1882). This effect is quantified by the Seebeck coefficient, α, as represented in eq. 1:

$$\alpha = \frac{\Delta V}{\Delta T} \ (VK^{-1}) \tag{1}$$

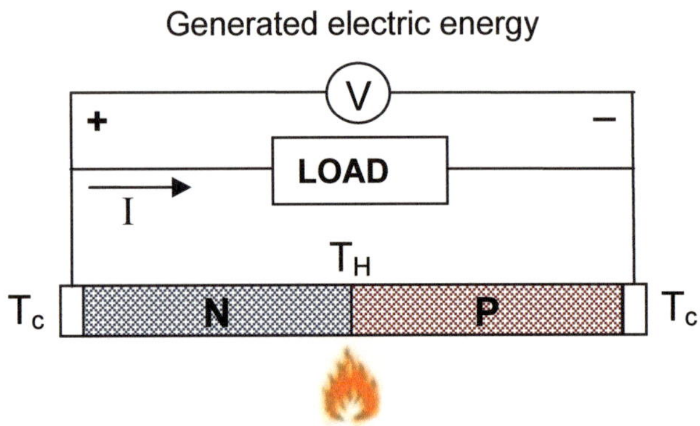

Fig. 1. In the Seebeck effect, a temperature difference between the junctions of two different materials makes an electric voltage to arise.

The Seebeck effect is used for two types of applications: temperature sensors and thermoelectric generators.
In the Peltier effect, when a current flow through the junction of two different materials, heat is absorbed or released in the junction, depending on the current direction (Peltier, 1834). This effect is quantified by the Peltier coefficient, π, related with the Seebeck coefficient (eq. 2):

$$\pi = \alpha T \tag{2}$$

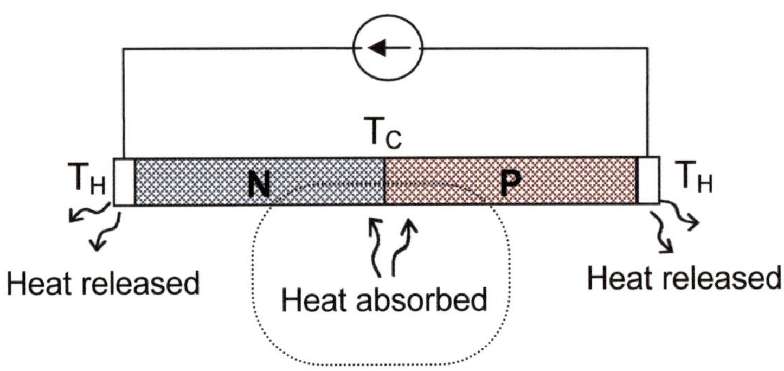

Fig. 2. In the Peltier effect, when a current flows through the junction of two different materials, heat is absorbed or released in the junctions.

The heat absorbed (Q_C) in the centre junction of figure 2, by Peltier effect can be calculated with eq 3.

$$Q_c = \left(\alpha_p - \alpha_n\right) T_c I \tag{3}$$

Due to the current flowing in N and P materials and the interfaces between materials (contacts), heat is generated by joule effect (Q_J). Eq. 4 calculates the total heat generated, since R represents the total resistance of the Peltier device.

$$Q_J = RI^2 \tag{4}$$

The complete model of a Peltier device (Wijngaards, 2000) can be considered as in figure 3. The electrical model (on the left of figure 3), includes the electrical equivalent resistance of the device (R), a voltage source that provides power to the device and a voltage source modelling the Seebeck effect of junctions, $(\alpha_p - \alpha_n)(T_h - T_c)$. The resultant current is I_e. On the right side of figure 3, the, thermal model is presented. The two current sources, $(\alpha_p - \alpha_n)I_eT_c$ and $(\alpha_p - \alpha_n)I_eT_h$, represent the cooling and heating by Peltier effect, respectively. The capacitors $C_{t,c}$ and $C_{t,h}$ are the heat capacity of on cold side and hot side and resistances $R_{t,c}$ and $R_{t,h}$ are the losses by convection and radiation to ambient temperature, T_a. $R_{t,h}$ is usually very small, and $T_h \approx T_a$. $R_{t,d}$ represents half of the thermal resistance between hot side and cold side. Q_j and Q_{jc} represent the heating by Joule, respectively on the thermoelectric materials and contacts. The load applied on the cold side of the device is represented by Q_L.

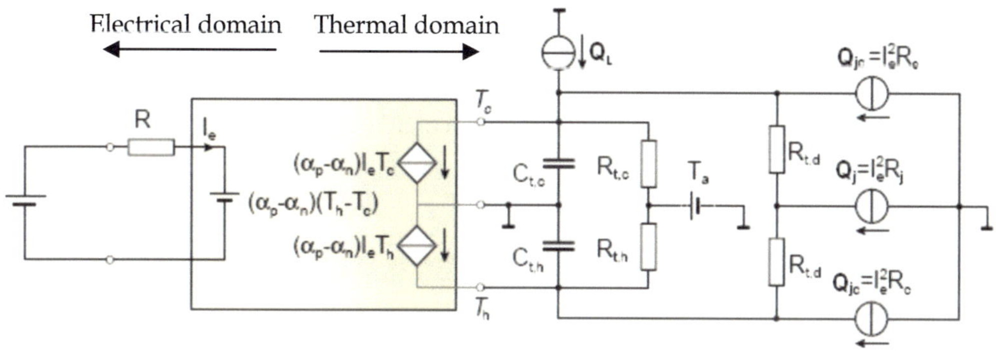

Fig. 3. Complete model of a Peltier microcooler, including electrical and thermal domains.

The same model can be used in an electrical power generator application, based on Seebeck effect, as figure 1. Instead of a voltage source (on the left side of the model in figure 3), a load is connected and the heat to convert to electrical energy (Q_L) is applied to the hot side of the device (T_h). The voltage generated is proportional to the temperature difference: $V=(\alpha_p - \alpha_n)(T_h - T_c)$.

Once each thermoelectric pair can produce a voltage near 400 μVK^{-1}, many pairs, connected in series, are necessary to generate a usable voltage. The maximum power in a thermoelectric generator, calculated with eq 5, is obtained when the load resistance equals the internal resistance (R).

$$P_{MAX} = \frac{V_{OUT}^2}{4R} = \frac{\left(n(\alpha_p - \alpha_n)\Delta T\right)^2}{4n\left(R_n + R_p + R_j + 4R_c\right)} \tag{5}$$

where n is the number of elements (pairs of thermoelectric p-n junctions), α is the Seebeck coefficient, ΔT is the temperature difference between the hot side and cold side of

thermoelectric elements (T_h- T_c) and R is the electric resistance. The indexes p and n refer to p-type and n-type materials respectively and the indexes j and c refer to materials of contacts and the contact itself. In order to obtain the maximum power, it is also important do match the thermal resistance of the generator with the heat sink (on the cold side) and hot object (in the hot side), not represented in the previous equation. In several applications, it is also important to analyze the impact of the generator in the temperature of the hot object. If a human-body generator is designed, it will not suit comfortable if much thermal power is absorbed from the skin (the sensation of cold will be noticed). By the other hand, when designing a thermoelectric generator for waste heat recovering (ex. recovering heat from a laptop CPU), an increase of temperature could occur where the heat is generated.
The coefficient of performance (COP) of the Peltier coolers is four to five times below to those found in conventional coolers (based on the Carnot cycle). Additionally, the unitary limit of the figure-of-merit (ZT - a performance measure of TE materials) it was seemed as an impossible barrier to pass, but also unexplainable. Bismuth telluride (Bi_2Te_3) and antimony telluride (Sb_2Te_3) compounds, were known for decades as the best thermoelectric materials at room temperature. Figure-of-merit is calculated by eq. 6.

$$Z = \frac{\alpha^2}{\rho k} \tag{6}$$

where α [μVK^{-1}] is the Seebeck coeficient, ρ [Ωm] is the electric resistivity and κ [$Wm^{-1}K^{-1}$] is the thermal conductivity. Figure-of-merit can also be calculated for a specific temperature, including T (absolute temperature) in the previous equation, resulting ZT:

$$ZT = \frac{\alpha^2}{\rho k} T \tag{7}$$

In thermoelectric generation applications, the power factor is sometimes used instead of figure-of-merit:

$$PF = \frac{\alpha^2}{\rho} \tag{8}$$

3. Materials for thermoelectric applications

Despite the continuous efforts in the search of an adequate material for fabrication of Peltier effect devices, more than 50 years ago that the value close to one of the figure-of-merit (ZT) seems to appear as a goal that can not be overtaken at room temperature. A good thermoelectric material must have high Seebeck coefficient, low electric resistivity and low thermal conductivity. But these three parameters are correlated. A material with low electric resistivity (a metal for example) frequently has a high thermal conductivity. The thermal conductivity based in the electronic transport (κe), which it is the dominant mechanism of thermal conduction in metals, is related with the electric conductivity (σ) by the Wiedemann-Franzem law, where L is the Lorenz number and T the temperature:

$$\frac{\kappa_e}{\sigma} = LT. \tag{9}$$

Theoretically, the Lorenz number is equal to:

$$L = \frac{\pi^2}{3}\left(\frac{k_B}{e}\right)^2 \approx 2.44 \times 10^{-8}\ \mathrm{W\Omega K^{-2}} \tag{10}$$

where k_B is the Boltzmann constant and e is the electron charge.
Figure 4 shows the figure-of-merit (Z) calculated for different materials at different temperatures.

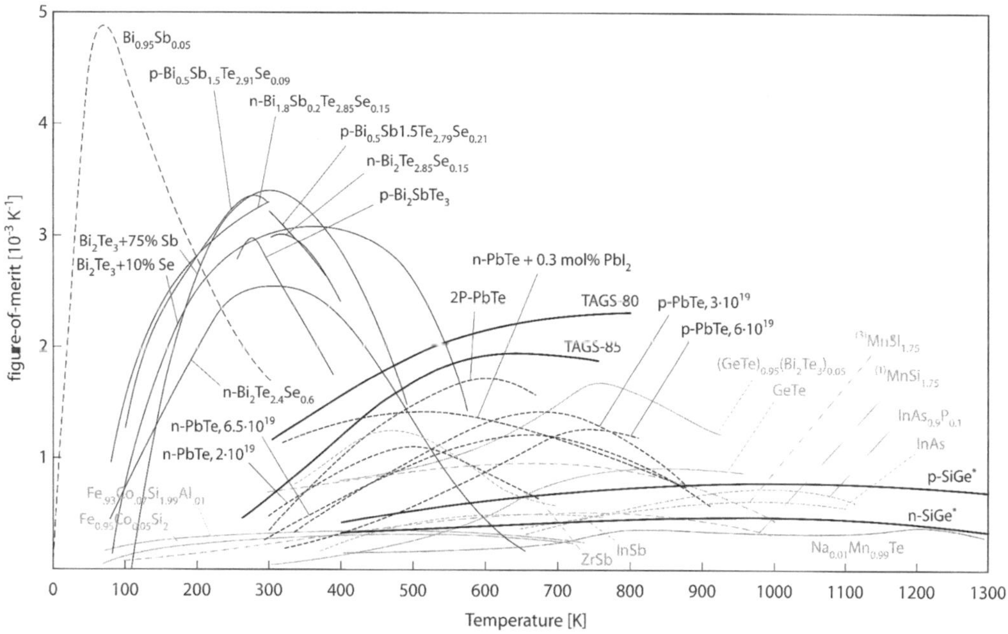

Fig. 4. Figure-of-merit (Z) calculated for different materials at different temperatures (Wijngaards, 2003).

Of the great number of materials investigated, those based on bismuth telluride, lead telluride and silicon-germanium alloys emerged as the best for operating at temperatures near 300 K, 900 K and 1400 K respectively.
Near the room temperature (250-350 K), tellurium (Te), bismuth (Bi), antimony (Sb) and selenium (Se) composites show the highest figure-of-merit values. For this reason, they are used in many of the commercial Peltier devices. The thermoelectric properties at room temperature of some of these materials are displayed in table 1.
For operation at temperatures around 800 K, lead antimony telluride shows the highest figure-of-merit. A ZT value around 1 was reported at 800 K (Fano, 1997). However, there are environmental restrictions to the use of lead. Silicon-Germanium is a candidate material for operation at temperatures above 1000 K. A figure-of-merit around unity was achieved at 1200 K (Vining, 1997). These materials also have the advantage of easy integration with microelectronics.

Material	*Symbol*	*Seebeck coefficient* α *(μVK⁻¹)*	*Resistivity* ρ *(μΩm)*	*Thermal conductivity* κ *(Wm⁻¹K⁻¹)*	*Figure of merit ZT*	*Temperature K*
Nickel	Ni	-18	0.070	91	0.015	300
Chromium	Cr	18	0.13	94	0.008	300
Bismuth	Bi	-60	1.15	8.4	0.110	300
Antimony	Sb	40	0.42	18.5	0.062	300
Silicon-Germanium (n)	SiGe	-242	17.8	4.2	0.94	1200
Silicon-Germanium (p)	SiGe	240	31.9	4.38	0.50	1200
Bismuth telluride (n)	Bi_2Te_3	-240	10	2.02	0.86	300
Antimony telluride (p)	Sb_2Te_3	92	3.23	1.63	0.48	300

Table 1. Thermoelectric properties of some materials.

4. Quantum confinement in thermoelectricity

There are a lot of attempts to produce thermoelectric materials with ZT greater than one. Nevertheless, the best commercial thermoelectric modules, fabricated from bismuth, antimony and tellurium compounds, have ZT close to one. This is mainly due to the fact that in conventional 3D crystalline systems the Seebeck coefficient (α), the electrical conductivity (σ) and the thermal conductivity (κ) are interrelated, being difficult if not impossible to control each factor independently in order to improve ZT (Bottner, 2006; Bell, 2008). An increase of α, usually results in a decrease of σ. By the other hand, a decrease of σ leads to a decrease of the electronic contribution to κ. However, if the dimensions of the material decrease, a scale factor becomes available for the control of material properties (Hicks, 2003). This phenomenon is due to the reduction of the 3D solid crystalline structures to 2D superlattices (figure 5), 1D nanowires, or quantum dots, introducing new forms to control α, σ or κ more independently. The introduction of many interfaces in the structure can scatter phonons more effectively than electrons and allows enhanced ZT in such nanostructured materials. Recent work with PbTe (Harman, 2002), SiGe (Caylor, 2007) and BiSbTe (Bottner, 2006; Venkatasubramanian, 1992) superlattices demonstrated an enhancement of ZT. ZT=2.4 and ZT=1.4 were measured in p-type and n-type Bi/Sb/Te superlattices (Venkatasubramanian , 2001), respectively.

The use of thin-film processes in thermoelectric structures limits the thickness of deposited films to few micrometers. Using this thickness, the achieved heat-flow density has higher value, compared with traditional large scale devices. If 10 Wcm^{-2} can be found in typical large-scale devices, 500 Wcm^{-2} could be supported in a thin-film device. However, this density could not be attended with conventional heatsinks. For lower density applications, efforts are also being done to achieve bulk materials (rather than films) with increased figure-of-merit. A periodic structure is the major mechanism to reduce thermal conductivity and support the enhanced figure-of-merit in superlattices. However, nanocomposites become a natural step for extending the success in superlattices to more scalable materials. Randomly distributed nanostructures in nanocomposite materials (figure 6) can lead to a reduction in the thermal conductivity below that of an alloy of the same overall chemical stoichiometry (Dresselhaus, 2007). These materials can be prepared by either wet-chemistry, ball-milling, or by inert-gas condensation methods. Nanometer or micrometer sized particles are then hotpressed to obtain dense and mechanically strong, bulk nanocomposites.

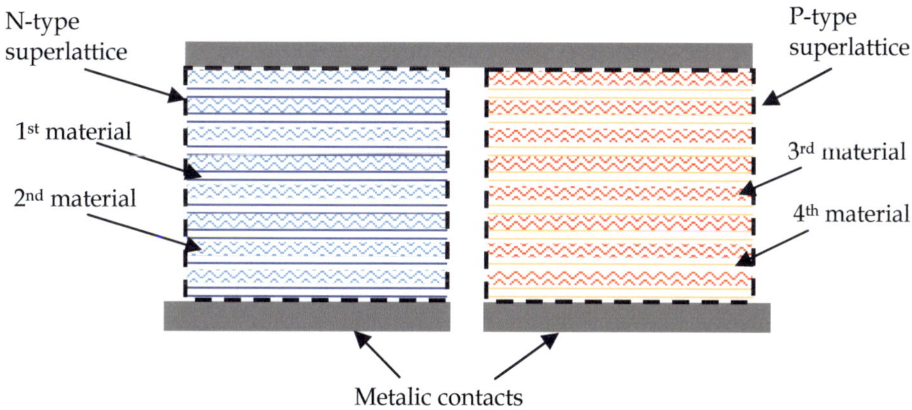

Fig. 5. Thermoelectric pair with superlattice materials. Each material is composed by alternating layers of two different materials, whose thickness is in the range of tens of nanometers.

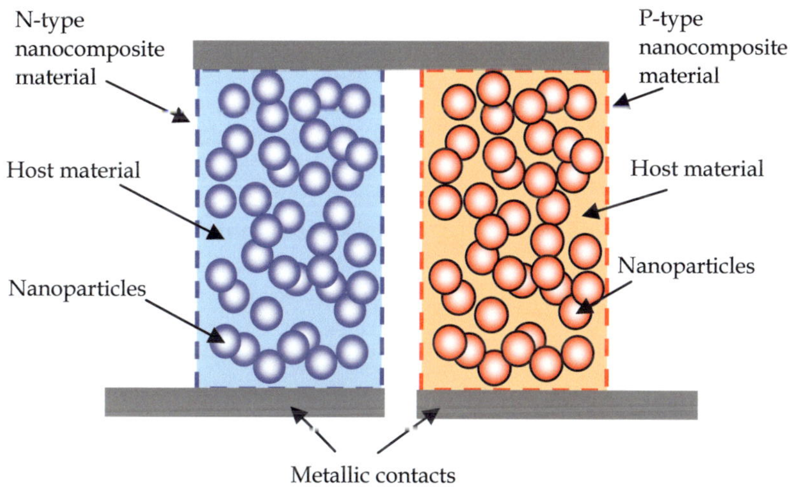

Fig. 6. Thermoelectric pair with nanocomposite materials.

There is also certain optimism concerning the materials of the group of clathrates, which create crystals with nanocages and whose thermal conductivity can be reduced if an atom of an heavy element is placed inside of cages.

The energy and environmental circumstances have relaunched the current research on these materials significantly. The results presented by a group of the University of Aarhus, the Copenhagen University and the Technical University of Denmark are part of this new wave and should help to accelerate research in the world. Their study describes why some materials may have very low thermal conductivity without degrading their electrical properties. Their research work has focused on the properties of one of the thermoelectric materials of the most promising family of clathrates, which the crystal is filled with nanocages. By placing a heavy atom in the heart of each nanocage, it is possible to reduce the ability of the crystal conduct heat. The research team thought that the random movements of atoms in the cage were responsible for the phenomenon. They used the technique of neutron scattering which allows the observation of the movements of atoms within the material. They understood that the thermoelectric properties were determined by the global movement of nanocage structure, which is influenced by the heavy atom therein (Christensen, 2008).

5. Thin-film fabrication

A single junction of Bi_2Te_3-Sb_2Te_3 thermocouple has a Seebeck voltage of only 400 μVK^{-1}. To achieve a usable voltage in generator devices, more than 4000 thermocouples must be connected in series. If these 4000 thermocouples are to be fitted in a 1 cm^2 device, each thermocouple is about 100 μm × 200 μm. The fabrication methods used in macro-sized TE devices cannot be used in the fabrication of these micro-devices. In these devices, microsystems technology should be used instead. Materials can be deposited by thin-film deposition processes (physical and chemical vapour deposition or electrochemical deposition). Some techniques were tried before for the deposition of Bi/Sb/Te thin-films. Electrochemical deposition (ECD), metal-organic chemical vapour deposition (MOCVD), pulsed laser deposition (PLD), sputtering and thermal evaporation are some examples. Independently of the technique used, a good control of film composition and crystalline structure is very important to fabricate films with high figure-of-merit. Previous research (Goncalves, 2009), demonstrated the optimum composition to maximize figure-of-merit. A tellurium content in the range 60%-65% can maximize figure-of-merit. When evaporating directly the compounds (either Bi_2Te_3 or $Bi_xSb_{2-x}Te_3$), these elements decompose and the final composition of the deposited film does not match the composition of the initial target, due to different vapour pressure of each element (Bi, Sb or Te). Moreover, when thicker films are deposited, the composition differs from surface to deep film layers (Silva, 2005). This effect is more evident in thermal evaporation, since the increase in temperature promotes de decomposition of source materials. To overcome this problem, co-deposition systems (either thermal co-evaporation (Goncalves, 2007) or co-sputtering (Kim, 2006; Bottner, 2004)) are usually used, and the deposition rate of each element (Bi, Sb or Te) is controlled independently, in order to obtain the final optimal composition. The power factor of films deposited by co-evaporation, as function of composition (measured by EDX) and substrate temperature is presented in figure 7.

The importance of crystalline structure in figure-of-merit was also demonstrated before. The structure of these films changes from amorphous to polycrystalline. Films with a more crystalline structure have usually low electrical resistivity. The polycrystalline structure of these films also decreases the thermal conductivity (compared with a single crystal) (Scherrer, 1997) thus increasing the figure-of-merit. The crystalline structure can be controlled by the substrate temperature during deposition or with annealing cycles after deposition. However, due to different vapour pressure of tellurium, bismuth or antimony, the composition of the films can change with heating, resulting in films poor in tellurium. The influence of substrate temperature in films deposited by co-evaporation is presented in Fig 7. A low deposition rate (bellow 2 μm/h) also allows an appropriate crystallization, resulting in higher figure-of-merit. This low deposition rate limits the thickness of film that can be deposited. Using co-sputtering or ECD, higher deposition rate can be obtained. IPM (Bottner, 2004) reported a deposition rate of 5 μm/h using co-sputtering and JPL (Fleurial, 2003) fabricated a device with thermoelectric columns 20 μm high by ECD. Table 2 compares the thermoelectric properties of Bi_2Te_3 and $Bi_xSb_{2-x}Te_3$ films fabricated by different techniques.

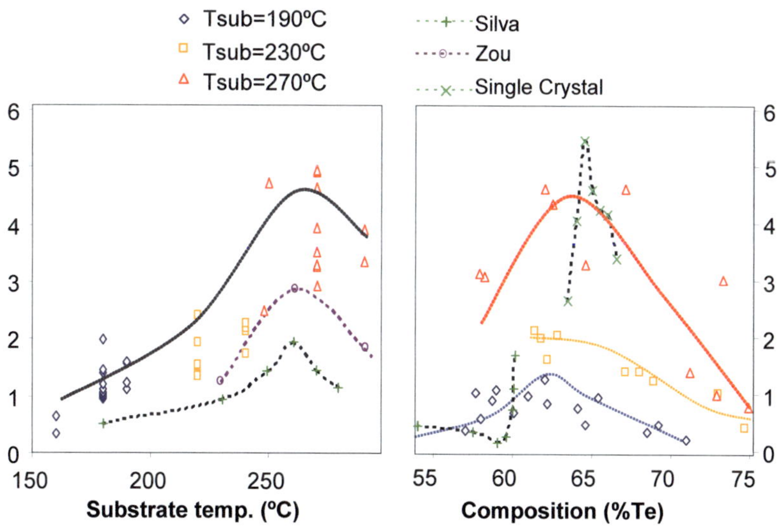

Fig. 7. The power factor of films deposited by co-evaporation, as functions of substrate temperature and composition (measured by EDX). Results from other authors and from single crystal are also presented.

Material		*Deposition technique*	*Seebeck* α *(μVK^{-1})*	*Resistivity* ρ *($\mu\Omega m$)*	*Power factor* $10^{-3}WK^{-2}m^{-1}$	*Fig. of merit Z* $\times 10^{-3}K^{-1}$	*Reference*	*Obs*
Bi_2Te_3	n	Co-evaporation	-220	10.6	4.57	3.03	Goncalves, 2009	
Sb_2Te_3	p	Co-evaporation	188	12.6	2.81	1.87	Goncalves, 2007	
Bi_2Te_3	n	Electrochemical	-60	10	0.36	-	Lim, 2002	
Bi_2Te_3	n	MOCVD	-210	12	3.7	2.48	Giani, 1999	(1)
Sb_2Te_3	p	MOCVD	-110	3.5	3.46	-	Giani, 1999	
Bi_2Te_3	p	MOCVD	190	78	0.46	2.5	Giani, 1997	(1)
Bi_2Te_3	n	MOCVD	-218	6.9	6.9	-	Boulouz, 1998	
$Bi_{0.5}Sb_{1.5}Te_3$	p	Flash	230	17	3.1	2.9	Volklein, 1990	
$Bi_2Te_{2.72}Se_{0.3}$	n	Flash	-200	15	2.7	-	Foucaran, 1998	
$Bi_{0.5}Sb_{1.5}Te_3$	p	Flash	240	12	4.8	-	Foucaran, 1998	
$Bi_{1.8}Sb_{0.2}Te_{2.7}Se_{0.3}$	n	Sputtering	-235	47	1.2	-	Kessler, 2003	(2)
Bi_2Te_3	n	Co-Sputtering	-160	16.3	1.6	-	Bootner, 2004	(3)
$(BiSb)_2Te_3$	p	Co-Sputtering	175	12.1	2.5	-	Bootner, 2004	(3)
$Bi_2Se_{0.3}Te_{2.7}$	n	Sputtering	-160	20	1.3	-	Stordeur, 1997	
$Bi_{0.5}Sb_{1.5}Te_3$	p	Sputtering	210	25	1.8	-	Stordeur, 1997	
Bi_2Te_3	n	Co-Sputtering	-55	10	0.3	-	Kim, 2006	
Bi_2Te_3	n	Co-evaporation	-228	13.0	4.0	2.7	Zou, 2001	(1)
Sb_2Te_3	p	Co-evaporation	171	10.4	2.8	1.76	Zou, 2001	(1)
Bi_2Te_3	n	Co-evaporation	-228	28.3	1.8	-	Silva, 2005	
Sb_2Te_3	p	Co-evaporation	149	12.5	1.78	-	Silva, 2005	

Obs: (1) Z estimated by the author.
(2) Doped with CuBr.
(3) The power factor of de 3×10^{-3} $WK^{-2}m^{-1}$ and 4×10^{-3} $WK^{-2}m^{-1}$, respectively for type n and type p was reported latter by the same authors (Bottner, 2007) but no reference of other thermoelectric properties was found.

Table 2. Properties of selected Bi_2Te_3 and $Bi_xSb_{2-x}Te_3$ films

6. Patterning of devices

Common techniques used in MEMS fabrication, namely wet-etching, lift-off (with SU-8 photoresist), Reactive Ion Etching (RIE) and Lithography-Electroplating-Molding (LIGA) were tried before in the fabrication of thermoelectric microstructures.
IPM (Bottner, 2004) used RIE techniques to pattern thick films of Bi,Sb,Te materials, using photoresist as an etching mask. Two wafers with patterned thermoelectric materials were soldered to create the columnar thermoelectric device. Each wafer contains n-type or p-type materials, deposited on top of metal contacts and a soldering material deposited on top. The wafers are then aligned and soldered. This process allows the deposition of thermoelectric materials with crystalline structure by heating the substrate during the deposition of thermoelectric materials.
The JPL laboratory (Snyder, 2003) used a MEMS like process, LIGA, to fabricate micro-columns of TE materials. Gold-chromium contacts were deposited and patterned on the substrate. Thick photoresist was patterned to create holes were TE materials were deposited by ECD. A gold-nickel layer was then deposited and patterned over the structures to create top contacts. Photoresist, gold and chromium layer were etched, creating the complete device. By this process, height columns can be created, however the figure of merit of thermoelectric materials deposited by ECD is low.
Lift-off can also be used in thermoelectric materials. Photoresist is spun cast and patterned to define the lift-off pattern for thermoelectric materials that will be deposited on top. The photoresist is then removed, removing also the TE material on top of it and creating the structures. The process is repeated for each thermoelectric material. The technique was applied by Silva (Silva, 2005), using SU-8 photoresist and thermal co-evaporated Bi_2Te_3 and Sb_2Te_3 thin-films. Yield of this process was low, in particular with small TE elements (7 µm × 7 µm). Due to temperature limit of photoresist, substrate cannot be heated above 170 °C during the deposition of TE materials and the figure-of-merit is lower than these obtained at higher substrate temperatures.
Shafai (Shafai, 2001) reported the possibility of use nitric acid (HNO_3) and hydrochloric acid (HCl) diluted in water (H_2O) for etching Bi_2Te_3, but his work was not extended to full characterize this process, or apply it to other tellurium compounds. Recent work from Sedky (Sedky, 2009) also proposed suspended Bi_2Te_3 microstructures fabricated by wet-etching. Goncalves (Goncalves, 2007) deposited thin-films of Bi_2Te_3 and Sb_2Te_3 (1 µm thick) on polyimide substrates, by thermal co-evaporation. Transene's PKP negative photoresist was applied on the surface and test structures were patterned by photolithography. Different etching solutions were prepared (Goncalves, 2008) using water, pure HNO_3 and 37% HCl dil. in water and the effect of etchant composition in etch rate and final result was evaluated. Figure 8 plots the etch rate of Bi_2Te_3 and Sb_2Te_3 films in (1-x)HNO_3:(x)HCl solution (diluted 70% in water, in volume).
Higher percent of HCl (%HCl / %HNO_3 > 0.5) induces cracking of the film and peeling occurs. Using only HNO_3 (diluted at 70% in water) Bi_2Te_3 is etched at ≈300 nm/sec and the Sb_2Te_3 etch rate is below 6 nm/sec. This difference of 50× can be useful to pattern devices with both materials, etching Bi_2Te_3 with HNO_3 without etching Sb_2Te_3 films. However, this method cannot be used with $Bi_xSb_{2-x}Te_3$ instead of Sb_2Te_3. The behavior of $Bi_xSb_{2-x}Te_3$ is equivalent of Bi_2Te_3, even with small percentage of Bi in composition. Figure 9 plots the influence of etchant dilution (in water) in etch rate, respectively in Bi_2Te_3 and Sb_2Te_3 films.

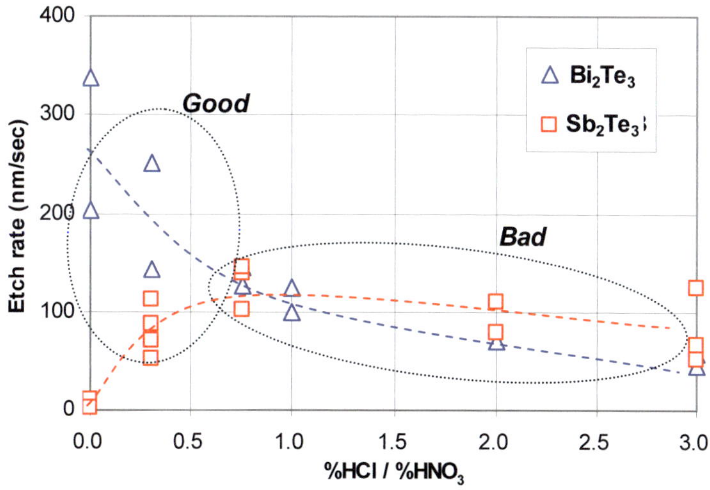

Fig. 8. Etch rate of Bi_2Te_3 and Sb_2Te_3 films in (1-x)HNO_3:(x)HCl solution (diluted 70% in water, in volume).

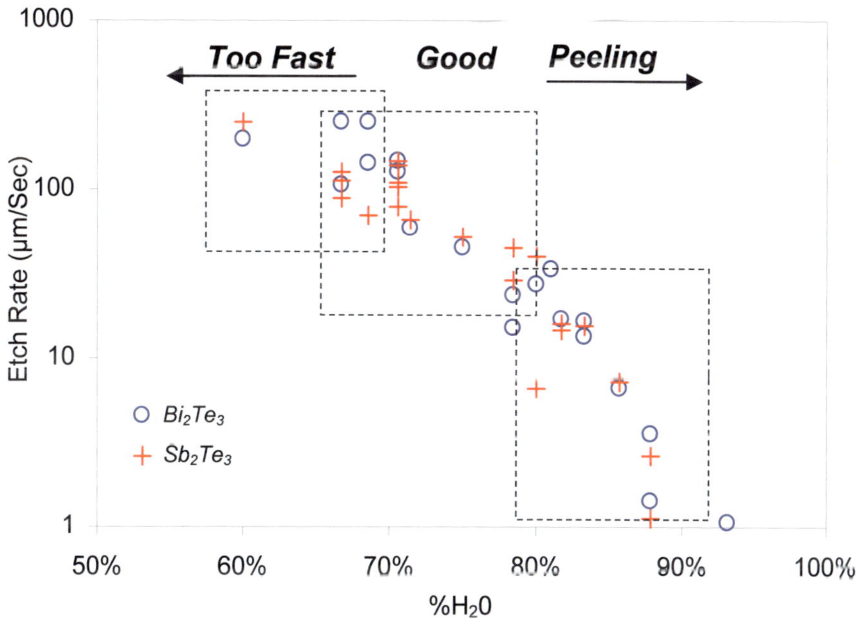

Fig. 9. Influence of dilution (in water) in Etch rate of Bi_2Te_3 and Sb_2Te_3 films in 10:3 HNO_3:HCl solution.

Etchant of composition with dilution of 70% produces the best results. With dilution above 80% in water, the etch rate is very low and peeling of the film occurs. With dilutions below 60%, the etch rate is very high and becomes difficult to control de etch time. Table 3 shows the etch rate of Bi_2Te_3, Sb_2Te_3, chromium and aluminum in HNO_3:HCl:H_2O, HNO_3, aluminum etchant (Transene type A) and chromium etchant (Transene 1020). The selectivity between different etchants in different materials allows different possibilities to fabricate a complete device. Figure 10 shows the patterned structures used in etchant evaluation.

Etchant \ Material	Bi_2Te_3	Sb_2Te_3	Aluminum	Nickel
Al - Transene type A	8	5	10-80	< 0.1
Cr - Transene 1020	≈ 20	<1	-	10-40
$3HNO_3$:1HCl (dil 70% H_20)	2000	800	< 2	< 0.2
HNO_3 (dil 70% H20)	2500	50	< 0.1	< 0.1

Table 3. Summary of etch rates (Å /sec).

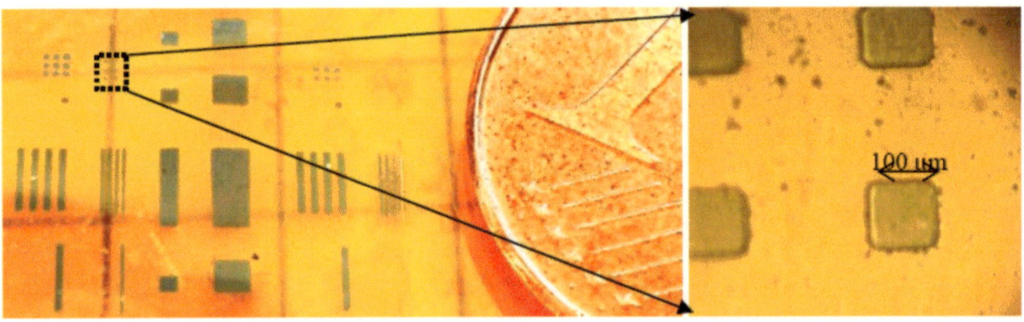

Fig. 10. Test structures of thermoelectric films patterned by chemical etching.

7. Applications

Thermoelectric materials have unique properties that make them useful to convert thermal energy into electric energy and vice-versa. For this propose, beyond a large Seebeck coefficient, they must have high electrical conductivity and low thermal conductivity. Despite this two properties being related, Bi/Sb/Te compounds are the best materials for thermoelectric applications at room temperature. This applications fall in two main categories: Cooling and electric energy generation.

In recent years, the available air conditioners and refrigerators have become a way of life for millions of people around the world. At the same time, energy costs and environmental regulations regarding the manufacture and release of CFCs are also increasing. These facts are encouraging manufacturers and their customers to seek alternatives to conventional refrigeration technology. Despite the performance of thermoelectric cooling being far from Carnot cycle compressors, some specific application requiring low maintenance, long life with no moving parts are using thermoelectric refrigeration.

Regarding generator applications, the use of thermoelectric devices in vehicles is currently being addressed. Thermoelectric converters can be used in cars, in order to make the engines most efficient. A conventional combustion engine wastes about 80% of the energy of the fuel under the form of heat. The thermoelectric devices can be directly used to generate electric energy from this wasted heat. A fuel consumption reduction of 10% is achievable, which represents a significant impact in the global energy spent in transport of people and goods all over the world. Using thermoelectric materials, this heat can be used to produce electrical energy, which in turn can be used to charge the car battery.

Small sized applications, from few cubic-centimetres to few cubic micrometers are not realizable with Carnot cycle compressors. In these low-power applications, thermoelectric devices are applied with advantages. The same thermoelectric principle can be used to build cooling systems inside microchips, optimizing the heat sink capability. These small devices can also be used to control temperature of sensible electronic circuits.
The micro thermoelectric generators have applications in energy harvesting microsystems. From low temperature gradients found in human-body or house environments, energy can be generated to power wireless devices. Self-powered wrist-rings for watch or sensor applications or thermoelectric bolts, that can generate energy wherever they attach, are being developed to power-up microwatt electronic circuits.

8. References

Bell, L. (2008). Cooling, heating, generating power, and recovering waste heat with thermoelectric systems. *Science*, 321, pp. 1457-1461.

Bottner, H. et al (2004). New thermoelectric components using microsystem technologies, *Journal of Microelectromechanical System* 13 Issue 3, pp. 414 - 420.

Bottner, H. et al (2007). New high density micro structured thermogenerators for stand alone sensor systems, *in Proc. International Conference on Thermoelectrics ICT'07*, Korea.

Boulouz, A. et al (1998). Preparation and characterization of MOCVD bismuth telluride thin films, *J. Crystal Growth* pp. 194 336.

Dresselhaus M. S. et al (2007). New Directions for Low-Dimensional Thermoelectric Materials, *Advanced Materials* 19, 2007

Fano, V. (1987). CRC handbook of thermoelectrics, edited by D.M. Rowe, 261.

Foucaran, A. et al (1998). Flash evaporated layers of (Bi2Te3–Bi2Se3)(N) and (Bi2Te3–Sb2Te3)(P), *Materials Science and Engineering B* 52, pp. 154–161.

Giani, A. et al (1997). Bi2Te3 films grown by MOCVD, *Thin Solid Films*, Vol 303 (1997) 1-3.

Giani, A. et al (1999). Growth of Bi2Te3 and Sb2Te3 thin films by MOCVD, *Materials Science and Engineering B* 64, pp. 19–24.

Goncalves L. M. et al, (2007). Fabrication of flexible thermoelectric microcoolers using planar thin-film technologies, *Journal of Microelectromechanical Systems* 17.

Goncalves, L. M. et al (2008). Thermoelectric Micro Converters for Cooling and Energy Scavenging Systems" *Journal of Microelectromechanical Systems* 18.

Goncalves L. M. et al (2009). *Thin Solid Films*, In press.

Harman TC, Taylor PJ, Walsh MP, LaForge BE, (2002). Quantum Dot Superlattice Thermoelectric Materials and Devices, *Science* Vol. 297, no. 5590, pp. 2229 - 2232

Hicks D. and Dresselhaus M. S. (1993). Effect of quantum-well structures on the thermoelectric figure of merit. *Physical Review B: Condensed Matter*, 47, 12727-12731

Kessler, E. et al (2003). Thin-film infrared thermopile sensors with thermoelectric high-effective materials combination. *Proceedings of Sensor 2003*, 11th International Conference, Vol. II, Nürnberg 249-254.

Kim, Ding-ho et al (2006). Effect of deposition temperature on the structural and thermoelectric properties of bismuth telluride thin films grown by co-sputtering. *Thin Solid Films*, 510 148-153.

Lim, J.R. et al (2002). Thermoelectric Microdevice Fabrication Process and Evaluation at the Jet Propulsion Laboratory (JPL), *in Proc. International Conference on Thermoelectrics.*

Mogens Christensen et al (2008). Voided crossing of rattler modes in thermoelectric materials, *Nature Materials*, Vol. 7, (10) pp. 811 - 815.

Peltier, J. C. (1834). Nouvelles experiences sur la caloricité des courans électriques. *Annales de Chimie et the Physique*, LVI 56, pp. 371-386.

Scherrer, H. and Scherrer, S. (1987). CRC handbook of thermoelectrics, edited by D.M. Rowe, pp. 211-237.

Sedky, Sherif et al (2009). Bi2Te3 as an active material for mems based devices fabricated at room temperature, *in Proc. International Conference Transducers09*, Denver, Colorado USA

Seebeck, T. I. (1822). Magnetische polarisation der metalle und erze durch temperatur-differenz. *Abhandlungen der Deutschen Akademie der Wissenschaften zu Berlin*, pp. 265-373.

Shafai, C. and Brett, M.J. (2001). Optimization of Bi2Te3 thin films for microintegrated Peltier heat pumps, *Journal of Vaccum Sci Technol A*, Vol. 17- 1 pp. 305-309.

Silva, Luciana; Kaviany, Massoud and Uher, Citrad (2005), *Journal of Applied Physics* 97, pp. 114903.

Snyder, G. Jeffrey et al (2003). Thermoelectric microdevice fabricated by a MEMS-like electrochemical process, *Nature Materials* 2 pp. 528.

Stordeur, Matthias and Stark, Ingo (1997). Low Power Thermoelectric Generator - self-sufficient energy supply for micro systems. *16th International Conference on Thermoelectrics*.

Venkatasubramanian, R. et al. (1992). *in Proc. 1st Natl Thermogenic Cooler Workshop*. Center for Night Vision and Electro-Optics, Fort Belvoir, VA, pp. 196-231.

Venkatasubramanian, R. et al. (2001). Thin-film thermoelectric devices with high room-temperature figures of merit. *Nature*, pp. 413.

Vining, Cronin B. (1987). CRC handbook of thermoelectrics, edited by D.M. Rowe, 329.

Völklein, F. et al (1990). Transport properties of flash-evaporated (Bi1-xSbx)2Te3 films: Optimization of film properties, *Thin Solid Films* 187 pp. 253-262.

Wijngaards D.D.L.; Cretu E.; Kong S.H. and Wolffenbuttel R.F (2000). Modeling of integrated polySiGe Peltier elements, *Sixth THERMINIC Workshop*, Budapest, Hungary, 24–27 September (2000) 2 pp. 75a–275d.

Wijngaards, Davey (2003). Lateral on-chip integrated Peltier elements based on polycrystalline silicon germanium, *Phd Thesis*, Tu Delft.

Zou, Helin; Rowe, D.M. and Min, Gao (2001).Growth of p- and n-type bismuth telluride thin films by co-evaporation. *Journal of Crystal Growth*, Vol. 222 (2001) pp. 82-87.

20

Ppt-level Detection of Aqueous Benzene with a Portable Sensor based on Bubbling Extraction and UV Spectroscopy

Serge Camou, Akira Shimizu, Tsutomu Horiuchi and Tsuneyuki Haga
NTT Microsystem Integration laboratories, Microsensor research group, NTT Corporation Japan

1. Introduction

Benzene (EPA, 1993), sometimes also referred to as part of BTX or BTEX, which stands for benzene, toluene, ethylbenzene, and xylenes, is widely used in industrial activities despite being a known carcinogen that can easily contaminate different media (gas, water, and soil). It can be inhaled, ingested, and absorbed through the skin and therefore represents a potential threat to human health, even at trace levels. With respect to benzene toxicity, national regulations have been established in an effort to minimize its impact on human health. Water regulations cover drinking water and wastewater, with levels varying from one country to another but falling within the same order of magnitude. Wastewater regulations give the maximum concentration allowed for disposal, about a few hundred parts per billion (ppb) (Volume) (110 ppbV in Japan (Ministry of the Environment in Japan, 2008)). Below this level, the impact on the environment is assumed to be negligible. This level is sometimes referred to as the alarm level. Drinking water regulations, which are aimed at water for human consumption, stipulate permissible levels in the low ppbV range, two orders of magnitude lower than for wastewater (11, 5 and 1 ppbV in Japan, America, and Europe, respectively (Ministry of Health, Labour and Welfare in Japan, 2003) (EPA, 2006) (European Council, 1998)).

Laboratory procedures for detecting benzene concentrations in water involve several consecutive steps: (i) extraction by inert-gas stripping or heating to transfer benzene to the vapor phase, (ii) trapping to enhance the concentration, and (iii) detection by gas chromatography combined with either flame ionization, photo-induced detection or mass-spectrometry (respectively GC/FID, GC/PID, and GC/MS) (Martinez *et al.*, 2002) (Serrano & Gallego, 2004) (Richardson & Ternes, 2005). The detected thresholds far exceed the requirements, but the procedures are complex and involve the use of expensive equipment. The required apparatuses are also not compatible with the requirements for on-site measurements, which means samples have to be collected in the field and sent to a laboratory for analysis (Richardson & Ternes, 2005). To prevent contamination of samples or changes in their characteristics due to long contact-exposure to the container (with transportation time to the laboratory varying from a few hours to a few days), complex and error-prone procedures have been established (Namiesnik *et al.*, 2005), which, depending mainly on the potential contaminants, include choosing an appropriate container material, dilution with organic solvent, and avoiding headspace. Recent disasters in 2005/2006 (chemical plant explosions in China; hurricane Katrina in America) point to an urgent need for on-site, real-time monitoring of pollutant contamination to prevent dramatic impacts on the local population's health, making response time a crucial factor.

A suitable on-site sensor should then exhibit robustness, sensitivity in the low-ppbV range for drinking water regulation levels, and response time of about a few minutes or less. A few studies have examined portable sensors dedicated to on-site benzenic-compounds measurements from water samples. Those devices use alternative detection techniques, and, though the sensitivity is relatively low, exhibit several advantages over conventional methods in cost, protocol and data analysis, and size. In addition, they do not require any inert gases, such as the carrier gas used in gas chromatography-based measurements, which improves portability.

These sensors can be classified into direct measurements systems, which directly measure pollutant concentrations in the liquid phase, and indirect measurements ones, which require pre-conditioning of a sample before its characterization.

Among the direct methods, mid-infrared (MIR) evanescent field spectroscopy has been largely utilized. The sensing element is a polymer-coated optical fiber or an attenuated-total-reflection (ATR) crystal. The polymer coating provides selective enrichment of compounds to be detected within the depth penetration of evanescent wave in the aqueous media. Silver halide fibers and various polymer coatings have been used to measure hydrocarbons in water (Hahn *et al.*, 2001) (Krska *et al.*, 1993) (Steiner *et al.*, 2003) (Beyer *et al.*, 2003), though the limit of detection (LOD) is in the low ppmV range and measurement time is longer than an hour. Systems based on coated quartz-glass optical fibers and detection in near infrared (NIR) spectral range have been reported (Burck *et al.*, 2001) (Zimmermann *et al.*, 1997). A recent system on ATR crystal (Karlowatz *et al.*, 2004) has demonstrated a LOD to benzene of about 45 ppbV with a 20-min measurement cycle. Although the simultaneous detection of five compounds has been demonstrated at low concentration levels, consecutive measurements using the same apparatus may take longer because the equilibration time to water diffusion through the coating (about several hours) for stabilizing the baseline is not included in the measurement time and because enrichment reversibility, i.e., the refreshment of sensing system, is a slow process limited by the partition coefficient and diffusion through the polymer coating. As a potential improvement of evanescent wave spectroscopy, Tobiska *et al.* (Tobiska *et al.*, 1998) proposed bending the fibers to increase the evanescent field intensity. Mohacsi *et al.* (Mohacsi *et al.*, 2001) described a sensor based on photo-acoustic detection principle with a 300-ppbV LOD and a 40-min response time. To achieve high sensitivity, aromatic compounds are detected in the vapor phase after the molecules have been selectively transferred from the liquid phase through a semi-permeable membrane surrounding the acoustic cavity. Vogt *et al.* (Vogt *et al.*, 2000) proposed direct and simultaneous detection of hydrocarbons in water with a system based on UV derivative spectrometry. This approach offers a fast response time of few minutes and a LOD to benzene below 50 ppbV. However, turbidity is expected to strongly bias the results, which limits the applicability of such a sensor to specific on-site measurements.

In comparison with direct measurements, indirect ones require sample extraction from the liquid phase. Yang *et al.* combined gas-stripping (Yang & Her, 1999) or headspace (Yang & Tsai, 2002) techniques to transfer aromatic compounds from the liquid to the vapor phase and performed detection by MIR evanescent wave spectroscopy. The reported LOD is in the few-hundred ppbV range. Measurements are faster, typically less than thirty minutes all steps included, from sample introduction to complete regeneration, but the sensor gains in complexity because of the large number of connected elements. Solid phase micro extraction (SPME) combined with IR (Heglund & Tilotta, 1996) or UV (Lamotte *et al.*, 2002) spectroscopy has also been used to quantitatively and qualitatively determine benzenic compounds diluted in water. Since only the disposable SPME matrix is in direct contact with the sample, the contamination risk to the re-usable detection setup is almost negligible, making the regeneration step unnecessary. However, because enrichment is based on diffusion through the SPME matrix, the sensitivity is directly proportional to exposure time. In order to reach a LOD in the hundreds of ppbV range, enrichment time exceeding one hour is then common.

Field-test validation has been carried out successfully for high pollutant concentration levels (Krska *et al.*, 1993) (Steiner *et al.*, 2003), indicating the proposed sensors offer alternatives to laboratory equipment at the alarm-level threshold. However, with the drinking water regulation level as the final target, none of those devices fulfill the requirements in terms of sensitivity.

This chapter then describes the design of portable aqueous benzene sensor based on UV-spectroscopy with a sensitivity in the pptV range. Starting from our portable BTX sensor dedicated to air monitoring (Ueno *et al.*, 2001) (Camou *et al.*, 2006, a), we extended its use to aqueous sample solution by adding an extraction module into the inlet to transfer the benzene compound from the liquid to the vapor phase (Camou *et al.*, 2008). However, the concentration cell developed for the airborne system, which provides benzene concentration enhancement prior to the detection cell, could not be used. As a consequence, the sensor demonstrated a LOD in the hundred ppbV range, several times higher than the requirements (Camou *et al.*, 2008). Since the concentration stage is a key component for achieving high sensitivity, we developed a new one, taking into account the carrier gas specific properties after extraction, and demonstrated pptV range measurements. These results open the door to potential on-site and high sensitive measurements of aqueous benzene concentrations and make the presented sensor a viable alternative to standard methods.

2. Airborne benzene portable system

Figure 1 shows the schematic view of our portable airborne BTX sensor, which is mainly composed of detection and concentration cells connected to external elements.

As the detection principle, we chose UV spectroscopy because it offers several advantages for on-site and real-time monitoring: portability and simultaneous detection of several compounds if the wavelength range has been selected appropriately. The output spectrum results from summation of all contributions of solutes absorption weighted by ponderation coefficients proportional to their respective concentration. This method can therefore provide qualitative and quantitative measurements of complex mixtures if every compound that significantly contributes to the final spectrum has been identified and their reference spectra are taken into account in the data analysis process (Ueno *et al.*, 2002). Regarding the three BTX compounds, they exhibit specific and characteristic absorption spectra in the 235 - 275 nm range, whereas water (and especially water vapor), as a potential interferant, shows no peak absorption.

For BTX mixtures in ambient air, we previously demonstrated the accuracy of UV spectroscopy detection in the low-concentration range when the micromachined detection

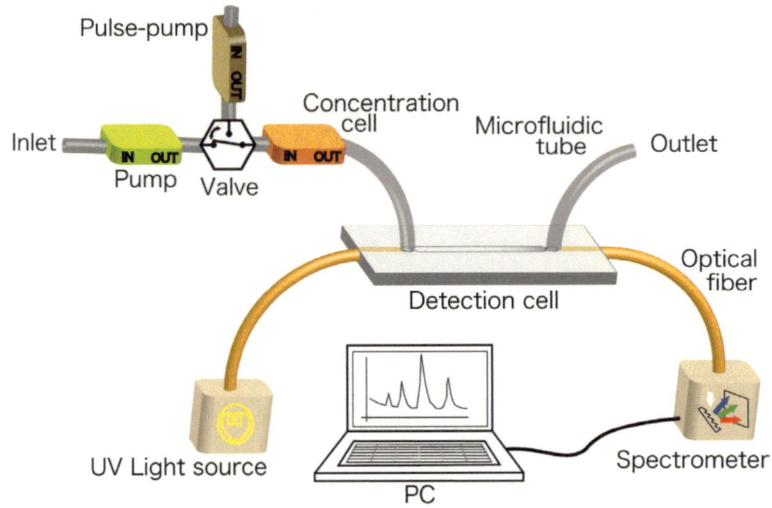

Fig. 1. Schematic view of the airborne BTX portable monitoring system, including the concentration and detection cells with pumps and a few external elements.

cell shown in Fig. 2 was used. The cell's fabrication process is described in detail elsewhere (Camou *et al.*, 2006, b), but can be briefly summarized as follows. A hollow fiber whose inner side-wall has been uniformly covered with an aluminum reflective layer (Souken Co., 2005) acts as an optical waveguide. It is sandwiched between two patterned Pyrex glass wafers, which are anodic-bonded to each other through a silicon nitride thin film. A multi-step dicing process produces trenches with various depths and shapes, providing excellent optical alignment between optical fibers and the hollow fiber in the plane perpendicular to the optical axis. Finally, sealing the different parts with UV resin yields good mechanical stability and prevents gas leaks. By separating mechanical parts (alignment of optical fibers in regards to the hollow fiber, microfluidic connections, etc) from optical parts (waveguide optical efficiency), we were able to fabricate a 10-cm-light-path detection cell with high coupling efficiency using conventional equipment for four-inch wafer processing.

The fabrication process for the concentration cell shown in Fig. 2, which is also described elsewhere, used classical microfabrication processes combined with sand blasting for the patterning of a trench with arbitrary shape in the Pyrex glass. After completing the concentration cell, the adsorbent material is inserted through the inlet by aspiration from the outlet.

The measurement sequence can then be described as follows. First, the ambient air is pumped at a high flow rate through the concentration cell during the so-called concentration time. The adsorbent, characterized by its high active surface/volume ratio, then selectively adsorbs the BTX compounds at its surface.

Meanwhile, since most of the BTX compounds are trapped by the adsorbent, spectrometer calibration measurements are performed. At the end of concentration time, the pump stops, and BTX compounds trapped by the adsorbent are desorbed by heating the concentration cell. After switching the valve, the pulse-pump precisely transfers the resulting high-density sample gas from the concentration cell to the detection cell, where simultaneous and efficient detection of BTX compounds are performed by UV-absorption-based measurements. Among the several benzene absorption peaks in the studied wavelenght range, we chose the main peak located at 253 nm to perform quantitative measurements.

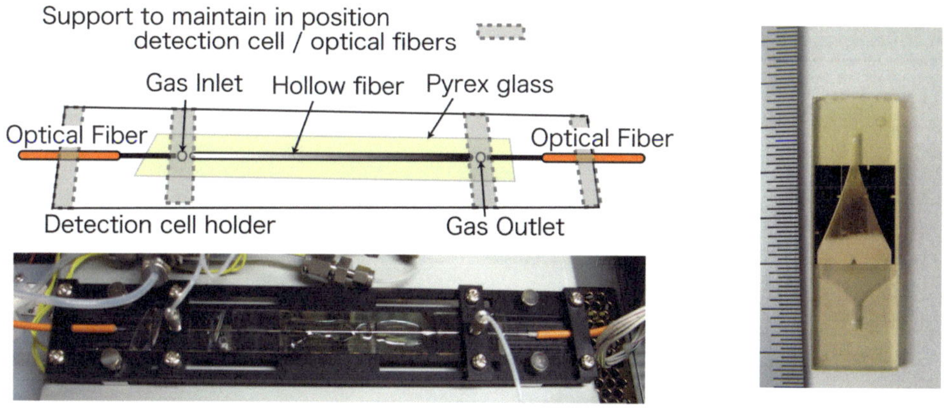

Fig. 2. Pictures of (left) the detection cell (10-cm-long light path) on its holder and (right) a micro-fabricated concentration cell half-filled with adsorbent (platinum heater electrodes on the rear side).

Figure 3 shows the sensor response versus benzene calibrated gas concentration for concentrations varying from 0 to 10 ppbM and with a concentration time of fifty minutes. The sensor exhibits linear response over more than one order of magnitude, with a detection limit of about 1 ppbM. These characteristics are consistent with our primary objectives, but improvements in the detection limit are still necessary in order to really claim accurate and reproducible measurements at 1 ppbM benzene concentration.
Meanwhile, due to the high toxicity of benzene diluted in water despite its low solubility, we also started the development of a portable, high-sensitivity aqueous benzene sensor based on our technology.

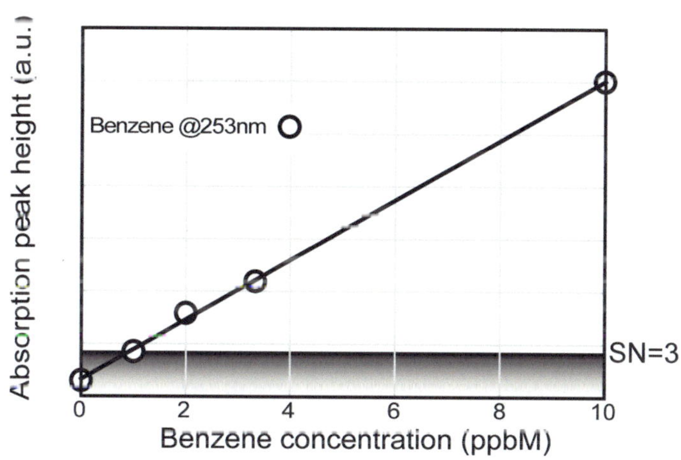

Fig. 3. Sensor response versus benzene concentration of calibrated gas mixture in dried nitrogen.

3. Extension of portable sensor usage to aqueous samples

The airborne system was designed to deal with gaseous samples. To extend the portable sensor's use to aqueous sample, an extraction module based on a bubbling method - sometimes also referred to as gas stripping- was added as shown in Fig. 4.

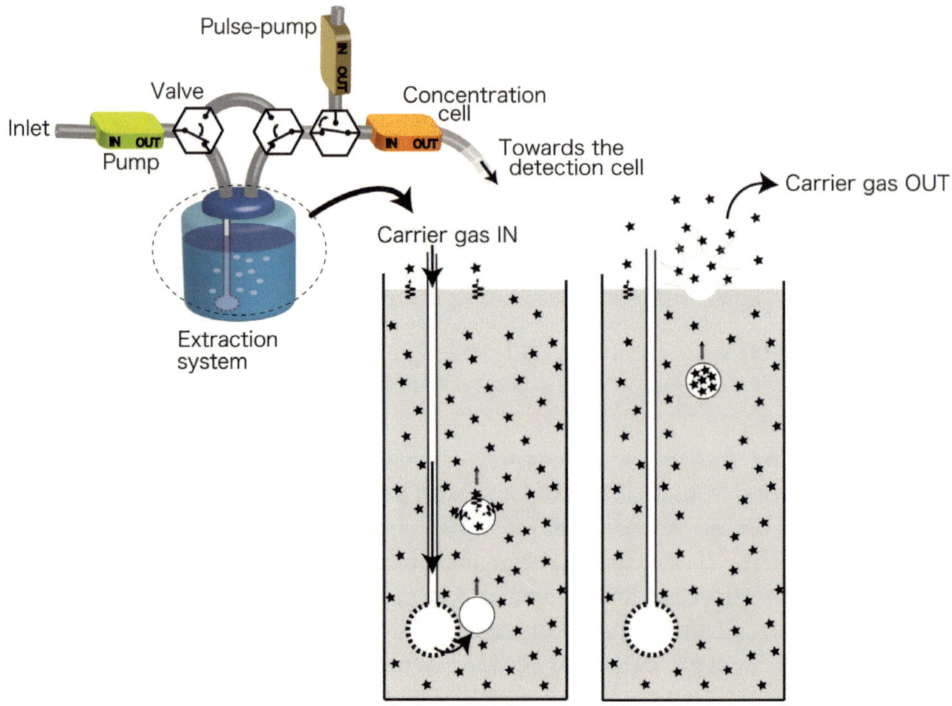

Fig. 4. Schematic view explaining the basic concept of aqueous benzene extraction by bubbling, with the black stars representing benzene molecules.

3.1 Extraction system: transfer of benzene from liquid to gas phase

Thanks to two compound-specific properties of benzene -a low boiling point (80.1 degrees C) and a low solubility in water (1.79 g/l at 25 degrees C) (EPA, 1993)- bubbling is a very efficient way to extract benzene diluted in aqueous solution. Standard methods based on chromatography sometimes use bubbling extraction, where an inert gas flows through the sample solution. This is known as "purging" from the so-called "purge and trap" technique (EPA,, 2003). This method, also used by Yang *et al.* (Yang & Her, 1999), does not require any additional components (actuators, heaters, etc) and provides easy and robust benzene extraction from aqueous samples.

The bubbling method is briefly described as follows. Due to the low solute-solvent interaction of benzene diluted in water, benzene naturally evaporates at the air-water interface. By generating air bubbles inside the liquid, we can greatly increase the air-liquid interface area so that gas exchange can proceed at a faster rate. Since the main gas exchange is considered to be benzene evaporation, the bubbles take in a large amount of benzene from the solution while rising and then release it in the gas volume over the sample solution.

This sample gas is then transferred to the next stages, where its benzene concentration is finally determined by UV spectroscopy.
However, a major drawback of bubbling extraction is a lack of selectivity. When air is bubbled through an aqueous sample, most diluted compounds as well as water itself will evaporate with a compound-specific efficiency and flow through the concentration/ detection cell. As a result, the sample gas measured will also contain a high relative humidity (RH) ratio, estimated to be over 90%.

3.2 Without concentration stage

Figure 5 shows the set-up without a concentration stage and the corresponding measurement sequence. While the spectrophotometer calibration measurements are performed, ambient air flows through the detection cell thanks to the valve positioning.

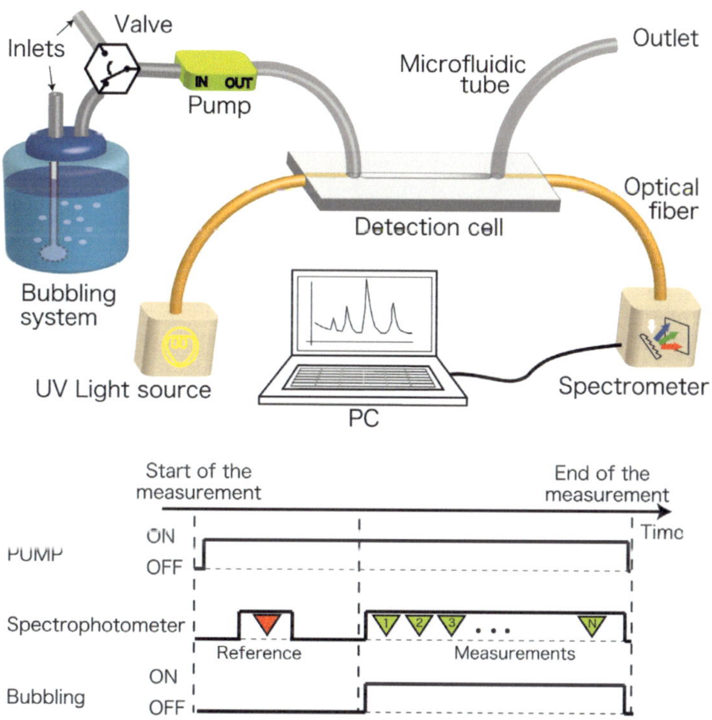

Fig. 5. Schematic view of the experimental set-up (top) and the corresponding measurement sequence (bottom)

Then, the valve switches and simultaneously bubbling through the sample solution occurs and spectra acquisition starts. As the benzene compound is extracted from the liquid phase (with a certain efficiency) and passes through the optical detection cell mixed with the carrier gas (ambient air), spectra in the specific UV range are captured continuously and the results are displayed on the computer screen in real time.
This set-up has several advantages compared to previously reported technologies. The proposed system does not require any enrichment or regeneration. This reduces the measurement time to less than 10 minutes, from sample collection to the reading of the

results on the computer display. Also, the UV absorption-based method is a relative measurement. Therefore, by using ambient air as the carrier gas for both spectrometer calibration and bubbling extraction, the contribution from compounds in ambient air should then cancel each other out and leave the output spectra free of interference.

With measurement every 80 seconds, plotting the peak heigh at 253 nm versus time leads to the characteristic response shown in left part of Fig. 6. After a fast increase of benzene absorption within the first two measurements and a narrow plateau, the absorption signature of benzene decreased at a stable and slow pace. Using the plateau value as the measurement, we then investigated the sensor response to benzene solution whose concentrations vary from 0.1 to 3 ppmV (right, Fig. 6).

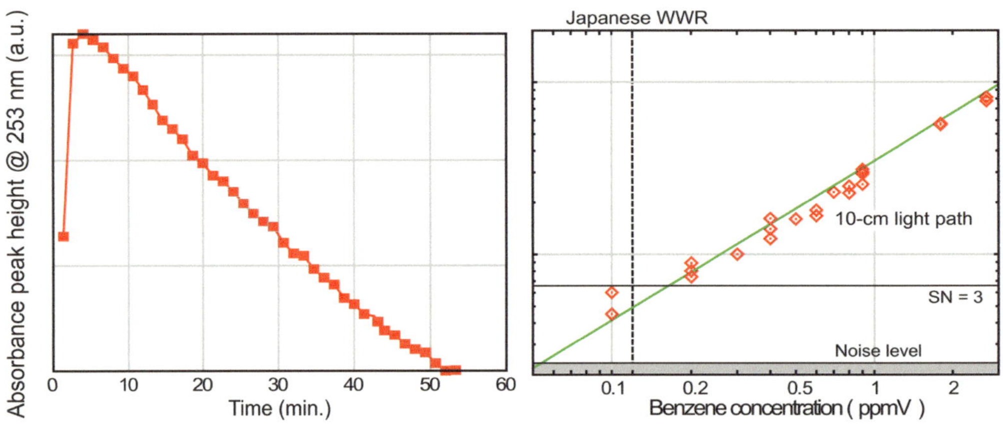

Fig. 6. Benzene main peak absorption amplitude at 253 nm versus time (left) and sensor response versus aqueous benzene concentration (right) (Camou *et al.*, 2008)

First, the sensor exhibits a linear response in the 0.2 - 3 ppmV range, which is more than one order of magnitude. This range may be extendible; however the high concentration range is not of much interest regarding the final application.

Concerning the actual detection limit, with a S/N ratio of 3 as our minimum requirement for extracting reliable data from the spectra, the LOD was estimated to be about 150 ppbV.

Experiments at lower concentrations were carried out, but the low S/N prevented a precise estimation of peak heights, as the points slightly diverged from the linear fit in Fig. 6.

As a result, with a LOD in the 150-ppbV range and ease of operation, this sensor represents an alternative suitable for on-site alarm-level measurements. In comparison to other relevant sensors, it exhibits similar sensitivity to benzene, while the time needed per measurements is shortened to just 10 minutes.

Nevertheless, the drinking-water regulation levels remain about two orders of magnitude below the reported LOD (5 ppbV in both America and Japan; 1 ppbV in Europe), and gaining two orders of magnitude in sensitivity requires drastic modifications. Previously, in the framework of the air monitoring system, the concentration stage led to a gain in sensitivity of about three to four orders of magnitude. The concentration stage then remains the key technology for reaching the drinking-water regulatory levels, typically in the high pptV levels.

3.3 Results with airborne system

We first equipped the airborne system (Fig. 1) with bubbling-extraction module (Fig. 4) and tested its ability to detect high-aqueous-benzene-concentration [several ppmV] solutions. During these preliminary experiments, two problems occurred that allowed us to identify the major difference between airborne and aqueous benzene detection: the carrier gas relative humidity.

Figure 7 depicts the first five consecutive raw spectra obtained in this experiment. Despite absorption peaks at wavelengths characteristics of benzene compounds, the raw spectra exhibited unusual shapes. After a positive first response characterized by peak amplitudes drastically lower than first expected, the second spectrum exhibited no peak, and from the third spectrum, the benzene signature appeared negative. A decrease of the absorption peak amplitude with time, where time is equivalent to increasing spectrum number, is normal due to the diffusion of concentrated gas within the fluidic system. As a result, when the benzene molecules remaining within the optical light path lead to absorption below the noise level, the output and reference spectra become comparable and free of absorption due to benzene. Thus, the raw spectra tend to a flat baseline around 0 as the spectrum number increases. The results of Fig. 7 then clearly indicated that the reference spectrum had been corrupted. In fact, silicate adsorbent active sites also exhibit strong affinity to water molecules, even stronger than that to benzene molecules. Thus, water molecules have the ability to remove and replace benzene molecules adsorbed at the active sites while the reverse reaction can hardly occur. As a consequence, competition between water and benzene molecules resutls in random release of benzene molecules in the carrier gas that can potentially alter the reference spectrum. This explanation elucidates the negative peaks seen in Fig. 7 and also corroborates some previous results.

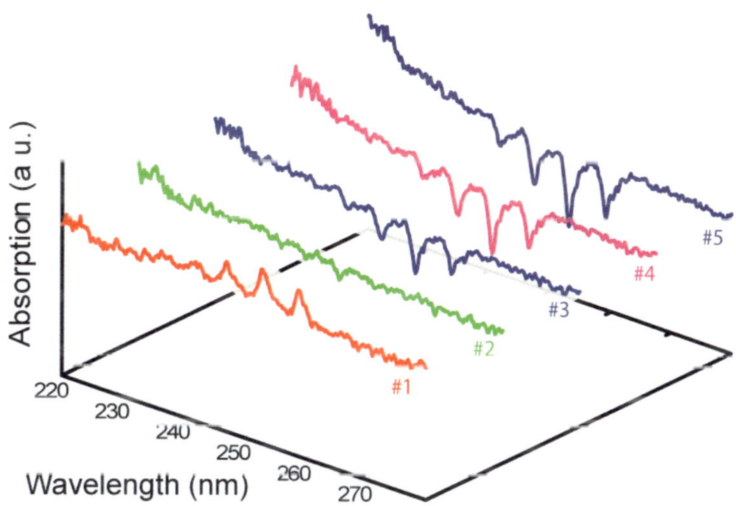

Fig. 7. First five consecutive raw spectra obtained with sensor combining bubbling extraction and a concentration cell filled with silicate adsorbent.

Second, while carrier gas containing benzene molecules flowed through the concentration cell, we observed a continuous decrease of the flow rate. Furthermore, at the end of this experiment, the recovery/refresh process of the adsorbent required an unusually large number of thermal cycles at high temperature before the flow rate would return to reference value. These phenomena come from capillary condensation of water, which is due to the combination of carrier gas containing high RH levels with the nano-sized structure of silicate adsorbent.

The high RH value of carrier gas after the extraction exerts two side effects that make the approach inefficient. However, replacing the silicate adsorbent with another compound is not sufficient because of the capillary condensation. Indeed, we are facing here two contradictory properties. The active sites of the adsorbent are localized at its surface. In order to optimize the number of active sites within a finite volume, high surface/volume ratios are a prerequisite for an efficient concentration enhancement. As a consequence, all the potential adsorbents exhibit dense structures through pores whose sizes vary greatly but always in the range where 90% RH carrier gas leads to capillary condensation. Whatever adsorbent material we use, the relative humidity of carrier gas flowing through the concentration cell should exhibit lower levels. Extraction systems based on different mechanism and with lower efficiency have been tested, but the RH levels of the carrier gas remained over 80%, leaving the issue intact.

3.4 New concentration stage

We then developed a new concentration stage that could provide benzene concentration enhancement prior to the detection cell despite interference due to water molecules. As shown in Fig. 8, the new stage is composed of a passive drying system (Nafion tube), which decrease RH to a level where capillary condensation doesn't occur, and a concentration cell filled with a zeolite adsorbent, which exhibits weaker affinity to water molecules.

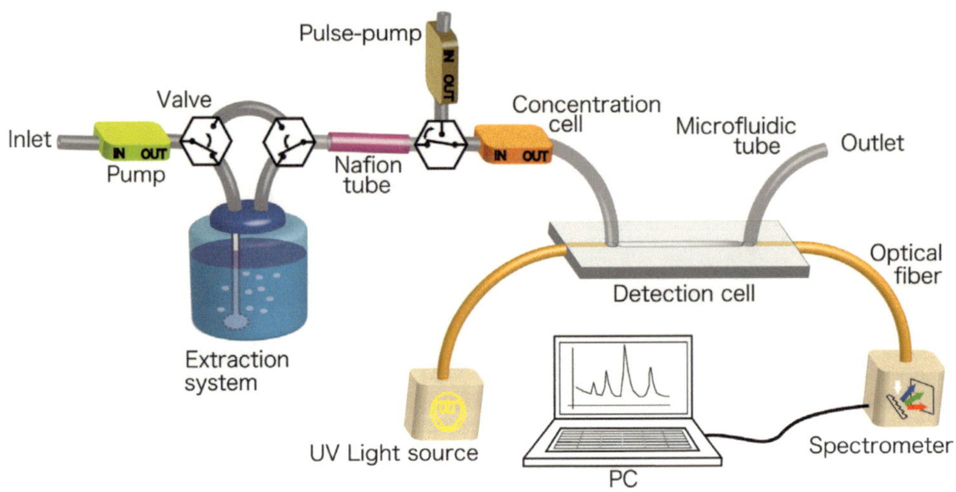

Fig. 8. Schematic view of the high-sensitivity aqueous benzene sensor with an extraction, concentration, and detection stages.

Adsorbent

Over the past few years, several compounds have been investigated as potential adsorbents for air monitoring system: MCM-41, SBA-15, and SBA-16 as mesoporous silicates, and ZSM5 as zeolite. Mesoporous silicates offer huge surface areas, large pore sizes, and regular and well-ordered structures. Regarding the benzene adsorption, several papers dealing with the optimization of pore sizes via the synthesis process have pointed to SBA-16 as the best alternative (Ueno *et al.*, 2005). Meanwhile, the ZSM5 zeolite has been widely used for gas separation, including for that of volatile organic compounds, which makes it a potential adsorbent.

With benzene/dry nitrogen calibrated gas samples, the performance of SBA-16 and ZSM5 materials as benzene adsorbent were measured under similar experimental conditions (Fig. 9). SBA-16 exhibit saturation after few minutes, while ZSM5, despite a gradual slope decrease with time, does not within the first 70 minutes. This indicates that concentration efficiency with SBA-16 is better at short concentration time, but the tendency reverses as the concentration time increases (Fig. 9).

Thus, despite a lower efficiency within the first 10 minutes of concentration time, both SBA-16 and ZSM5 remain potential candidates as adsorbents. However, SBA-16 exhibits stronger affinity to water molecules than ZSM5. Thereafter, we then exclusively used ZSM5 as adsorbing material.

Drying system

Prior to the concentration cell, we tested two passive methods based on Nafion tube technology to get rid of some of the RH content of the carrier gas. Explanations about the basic concept can be found in the product datasheet:

"When gas containing water vapor passes through Nafion tubing, the water is absorbed by and moves though the walls of the tubing, evaporating into the surrounding air in a process called pervaporation. The reaction is driven by the humidity gradient until equilibrium is reached".

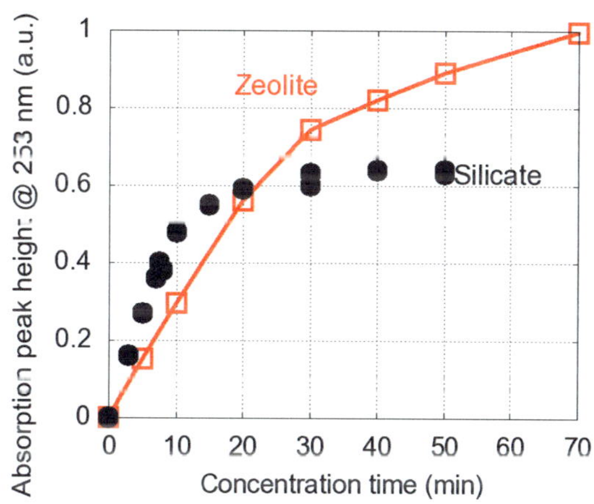

Fig. 9. Concentration efficiency at 10 ppbV calibrated gas with SBA and ZSM5 adsorbing materials versus concentration time.

The other components, such as BTX gases, are totally retained, and the benzene concentration of the carrier gas remains unaffected despite the gas's flowing through the Nafion tube. This reaction is quite fast, and a 30-cm long tube operating at flow rates of about a few ten standard cubic centimeter per minute (sccm) provides sufficient space for reaching equilibrium at the outlet. Moreover, the Nafion tube requires no maintenance or source of energy and is resistant to most chemicals. Its robustness and maintenance-free characteristic then makes it particularly suitable for on-site experiments.

Nevertheless, this system doesn't provide any active control of the sample gas RH, and the carrier gas humidity level depends directly on the ambient RH level. All the experiments were performed in a clean-room environment, with the RH parameter effectively monitored and controlled at levels within the 45% +/- 15% range. As a consequence, the Nafion tube leads to a decrease of the carrier gas RH from 90 to a mean value of 45% in our experiments (Fig. 10).

If more precise control of the RH is needed (such as when operating in a very humid environment, such as summer in Tokyo, Japan), similar technology, which we call a "Nafion box", enables drying of sample gas down to the dew point of -30/-40 degree C whatever the ambient air RH is. The Nafion tube is encapsulated inside a closed container filled with desiccant (millimeter-sized spheres) that provides a dry local environment by absorbing almost all the water molecules within the container volume. Then, after reaching equilibrium, the carrier gas also exhibits a very low RH level at the tube outlet. However, this device requires regular maintenance to replace or refresh the spheres (re-activation of the used spheres by thermal desorption), which remains a significant drawback.

The two drying systems were tested and their performance compared using the same pre-concentration cell filled with zeolite adsorbent and using equivalent benzene/water sample solution. The results did show some measurable differences, given that all the experiments were performed in a clean-room environment (RH: 45% +/- 15%). However, the configuration with just the Nafion tube enables sensitivity levels far below the requirements, making any further improvement unnecessary. In order to get a maintenance-free set-up, we then used exclusively that set-up for the next experiments.

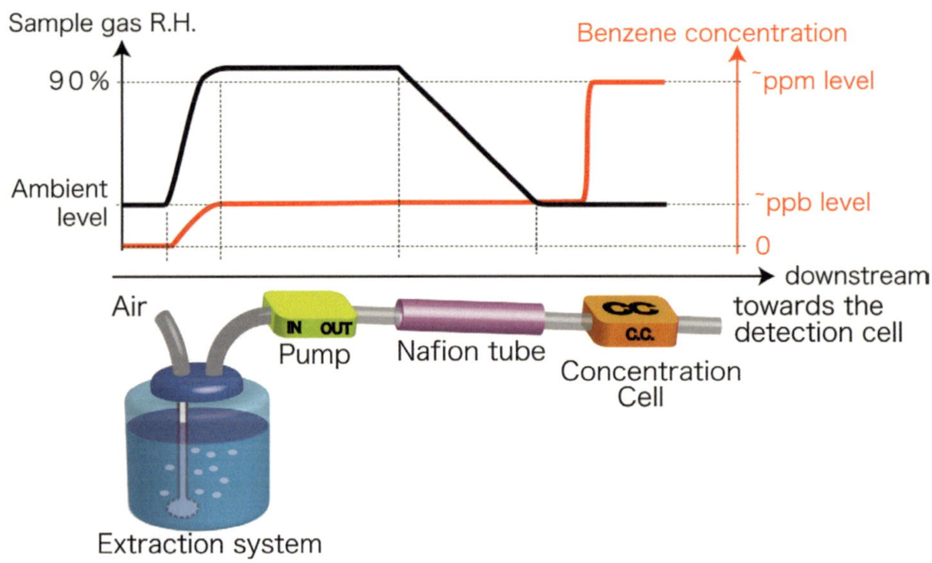

Fig. 10. Evolution of carrier gas content in terms of RH and benzene concentration along the microfluidic channel.

4. Portable aqueous benzene characterization

4.1 Sensitivity

Experiments were first performed with a 5-min concentration time and aqueous benzene concentration varying within the low ppbV range. To assess the improvement in terms of sensitivity, we plot in Fig. 11 the benzene peak absorption amplitude as function of aqueous benzene concentration for the two experimental set-ups, *i.e.,* with and without the concentration stage.

First, the concentration stage doesn't deteriorate the sensor's response, which remains linear in both cases with a comparable slope. For both set-ups, the linear range may extend for concentrations higher than those shown in Fig. 11. However, the lack of experimental data at higher concentration levels does not allow us to assess the linear range upper limit with certainty.

Furthermore, the use of the concentration stage leads to an overall shift of the response of about more than 2 orders of magnitude towards the low concentration levels. This huge improvement yields a detection limit of about 300 pptV, which is five hundred times below the previously reported LOD and more than ten times below the regulatory levels in both Japan and America (11 and 5 ppbV respectively).

In summary, the concentration cell leads to subsequent sensitivity improvement that enables us to clear the drinking water regulatory levels, while the response remains linear over more than two orders of magnitude. Furthermore, due to the 5-min concentration time, one measurement takes less than ten minutes, including all the steps required to calibrate the spectrophotometer.

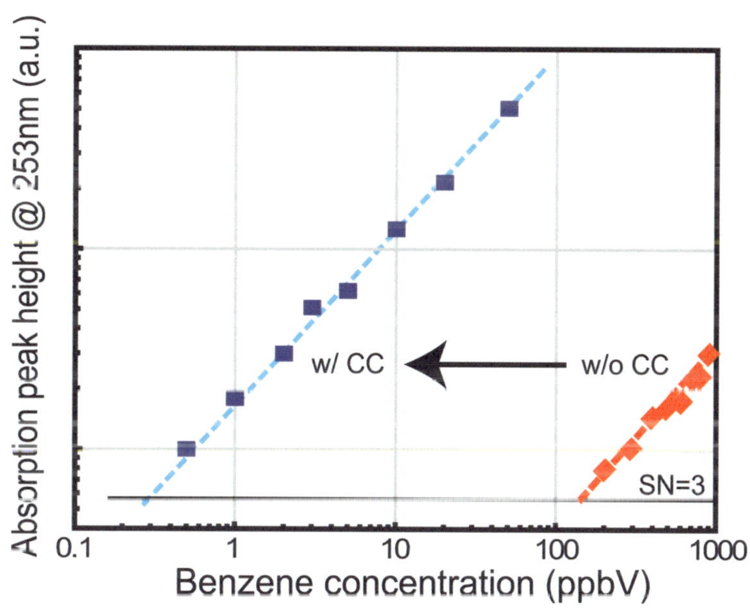

Fig. 11. Aqueous benzene sensor response around the ppbV benzene concentration range with (blue squares) and without (red diamonds) the concentration stage.

4.2 Concentration time

The previous results were all otained with a 5-min concentration time, which provides enough sensitivity improvement to clear the regulatory levels without increasing the measurement time too much. However, the concentration time directly influences the sensitivity. From a theoretical point of view, there should be a linear relationship between concentration enhancement and concentration time. However, depending on the adsorbing material, we already demonstrated saturation or a decrease of the slope as the concentration time increases (Fig. 9). A series of experiments with 4-ppbV benzene solution and bubbling extraction were then carried out to evaluate the concentration efficiency profile of zeolite adsorbent versus concentration time (Fig. 12).

As shown in Fig. 12, from 0 to 20 min., the absorption peak amplitude due to benzene compound linearly increases, pointing to a linear increase of the accumulated benzene molecule versus time. However, with concentration time exceeding 20 minutes, the signal saturates. The blue square is from the results at 4-ppbV concentration shown in Fig. 11, and demonstrates consistency between the two sets of experiments, and that increasing the concentration time within the linear range will result in linear improvement of the LOD reported earlier. We can therefore expect a further improvement of sensitivity by a factor of three, leading to a LOD down to less than 100 pptV.

In terms of the response profile, the results in Fig. 12 are quite different from those in Fig. 9. Nevertheless, the benzene concentration profile of carrier gas differs in the two cases.

Figure 13 summarizes the tendencies: with the air monitoring system, a steady carrier gas concentration leads to a gradual decrease of the concentration efficiency, while for aqueous sample, the time-dependent carrier gas benzene concentration results in saturation of the corresponding concentration efficiency. Actually, with the air monitoring system, we used

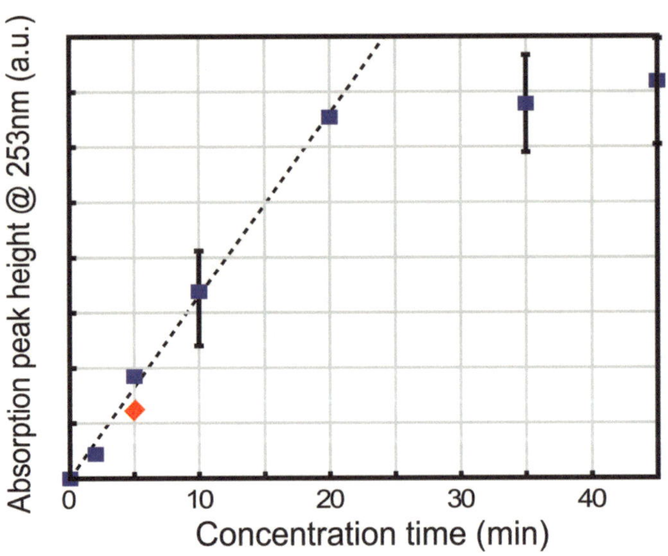

Fig. 12. Concentration efficiency of a cell filled with zeolite adsorbent versus concentration time.

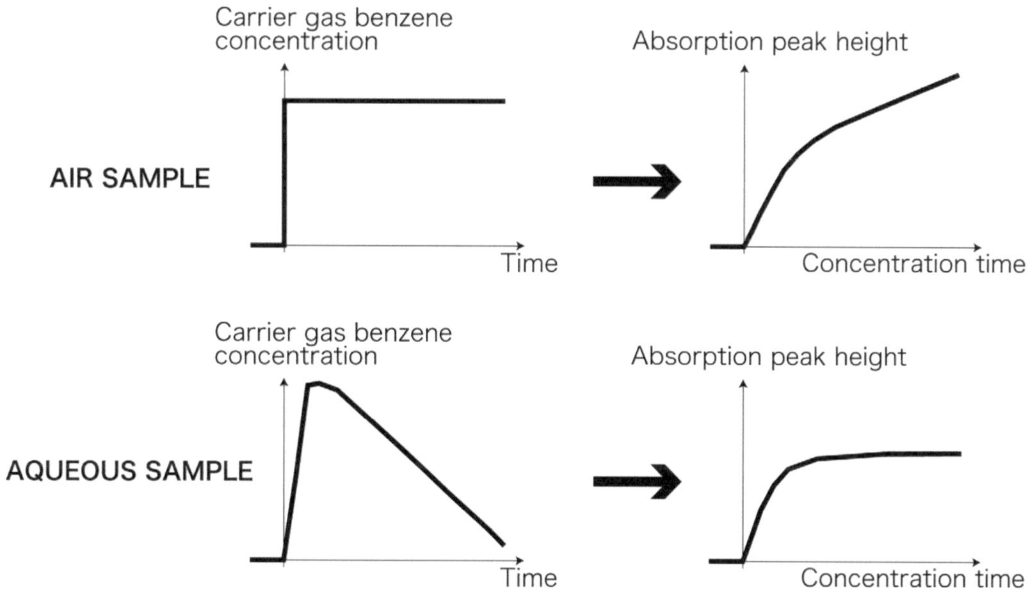

Fig. 13. Comparison of response of concentration cell filled with zeolite (right) for two carrier gas concentration profiles (left).

calibrated benzene sample gas mixed with dry nitrogen as the carrier gas. As a result, the RH remained very low, and the benzene concentration stayed constant during the entire measurement process. In comparison, the carrier gas RH after extraction/passive drying tube exhibits RH levels of about 45% and the benzene concentration exhibits a time-dependent profile.

However, as explained earlier in section 3-4, drying the carrier gas after extraction to very low RH levels leads to noticeable improvement of about 20%, but it doesn't change the overall tendency. Independantly of the RH difference, the saturation with aqueous measurements then may be seen as a more drastic decrease of the slope as the carrier gas concentration also decreases with time.

4.3 Selectivity

All the results presented earlier were obtained with pure benzene solutions we prepared at desired concentrations. However, the main source of environmental contamination has been identified as gasoline pollution, where benzene toluene and xylene are mixed with other compounds. Figure 14 shows the absorption spectra of benzene, toluene, and o-xylene, as three compounds diluted in commercially available gasoline. Due to the severe toxicity of benzene, drastic regulations have been set for the benzene concentration in gasoline. Nowadays, gasoline is composed of about 5% benzene, 35% toluene, and other compounds that may include o-xylene at lower concentration levels. Therefore, the main contamination source has a toluene concentration about seven-fold higher than that of benzene on average, though toluene absorption spectrum exhibits peaks in the same area as benzene (areas in grey in Fig. 14). Nevertheless, benzene is the only compound subject to mandatory regulation and thus the only one requiring a direct measurement procedure. Theoretically, if the analysis takes into account all the reference spectra of compounds in

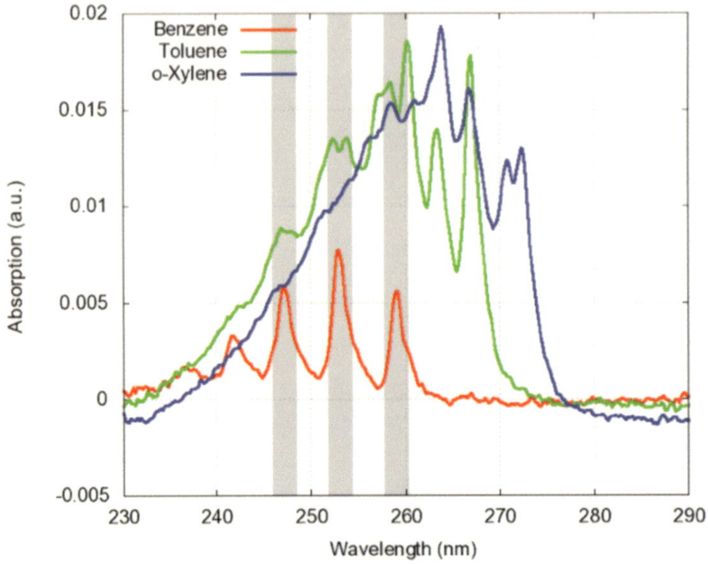

Fig. 14. Reference absorption spectra of benzene, toluene and o-xylene in the 230-290-nm wavelength range.

solution that aborb in the studied wavelength range, accurate and simultaneous quantitative measurements of several compounds from one spectrum should be possible. However, in practice, such a database of reference spectra including all potential contaminants remains an ideal, making the separation efficiency a valuable characteristic. By analogy with, for example, the gas chromatographic column prior to mass-spectrometer, efficient separation should bring to the detecting area all compounds successively, one by one, preventing overlap and interference between two or more solutes. Thus, unidentified compounds should be separated from the compounds of interest and detection of each compound done at maximum sensitivity, despite huge variation in concentrations among all the solutes.

Our sensor is composed of extraction and concentration stages, which may both result in selectivity. Nevertheless, the selectivity coming from the concentration cell remains negligible due to our thermal cycle characteristic. In our experiments, we quickly heated the adsorbent to temperatures far above the level at which benzene desorption occurs. This procedure then guaranties the best sensitivity because all of the adsorbed molecules are released simultaneously, within as small a carrier gas volume as possible. However, the three BTX compounds exhibit quite close desorption temperature. As a result, despite a chromatographic desorption process for adsorbed molecules, the fast increase of temperature yields the almost simultaneous release of adsorbed BTX compounds, cancelling the chromatographic effect. In what follows, we therefore focus exclusively on the bubbling extraction method.

Measurement of a benzene/toluene/o-xylene solution in water at 0.45/3/3 ppmV concentrations, respectively, was then performed without any concentration stage (Fig. 6).

Figure 15 shows the first eight consecutive output raw spectra. The response exhibits the typical "bubbling-like" profile as mentioned previously (Fig. 7), with an initial rapid increase followed by a constant and slow decrease of the absorption amplitude.

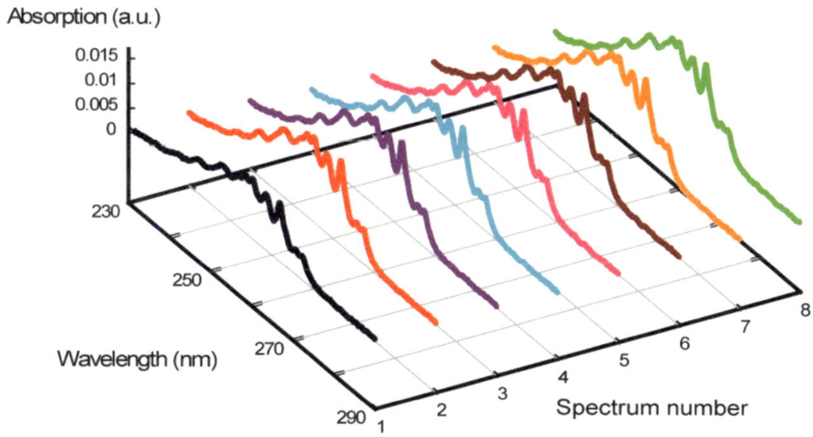

Fig. 15. Raw absorption spectra versus time with bubbling extraction.

In order to evaluate the contribution of each compounds independently, we then performed manually a fit of the experimental data from the reference spectra shown in Fig. 14 weighted by coefficients. Thus,

$$RawSpec = a.BenzRS + b.TolRS + c.oXylRS \tag{1}$$

where *RawSpec* stands for the raw output spectrum, *BenzRS*, *TolRS,* and *oXylRS* for reference spectra of benzene, toluene, and o-xylene, respectively, and *a,b,* and *c* are linear coefficients determined manually.

Results corresponding to the third spectrum are summarized in Fig. 16, which includes three different graphs: the experimental data and the spectrum built from the fitting process (top); the experimental raw data and the three BTX contributions pondered by fitted coefficients and plotted separately (middle); the experimental data and the spectrum built from the fitting process without the benzene contribution (bottom).

As shown in top graph of Fig. 16, we could reach a good correlation between the experimental spectrum and the reconstructed one obtained from the manual fitting procedure. Despite the noise background slightly diverging at higher wavelenght, the two curves are almost perfectly super-imposed in the peak area.

When the three contributions from the reconstructed spectrum are plotted separately (middle, Fig. 16), the benzene contribution remains comparatively low, with a ponderation coefficient approximately six times lower that those of toluene and o-xylene. This ratio is similar to the concentration differences between the three compounds at which the sample solution was prepared. Furthermore, the same procedure has been utilized with later output spectra (not shown). It was found that despite an overall decrease of the absorption peak amplitudes as the spectrum rank increases, the ratio between the three compounds from the manual fit remains constant. With the exact same extraction profile for the three compounds (tendancy similar to Fig. 7 and amplitude proportional to the compound concentration in the feed solution), this extraction method provides no specificity and operates with the same efficiency on the three BTX compounds.

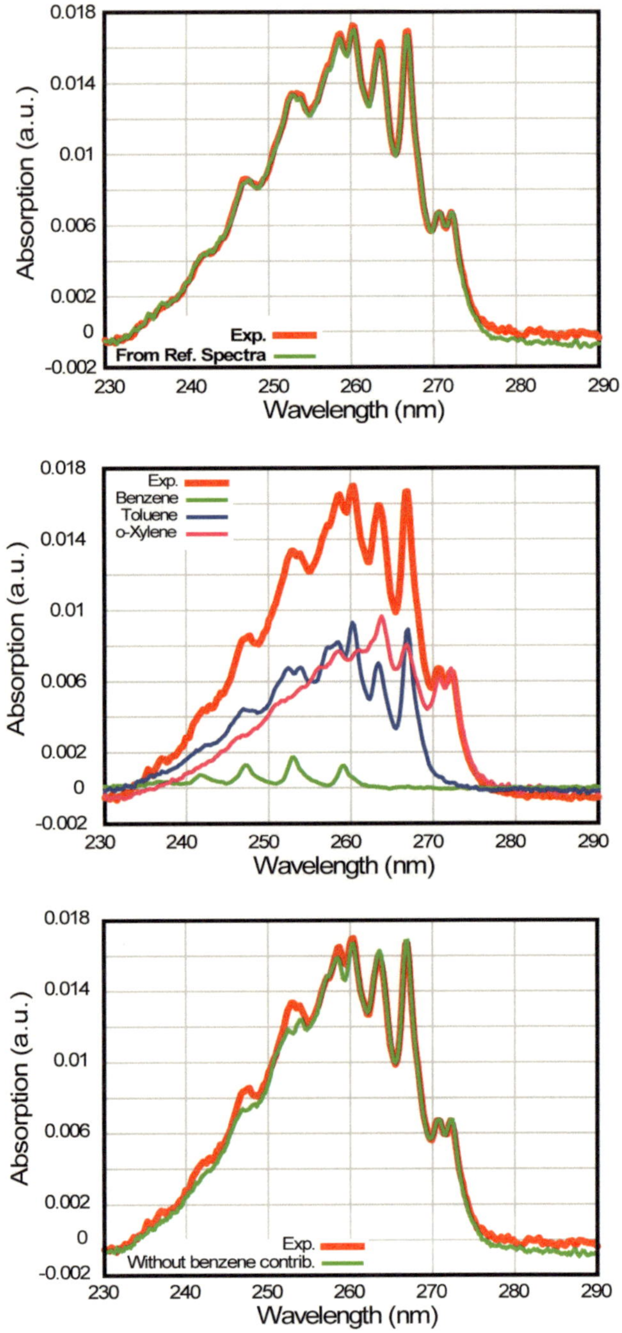

Fig. 16. Estimation of the respective contributions of benzene, toluene, and o-xylene for the third spectrum with bubbling extraction.

When removing the benzene contribution from the reconstructed spectrum (bottom, Fig. 16), the two curves slightly and locally diverge, but the difference remains quite small compared to the overall signal amplitude and shape. The overlap of absorption bands, especially between benzene and toluene, leads to interference that can potentially disturb the precise estimation of benzene contribution. In practice, determination of benzene still remains possible, but the task may be quite difficult due to the background level (background including toluene at a concentration seven-times higher) and potential interference from unknown compounds dissolved in the feed solution.

5. Summary and future work

We described in this chapter a portable aqueous benzene sensor that combines bubbling extraction and concentration and detection stages. The bubbling module extracts several compounds simultaneously from the liquid to the vapor phase, while the performance of the concentration stage prior to detection cell leads to high sensitivity. We then demonstrated a LOD about 300 pptV, far below the requirements with a ten minutes measurement time. Furthermore, the sensor response remains linear over more than two orders of magnitude. Systematic studies of concentration time also demonstrated that this sensor allows some flexibility for finding the appropriate compromise between sensitivity/measurement time depending on the application requirements. All the measurements were performed in a controlled atmosphere with RH levels of around 45%. When the RH of ambient air may become ploblematic, the moisture exchanger tube should be replaced with the drying box, which provides active and efficient control over the carrier gas RH. Though a system with the drying box requires more often maintenance, it provides a sensor unit the proper on-site conditions without any limitation in terms of ambient air RH. The sensor then represents a potential alternative to bulky standard equipment as an on-site early alert system.

However, some issues remain for future development of our sensor. As discussed earlier, considering the main contamination source to be a gasoline spill, the sensor should exhibit specificity in order to separate benzene and toluene at the detection stage. Thanks to the concentration stage, we have achieved LOD levels far below the requirement. The margin we got about the sensitivity allows some degree of freedom for improving the selectivity. As a consequence, another chromatographic extraction method may represent a good compromise by providing better selectivity despite worsened sensitivity, but still in the pptV range.

Regarding the final application of this sensor, we will also focus our efforts on the development of an in-line and continuous aqueous benzene extraction system. Right now, the portable sensor enables one to perform on-site high-sensitivity measurements. However, an operator must still take a sample of the liquid and transfer it to the extraction tank, as is the case for measurements based on standard techniques. Due to the limited number of skilled operators and the huge number of sites to be monitored, the frequency of benzene monitoring is calculated from previous measurement campain results and the potential risk/impact of a benzene contamination. As a consequence, the time between two consecutive measurements at a specific site may vary from days to months. In order to detect benzene contamination at a very early stage, a drastic reduction of this delay is a real need that only continuous and operator-free measuring devices can fulfil. A sensor combining high-sensitivity with continuous measuring sequence may then result in significant advances towards the supply of safe drinking water.

7. Acknowledgments

This work is based on the concept of a portable airborne BTX sensor system developed by Drs. Y. Ueno and O. Niwa, and Mr. A. Tate in the early stage of the project. The authors would like to thank for their contributions in proving the concept.

8. References

Beyer, T.; Hahn, P.; Hartwig, S.; Konz, W.; Scharring, S.; Katzir, A.; Steiner, H.; Jakush, M.; Kraft, M. & Mizaikoff, B. (2003). Mini spectrometer with silver halide sensor fiber for in situ detection of chlorinated hydrocarbons. *Sensors and Actuators B*, 90, 2003, pp. 319-323

Burck, J.; Schagenhof, M.; Roth, S. & Mathieu, H. (2001). Kinetic evaluation method for SPME-NIR measurements of analytes with long equilibration time. *Field Anal. Chem. Techn.*, 5(3), 2001, pp. 131-142

Camou, S.; Horiuchi, T. & Haga, T. (2006),a. Ppb level benzene gas detection by portable BTX sensor based on integrated hollow fiber detection cell. *IEEE Sensors 5th. Proceedings.* 2006. pp. 73, Daegu (South-Korea)

Camou, S.; Horiuchi, T. & Haga, T. (2006),b. Absorption detection cell fabrication based on aluminum coated hollow fiber: application to airborne benzene measurements. *,Eurosensors XX', Proceedings.* 2006. pp. 42-43. Goteborg (Sweden)

Camou, S.; Shimizu, A.; Horiuchi, T. & Haga, T. (2008). ppb-Level detection of benzene diluted in water with portable device based on bubbling extraction and UV spectroscopy. *Sensors and Actuators B*, 2008, 132, pp. 601-607

EPA (1993). 1993 Motor Vehicle - Related Air Toxics Study - Chapters 5 -7. USA, available at: <http://www.epa.gov/otaq/regs/toxics/airtox1b.pdf>

EPA (2003). Method 5030C, Purge-and-trap for aqueous samples. Revision 3 May 2003, USA, available at: <http://www.epa.gov/epaoswer/hazwaste/test/pdfs/5030c.pdf>

EPA (2006). Drinking water contaminants. Revision 28 November 2006, USA, available at: <http://www.epa.gov/safewater/contaminants/index.html#organic>

European Council (1998). Directive 98/83/EC of the Council of 3 November 1998. *Official Journal of the European Communities 330*, 05.12.1998

Hahn, P.; Tacke, M.; Jakusch, M.; Mizaikoff, B.; Spector, O. & Katzir, A. (2001) Detection of hydrocarbons in water by MIR evanescent-wave spectroscopy with flattened silver halide fibers. *Applied Spectros.*, vol. 55, 1, 2001, pp. 39-43

Heglund, D.L. & Tilotta, D.C. (1996). Determination of volatile organic compounds in water by solid phase microextraction and infrared spectroscopy. *Environ. Sci. Technol.*, 30, 1996, pp. 1212-1219

Karlowatz, M.; Kraft, M. & Mizaikoff, B. (2004). Simultaneous quantitative determination of benzene, toluene, and xylenes in water using mid-infrared evanescent field spectroscopy. *Anal. Chem.*, 2004, 76, pp. 2633-2648

Krska, R.; Taga, K. & Kellner, R. (1993). New IR fiber-optic chemical sensor for in situ measurements of chlorinated hydrocarbons in water. *Applied Spectros.*, vol. 47, 9, 1993, pp. 1484-1487

Lamotte, M.; Fornier de Violet, F.; Garrigues, P. & Hardy, M. (2002). Evaluation of the possibility of detecting benzenic pollutants by direct spectrophotometry on PDMS solid absorbent. *Anal. Bioanal. Chem.*, 372, 2002, pp. 169-173

Martinez, E.; Lacorte, S.; Llobet, I.; Viana, P. & Barcelo, D. (2002). Multicomponent analysis of volatile organic compounds in water by automated purge and trap coupled to gas chromatography-mass spectrometry. *J. Chromatogr. A*, 959, 2002, pp. 181-190

Ministry of Health, Labour and Welfare in Japan (2003). Drinking water regulation levels. Revision 30 May 2003, Japan, available at (in Japanese): http://www.mhlw.go.jp/topics/bukyoku/kenkou/suido/kijun/dl/syourei.pdf

Ministry of the Environment in Japan (2008). Wastewater regulation levels. Revision 30 September 2008, Japan, available at (in Japanese): http://law.e-gov.go.jp/htmldata/S46/S46F03101000035.html

Mohacsi, A.; Bozoki, Z. & Niessner R. (2001). Direct diffusion sampling-based photoacoustic cell for in situ and on-line monitoring of benzene and toluene concentrations in water. *Sensors and Actuators B*, 79, 2001, pp. 127-131

Namiesnik, J.; Zabiegala, B.; Kot-Wasik, A.; Partyka, M. & Wasik, A. (2005) Passive sampling and/or extraction techniques in environmental analysis: a review. *Anal. Bioanal. Chem.*, 381, 2005, pp. 279-301

Richardson, S.D. & Ternes, T.A. (2005). Water analysis: emerging contaminants and current issues. *Anal. Chem.*, 2005, 77, pp. 3807 3838

Serrano, A. & Gallego, M. (2004). Direct screening amd confirmation of benzene, toluene, ethylbenzene and xylenes in water. *J. Chormatogr. A*, 1045, 2004, pp. 181-188

Souken Co., Ltd. (2005). Beam Homogenizer and Aluminum Hollow Fiber catalogue 2005.

Steiner, H.; Jakusch, M.; Kraft, M.; Karlowatz, M.; Baumann, T.; Niessner, R.; Konz, W.; Brandenburg, A.; Michel, K.; Boussard-Pledel, C.; Bureau, B.; Lucas, J.; Reichlin, Y.; Katzir, A.; Fleichmann, N.; Staubmann, K.; Allabashi, R.; Bayona, J.M. & Mizaikoff, B. (2003) In situ sensing of volatile organic compounds in groundwater: first field tests of a mid-infraredfiber-optic sensing system. *Applied Spectros.*, vol. 57, 6, 2003, pp. 607-613

Tobiska, P.; Chomat, M.; Matejec, V.; Berkova, D. & Huttel, I. (1998). Investigation of fiber-optic evanescent-wave sensors for detection of liquid hydrocarbons. *Sensors and Actuators B*, 51, 1998, pp. 152-158

Ueno, Y.; Horiuchi, T.; Morimoto, T. & Niwa, O. (2001). Microfluidic device for BTEX airborne detection. *Anal. Chem.*, 2001, 73, pp. 4688-4693

Ueno, Y.; Horiuchi, T.; Tomita, M. & Niwa, O. (2002). Separate detection of BTX mixture gas by a microfluidic device using a function of nanosized pores of mesoporous silica adsorbent. *Anal. Chem.*, 2002, 74, pp. 5257-5262

Ueno, Y.; Tate, A.; Niwa, O.; Zhou, H-S.; Yamada, T. & Honma, I. (2005). High benzene selectivity of mesoporous silicate for BTX gas sensing microfluidic devices. *Anal. Bioanal. Chem.*, 2005, 382, pp. 804-809

Vogt, F.; Tacke, M.; Jakush, M. & Mizaikoff, B. (2000). A UV spectroscopic method for monitoring aromatic hydrocarbons dissolved in water. *Anal. Chim. Acta*, 422, 2000, pp. 1887-198

Yang, J. & Her, J-W. (1999). Gas-assisted IR-ATR probe for detection of volatile compounds in aqueous solutions. *Anal. Chem.*, 1999, 71, pp. 1773-1779

Yang, J. & Tsai, S-S. (2002). Cooled internal reflection element for infrared chemical sensing of volatile to semi-volatile organic compounds in the headspace of aqueous solutions. *Anal. Chim. Acta*, 462, 2002, pp. 235-244

Zimmermann, B.; Burck, J. & Ache, H-J. (1997). Studies on siloxane polymers for NIR-evanescent wave absorbance sensors. *Sensors and Actuators B*, 41, 1997, pp. 45-54

21

CMOS Readout Circuit Developments for Ion Sensitive Field Effect Transistor Based Sensor Applications

Wen-Yaw Chung[1], Febus Reidj G. Cruz[1], Chung-Huang Yang[2], Fu-Shun He[1], Tai-Tsun Liu[1], Dorota G. Pijanowska[3], Wladyslaw Torbicz[3], Piotr B. Grabiec[4] and Bohdan Jarosewicz[4]
[1]Electronic Engineering Department, Chung-Yuan Christian University,
[2]Electronic Engineering Department, Vanung University,
[3]Institute of Biocybernetics and Biomedical Engineering, PAS, Warsaw,
[4]Institute of Electron Technology, Warsaw,
[1,2]Taiwan, R.O.C.,
[3,4]Poland

1. Introduction

Biomimetic devices have become more and more important in modern life where populations are aging; and the applications of electronic tongue system to water quality and environmental monitoring have become a significant field all over the world. Electronic tongue system uses sensor arrays and signal processing techniques such as identification, classification and recognition for quantitative multi-component analysis and for artificial assessment of taste and flavor of various liquids (Cjosek & Wroblewsk, 2007). Ion Sensitive Field Effect Transistor (ISFET), an electrochemical and potential type sensor, has served as excellent candidate for various electronic tongue applications.

The ISFET, invented by Bergveld in 1970, is a solid-state device that combines a chemically sensitive membrane with a MOS type field-effect transistor (Bergveld, 1970). Due to its small size, rapid pH response and rugged solid-state construction, the ISFET exhibits a number of advantages over conventional pH-glass electrodes. ISFET has been extensively studied in past 36 years (Bergveld, 1991 and 2003; Garde et al., 1995). The current status and trends of main ISFET-based research, shown in Fig.1, are (1) single and sensor array applications, (2) ISFET micro-system fabrication in a standard CMOS technology, and (3) diversified ISFET-based biosensor development. For example, the ISFET research topics in Taiwan for the past ten years (Yin et al., 2001; Chin et al., 2001; Chung et al., 2004, and 2008) are focused on the study of new sensing material, on fabrication technology and device structure development, on diversified field applications, on the study and improvement for non-ideal characteristics, and on new readout circuit development. Based on our previous researches, the key problems in readout circuit development are due to the inherent characteristics of ISFET and to the body effect caused by common substrate of sensor array applications. The inherent characteristics of ISFET, like time drift and temperature dependency, cause drawbacks on ISFET continuous-mode monitoring applications. Furthermore, the conventional floating-source constant-voltage and constant-current circuit (Caras & Janata, 1980) in Fig. 2 faces problems including noise interference, requirement of two external current sources and body effect. In order to solve the aforementioned problems, this chapter focuses on developing a series of improved readout circuit techniques that enhances the performance of ISFET and demonstrates their pH sensing capability for environmental monitoring.

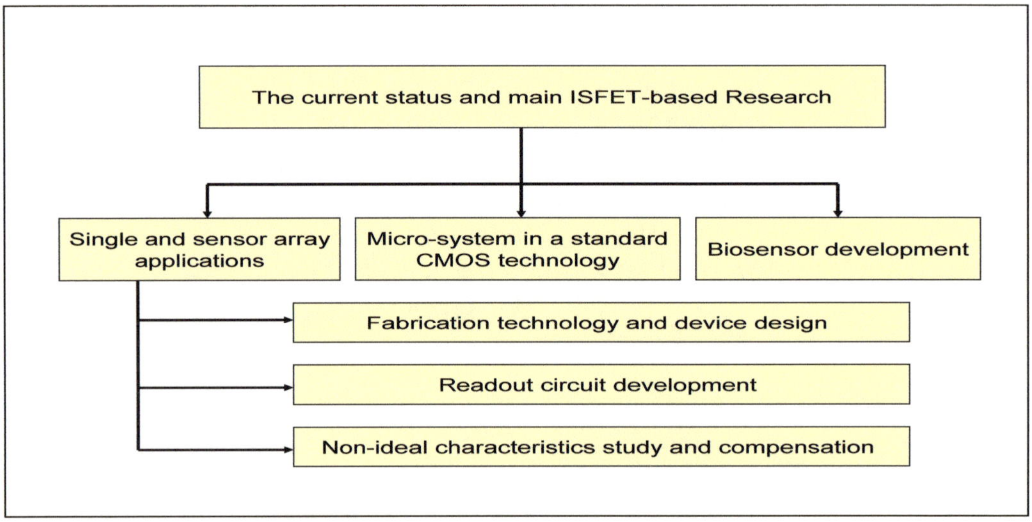

Fig. 1. The current status and main ISFET-based research

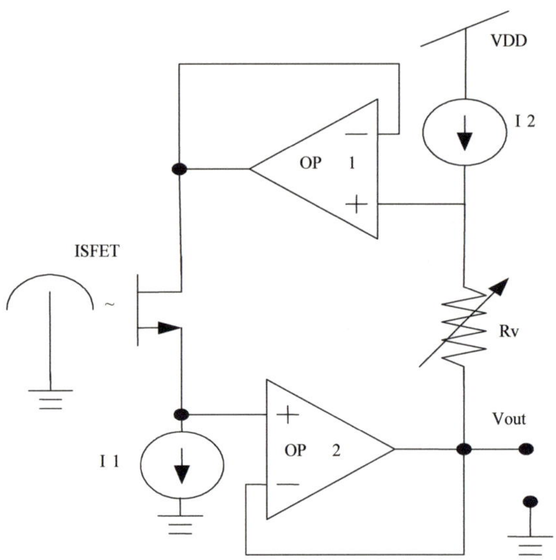

Fig. 2. A conventional floating source constant-voltage constant-current circuit (Caras & Janata, 1980)

Section 2 of this chapter explores the main concerns on ISFET device structure, operation and its stable signal readout circuit design. A bridge-type floating source circuit is developed for ISFET-based single and sensor array applications.

In order to investigate the performance of readout circuit due to the non-ideal characteristics of ISFET such as drift response, **Section 3** develops a behavioral macro model for a depletion-mode ISFET with a silicon nitride gate insulator.

Fig. 3 gives a typical measured data for Si_3N_4-gate ISFET at different pH buffer solutions. The temperature dependency may cause around 15% error in pH reading in real applications. **Section 4** demonstrates and investigates a V_{TH} extractor circuit that provides sensitive measurements with improved temperature compensation. This circuit uses Si_3N_4-gate ISFET and depletion-type MOSFET sensor pairs that are fabricated on the same wafer.

Section 5 develops a new readout circuit that improves the performance parameters, including stability of readout circuit, dependency of temperature, and wide-usage for sensor array applications. A bridge-type floating source circuit with body-effect reduction has been developed for capturing more accurate threshold voltage variation which is corresponding to different H+ concentrations. The presented readout circuit interface improves the accuracy of pH measurements, while maintaining operation at constant drain-source voltage and current condition.

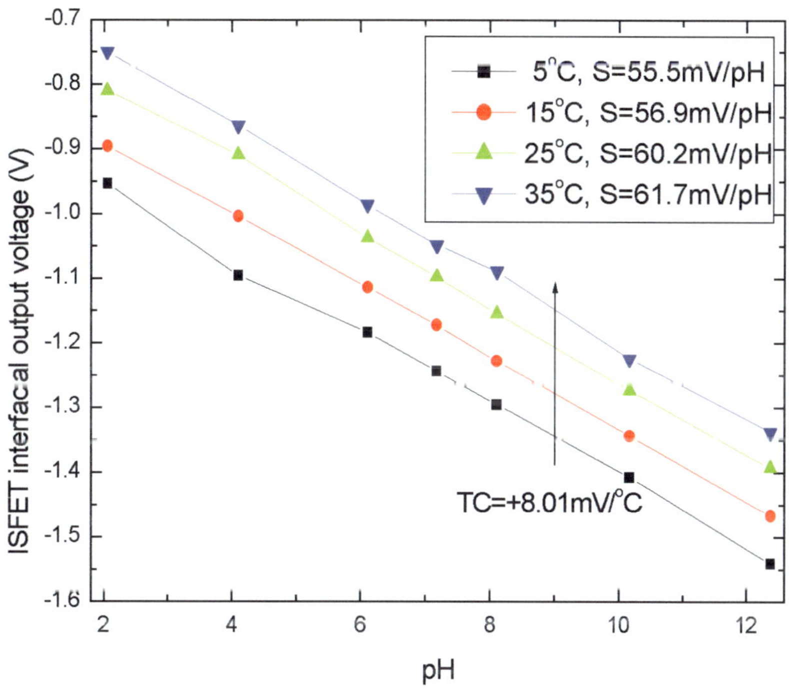

Fig. 3. Measured data versus different pH buffer solution with temperature variation.

2. ISFET operation and its signal readout

ISFET-based potentiometric transducers have created valuable applications in biomedical data acquisition and environmental monitoring. Two basic Si_3N_4-gate ISFETs are depicted in Fig. 4. In Fig. 4(a) is a simple ISFET device structure that is compatible to a standard p-substrate CMOS process, while in Fig. 4(b) is an n-substrate/p-well/n-type ISFET which have a better performance in sensor array application because of isolated p-well structure.

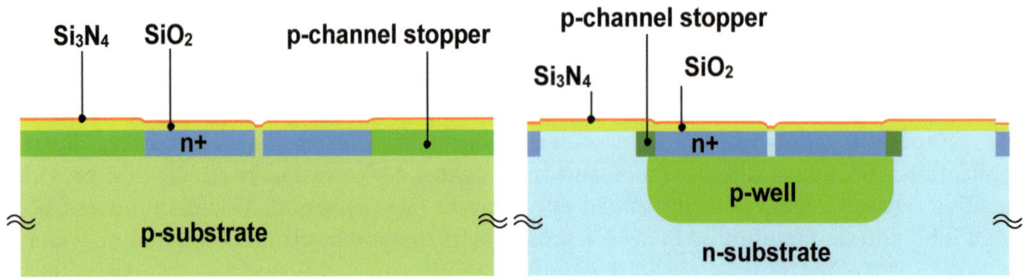

Fig. 4. (a) p-substrate/n-type ISFET; (b) n-substrate/p-well/n-type ISFET

The model of a conventional MOSFET device can also define an ISFET sensor (Bergveld, 1970) as in (2.1). The only difference is that the threshold voltage of MOSFET is replaced by the threshold voltage of ISFET.

$$I_{DS} = \mu_n C_{ox} \frac{W}{L}\left[(V_{GS} - V_{TH(ISFET)})V_{DS} - \frac{1}{2}V_{DS}{}^2 \right] \tag{2.1}$$

The I_{DS} is drain current, μ_n is mobility of electron carriers in semiconductor layer, C_{ox} is oxide capacitance density, W/L is device aspect ratio, V_{GS} is gate-source voltage, V_{DS} is drain-source voltage, and $V_{TH(ISFET)}$ is the threshold voltage of ISFET.
With gate region exposed to the chemical solution, the threshold voltage of ISFET changes accordingly with the activity of ions in the chemical solution. This electrochemical phenomenon is defined by Nernst for single-ion, e.g., hydrogen in (2.2) and (2.3).

$$V_{TH(ISFET)} = V_{TH(MOSFET)} - V_{CHEMICAL} \tag{2.2}$$

$$V_{CHEMICAL} = E_i + \frac{RT}{n_i F}\ln(a_i) \tag{2.3}$$

The $V_{TH(ISFET)}$ is a combined outcome of $V_{TH(MOSFET)}$ and $V_{CHEMICAL}$. The $V_{TH(MOSFET)}$ is threshold voltage of inherent MOSFET structure in ISFET, $V_{CHEMICAL}$ is electrochemically induced voltage in the threshold voltage of ISFET, E_i is chemical constant, R is gas constant, T is absolute temperature in Kelvin, F is Faraday constant, n_i is charge of ion i, and a_i is ion activity of ion i. To include the effect of ion activity to ISFET electrical characteristics, the Nernst model is added to the ISFET equation as in (2.4).

$$I_{DS} = \mu_n C_{ox} \frac{W}{L} V_{DS}\left[V_{GS} - V_{TH(MOSFET)} + \left[E_i + \frac{RT}{n_i F}\ln(a_i) \right] - \frac{1}{2}V_{DS} \right] \tag{2.4}$$

When ISFET is connected to a readout circuit, the output voltage V_{OUT} is usually the gate-source voltage V_{GS} of ISFET. From (2.5), the V_{GS} of ISFET is proportional to the logarithmic function of ion activity a_i. Hence, the V_{OUT} of readout circuit reflects the ion activity.

$$V_{GS} = \frac{I_{DS}}{\mu_n C_{ox} \frac{W}{L} V_{DS}} + V_{TH(MOSFET)} - E_i - \frac{RT}{n_i F}\ln(a_i) + \frac{1}{2}V_{DS} \quad (2.5)$$

In order to monitor the change of ion activity, ISFET should operate in the linear region ($V_{DS}<V_{GS}-V_{TH}$) and should maintain constant voltage constant current mode (CVCC). These conditions make the gate-source voltage (V_{GS}) proportional to the internal threshold voltage. We developed a more stable bridge-type readout circuit for ISFET pH sensor, shown in Fig.5. This circuit provides low drain-source voltage (V_{DS}) to ensure the linear operating condition of ISFET and maintain CVCC mode (V_{DS} =0.5V, I_{DS}=100uA) so that the gate-source voltage, which is the output voltage (*OUT*) of the circuit, becomes proportional to the threshold voltage of ISFET. Hence, the output voltage also becomes proportional with the pH concentration of solution. Equations (2.6) to (2.8) show the basic concept of circuit operation, where V_1 is a voltage drop across R_1 and R_2.

$$V_{DS} = V_{R2} = V_1\left[\frac{R_2}{R_1 + R_2}\right] \quad (2.6)$$

$$I_{DS} = I_{R3} = \frac{V_1 - V_{DS}}{R_3} \quad (2.7)$$

$$\begin{aligned} OUT &= V_{GS} = f(V_{TH(ISFET)}) = f(pH) \\ OUT &= V_{RE} - V_{OUT}; V_{RE} = 0V = grounded \\ OUT &= -V_{OUT} \end{aligned} \quad (2.8)$$

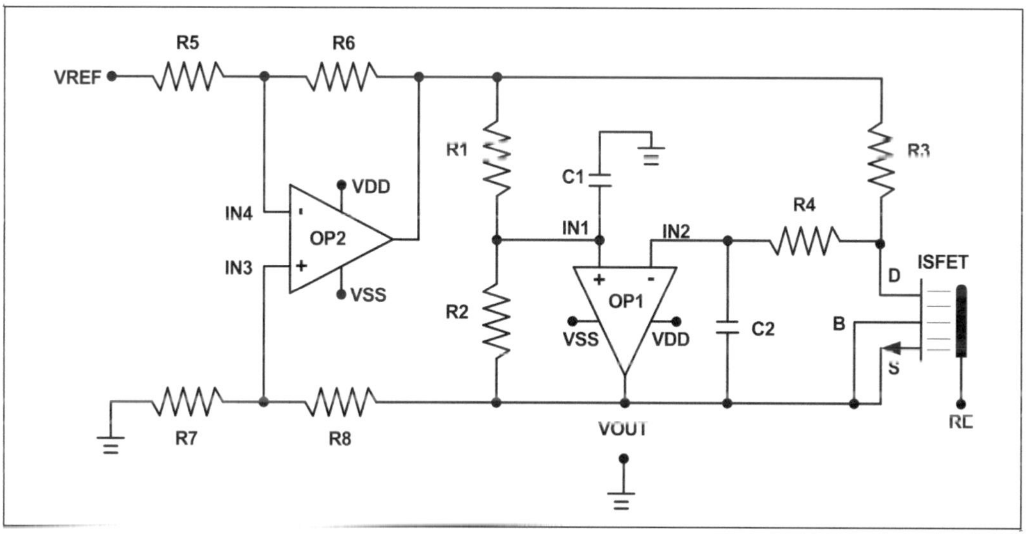

Fig. 5. The schematic diagram of ISFET bridge-type readout circuit

To enhance the signal to noise ratio, the readout circuit incorporated two low pass filters (LPF) for removing noise signals from the power supply and the ISFET itself, namely from the external electromagnetic field interference or the fluid fluctuation. One LPF is formed by

R_1, R_2 and C_1, and the other LPF is provided by R_3, R_4, R_{DS} and C_2. The pass band edge f_P is set by (2.9):

$$f_P = \frac{1}{2\pi(R_4 + R_3 \,||\, R_{DS})C_2} = \frac{1}{2\pi(R_1 \,||\, R_2)C_1} \tag{2.9}$$

The ISFET bridge-type readout circuit shown in Fig. 5 can be extended for sensor array applications. Instead of using a single ISFET sensor, a parallel configuration of ISFET sensors with respective analogue switches in each leg has been designed as in Fig. 6. The key concern of this design is the response time and linearity of analogue switches.

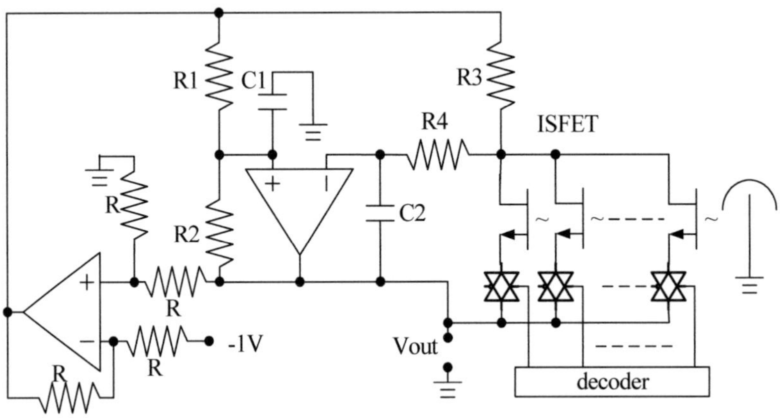

Fig. 6. The schematic diagram of bridge-type readout circuit for ISFET sensor arrays

3. Modelling the ISFET drift effects

For long-term monitoring, the drift effect of ISFET sensor is frequently observed which can last up to several hours. Studies have indicated that a drift effect of ISFET limits the measurement accuracy and quality of water monitoring.

Electronic circuit simulation programs such as SPICE (Simulation Program with Integrated-Circuit Emphasis), which were originally developed for designing and simulating electronic circuits, can also be adapted to design silicon-based chemical- and bio-sensors micro-system. Researches to model the ISFET was carried out in two main ways: (a) development of physical-chemical models (Jamasb et al., 2000; Kuhnhold et al., 2000; Chou et al., 2000) and (b) investigation of electronic circuits by SPICE built-in models or macro models (Martinoia et al., 1999; Lauwers et al., 2001). In order to include the drift effect of ISFET for circuit simulation, we developed a behaviour model which can be used in circuit design using SPICE simulator.

3.1 Drift effect of ISFET

Based on the CVCC and constant temperature conditions, ISFET drift is defined as the shift of dV_{GS}/dt. Previous works reported that the non-ideal effects of the ISFET are modelled by both responses of buried sites and the surface oxidation of silicon nitride (Bousse et al., 1990; Kuhnhold et al., 2000). The sensor output signal is influenced by both fast and slow responses. The fast time dependence is caused by the surface oxidation of silicon nitride, while the response due to buried sites mainly affects the slow pH response. So, the effective ISFET threshold voltage, V_{TH}^*, is given with time dependence by equation (3.1) (Liao, 2000):

$$V_{TH}^* = V_{TH}(0) + \Delta V_{THFS}(t) + \Delta V_{TDFT}(t) \tag{3.1}$$

Where $V_{TH}(0)$ is the original threshold voltage of ISFET at time $t=0$, $\Delta V_{THFS}(t)$ is the contribution of the drift effect by fast and slow responses, and $\Delta V_{TDFT}(t)$ is the drift-induced threshold voltage variation during the overall time interval of interest. The $\Delta V_{THFS}(t)$ can be expressed as (3.2):

$$\Delta V_{THFS}(t) = fm \times (1 - e^{-t/\tau f}) + sm \times (1 - e^{-t/\tau s}) \tag{3.2}$$

Where fm is the maximum shift of threshold voltage due to the fast time response, sm is the maximum shift of threshold voltage caused by the slow time response, τf and τs are the time constants of fast and slow responses. The typical values for τf and τs are several seconds and 2 to 3 hours, respectively. In addition, the $\Delta V_{TDFT}(t)$ can be modeled by (3.3):

$$\Delta V_{TDFT}(t) = dm \times (1 - e^{-t/\tau ov}) \tag{3.3}$$

Where τov is the time constant of overall time interval of interest, and dm is the maximum drift during a long period of measurement. The drift rate can be defined by (3.4):

$$Drift\ rate = \frac{d\Delta V_{TDFT}(t)}{dt} = \frac{dm}{\tau ov} e^{-t/\tau ov} \tag{3.4}$$

Thereby, a constant drift rate of $dm/\tau ov$ and a drift rate of 0 mV/hour can be given for $t \ll \tau$ and $t \to \infty$, respectively. From equations (3.1) to (3.4), the overall drift-induced threshold voltage variation can be concluded in (3.5):

$$\Delta V_{Toverall} = fm \times [1 - \exp(-\frac{t_f}{\tau f})] + sm \times [1 - \exp(-\frac{t_s}{\tau s})] + dm \times [1 - \exp(-\frac{t_d}{\tau ov})] \tag{3.5}$$

In 2000, Chou et al. reported that the drift rate increases as the temperature rises (Chou et al., 2000) according to the following relation in (3.6):

$$\Delta V_{d,Temp} = c_{T1} \times \exp(-c_{T2}/T) \tag{3.6}$$

Where T is the operating temperature, C_{T1} and C_{T2} are the coefficients of drift rate against temperature for different sensors. The ratio of operating temperature T to room temperature, i.e., 25°C is described in (3.7), where $driftT$ is the drift due to temperature variation.

$$driftT = \Delta V_{THoverall} \times \exp(-\frac{(25 - T) \times c_{T2}}{25 \times T}) \tag{3.7}$$

In addition, drift rate is linearly proportional to pH value as described in (3.8):

$$\Delta V_{d,pH} = c_{pH} \times pH \tag{3.8}$$

Where c_{pH} is the coefficient of drift rate versus pH for different sensors; the ratio of pH to pH7 is described as (3.9)

$$driftpH = \Delta V_{Toverall} \times (1 - \frac{c_{pH} \times (7 - pH)}{drift7}) \tag{3.9}$$

Where *driftpH* is the drift rate that changes with pH, and *drift7* is the drift rate at pH7. Finally, considering the dependence on temperature and pH value, the expression for the drift rate is provided in (3.10):

$$\Delta V^{*}_{TH,Temp,pH} = \Delta V_{THoverall} \times [1 - \frac{c_{pH} \times (7 - pH)}{drift_{pH7,25C}}] \times \exp\left\{-[\frac{(25 - T)}{25T} c_{T2}]\right\} \tag{3.10}$$

Where $\Delta V_{THoverall}$ is the drift rate at pH=7 and 25°C.

This study evaluated the n-channel p-well depletion-mode Si_3N_4-gate ISFET sensor with W/L=600μm/15μm and fabricated by the Institute of Electron Technology, Poland. The physical layout of this ISFET is provided in Fig. 7(a). Previous research (Martinoia & Massobrio, 2000) together with the approach we have formulated in equation (3.10), leads to the ISFET equivalent circuit shown in Fig. 7(b). The E_{ref} is the potential of reference electrode, C_{Gouy} and C_{Helm} are the Gouy-Chapman and Helmholtz capacitances, and E_{drift} models the drift response. The HSPICE-compatible macro model in Fig. 7(b) and Fig. 7(c) characterizes the ISFET as two stages: an electronic stage and an electrochemical stage. The nodes 1, D, S, and B stand for the reference electrode, drain, source, and bulk connections, respectively.

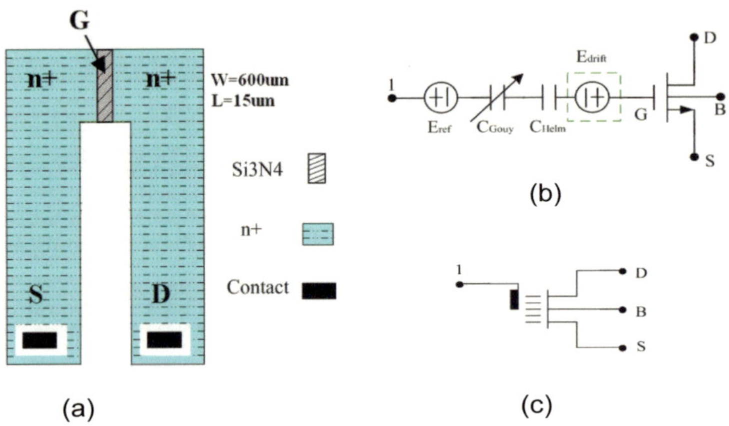

Fig. 7. (a) Physical layout of ISFET sensor, (b) equivalent circuit of the ISFET including drift effect, (c) HSPICE-compatible macro model connections

3.2 Experimental set-up and results

Based on the physical layout of ISFET sensor in Fig. 7(a), the electrical drain and source contacts are not close to the actual transistor drain and source terminals. Thus, a significant internal series drain resistance, series source resistance, and parasitic capacitances have to be considered.

In the following simulations and experiments, the ISFET is biased on a constant drain-source voltage of 0.5V and a constant drain current of 100μA (CVCC). The simulations were investigated using the developed ISFET macro model with incorporated drift effect, and using the designed bridge-type readout circuit mentioned in **Section 2**. The experiments used the standard buffer solutions purchased from Riedel-de Haen (Germany) with pH levels from 2 to 12. The measurements were conducted in a temperature-controlled system illustrated in Fig. 8.

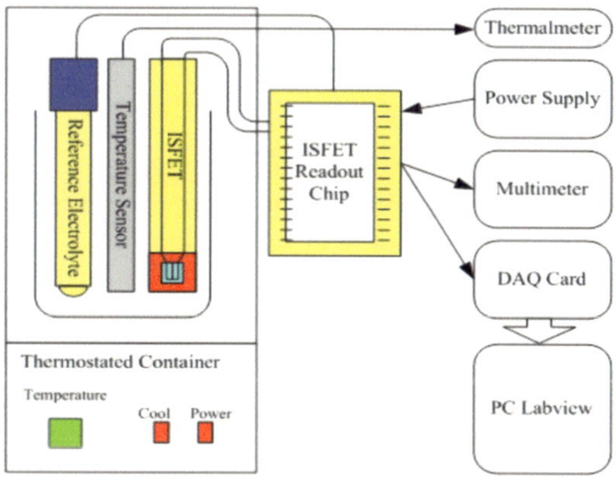

Fig. 8. Experimental set-up

The rail-to-rail op amp meets the design specifications to serve as a basic building block of the proposed bridge-type floating source readout circuit. The graph in Fig. 9 shows the dependence of ISFET source terminal potential over pH range of 2 to 12. The calculated slope for the curve is -46.34mV/pH and it presents a linear function with very high correlation coefficient of 0.998.

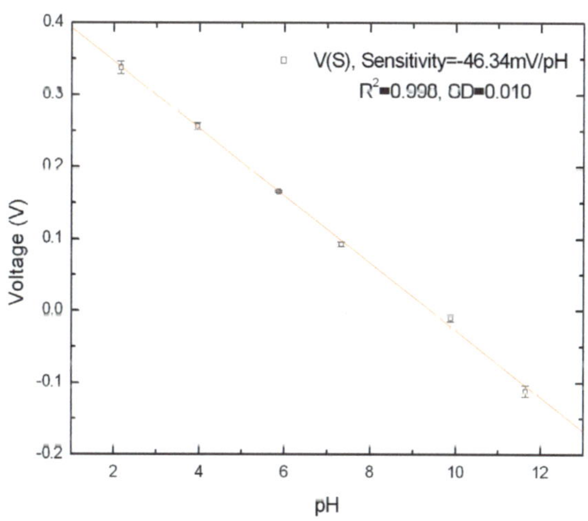

Fig. 9. ISFET readout voltage, VOUT in the pH range of 2 to 12

In order to evaluate the stability of the same ISFET sensor, the base line drift was measured twice using a standard buffer solution of pH 7. The drift rates of ISFET were evaluated after initial time of stabilization as a linear change of V_{GS} per time unit, and the drift rate is called the drift coefficient, c_d (mV/hr). Fig. 10 shows the time response of ISFET for 18-hour time period operated in pH 7 at a controlled temperature of 25°C. The c_d values were calculated for experimental data after 4 hours of conditioning. The respective c_d for Test 1 and Test 2 are -1.44mV/hr and -1.34mV/hr with a standard deviation of 0.00838.

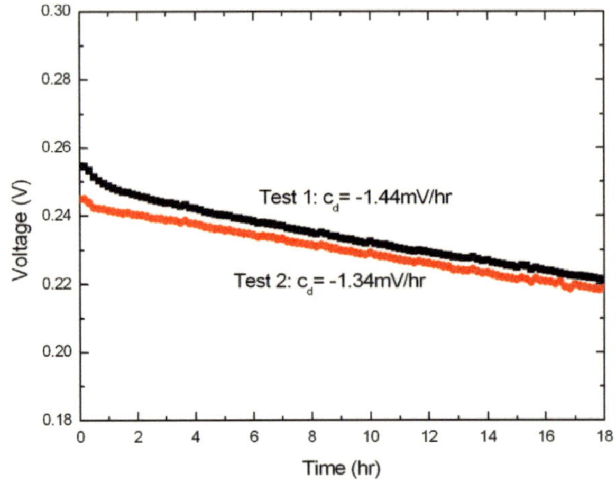

Fig. 10. The stability of ISFET in drift test

Like the time drift effect, the hysteresis also limits the accuracy of ISFET pH measurements. The hysteresis is a good marker to evaluate the reproducibility of the devices. In experiments, hysteresis tests are performed by titration in both direction, acidic or basic first and then in opposite direction to close the pH loop. Fig. 11 shows the time response of ISFET that was conditioned at pH 7 buffer solution and that went through titration with direction of pH 7→pH 12→pH 2→pH 7 in 10-minute step. Table 1 provides the parameter of Si_3N_4-gate ISFET for a statistic group of 30 ISFET sensors. The width of hysteresis is the difference in the ISFET response at pH 6.0. The results present a good stability and good reproducibility of ISFET used. Based on the measurements, we conclude with the ISFET specifications in Table 2.

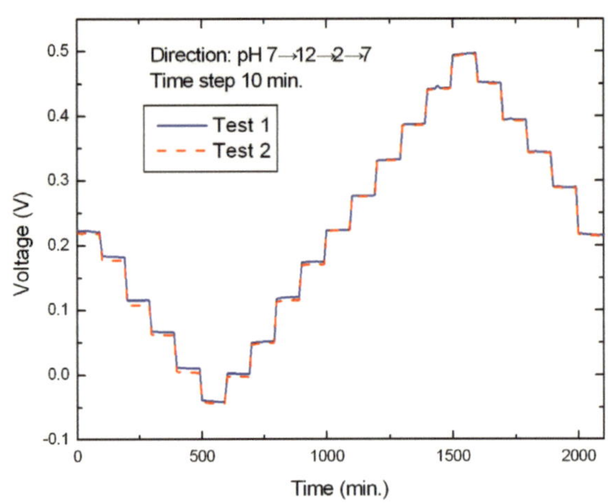

Fig. 11. Time response of reproducibility

Parameter	Range	Mean value	Standard deviation
Sensitivity (mV/pH)	-45.86~-47.12	-46.34	0.010
Correlation	0.996~0.9988	-	-
Drift (mV/hr)	-0.52~-1.71	-1.10	0.52
Hysteresis (mV)	9.22~13.0	11.1	2.68

Table 1. Parameter of Si_3N_4 gate ISFET

Item	Range
Bias condition	0.5V, 100μA
Temperature	5 ~40°C
Aspect ratio	600μm/15μm
Temperature coefficient	-0.9±0.5mV/°C
Drift coefficient	-1±0.5mV/hr
Hystersis	10±5mV
Sensitivity	-46±1mV/pH

Table 2. Specification of ISFET macro model

The ISFET drift data at room temperature was collected using the designed bridge-type readout circuit. The ISFET gate was applied using a commercial Ag/AgCl reference electrode immersed in a pH 7 buffer solution. Drift measurements and model parameter extraction were performed on a total of eight ISFET devices. The known and extracted parameters obtained on four devices, are given in Table 3. The modeled-versus-measured fit for an 18-hour time period depicted in Fig. 12 is characterized by an RMS error of 2.2%.

Parameter	Extracted value
fm	37.69mV
sm	33.28mV
dm	25.36mV
C_{T2}	60
τf	0.01 hour
τs	2 hour
τ	16 hour

Table 3. Extracted parameters for drift of Si_3N_4 ISFET

Dealing with the correlation of drift rate at different pH buffer solution, the measurement was done over a pH range of 2 to 12. The dependence of drift response on pH value between modeled and measured fit is shown in Fig. 13. The RMS error is 2.4%. The results show that the drift rate becomes larger as pH rises. Fig. 14 presents the measured drift rate at pH 7 as temperature varies from 5°C to 35°C. The temperature dependence on drift rate indicates an RMS error of 6.6% between modeled and measured fit.

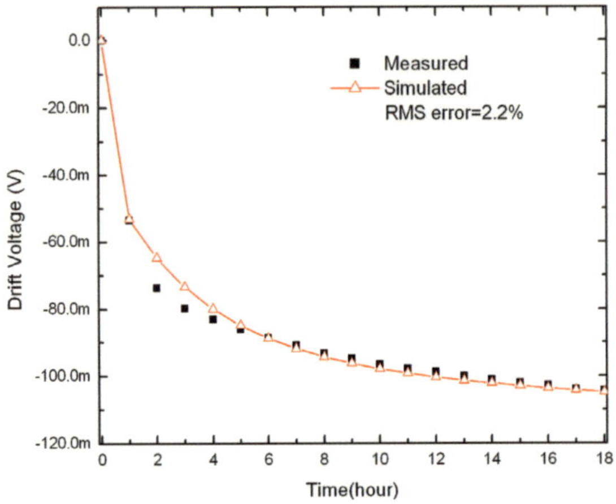

Fig. 12. ISFET drift characteristics at pH=7, 25°C

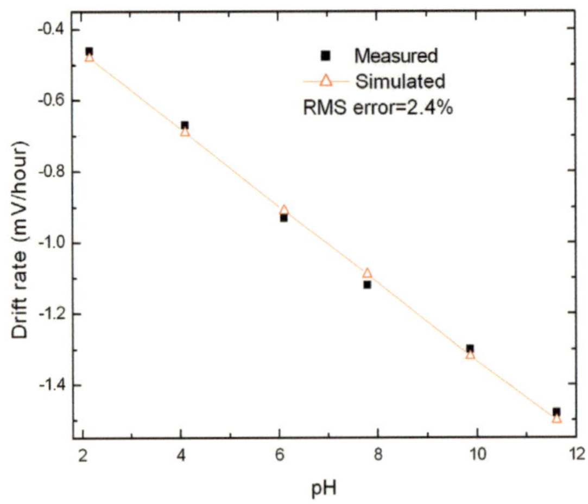

Fig. 13. Dependence of drift response on pH value

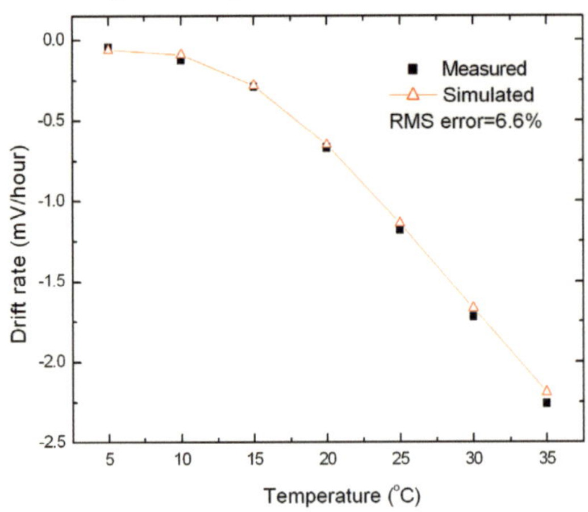

Fig. 14. Temperature dependence of drift rate

In conclusion, an HSPICE-compatible macro model that considers the drift response of pH ISFET has been presented. The modeled-versus-measured fit of the dependence of drift rate with long time period, temperature and pH variations present an RMS error of 2.2%, 2.4% and 6.6% respectively. The developed macro model can be adapted to speed up the design of silicon-based chemical- and bio-sensor micro-systems.

4. Temperature dependency of ISFET and its compensation

We developed a readout circuit that improved the performance of ISFET against non-ideal effects such as temperature dependency, time drift and hysteresis. The design concerns were also inspired by real application requirements for simpler and lower power consumption of the sensing system. Based on the concept of threshold voltage extractor (Wang, 1992), shown in Fig. 15, we developed a new ion sensing and interfacing circuitry with temperature compensation.

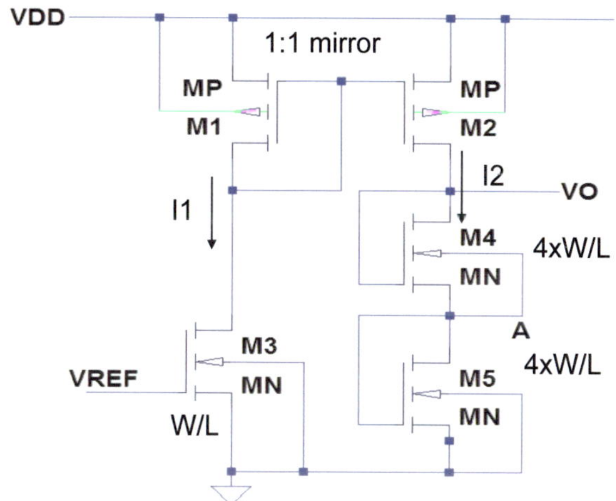

Fig. 15. Four-terminal extractor circuit (Wang, 1992)

With a 1:1 current mirror in M1 and M2, with specific W/L sizes of M3, M4 and M5, and with I_{DM3} = $I1$ = I_{DM4} = I_{DM5} = $I2$, the output voltage Vo of extractor circuit yields the threshold potential V_{TH} as expressed in equations (4.1) to (4.4).

$$I_{DM3} = \frac{1}{2}\mu_n C_{ox}\frac{W}{L}\left(V_{REF} - V_{TH}\right)^2 \tag{4.1}$$

$$I_{DM4} = (4)\frac{1}{2}\mu_n C_{ox}\frac{W}{L}\left(\frac{Vo}{2} - V_{TH}\right)^2 \tag{4.2}$$

$$\frac{1}{2}\mu_n C_{ox}\frac{W}{L}(V_{REF} - V_{TH})^2 = (4)\frac{1}{2}\mu_n C_{ox}\frac{W}{L}(\frac{Vo}{2} - V_{TH})^2 \tag{4.3}$$

$$V_o = V_{REF} - V_{TH} \tag{4.4}$$

Taking advantage of compatible CMOS process, the sensors that include ISFET and depletion-type Al-gate DMOSFET devices DM1 and DM2 were fabricated on the same wafer. Fig. 16 gives the complete design of a novel readout circuit of ISFET with temperature compensation based on the V_{TH} extractor circuits. It consists of blocks (a), (b), and (c). Block (a) is the ISFET pH sensing circuit. The gate and the drain of ISFET are connected together to make sure that it operates in saturation. The saturation drain-source current is set at 50μA because of the stable characteristics of ISFET in this value. With a fix drain-source current, the gate-source voltage of ISFET varies directly with its threshold voltage in the saturation region according to equation (4.1). As a result, the voltage V_{ISFET} is proportional with pH levels. However, the voltage V_{ISFET} exhibits temperature dependency with $V_{ISFET} = V_{pH} + V_{TEMP}$. Block (b) is the DMOSFET temperature sensing circuit. The voltage V_{TEMP} is proportional to the temperature value in DMOSFET. Block (c) is the output differential stage formed by M12 and M17 and served as subtractor circuit. The common temperature dependency of ISFET and DMOSFET is cancelled, and therefore the output voltage *VCOMP* is temperature compensated and is equal only to V_{pH}. Simulation results in Fig. 17(a) illustrate the voltage V_{ISFET} without temperature compensation and in Fig. 17(b) show the voltage *VCOMP* with temperature compensation.

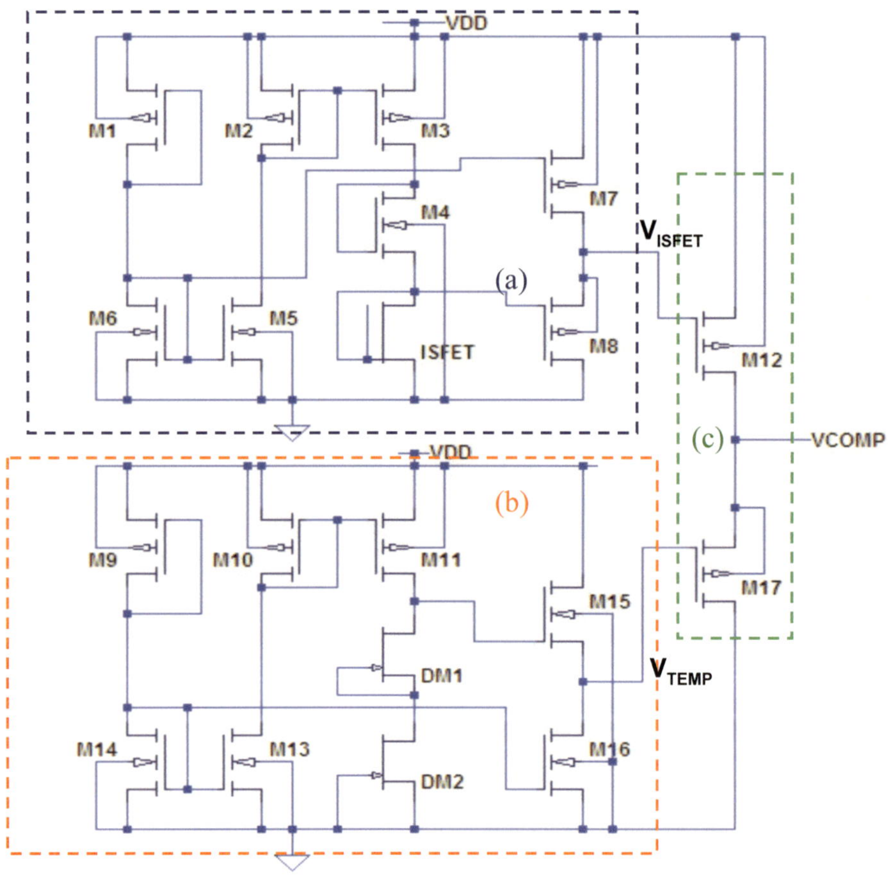

Fig. 16. V_{TH} extractor based readout circuit with temperature compensation

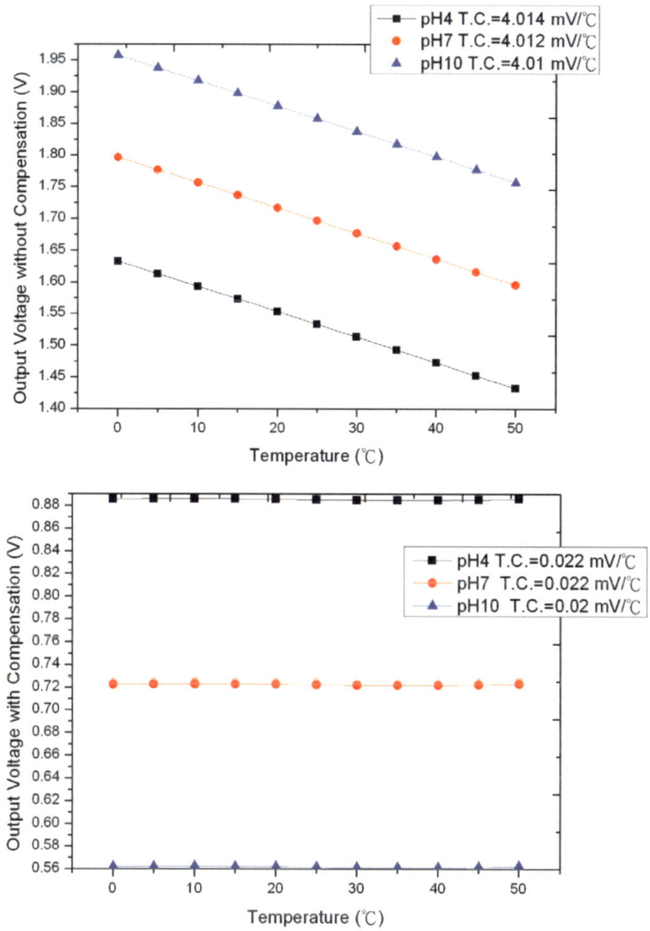

Fig. 17. (a) Output voltage without compensation, (b) output voltage with compensation

5. Body effect and its reduction technique

We have discussed the readout circuit techniques to improve the performance of ISFET over its inherent and non-ideal characteristics such as temperature-dependency and long-term time drift effects. To expand the benefits of most CMOS standard technologies, recent works have projected to integrate ISFET and interface electronics on the same chip (Wong & White, 1989; Ravczzi & Conci, 1998; Bausells et al, 1999; Palan et al, 1999; Chin et al, 2001). Because of low drift and high mobility properties of carriers, the n-channel ISFET devices are generally used. In most of current CMOS processes, the NMOS device is fabricated into a p-type substrate that is globally and constantly grounded to the most negative supply in the system. Thus, the above-mentioned interface circuits suffer from the problem where the substrate potential greatly influences the device characteristics in ISFET-based integrations. In 2004 Morgenshtein et al. presented a novel technique, which allows body effect elimination of readout interface in CMOS ISFET-based micro-systems (Morgenshtein et al, 2004). However, in this case a portion of the architecture of the ISFET does not have a constant current and voltage bias.

This section presents our approach of enhancing the accuracy of ISFET measurements using a body effect reduction technique while maintaining constant drain-source voltage and current. With a differential configuration of amplifier circuit, this design technique generates an output signal independent of temperature and long-term drift. In addition, a voltage-controlled DC offset error compensation circuit modulates the extracted signal to the desired DC level for the A/D converter for each sensor. Simulation and experimental results demonstrate the effectiveness of the instrumentation system for monolithic ISFET integration in CMOS technology.

5.1 Body effect in ISFETs

In Complementary MOS (CMOS) technologies, both NMOS and PMOS transistors must be fabricated into the same "local substrate". In current CMOS processes, the NMOS device is fabricated into a p-type substrate that is why the substrate potential greatly influences the device characteristics. Usually the substrate of NMOS transistors is connected to the most negative supply in the system. Thus, in typical MOS operations the S/D junction diodes must be reverse-biased. It can be proved that with body effect, the threshold voltage is expressed as (5.1) and (5.2):

$$V_{TH} = V_{TH0} + \gamma(\sqrt{|2\phi_F + V_{SB}|} - \sqrt{|2\phi_F|}) \tag{5.1}$$

$$V_{TH0} = \phi_{MS} + 2\phi_F + \frac{Q_{dep}}{C_{ox}} \tag{5.2}$$

Where V_{TH0} is the threshold voltage when there is no body effect, ϕ_F is the Fermi potential, γ denotes the body effect coefficient, and V_{SB} is the source-bulk potential difference.

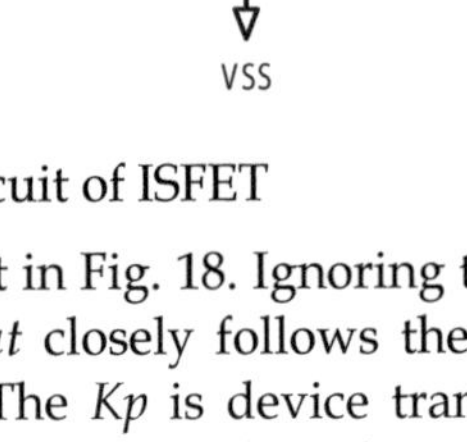

Fig. 18. Source-follower readout circuit of ISFET

Consider the source-follower circuit in Fig. 18. Ignoring the body effect, as the input voltage *Vin* varies, the output voltage *Vout* closely follows the input variation because the drain current remains equal to the I_{DS}. The *Kp* is device transconductance factor, V_{DS} is drain-source voltage, and $V_{TH}^* = V_{TH} - EPH$ is threshold voltage of ISFET from threshold voltage of FET V_{TH} and from interface potential *EPH* between sensing membrane and buffer solution.

$$\begin{aligned} I_{DS} &= K_p[(V_{GS} - V_{TH}^*) - \frac{V_{DS}}{2}]V_{DS} \\ &= K_p[(V_{in} - V_{out} - V_{TH}^*) - \frac{V_{DS}}{2}]V_{DS} \end{aligned} \tag{5.3}$$

Suppose the substrate is connected to the most negative supply and the body effect is significant in Fig. 18 circuit. As *Vin* increases, *Vout* becomes more positive and the potential difference between the source and the bulk increases. This condition raises the V_{TH} in ISFET. Therefore, the equation (5.3) implies that V_{in}-V_{out} must increase to maintain a constant I_{DS}. Moreover, a non-zero V_{SB} contributes a parasitic change in V_{TH}* that is not due to variation of ion concentration.

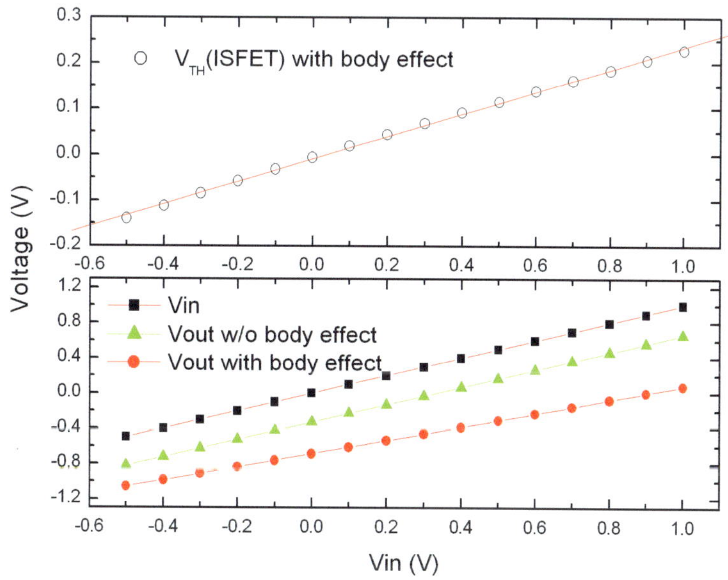

Fig. 19. Vin and Vout of source-follower readout, with and without (w/o) body effect.

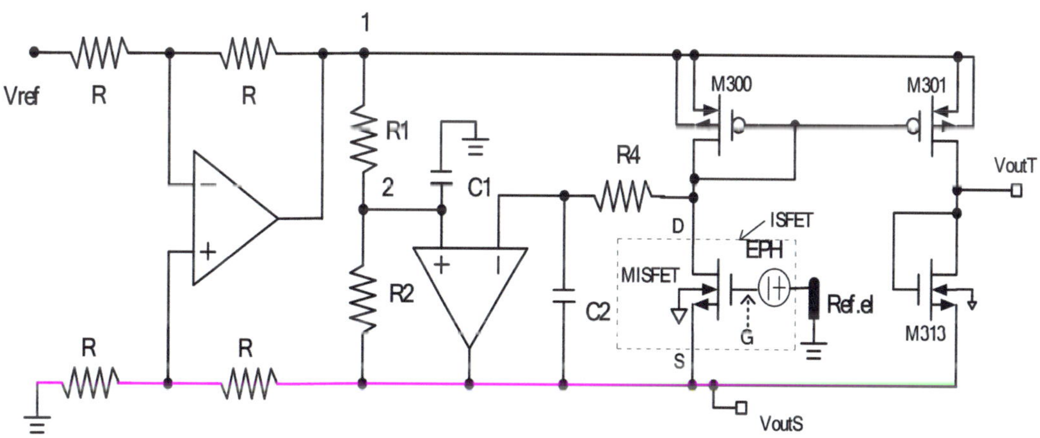

Fig. 20. Schematic diagram of the body-effect reduction readout interface

In order to reduce the body effect in ISFET, the circuitry described in Fig. 5 was modified with a current mirror as shown in Fig. 20. The two devices, MISFET and M313, carry equal drain currents under the influence of body effect (Wade & Tadokoro, 2002). To simplify the analysis, we employ MOSFET in the saturation region. Hence,

$$I_{DS,misfet} = I_{DS,m313}$$
$$\Rightarrow K_p(V_{GS,misfet} - V_{TH,misfet})^2 = K_p(V_{GS,m313} - V_{TH,m313})^2 \quad (5.4)$$

Assume that MISFET and M313 are matched in the same p-type substrate, the extracted signal *VoutT* is equal to the electrolyte-insulator interface potential *EPH* and is independent of the body effect.

$$V_{GS,misfet} - V_{GS,m313} = V_{TH,misfet} - V_{TH,m313}$$
$$\Rightarrow -EPH - VoutS - (VoutT - VoutS)$$
$$= V_{TH0,misfet} + \gamma(\sqrt{2\phi_f + V_{SB,misfet}} - \sqrt{2\phi_f}) - [V_{TH0,m313} + \gamma(\sqrt{2\phi_f + V_{SB,m313}} - \sqrt{2\phi_f})] \quad (5.5)$$
$$\Rightarrow VoutT = -EPH$$

In general, single ISFET interface circuits do not offer any degree of compensation for temperature dependency or long-term drift. The body-effect reduction readout interface in Fig. 20 accompanies a performance enhancement circuit in Fig. 21. Assume that resistances R are perfectly matched with one another and that op amp have infinite CMRR in the differential amplifier circuit. With *VoutS* and *VoutT* having equal temperature dependency, equal long-term drift as well as common noise, the output signal *VoutU* in (5.6) becomes unaffected by temperature, long-term drift and common noise.

$$VoutU = VoutT - VoutS \quad (5.6)$$

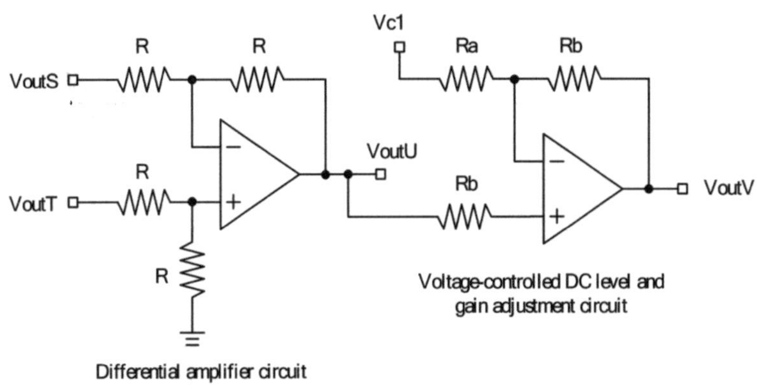

Fig. 21. Differential circuit used for ISFET performance enhancement

In (5.7) is the output signal *VoutV* of voltage-controlled DC offset error compensation and gain adjustment circuit. The first term generates an output signal that is independent from thermal and long-term drift effects, while the second term modulates the extracted signal to the desired DC level for the A/D converter of each sensor.

$$VoutV = (1 + \frac{R_b}{R_a})VoutU - \frac{R_b}{R_a}V_{c1} \quad (5.7)$$

5.2 On-chip circuit implementation and results

A photomicrograph of the realized bridge-type ISFET readout circuit with band-gap reference voltage generator in Fig. 22 was fabricated in Taiwan Semiconductor Manufacturing Company using TSMC 0.35 μm CMOS technology. (BFDSF stands for bridge-type floating drain source follower). The core die size is around 963 × 892 μm². Fig.23 shows the photomicrograph of the total body-effect reduction and performance enhanced circuitry (9.1 x 9.1 mm²).

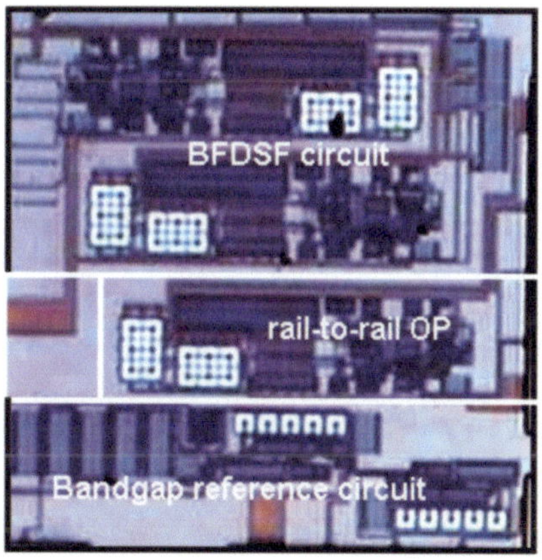

Fig. 22. Photomicrograph of the realized BFDSF circuit (963 × 892 μm²)

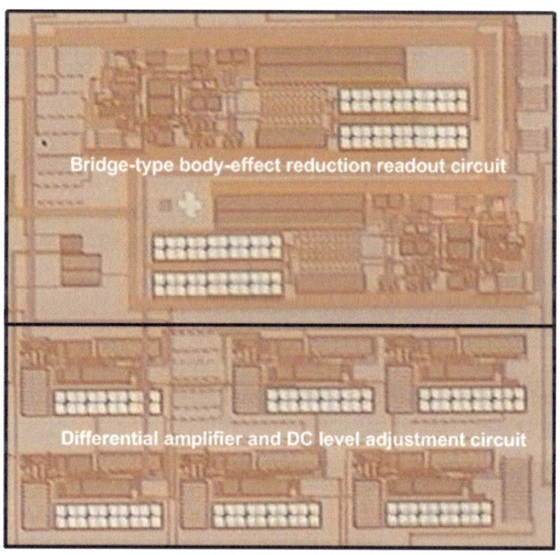

Fig. 23. Photomicrograph of the total body-effect reduction and performance enhanced circuitry (9.1 x 9.1 mm²)

Table 4 depicts the measured bias voltage and current for Al_2O_3-gate and Si_3N_4-gate ISFET sensors operated from pH 2 to pH 12. The inaccuracy is only 0.2% in the drain current and only 0.006% in the drain-source voltage, and is attributed to the process variation on circuit resistors. The results show very small variations in the ISFET bias voltage and current and prove that the readout circuit maintains a stable operating point with different pH value and different sensor.

Both	Al_2O_3 ISFET		Si_3N_4 ISFET	
pH	Ids(μA)	Vds(V)	Ids(μA)	Vds(V)
2.001	100.2	0.505	100.2	0.505
3.007	100.2	0.505	100.2	0.505
4.007	100.2	0.505	100.2	0.505
5.001	100.2	0.505	100.2	0.505
6.006	100.2	0.505	100.2	0.505
7.006	100.2	0.505	100.2	0.505
8.015	100.2	0.505	100.2	0.506
8.999	100.2	0.505	100.2	0.506
10.035	100.2	0.505	100.2	0.506
11.011	100.2	0.505	100.2	0.506
12.050	100.1	0.505	100.2	0.506

Table 4. Measurement of bias voltage and current for Al_2O_3-gate and Si_3N_4-gate ISFET sensors operating in buffer solutions of different pH

Fig. 24 show the potentials of four terminals in the body-effect reduction circuit over pH range of 2 to 12, namely *VoutS*, *VoutT*, *VoutU* and *VoutV*. Refer to Fig. 20 and Fig. 21. The bulk of ISFET was connected to the most negative supply with VSS=-1.65V. The calculated slopes for graphs (a), (b), (c), and (d) are -40.06mV/pH, -48.43mV/pH, -8.22mV/pH and -50.68mV/pH for terminals *VoutS*, *VoutT*, *VoutU* and *VoutV* respectively. The increase of slope for curve (b) compared to curve (a) demonstrates the improvement that resulted from the reduction of body effect. The experimental data presented here correlates well with the simulation data presented in Fig.19.

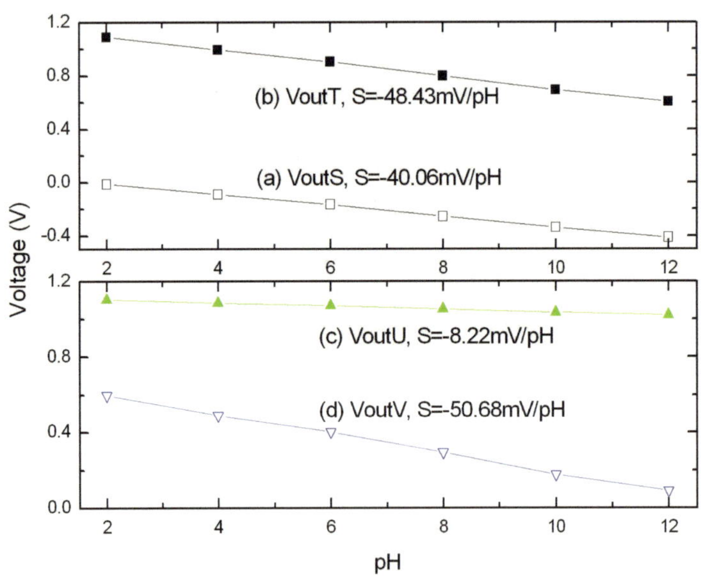

Fig. 24. pH response of ISFET operating in the body-effect reduction circuit with the ISFET bulk connected to VSS

5.3 On-board prototyping

Considering the practical applications, Fig. 25 and Fig. 26 give the system diagram and initial prototype of pH meter using separate on-board modules for the readout circuit and the microcontroller unit (MCU). The calibration and measurement routines are coded inside the MPC82G516A MCU. The experimental readings of this pH meter prototype agree with that of commercial ISFET pH meter KS701 (Shindengen Co., Japan) and measures from pH2 to pH12 with 0.1 pH resolution.

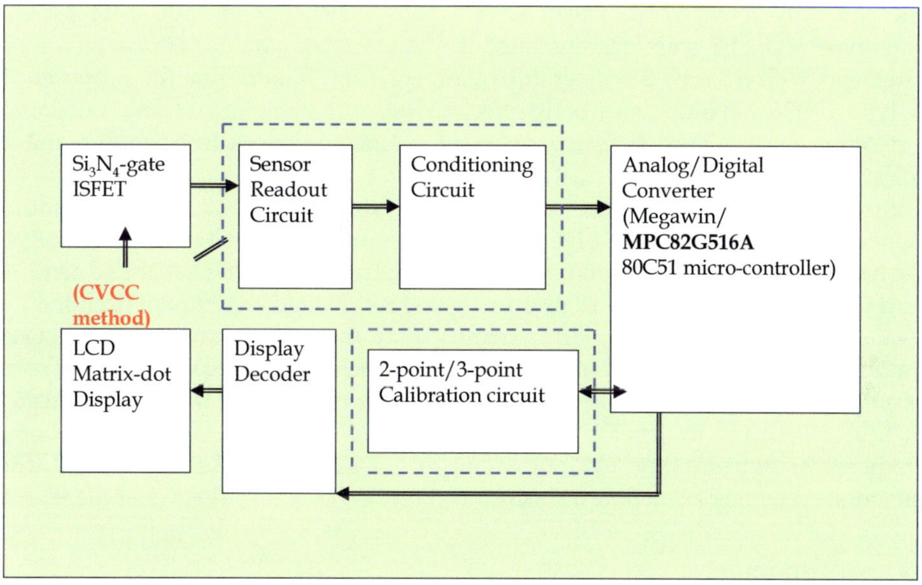

Fig. 25. System block diagram of a prototype of pH meter

Fig. 26. PCB-based hardware implementation of a pH-meter prototype

6. Conclusion and future works

This chapter explored the characteristics and the non-ideal parameters of ISFET that were important to the practical and long-term sensing applications of ISFET. This chapter also presented series of improved readout circuit techniques that enhanced the performance of ISFET and demonstrated the pH sensing capability of ISFET for environmental monitoring. The SPICE-based drift model of ISFET developed in this chapter can be used for further ISFET-based sensor interface circuit designs. With the advantage of compatible CMOS process and only fewer mask steps, sensor pairs consisting of Si_3N_4-gate ISFET and depletion-type MOSFET were demonstrated in V_{TH} extractor circuit that provided sensitive measurements with improved temperature compensation. In addition, the proposed ISFET bridge-type CVCC circuitry with body-effect reduction technique not only enhanced the noise rejection performance but also removed the interferences from source and drain terminals.

For future works, the multi-ion sensing based on ISFET sensor arrays and their corresponding signal processing algorithms such as independent component analysis or blind source separation will be continuously studied. In addition, the integrated sensors in a standard CMOS process will be further investigated for diversified field applications.

In conclusion, CMOS technology and circuitry play more important roles on biosensor applications especially in the field of sensor interface design and development. The response of biosensor can be potential, current and impedance changes. Thus, the systematic and hierarchical approaches to develop more advanced electronic tongue using potentiometric, amperometric or impedimetric readout circuit techniques should be emphasized through the collaboration among academic, industrial and research organizations over the world.

7. Acknowledgement

This work was supported by the National Science Council, Republic of China, under multi-year contracts of NSC: 98-2221-E033-050; and the long-term, bilateral Polish-Taiwanese joint research projects with the Institute of Biocybernetics and Biomedical Engineering, Polish Academy of Sciences and the Institute of Electron Technology, Poland. Moreover, the authors would also like to thank the National Chip Implementation Center (CIC), and Taiwan Semiconductor Manufacturing Company (TSMC), Taiwan for technical support and chip fabrication service.

8. References

Bausells, J.; Carrabina, J.; Errachid, A.; Merlos, A. (1999) Ion-sensitive field effect transistors fabricated in a commercial CMOS technology, *Sensors and Actuators B*, Vol. 57, 1999 56-62

Bergveld, P. (1970). Development of an ion sensitive solid-state device for neurophysiological measurement, *IEEE Trans. Biomed. Eng.*, Vol. 17, 1970, 70-71

Bergveld, P. (1991). Future applications of ISFETs, *Sensors and Actuators B*, Vol. 4, 1990, 125-133

Bergveld, P. (2003). Thirty years of ISFETOLOGY-What happened in the past 30 years and what may happen in the next 30 years, *Sensors and Actuators B*, Vol. 88, 2003, 1-20

Bousse, L.; Hafeman, D.; Tran, N. (1990). Time-dependence of the chemical response of silicon nitride surface, *Sensors and Actuators B*, Vol. 1, 1990, 361-367

Chin, Y.; Chou, J.; Sun, T. ; Chung, W. ; Hsiung, S. (2001). A novel pH sensitive ISFET with on chip temperature sensing using CMOS standard process, *Sensors and Actuators B*, Vol. 76, 2001, 582-593

Chou, J. & Hsiao, C. (2000). Drift behavior of ISFETs with a-Si: H-SiO_2 gate insulator, *Materials Chemistry and Physics*, Vol. 63, 2000, 270-273

Chung, W.; Yang, C.; Pijanowska, G.; Krzyskow, A. (2004). ISFET interface circuit embedded with noise rejection capability, *Electronics Letters*, Vol. 40, 2004, 1115-1116

Chung, W.; Chang, K.; Hong, D.; Cheng, C.; Cruz, F.; Liu, T.; Yang, C.; Chiang, J.; Pijanowska, G.; Dawgul, M.; Torbicz, W.; Grabiec, P.; Jarosewicz, B. (2008). An electronic tongue system design using ion sensitive field effect transistors and their interfacing circuit techniques, *Proceedings of the 17th Biennial University/Government /Industry Micro-Nano Symposium*, pp. 44-48, ISBN, 978-1-4244-2484-9, Louisville, KY, USA, July, 2008, IEEE, Louisville, KY

Ciosek, P. & Wroblewski, W. (2007). Sensor array for liquid sensing - electronic tongue systems, *Analyst*, Vol. 132, 2007, 963-978

Garde, A.; Alderman, J. & Lane, W. (1995). Development of a pH-sensitive ISFET suitable for fabrication in a volume production environment, *Sensors and Actuators B*, Vol. 27, 1995, 341-344

Jamasb, S.; Collins, S.; Smith, R. (2000). A physical model for drift in pH ISFETs, *Sensors and Actuators B*, Vol. 49, 2000, 146-155

Kuhnhold, R. & Ryssel, H. (2000). Modeling the pH response of silicon nitride ISFET devices, *Sensors and Actuators B*, Vol. 68, 2000, 307-312

Lauwers, E.; Suls, J.; Gumbrecht, W.; Maes, D.; Gielen, G.; Sansen, W. (2001). A CMOS multiparameter biochemical microsensor with temperature control and signal interfacing, *IEEE J. Solid State Circuits*, Vol. 36, 2001, 2030-2038

Liao, H. (2000). Novel calibration and compensation technique of circuit design for Biosensor, *Master Thesis*, Chung Yuan Christian University, Chung-Li, Taiwan, 2000.

Martinoia, S.; Lorenzelli, L.; Massobrio, G.; Margesin, B.; Lui, A. (1999) A CAD system for developing chemical sensor-based microsystems with an ISFET-CMOS compatible technology, *Sensors and Materials*, Vol. 11, 1999, 32-49

Martinoia, S. & Massobrio, G. (2000). A behavioral macromodel of the ISFET in SPICE, *Sensors and Actuators B*, Vol. 62, 2000, 182-189.

Morgenshtein, A.; Sudakov-Boreysha, L.; Dinnar, U.; Jakobson, C.; Nemirovsky, Y. (2004) CMOS readout circuitry for ISFET microsystems, *Sensors and Actuators B*, Vol. 97 2004, 122-131

Palan, B.; Santos, F.; Courtois, B.; Husak, M.; (1999) Fundamental noise limits of ISFET-based microsystems, *Eurosensors*, Vol. 13, 1999, 169-172

Ravezzi, L. & Conci, P. (1998) ISFET sensor coupled with CMOS read-out circuit microsystem, *Electron Letters*, Vol. 34 , 1998, 2234-2235

Wada, K. & Tadokoro, Y. (2002) Design of a body-effect reduced-source follower and its application to linearization technique, *Proceedings of IEEE Int. Symposium on Circuits and Systems*, Vol. 3, 2002, 723–726

Wong, H. & White, H. (1989) A CMOS -integrated ISFET operational amplifier, chemical sensor employing differential sensing, IEEE Trans. Electron Devices, Vol. 36, 1989 479-487

Yin, L.; Chou, J.; Chung, W. ; Sun, T. ; Hsiung, S. (2001). Study of indium tin oxide thin film for separate extended gate ISFET, *Materials Chemistry and Physics,* Vol. 70, 2001, 12-16

22

Low-temperature Polymer Bonding Using Surface Hydrophilic Treatment for Chemical/bio Microchips

Hidetoshi Shinohara, Jun Mizuno and Shuichi Shoji
Major in Nano-science and Nano-engineering, Waseda University
Japan

1. Introduction

Polymer materials have been used for electronic, optical and bio micro/nano devices. Polymer device fabrication technologies based on replication methods including hot embossing (Becker & Heim, 2000; Park et al., 2003; Shinohara et al., 2007b), injection molding (Becker et al., 1986; Svedberg et al., 2003), ultraviolet (UV) imprinting (Haisma et al., 1996; Kawaguchi et al., 2007; Shinohara et al., 2008d) and casting (Duffy et al., 1998; Slentz et al., 2001) can reduce costs. Polymer bonding technologies have also been required for sealing or stacking the devices. Some examples of bonding methods have been reported, including thermal direct bonding (Spierings & Haisma, 1994; Chen et al., 2004; Shinohara et al., 2007b), solvent bonding (Wang et al., 2002; Lin et al., 2007), and bonding using other intermediate layer (Graß et al., 2001; Lei et al., 2004). Low-temperature bonding technologies are required with deformation of the previous surface structures as small as possible.
On the other hand, surface modification for biocompatibility is one of the most important processes for biochips. Polymer surface modification methods are classified into two categories. One is modification of the original surface (e.g., plasma treatment (Lianos et al., 1994; Kamińska et al., 2002; Chai et al., 2004, Lai et al., 2006), UV irradiation (Peeling & Clark, 1981; Murakami et al., 2003; Hozumi et al., 2004; Diaz-Quijada et al., 2007; Kim et al., 2009). The other is coating with other materials (Ratner, 1995; Oehr, 2003; Liu et al., 2004; Bi et al., 2006).
In this chapter, two low-temperature bonding technologies are described. Section 2 introduces low-temperature direct bonding methods of poly (methyl methacrylate) (PMMA) or cyclo-olefin polymer (COP), and their applications of microchannel devices. Section 3 describes surface hydrophilic treatment method using aromatic polyurea film, and bonding method using the polyurea film.

2. Low-temperature direct bonding of PMMA or COP

2.1 Surface pretreatment for low-temperature bonding

In our previous study, we developed a fabrication method for micro-scale flow devices by combining hot embossing and direct bonding techniques using a PMMA material. Direct bonding is superior to polymerize bonding or adhesive bonding because of its low optical loss in a bonded interface (Shinohara et al., 2007b). In this method, we fabricated flow channels around the glass transition temperature (T_g) of the material. Because of the applied pressure as well as heat during the direct bonding process, deformation of the channel was observed, although it was not a big problem in cell analysis. However, for single bio-molecule level analysis, which uses high-performance optical detection systems, high optical transparency of the material and nanometer-scale accuracy of the fabrication technologies are required.

In order to bond at lower than T_g, surface pretreatment was applied. Fig. 1 shows fabrication process of a polymer microchip using low-temperature direct bonding. First, silicon mold was fabricated by conventional photolithography and Deep-RIE (reactive ion etching) (Fig. 1 (a)). Microchannel patterns were formed by hot embossing (Fig. 1 (b)) (Shinohara et al., 2007b). After the microchannel plate and a lid were pretreated (Fig. 1 (c)), the microchannel was realized by the direct bonding (Fig. 1 (d)).

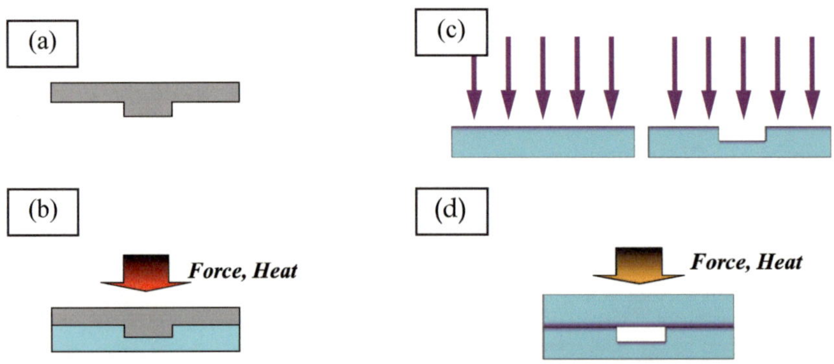

Fig. 1. Fabrication process of polymer microchip using low-temperature direct bonding (Shinohara et al., 2007a)

Examples of typical pretreatment methods are oxygen plasma, atmospheric-pressure oxygen plasma, UV/O_3, and VUV (vacuum UV) /O_3. Typical treatment conditions of the equipments were shown in Table 1.

Oxygen plasma was generated in a plasma activated bonding system (EVG810LT from EV Group Co.). Oxygen plasma can be generated between parallel electrodes in the vacuum chamber. Since the radiofrequency (397 kHz) was lower than that of other conventional plasma treatment systems (13.56 MHz or higher), the damage on the surfaces was expected to be smaller. Atmospheric-pressure oxygen plasma was generated by plasma cleaning unit (Aiplasma from Panasonic Electric Works, Ltd.), using dielectric-barrier discharge (Sawada, 2003). In this equipment, high-density active plasma can be expelled from a nozzle supplying mixed gas (98 % Ar and 2 % O_2) under atmospheric pressure. After oxygen plasma irradiation, the molecular bonds (e.g. C-H) on the polymer surface are expected to be dissociated and incorporated oxygen radicals. Polar oxidized components were increased because of the incorporation (Lianos et al., 1994; Chai et al., 2004). This surface state is considered to enhance the bonding reaction at the interface.

Condition	Oxygen plasma	Atmospheric plasma	UV/O_3	VUV/O_3
Gas	O_2	Ar 98%, O_2 2%	O_2	O_2
Power (W)	200	80	-	-
UV wavelength (nm)	-	-	185, 254	172
Chamber pressure (p) (MPa)	8.0×10^{-5}	0.1	0.1	5.0×10^{-2}
Exposure time (t)	30 sec	0.6 sec	20 min	30 min

Table 1. Typical treatment conditions of oxygen plasma, atmospheric-pressure oxygen plasma, UV/O_3, and VUV/O_3 (Shinohara et al., 2007a)

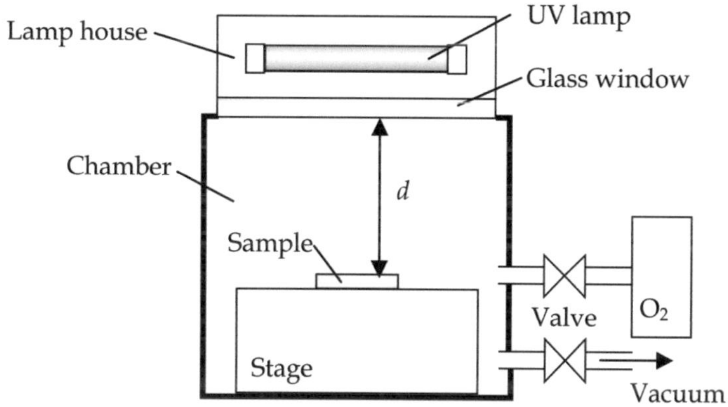

Fig. 2. Schematic diagram of VUV/O_3 equipment (Shinohara et al., 2008b)

The UV/O_3 system (NL-UV253 from Nippon Laser & Electronics Lab.,) has three low-pressure UV lamps that radiate 185 nm and 254 nm lights in wavelength. In the presence of O_2, the 185-nm UV is absorbed by O_2 to generate the atomic species in ground state O(^{3}P). O(^{3}P) can react with O_2 to form O_3. If this O_3 absorbs the 254-nm UV, excited oxygen atoms (O(^{1}D)) with 190 kJ/mol excitation energy are generated (Wang & Ray, 2000). The VUV/O_3 system (UER20-172 from Ushio Inc.) has a dielectric barrier discharge excimer lamp filled with Xe gas and radiates light of a central wavelength of 172 nm (VUV). The VUV/O_3 system is shown in Fig. 2. Oxygen gas was introduced into the chamber after evacuation. The VUV generates not only O_3 and O(^{1}D) in the same manner as the 185-nm and 254-nm UV lights, but is also absorbed directly by O_2 in the chamber to generate O(^{1}D) (Kaspar et al., 2003). The 172-nm UV light irradiance on the sample surface can be controlled by the oxygen pressure and the distance between the lamp window and the sample (d) (Hozumi et al., 2004; Shinohara et al., 2008b). In UV (VUV)/O_3 treatment, O(^{1}D) plays important roles on surface activation (Hozumi et al., 2004). Polar oxidized components were also increased as well as the oxygen plasma treatments (Peeling & Clark, 1981; Diaz-Quijada et al., 2007; Kim et al., 2009). Since absorption coefficient of O_2 at the 172-nm UV light are approximately 20 times greater than that at the 185-nm

(Watanabe et al., 1953), the efficiency of O(¹D) generated by VUV/O_3 treatment is better than that by UV/O_3. Thus, it is expected that the activation by the VUV/O_3 is more effective than that by UV/O_3. In addition, the UV light is expected to dissociate chemical bonds of polymer as C-C, C-O and C-H. Main or side chain cleavage of the polymer causes degradation of polymer so as to generate low-T_g layer on the surface (Truckenmüller et al., 2004). It is considered to be act as an adhesion layer for the direct bonding.

2.2 Bonding strength

Bonding strengths of PMMA plates (Acrylyte E IR from Mitsubishi Rayon Co., Ltd.) were measured by a tensile test method (Shinohara et al., 2007a). The results were shown in Fig. 3. In this figure, red broken lines indicate the values for direct bonding under temperature of 95 °C, pressure of 1.25 MPa and annealing time of 25 min, without any surface treatments. The bonding strengths were same or stronger than that bonded around T_g.

Bonding strengths of oxygen plasma-treated COP plates (Zeonex480 from Zeon Co.) measured by the tensile test were higher than 1 MPa. Bulk distraction was observed from the bonded sample after tensile test while no interface separation was observed. The bonding strengths of pretreated COP samples were also measured by razor blade method (Maszara et al., 1988). The bonding strength at room temperature was approximately 0.6 J/m^2. The strength was increased (~ 8 J/m^2) after annealing at 70 °C (Mizuno et al., 2005a).

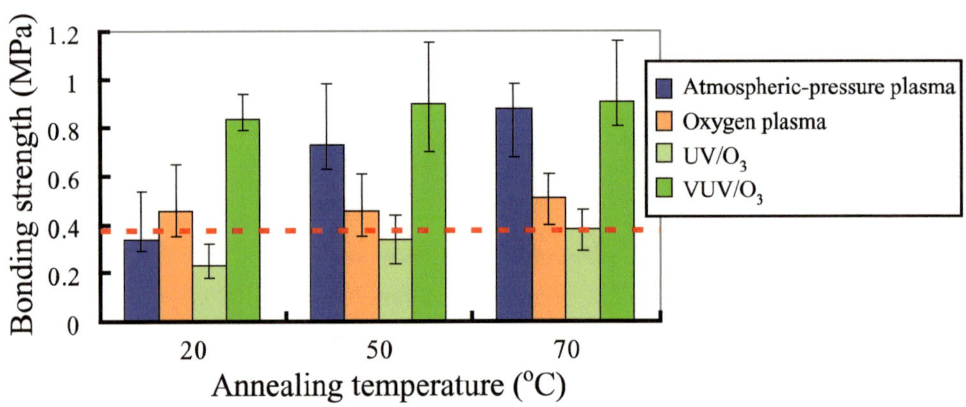

Fig. 3. Dependence of bonding strength of two PMMA plates on the annealing temperature (Shinohara et al., 2007a)

2.3 Shallow microchannel

A PMMA microchip which have fine channel of 5 μm in depth and 150 μm in width was fabricated by low-temperature direct bonding (bonding temperature of 75 °C) as shown in Fig. 4. (Shinohara et al., 2007a). The shallow microchannel was successfully fabricated without deformation, boids and leakages. To controlled conditions of surface treatment and bonding, the shallow microchannel can be also realized using COP materials (Shinohara et al., 2009b).

Fig. 5 shows a PMMA microchip which has two shallow dams of about 5 μm gaps (Shinohara et al., 2006). The dam structures were kept after low-temperature bonding. The flow behaviors of the dams were evaluated with fluorescent beads. Large microbeads (diameter: 5.7 μm) were completely trapped and filled between two dams, while small microbeads (diameter: 1.0 μm) were passed through the dams, as shown in Fig. 5 (c).

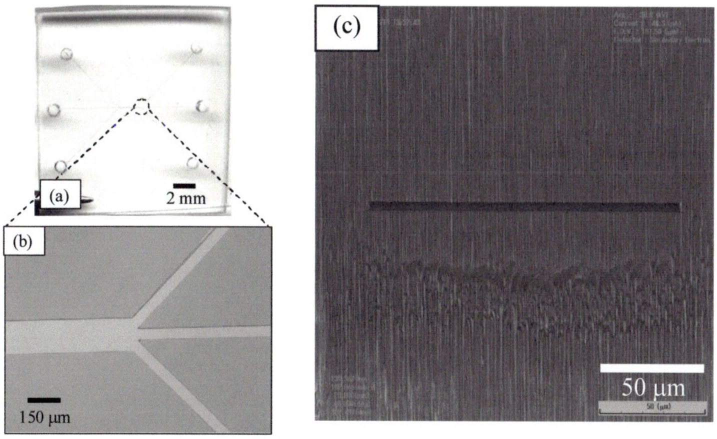

Fig. 4. A shallow PMMA microchip: (a) whole and (b) magnified view; (c) cross-section of a shallow microchannel (width: 150 μm, depth: 5 μm) (Shinohara et al., 2007a)

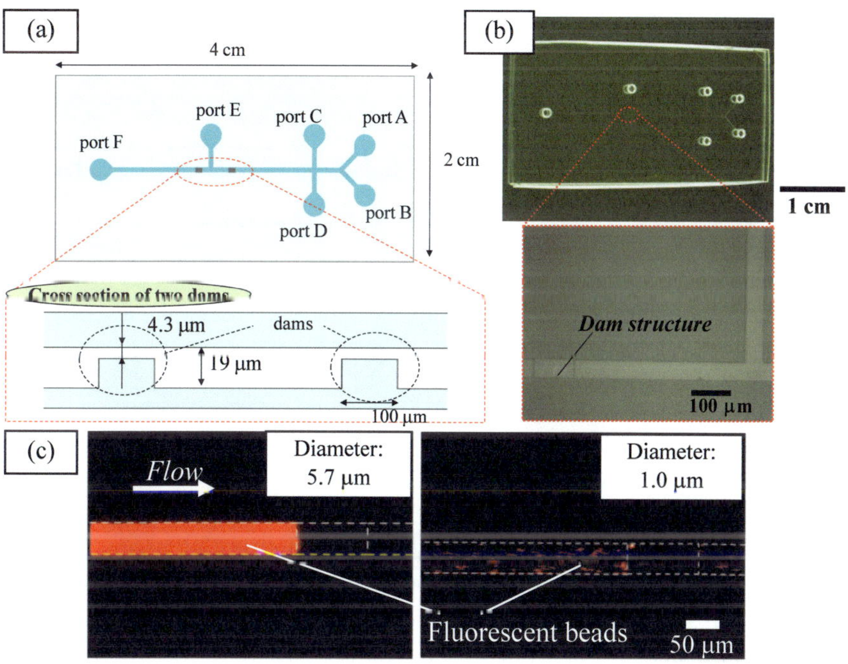

Fig. 5. A PMMA microchip which has two shallow dams of about 5 μm gaps: (a) design; (b) whole view and optical micrograph near a dam; (c) flow behaviour near a dam (Shinohara et al., 2006)

2.4 MCE-ESI-MS microchip

Mass spectrometry (MS) is one of the useful detection methods for microchip electrophoresis (MCE). The advantages of combining MCE and MS (MCE-MS) include high sensitivity, no need for the derivatization of samples and valuable for the analysis of complex mixtures such as biomedical samples. In many cases, the electrospray ionization (ESI) method is used as an interface of MCE-MS (MCE-ESI-MS). Tapered capillary of a spray nozzle was generally connected directly to the channel outlet (Li et al., 2000; Zhang et al, 2001, Tachibana et al., 2003; Tachibana et al., 2004). However, there are a few technical problems caused by the dead volume at a connecting joint between the spray nozzle and the microchip. Efficiency of the spray is strongly depends on the structure of the nozzle.

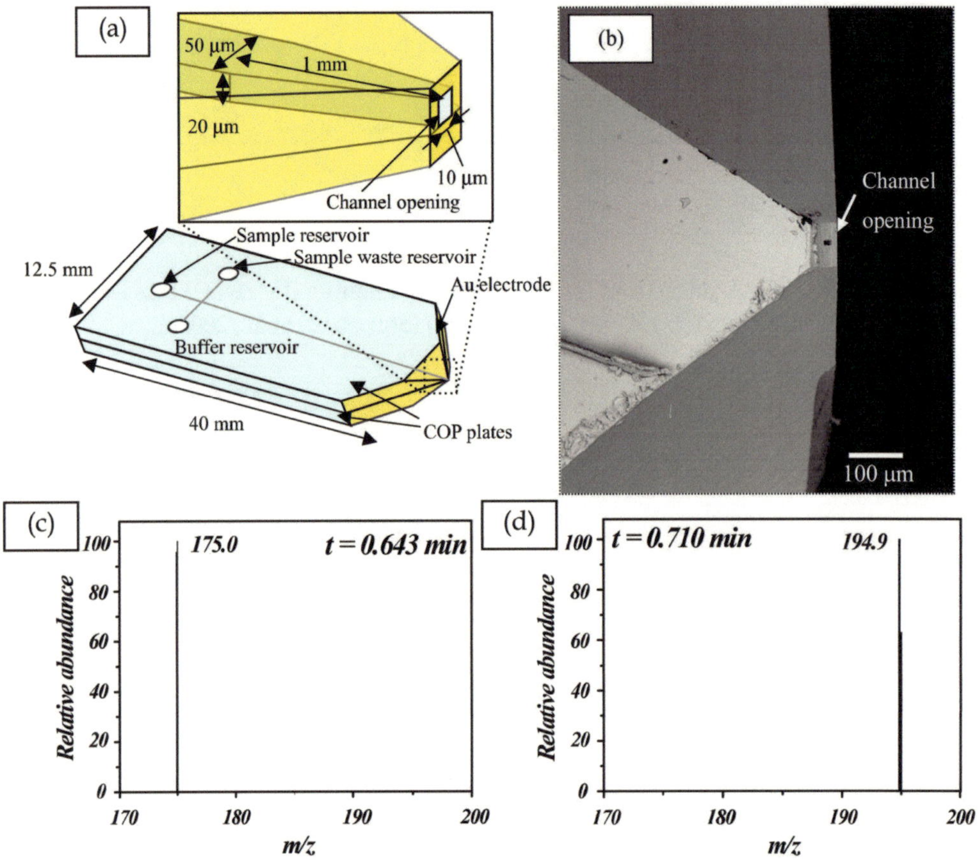

Fig. 6. A MCE-ESI-MS microchip made of two COP plates: (a) design; (b) SEM micrograph of the electrospray tip; MS spectra of (c) arginine and (d) caffeine (Shinohara et al., 2008a)

We developed a MCE-ESI-MS microchip made of two COP plates as shown in Fig. 6 (Shinohara et al., 2008a). An ESI emitter tip was fabricated directly on the opening of a separation channel by machining and electron beam evaporation of Au. Since the direct bonding is performed at the temperature lower than T_g, deformation of the channel structure was negligible. There was no crack at the bonded interface even after structuring the tip because of its sufficient bonding strength. Since the structure of the nano-electrospray tip enables neglected dead volume in the ESI interface, an efficient spray of a sample solution and higher separation efficiency are expected. The success rate of Taylor cone generation was increased with decreasing the tip angle (α). Arginine and caffeine were successfully separated and detected as $[M+H]^+$ in the MCE-ESI-MS analysis at $\alpha = 30$ °, the separation voltage for MCE of 1.3 kV, and the ESI voltage (potential difference between the nano-electrospray tip and the MS orifice) of 2.0 kV, as shown in Fig. 6 (c) and Fig. 6 (d).

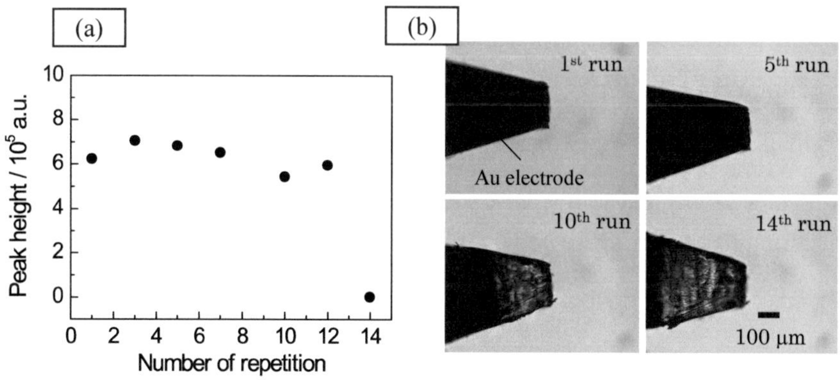

Fig. 7. Results of stability and reproducibility test: (a) reproducibility of the peak height detected as MS spectrum; (b) photomicrographs of the nano-electrospray tip after 1st, 5th, 10th, and 14th run (Shinohara et al., 2008c)

For stability and reproducibility test, MCE-ESI-MS analysis was carried out repeatedly, by using caffeine in 10 mM ammonium acetate as a sample solution (Shinohara et al., 2008c). A MCE-ESI-MS microchip was reused and the reproducibility of the peak heights detected as MS spectrum was observed. Fig. 7 (a) shows the peak heights at 1st, 3rd, 5th, 7th, 10th, 12th, and 14th run. Stable MS detection was achieved and reproducible peak heights were kept up to 13 times. The residual standard deviation (RSD) of the peak height was 9.4 %. At the 14th run, the peak was not detected. Fig. 7 (b) shows photomicrographs of the nano-electrospray tip after 1st, 5th, 10th, and 14th run. After 10th run, optical transparency of the tip was increased obviously. It is indicated that thickness of the Au film decreased. After 14th run, the decrease area was expanded, and deformation of the tip structure was observed. The obvious decrement of the peak at 14th run was caused by the deformation or damage of the Au electrode. The damages of the bonding interface were not observed. The Au thickness looked thinner; however, it was still remained on the COP tip. These results indicate that bonding strength of the COP plates and the adhesion strength of the Au film are strong enough. The stability and reproducibility of the fabricated nanospray tip is sufficient in practical use.

3. Low-temperature polymer bonding using polyurea film

3.1 Hydrophilic treatment of polyurea film using VUV/O_3

In our previous work, we fabricated and evaluated a blood analysis chip made of PMMA (Mizuno et al., 2005b; Shinohara et al., 2005). This chip has microchannel array, which equivalent diameter is 6 μm. When human whole blood is flowed into the microchannels, platelet aggregation was observed after channel passage due to activation of platelet. This

chip is used for the evaluations of the shear stress sensitivity of platelets, the adhesion of white blood cells and the hardness of red blood cells from blood transit time as well as the blood flow images (Kikuchi et al., 1992; Kikuchi et al., 1994). Hydrophilic treatment on the microchannels was required to flow the blood smoothly and not to adhesion of biomaterials. Direct hydrophilic treatment in section 2 was not sufficient because of low stability or low hydrophilicity on the treated surface (see Fig. 16). In this case, aromatic polyurea film coating was selected because of the advantages in visible transparency, non-toxicity, high purity and uniform film thickness (Shinohara et al., 2005). The aromatic polyurea film was prepared by vapor deposition polymerization of 4,4'-diaminodiphenyl methane (MDA) and 4,4'-diphenylmethane diisocyanate (MDI) (Takahashi et al., 1989) as shown in Fig. 8. After coating, highly hydrophilic surface was realized by annealing (50 - 150 °C) and exposing for O_3 at the same time under atmospheric pressure. This treated film had highly hydrophilic surface, water contact angle was smaller than 30 °, and hydrophilic surface was kept for long time (longer than a month) (Shinohara et al., 2005). However, the annealing process for hydrophilic treatment causes bending of the PMMA chip. On the other hand, the film surface was recovered to hydrophobic after washing by water. For reproducible measurements, improvement of the surface stability is required.
We improved the hydrophilic treatment of polyurea and removed the annealing process using VUV/O_3. The VUV/O_3 system used in section 2 was also used (see also Fig. 2). The polyurea surface is treated by the generated gases (O_3 and $O(^1D)$). Then, direct irradiation effect of the VUV light for surface modification is expected to be small in case of large *d*. The light intensity at the sample surface decreases because the VUV is absorbed by oxygen gas in the chamber. Therefore, O_3 and $O(^1D)$ are only generated near the lamp window, and these gases are spreaded over the chamber by diffusion. Since this treatment is carried out at room temperature, the deformation of the sample structure is negligible.

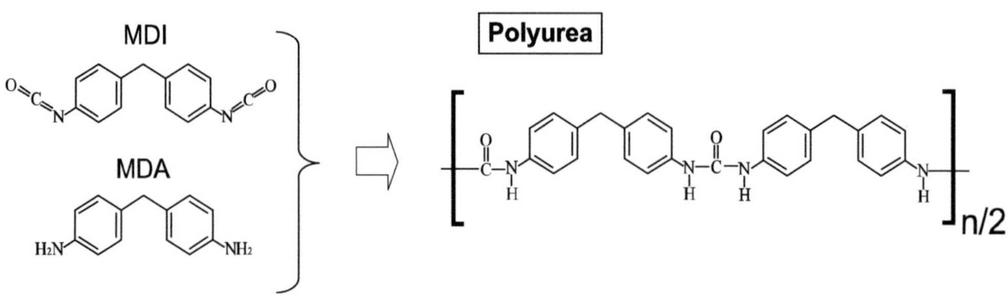

Fig. 8. Reaction scheme of aromatic polyurea

To evaluate the surface treatment effect, transit time of water contact angle after VUV/O_3 was measured under several conditions, as shown in Fig. 9 (Shinohara et al., 2008b). The untreated polyurea film has low hydrophilic surface, contact angle of about 80 °, while the treated films keep contact angles smaller than 45 ° for long time. Especially under the condition of chamber pressure (p) of 3.0 x 10^4 Pa, and exposure time (t) of 20 min, contact angle smaller than 20 ° was realized and kept about two months. Even after very hard condition of ultrasonic cleaning in de-ionized water for 3 min, contact angle of smaller than 40 ° was realized with the VUV/O_3-treated sample (Shinohara et al., 2008b). These results indicate that the VUV/O_3-treated polyurea was improved surface stability even after washing by water. In addition, the contact angle decreases with increasing the *d*, as shown in Fig. 10 (Shinohara et al., 2008b). Since the VUV light intensity decreases with distance from the light source, the direct irradiation effect of the VUV light (e.g., cross-linking (Sato et al., 1994), breakage of main polyurea structure) expected to be avoided.

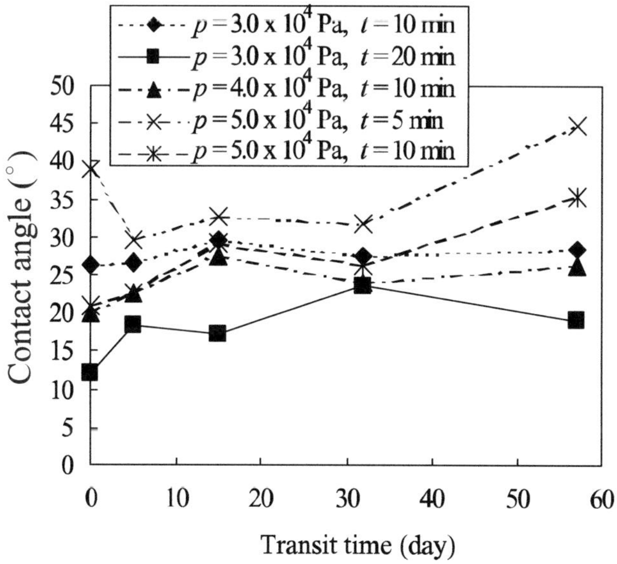

Fig. 9. Transit time of water contact angle on polyurea surface after VUV/O_3 treatment (d = 142 mm) (Shinohara et al., 2008b)

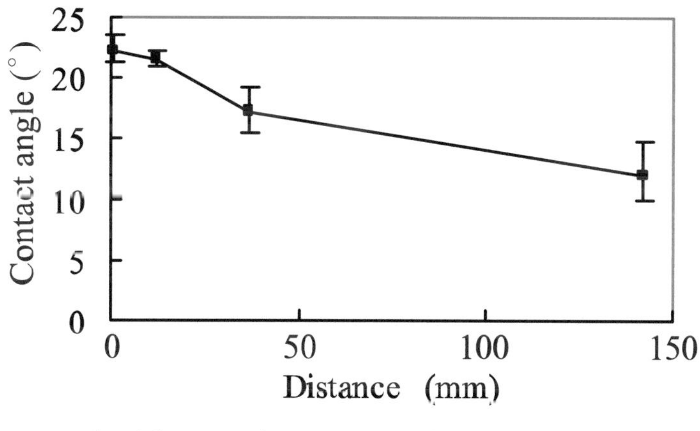

Fig. 10. Contact angle of de-ionized water versus distance between the lamp window and the sample (p = 3.0 x 10^4 Pa, t = 20 min) (Shinohara et al., 2008b)

The polyurea film was applied for PMMA blood analysis chip. As in the case of a conventional silicon chip (Kikuchi et al., 1992; Kikuchi et al., 1994), polyurea-coated PMMA chip was contacted with flat glass plate mechanically. The performance of the surface treatment was evaluated by actual human whole blood flow. The adhesion of platelets and white blood cells was significant in the case of a thermal-oxydized silicon chip (Fig. 11 (a)), while the PMMA chip coated polyurea film can reduce the adhesion of platelets and white blood cells (Fig. 11 (b)), even after ultrasonic cleaning in surfactant induced water (Fig. 11 (c)) (Shinohara et al., 2008b).

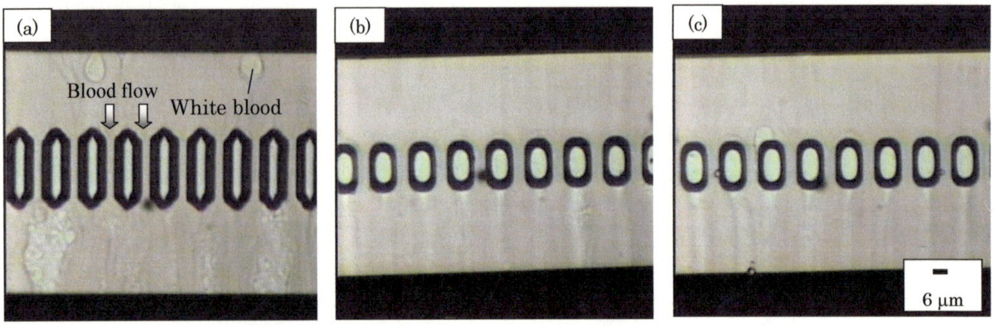

Fig. 11. Images of blood flow: (a) conventional chip made of Si for reference; (b) PMMA chip coated polyurea film; (c) reused PMMA chip after ultrasonic cleaning with surfactant-induced water (Shinohara et al., 2008b)

3.2 Thermal bonding using hydrophilic polyurea film

The hydrophilic polyurea film was used as intermediate bonding layers (Shinohara et al., 2009a). Fig. 12 shows a fabrication process of a microchip which has highly-hydrophilic microchannels. The polyurea was coated on the channel plate and the lid by vapor deposition polymerization (Fig. 12 (a)). Next, the polyurea-coated plates were treated with VUV/O_3 (Fig. 12 (b)). After VUV/O_3 treatment, the plates were brought into contact and then pressed (Fig. 12 (c)). The typical bonding temperature was 85 °C, and the pressure was 3 MPa for 20 min in the case of PMMA plates (Comoglass from Kuraray Co., Ltd.). Fig. 13 (a) and (b) shows a prototype PMMA microchip. Void-free structure was realized over the whole sample surface. Since the bonding temperature is lower than the Tg of the PMMA, negligible deformation of the channel structure is obtained. To observe its flow behavior, a 5-μL methylene blue aqueous solution droplet was applied onto a port (as indicated black arrow in Fig. 13 (a)) on the fabricated microchip (Shinohara et al., 2009a). Its flow behavior at the cross-junction is shown in Fig. 13 (c). All the microchannels were filled by capillary force. There was no leakage or obstacles to smooth fluidic flow at the bonded interface.

To evaluate the surface modification and annealing effect, contact angles of water (H_2O), glycerin ($C_3H_5(OH)_3$), formamide ($HCONH_2$) and diiodomethane (CH_2I_2) on the polyurea surface were measured (Shinohara et al., 2009a). The results were shown in Fig. 14. After the VUV/O_3 treatment, contact angles of water, glycerin, and formamide decreased dramatically, and the contact angles were kept even after annealing of 85 °C for 20 min. This result indicates that the highly hydrophilic surface of the microchannel was also realized after the above-mentioned bonding process.

In addition, surface free energy (γ_s), its polar (γ_s^p) and dispersive (γ_s^d) components ($\gamma_s = \gamma_s^p + \gamma_s^d$) were calculated using these contact angle results, according to Owens-Wendt theory (Owens & Wendt, 1969). The results were shown in Fig. 15 (Shinohara et al., 2009a). After VUV/O_3 treatment, the γ_s^p was increased significantly, while the γ_s^d was decreased. The result indicated that the additional new polar groups (e.g., OH, C=O, COOH) were created after the treatment. After annealing, the γ_s^p was decreased while the γ_s^d was increased. These results indicate two possibilities. One is that conformational transformations of the generated polar groups occurred. The other is that unreacted polymer tails (NH_2 or N=C=O) of polyurea were consumed by further polymerization during the annealing. In Fig. 8, the as-deposited polyurea film of only about five monomers (n = 5) is formed at room temperature (Wang et al., 1993). Further polymerization takes place (n > 5) when as-deposited films are annealed (without any surface treatment) by consuming the unreacted

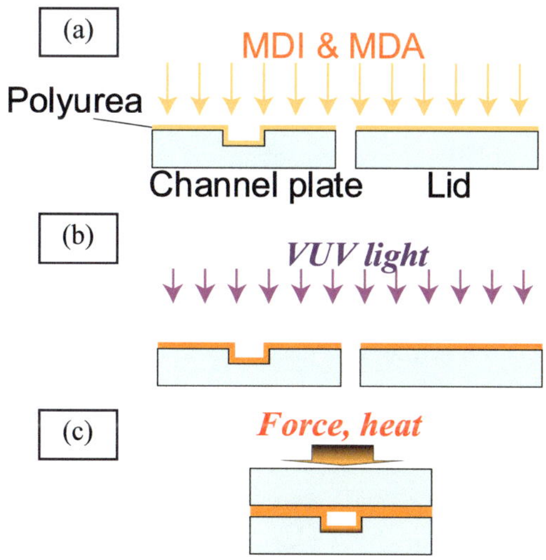

Fig. 12. Fabrication of a microchip which has highly-hydrophilic microchannels: (a) polyurea coating; (b) VUV/O_3 treatment; (c) thermal bonding (Shinohara et al., 2009a)

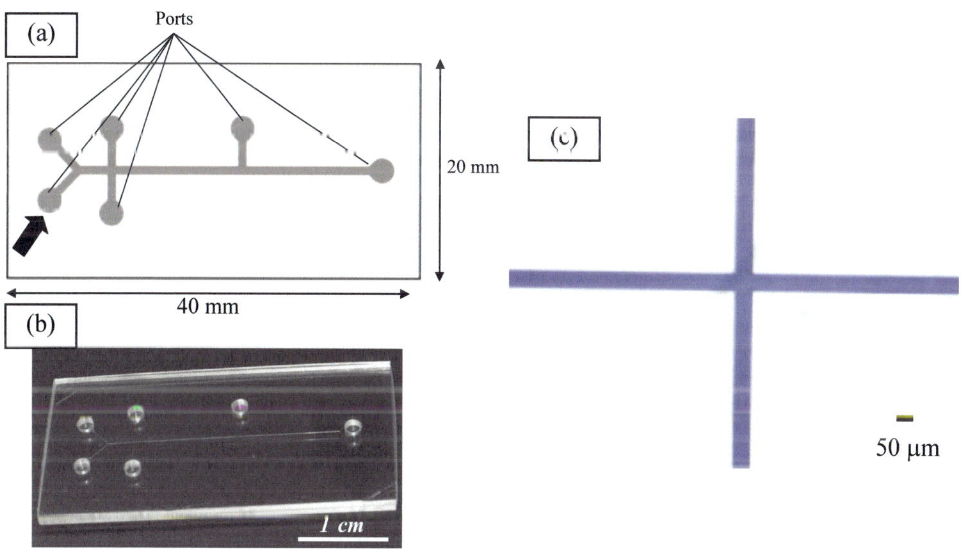

Fig. 13. Prototype PMMA microchip using polyurea film: (a) design; (b) whole view; (c) observation of flow behavior at the cross-junction (Shinohara et al., 2009a)

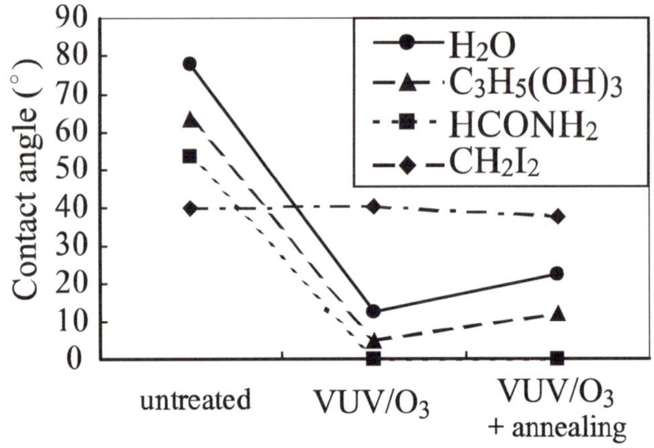

Fig. 14. Contact angles of water, glycerin, formamide, and diiodomethane on the polyurea surface before and after VUV/O_3 treatment (p = 3.0 × 10^4 Pa, t = 20 min, d = 142 mm) (Shinohara et al., 2009a)

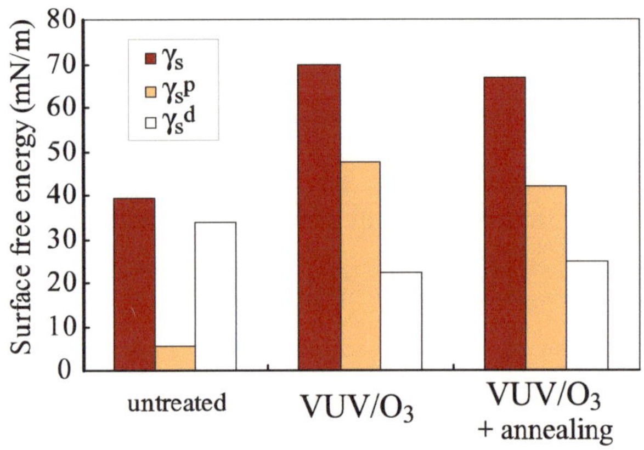

Fig. 15. Surface free energies of polyurea before and after VUV/O_3 treatment (Shinohara et al., 2009a)

polymer tails to form amid bonds (Takahashi et al., 1991). These transformations or polymerization could also have occurred at the interface of the two polyurea films during the bonding process.

To compare hydrophobic recovery with other low-temperature direct bonding, the water contact angle on the polyurea, the COP, and the PMMA surface before and after surface treatment, and after the treatment and annealing (at 85 °C for 20 min) were measured (Shinohara et al., 2009a). Oxygen plasma was selected for surface treatments of COP and PMMA. The results were shown in Fig. 16. In the case of the COP, a highly hydrophilic surface (~20 °) was realized after oxygen plasma treatment. However, the hydrophilic surface was not maintained after the annealing. In the case of the PMMA, the treatment effect was weak. From these results, the bonding using the polyurea as the intermediate layer is the best method from the hydrophilicity viewpoint.

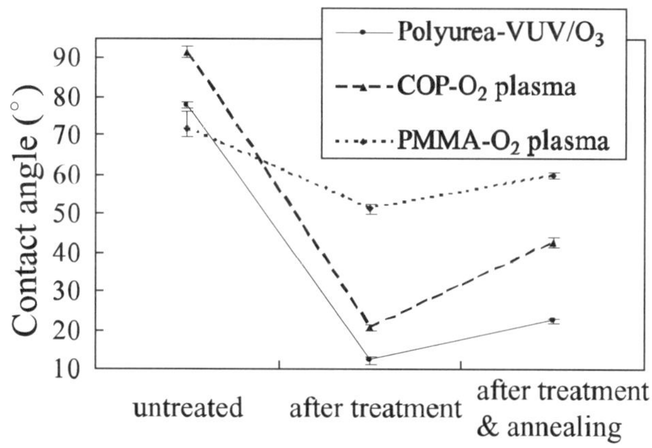

Fig. 16. Water contact angle in three conditions (untreated, after treatment, after treatment and annealing) on VUV/O_3-treated polyurea, oxygen plasma-treated COP (100 W, $p = 4.0 \times 10^{-5}$ MPa, $t = 30$ sec), and oxygen plasma-treated PMMA (200 W, $p = 0.8 \times 10^{-5}$ MPa, $t = 30$ sec) (Shinohara et al., 2009a)

4. Conclusion

In this chapter, two low-temperature bonding technologies, direct bonding of PMMA or COP, and bonding using surface hydrophilic polyurea film were described. The bonding was carried out at temperature lower than T_g of the polymer plates.

The low-temperature direct bonding was realized by surface pretreatment such as oxygen plasma, atmospheric-pressure oxygen plasma, UV/O_3, and VUV/O_3. Reasonable bonding strength was realized with negligible deformation. Shallow microchannels of about 5 mm gaps were successfully fabricated. By using this bonding technology, a MCE-ESI-MS microchip was developed. Arginine and caffeine were successfully separated and detected as $[M+H]^+$ in the MCE-ESI-MS analysis.

On the other hand, a novel hydrophilic treatment method in microchannel surface using aromatic polyurea was developed. The polyurea was changed highly hydrophilic (water contact angle < 20 °) after VUV/O_3 treatment, and the treated film kept highly hydrophilic surface for long time (~ 2 months). The polyurea film was applied for PMMA human blood analysis chip. The new chip can reduce the adhesion of platelets and white blood cells. The technology of the surface hydrophilic treatment of polyurea can be applied to low-temperature bonding. The VUV/O_3-treated polyurea film was used as intermediate bonding layers. The highly hydrophilic surface of the microchannel was retained after the thermal bonding process. There was no leakage or obstacles to smooth fluidic flow at the bonded interface. For actual micro-biochip fabrication with this method, the post-hydrophilic treatment after bonding process is expected unnecessary.

We are currently investigating these bonding mechanisms and optimizing these pretreatment conditions. In addition, these bonding methods will be applied to other polymer microchips.

5. Acknowledgments

This research was supported by the Grant-in-Aid for Specially Promoted Research "Establishment of Electrochemical Device Engineering", and the Waseda University Global COE Program "International Research and Education Center for Ambient SoC" sponsored by MEXT, Japan.

6. References

Becker, E. W.; Ehrfeld, W.; Hagmann, P.; Maner, A. & Münchmeyer, D. (1986). Fabrication of microstructures with high aspect ratios and great structural heights by synchrotron radiation lithography, galvanoforming, and plastic moulding (LIGA Process), *Microelectronic Engineering*, Vol. 4, No. 1, (May 1986) pp.35-56, ISSN 0167-9317

Becker, H. & Heim, U. (2000). Hot embossing as a method for the fabrication of polymer high aspect ratio structures, *Sensors and Actuators A: Physical*, Vol. 83, No. 1-3, (May 2000) pp.130-135, ISSN 0924-4247

Bi, H.; Zhong, W.; Meng, S.; Kong, J.; Yang, P. & B. Liu. (2006). Construction of a biomimetic surface on microfluidic chips for biofouling resistance, *Analytical Chemistry*, Vol. 78, No. 10, (May 2006) pp.3399-3405, ISSN 0003-2700

Chai, J.; Lu, F.; Li, B. & Kwok, D. Y. (2004). Wettability interpretation of oxygen plasma modified poly(methyl methacrylate), *Langmuir*, Vol. 20, No. 25, (December 2004) pp.10919-10927, ISSN 0743-7463

Chen, Z.; Gao, Y.; Lin, J.; Su, R. & Xie, Y. (2004). Vacuum-assisted thermal bonding of plastic capillary electrophoresis microchip imprinted with stainless steel template, *Journal of Chromatography A*, Vol. 1038, No. 1-2, (June 2004) pp.239–245, ISSN 0021-9673

Diaz-Quijada, G. A.; Peytavi, R.; Nantel, A.; Roy, E.; Bergeron, M. G.; Dumoulin M. M. & Veres, T. (2007). Surface modification of thermoplastics—towards the plastic biochip for high throughput screening devices, *Lab on a Chip*, Vol. 7, No. 7, (July 2007) pp.856-862, ISSN 1473-0197

Duffy, D. C.; McDonald, J. C.; Schueller, O. J. A. & Whitesides, G. M. (1998). Rapid prototyping of microfluidic systems in poly(dimethylsiloxane), *Analytical Chemistry*, Vol. 70, No. 23, (October 1998) pp.4974-4984, ISSN 0003-2700

Graß, B.; Neyer, A.; Jöhnck, M.; Siepe, D.; Eisenbeiß, F.; Weber, G. & Hergenröder, R. (2001). A new PMMA-microchip device for isotachophoresis with integrated conductivity detector, *Sensors and Actuators B: Chemical*, Vol. 72, No. 3, (February 2001) pp.249-258, ISSN 0925-4005

Haisma, J.; Verheijen, M.; Heuvel, K. V. D. & Berg, J. V. D. (1996). Mold-assisted nanolithography: a process for reliable pattern replication, *Journal of Vacuum Science and Technology B*, Vol. 14, No. 6, (November 1996) pp.4124-4128, ISSN 1071-1023

Hozumi, A.; Inagaki, H. & Kameyama, T. (2004). The hydrophilization of polystylene substrates by 172-nm vacuum ultraviolet light, *Journal of Colloid and Interface Science*, Vol. 278, No. 2, (October 2004) pp.383-392, ISSN 0021-9797

Kamińska, A.; Kaczmarek, H. & Kowalonek, J. (2002). The influence of side groups and polarity of polymers on the kind and effectiveness of their surface modification by air plasma action, *European Polymer Journal*, Vol. 38, No. 9, (September 2002) pp.1915-1919, ISSN 0014-3057

Kaspar, T.; Tuan, A.; Tonkyn, R.; Hess, W. P.; Rogers Jr., J. W. & Ono, Y. (2003). Role of O(1D) in the oxidation of Si(100), *Journal of Vacuum and Science Technology B*, Vol. 21, No. 2, (March 2003) pp.895-899, ISSN 0734-211X

Kawaguchi, Y.; Nonaka, F. & Sanada, Y. (2007). Fluorinated materials for UV nanoimprint lithography, *Microelectronic Engineering*, Vol. 84, No. 5-8, (May 2007) pp.973-976, ISSN 0167-9317

Kikuchi, Y.; Sato, K.; Ohki, H. & Kaneko, T. (1992). Optically accessible microchannels formed in a single-crystal silicon substrate for studies of blood rheology, *Microvascular Research*, Vol. 44, No. 2, (September 1992) pp.226-240, ISSN 0026-2862

Kikuchi, Y.; Sato, K. & Mizuguchi, Y. (1994). Modified cell-flow microchannels in a single-crystal silicon substrate and flow behavior of blood cells, *Microvascular Research*, Vol. 47, No. 1, (January 1994) pp.126-139, ISSN 0026-2862

Kim, Y.-J.; Taniguchi, Y.; Murase, K.; Taguchi, Y. & Sugimura, H. (2009). Vacuum ultraviolet-induced surface modification of cyclo-olefin polymer substrates for photochemical activation bonding, *Applied Surface Science*, Vol. 255, No. 6, (January 2009) pp.3648-3654, ISSN 0169-4332

Lai, J.; Sunderland, B.; Xue, J.; Yan, S.; Zhao, W.; Folkard, M.; Michael, B. D. & Wang, Y. (2006). Study on hydrophilicity of polymer surfaces improved by plasma treatment, *Applied Surface Science*, Vol. 252, No. 10, (March 2006) pp.3375–3379, ISSN 0169-4332

Lei, K. F.; Ahsan, S.; Budraa, N.; Li, W. J. & Mai, J. D. (2004). Microwave bonding of polymer-based substrates for potential encapsulated micro/nanofluidic device fabrication, *Sensors and Actuators A: Physical*, Vol. 114, No. 2-3, (September 2004) pp.340-346, ISSN 0924-4247

Li, J.; Wang, C.; Kelly, J. F.; Harrison, D. J. & Thibault, P. (2000). Rapid and sensitive separation of trace level protein digests using microfabricated devices coupled to a quadrupole - time-of-flight mass spectrometer, *Electrophoresis*, Vol. 21, No. 1, (January 2000) pp.198-210, ISSN 0173-0835

Lianos, L.; Parrat, D.; Hoc, T. Q. & Duc, T. M. (1994). Secondary ion mass spectrometry time of flight and in situ x-ray photoelectron spectroscopy studies of polymer surface modifications by a remote oxygen plasma treatment, *Journal of Vacuum and Science Technology A*, Vol. 12, No. 4, (July 1994) pp.2491-2498, ISSN 0734-2101

Lin, C. H.; Chao, C. H. & Lan, C. W. (2007). Low azeotropic solvent for bonding of PMMA microfluidic devices, *Sensors and Actuators B: Chemical*, Vol. 121, No. 2, (February 2007) pp.698-705, ISSN 0925-4005

Liu, J.; Pan, T.; Woolley, A. T. & Lee, M. L. (2004). Surface-modified poly(methyl methacrylate) capillary electrophoresis microchips for protein and peptide analysis, *Analytical Chemistry*, Vol. 76, No. 23, (December 2004) pp.6948-6955, ISSN 0003-2700

Maszara, W. P.; Goetz, G.; Caviglia, A. & McKitterick, J. B. (1988). Bonding of silicon wafers for silicon-on-insulator, *Journal of Applied Physics*, Vol. 64, No. 10, (November 1988) pp.4943-4950, ISSN 0021-8979

Mizuno, J.; Ishida, H.; Farrens, S.; Dragoi, V.; Shinohara, H.; Suzuki, T.; Ishizuka, M.; Glinsner, T.; Lindner, F. P. & Shoji, S. (2005a). Cyclo-olefin polymer direct bonding using low temperature plasma activation bonding, *Proceedings of the 13th International Conference on Solid-State Sensors, Actuators and Microsystems*

(Transducers'05), pp.1346-1349, ISBN 0-7803-8994-8, Seoul, Korea, June 2005, IEEE press, USA

Mizuno, J.; Shinohara, H.; Ishizuka, M.; Suzuki, T.; Tazaki, G.; Kirita, Y.; Nishi, T. & Shoji, S. (2005b). PMMA micro-channel array for blood analysis fabricated by hot embossing, *Proceedings of the 9th International Conference on Miniaturized Systems for Chemistry and Life Sciences (microTAS'05)*, pp.1340-1342, ISBN 0-9743611-1-9, Boston, USA, October 2005, Transducer Research Foundation, USA

Murakami, T. N.; Fukushima, Y.; Hirano, Y.; Tokuoka, Y.; Takahashi, M. & Kawashima, N. (2003). Surface modification of polystylene and poly(methyl methacrylate) by active oxygen treatment, *Colloids and Surfaces B: Biointerfaces*, Vol. 29, No. 2-3, (June 2003) pp.171-179, ISSN 0927-7765

Oehr, C. (2003). Plasma surface modification of polymers for biomedical use, *Nuclear Instruments and Methods in Physics Research Section B*, Vol. 208, (August 2003) pp.40-47, ISSN 0168-583X

Owens, D. K. & Wendt, R. C. (1969). Estimation of the surface free energy of polymers, *Journal of Applied Polymer Science*, Vol. 13, No. 8, (August 1969) pp.1741-1747, ISSN 0021-8995

Park, S. J.; Cho, K. S. & Choi, C. G. (2003). Effect of fluorine plasma treatment on PMMA and their application to passive optical waveguides, *Journal of Colloid and Interface Science*, Vol. 258, No. 2, (February 2003) pp.424-426, ISSN 0021-9797

Peeling, J. & Clark, D. T. (1981). ESCA study of the surface photo-oxidation of some non-aromatic polymers, *Polymer Degradation and Stability*, Vol. 3, No. 3, (May 1981) pp.177-185, ISSN 0141-3910

Ratner, B. D. (1995). Surface modification of polymers: chemical, biological and surface analytical challenges, *Biosensors and Bioelectronics*, Vol. 10, No. 9-10, (1995) pp.797-804, ISSN 0956-5663

Sato, M.; Iijima, M. & Takahashi, Y. (1994). Photoresist characteristics of polyurea films prepared by vapor deposition polymerization, *Japanese Journal of Applied Physics*, Vol. 33, No. 12A, (December 1994) pp.L1721-L1724, ISSN 0021-4922

Sawada, Y. (2003). Applications of atmospheric-pressure glow plasma, *Journal of Plasma and Fusion Research*, Vol. 79, No.10, (October 2003) pp. 1022-1028, ISSN 0918-7928 (in Japanese)

Shinohara, H.; Mizuno, J.; Tazaki, G.; Nishi, T.; Nakajima, M.; Takahashi, C. & Shoji, S. (2005). PU coated PMMA blood analysis chip, *Proceedings of the 5th International Symposium on MicroChemistry and Microsystems (ISMM'05)*, pp.114-115, Kyoto, Japan, December 2005, Society for Chemistry and Micro-Nano Systems, Japan

Shinohara, H.; Mizuno, J.; Kitagawa, F.; Otsuka, K. & Shoji, S. (2006). Fabrication of highly dimension controlled PMMA microchip by hot embossing and low temperature direct bonding, *Proceedings of the 10th International Conference on Miniaturized Systems for Chemistry and Life Sciences (microTAS'06)*, pp.158-160, ISBN 4-9903269-0-3-C3043, Tokyo, Japan, November 2006, Society for Chemistry and Micro-Nano Systems, Japan

Shinohara, H.; Mizuno, J. & Shoji, S. (2007a). Low temperature direct bonding of Poly(methyl methacrylate) for polymer microchips, *IEEJ Transactions on Electrical and Electronic Engineering*, Vol. 2, No. 3, (May 2007) pp.301-306, ISSN 1931-4973

Shinohara, H.; Mizuno, J. & Shoji, S. (2007b). Fabrication of a microchannel device by hot embossing and direct bonding of poly(methyl methacrylate), *Japanese Journal of Applied Physics*, Vol. 46, No. 6A, (June 2007) pp.3661-3664, ISSN 0021-4922

Shinohara, H.; Suzuki, T.; Kitagawa, F.; Mizuno, J.; Otsuka, K. & Shoji, S. (2008a). Polymer microchip integrated with nano-electrospray tip for electrophoresis-mass spectrometry, *Sensors and Actuators B: Chemical*, Vol. 132, No. 2, (June 2008) pp.368-373, ISSN 0925-4005

Shinohara, H.; Takahashi, Y.; Mizuno, J. & Shoji, S. (2008b). Surface hydrophilic treatment of polyurea film realized by vacuum ultraviolet light irradiation and its application for poly(methylmethacrylate) blood analysis chip, *Sensors and Actuators B: Chemical*, Vol. 132, No. 2, (June 2008) pp.374-379, ISSN 0925-4005

Shinohara, H.; Kitagawa, F.; Mizuno, J.; Otsuka, K. & Shoji, S. (2008c). Highly stable and reproducible cyclo-olefin polymer nano-electrospray tip for electrophoresis-mass spectrometry, *Proceedings of the 4th Asia-Pacific Conference on Transducers and Micro/Nano Technologies (APCOT'08)*, 2S45 (in CD-ROM), Tainan, Taiwan, June 2008

Shinohara, H.; Fukuhara, M.; Hirasawa, T.; Mizuno, J. & Shoji, S. (2008d). Fabrication of magnetic nanodots array using UV nanoimprint lithography and electrodeposition for high density patterned media, *Journal of Photopolymer Science and Technology*, Vol. 21, No. 4, (June 2008) pp.591-596, ISSN 0914-9244

Shinohara, H.; Takahashi, Y.; Mizuno, J. & Shoji, S. (2009a). Fabrication of post-hydrophilic treatment-free plastic biochip using polyurea film, *Sensors and Actuators A: Physical*, Vol. 154, No. 2, (September 2009) pp.187-191, ISSN 0924-4247

Shinohara, H; Mizuno, J. & Shoji, S. (2009b). Studies on low-temperature direct bonding methods of PMMA and COP ssing surface pretreatment, *Proceedings of International Conference on Electronics Packaging (ICEP'09)*, pp.628-631, Kyoto, Japan, April 2009, Japan Institute of Electronics Packaging, Japan

Slentz, B. E.; Penner, N. A.; Lugowska, E. & Regnier, F. (2001). Nanoliter capillary electrochromatography columns based on collocated monolithic support structures molded in poly(dimethylsiloxane), *Electrophoresis*, Vol. 22, No. 17, (October 2001) pp.3736-3743, ISSN 0173-0835

Spierings, G. A. C. M. & Haisma, J. (1994). Direct bonding of organic materials, *Applied Physics Letters*, Vol. 64, No. 24, (June 1994) pp.3246-3248, ISSN 0003-6951

Svedberg, M.; Pettersson, A.; Nilsson, S.; Bergquist, J.; Nyholm, L.; Nikolajeff, F. & Markides, K. (2003). Sheathless electrospray from polymer microchips, *Analytical Chemistry*, Vol. 75, No. 15, (July 2003) pp.3934-3940, ISSN 0003-2700

Tachibana, Y.; Otsuka, K.; Terabe, S.; Arai, A.; Suzuki, K. & Nakamura, S. (2003). Rubust and sinple interface for microchip electrophoresis-mass spectrometry, *Journal of Chromatography A*, Vol. 1011, No. 1-2, (September 2003) pp.181-192, ISSN 0021-9673

Tachibana, Y.; Otsuka, K.; Terabe, S.; Arai, A.; Suzuki, K. & Nakamura, S. (2004). Effects of the length and modification of the separation channel on microchip electrophoresis-mass spectrometry for analysis of bioactive compounds, *Journal of Chromatography A*, Vol. 1025, No. 2, (February 2004) pp.287-296, ISSN 0021-9673

Takahashi, Y.; Iijima, M. & Fukada, E. (1989). Pyroelectricity in poled thin films of aromatic polyurea prepared by vapor deposition polymerization, *Japanese Journal of Applied Physics*, Vol. 28, No. 12, (December 1989) pp.L2245-L2247, ISSN 0021-4922

Takahashi, Y.; Ukishima, S.; Iijima, M. & Fukada, E. (1991). Piezoelectric properties of thin films of aromatic polyurea prepared by vapor deposition polymerization, *Journal of Applied Physics*, Vol. 70, No. 11, (December 1991) pp.6983-6987, ISSN 0021-8979

Truckenmüller, R.; Henzi, P.; Herrmann, D.; Saile, V. & Schomburg, W. K. (2004). Bonding of polymer microstructures by UV irradiation and subsequent welding at low temperatures, Microsystem Technologies, Vol. 10, No. 5, (August 2004) pp.372-374, ISSN 0946-7076

Wang, J.; Pumera, M.; Chatrathi, M. P.; Escarpa, A.; Konrad, R.; Griebel, A.; Dörnger, W. & Löwe, H. (2002). Towards disposable lab-on-a-chip: poly(methylmethacrylate) microchip electrophoresis device with electrochemical detection, *Electrophoresis*, Vol. 23, No. 4, (February 2002) pp.596-601, ISSN 0173-0835

Wang, J. H. & Ray, M. B. (2000). Application of ultraviolet photooxidation to remove organic pollutants in the gas phase, *Separation and Purification Technology*, Vol. 19, No. 1-2, (June 2000) pp.11-20, ISSN 1383-5866

Wang, X. S.; Iijima, M.; Takahashi, Y. & Fukada, E. (1993). Dependence of piezoelectric and pyroelectric activities of aromatic polyurea thin films on monomer composition ratio, *Japanese Journal of Applied Physics*, Vol. 32, No. 6A, (June 1993) pp.2768-2773, ISSN 0021-4922

Watanabe, K.; Inn, E. C. Y. & Zelikoff, M. (1953). Absorption coefficients of oxygen in the vacuum ultraviolet, *Journal of Chemical Physics*, Vol. 21, No. 6, (June 1953) pp.1026-1030, ISSN 0021-9606

Zhang, B.; Foret, F. & Karger, B. (2001). High-throughput microfabricated CE/ESI-MS: automated sampling from a microwell plate, *Analytical Chemistry*, Vol. 73, No. 11, (June 2001) pp.2675-2681, ISSN 0003-2700